Springer
Tokyo
Berlin
Heidelberg
New York
Barcelona
Budapest
Hong Kong
London
Milan
Paris
Santa Clara
Singapore

Kiyosi Itô

N. Ikeda, S. Watanabe,
M. Fukushima, H. Kunita (Eds.)

Itô's Stochastic Calculus and Probability Theory

Springer

Nobuyuki Ikeda
Professor, Department of Computer Science, Ritsumeikan University,
Kusatsu, Shiga, 525-77 Japan

Shinzo Watanabe
Professor, Department of Mathematics, Graduate School of Science,
Kyoto University, Kyoto, 606-01 Japan

Masatoshi Fukushima
Professor, Department of Mathematics, Faculty of Fundamental Engineering,
Osaka University, Toyonaka, Osaka, 560 Japan

Hiroshi Kunita
Professor, Graduate School of Mathematics, Kyushu University,
Fukuoka, 812 Japan

ISBN 978-4-431-68534-0 ISBN 978-4-431-68532-6 (eBook)
DOI 10.1007/978-4-431-68532-6

Printed on acid-free paper

Typesetting: Camera-ready by editors using Springer TeX macropackage

Preface

The present volume is a tribute dedicated to Professor Kiyosi Itô on the occasion of his eightieth birthday. He was born in Mie Prefecture, Japan, on September 7, 1915. His paper in 1942, written in Japanese for the Journal of Pan Japan Mathematical Colloquium, Osaka University, marked the debut of Itô's theory of stochastic calculus. Since then this theory has played a crucial role in the development of probability theory and its applications over half a century. It has in fact provided us with an effective and indispensable method of constructing and analyzing probability models. His stochastic differential equations and basic formula, very well known as Itô's formula or Itô's lemma, are so fundamental that they have been used widely in science, even in finance theory by economists.

He has been a great teacher as well. Everyone who has had a chance to meet with him could never forget the happy experience of listening to his ideas and his way of thinking, which he has always eagerly and tirelessly tried to explain.

On behalf of all contributors of this volume, his colleagues, students and friends in the world-wide scientific community, we dedicate this volume to Professor Kiyosi Itô *with our deep admiration for his achievements in science and with our deep appreciation for his insight and ideas which he has so generously offered to all of us.*

We regret that we could not ask many of our colleagues and eminent probabilists to contribute their works because of the limitation of space; this would have taken several more volumes. We would like to thank our colleague Ichiro Shigekawa for his kind cooperation in editing this volume, and above all, for the painstaking job of unifying the style of the manuscripts for publication.

Nobuyuki Ikeda
Shinzo Watanabe
Masatoshi Fukushima
Hiroshi Kunita

April, 1996

Contents

Kiyosi Itô

More than a half century has passed since the first fundamental work of Kiyosi Itô on stochastic analysis appeared. In his first paper [1] [1] published in 1942, he succeeded in realizing an idea of Paul Lévy to describe the structure of Lévy processes (cadlag processes with independent increments, called $\mathcal{L}$-processes in [1] and differential processes in [2] and [12]) which is now known as *Lévy-Itô's canonical form of Lévy processes.* This was a remarkable achievement in stochastic analysis. Stochastic analysis, due originally to pioneering works by Paul Lévy and Norbert Wiener among others, is a method in probability theory to study probability models by an analysis on sample functions, being in contrast with the traditional analytical approach to study various distributions of probability models by means of analytical tools such as differential equations, Fourier analysis, Laplace transform or generating function methods, etc. Thus the Lévy-Itô canonical form of Lévy processes has provided us with a new proof and a deeper understanding of the Lévy-Kchinchin canonical form of infinitely divisible distributions, which, in analytical approaches, can be obtained by Fourier analysis or Hille-Yosida's theory of semigroups.

At almost the same time as he obtained the Lévy-Itô canonical form of Lévy processes, he applied this result to the study of sample functions of Markov processes. The analytical theory of Markov processes had been given already by Kolmogorov and then extended by Feller; they in fact obtained the fundamental integro-differential equations governing the transition probabilities. The equations of Kolmogorov and Feller have suggested that the tangent or the derivative of the time flow of transition probabilities is given by an infinitely divisible distribution. Then Itô proposed that for sample functions also, the tangent is given by a Lévy process which, due to his result, can be described in terms of a Wiener process and a Poisson random measure, so that the differential equations governing the sample functions of Markov processes can also be described by Wiener processes and Poisson random measures. As we can see in [2] and [12], Itô could realize this idea perfectly by introducing an original notion of stochastic integrals, so that the theory of stochastic differential equations was established for the first time. This is the very beginning of a really fruitful theory of stochastic analysis which has been one of the most important methods in probability theory and its applications over a half century.

We should perhaps follow the achievements of Itô in probability theory over fifty years after his first decisive contributions. To our great relief, however, the volume of *Kiyosi Itô Selected Papers*, edited by D. W. Stroock and S. R. S. Varadhan and published by Springer Verlag in 1987, has already fulfilled this purpose to a great extent. That volume contains a *Foreword* written by Itô himself, an autobiography, in which we can see how he began his study

---

[1] Numbers in brackets refer to the Bibliography of Kiyosi Itô

in probability theory and how and under what background and motivations it has been developed. We can also find in it an *Introduction* by the editors to the scientific achievements of Itô with excellent comments on distinguishing features shared by Itô's works, focussing on such samples as stochastic differential equations, stochastic integrals, one-dimensional diffusion processes and excursion point processes from his many important contributions. We give here only our comments on his recent works which are not contained in that volume.

His recent works [48–55] are devoted mainly to a mathematical foundation of *Malliavin's calculus*, a differential calculus for functions on Wiener space. Here we are happy to find once again beautiful and masterful treatments by Itô. Some of his ideas in these works have already stimulated works of younger generations; for example, his idea in [50] concerning regular versions and exceptional sets for a class of Wiener functionals has been used as a key by H. Sugita in his work on holomorphic Wiener functionals (H. Sugita: Regular version of holomorphic Wiener function, J. Math. Kyoto Univ. **34**(1994)).

We finally give a brief account on some remarkable features of Itô's approach to the Malliavin calculus. Firstly, he chooses as the basic Wiener space, functions on which the Malliavin calculus is to be applied, the space $(W_H, \mathcal{B}(W_H), \gamma)$ canonically associated to H, where H is a given real separable Hilbert space, W_H is the algebraic dual of H, i.e., the linear space formed of all linear functions on H, $\mathcal{B}(W_H)$ is the Kolmogorov σ-algebra on W_H and γ is the standard H-Gaussian measure realized on $(W_H, \mathcal{B}(W_H))$ by the Kolmogorov extension theorem. Usually in the Malliavin calculus, we choose an abstract Wiener space associated to H, but there is no canonical choice of it for given H. Secondly, his definition of Sobolev function spaces $\mathbf{D}_{p,\alpha}, 1 < p < \infty, \alpha \in \mathbf{R}$, is given in the following way: He first introduces the space $\mathbf{D}$ of generalized functionals as the algebraic dual of the space $\mathbf{P}$ of polynomial functionals on W_H. On the space $\mathbf{P}$, the (p, α)-Sobolev norm is defined through the Bessel potential operators acting on $\mathbf{P}$ and this norm naturally induces the (p, α)-Sobolev norm on $\mathbf{D}$ by the duality. Then the space $\mathbf{D}_{p,\alpha}$ is defined to be the Banach subspace of $\mathbf{D}$ formed of all elements having a finite (p, α)-norm. This is perhaps the most beautiful way of defining Sobolev spaces in which we can avoid such an abstract notion as completions of $\mathbf{P}$ by Sobolev norms. Thirdly, he introduces important notions of (p, α)-regular measures and Malliavin functions representing elements in $\mathbf{D}_{p,\alpha}$ for $\alpha \geq 0$. These notions replace the role of Malliavin's capacities and redefinitions or quasi-continuous modifications of elements in $\mathbf{D}_{p,\alpha}$ in a standard course of the Malliavin calculus. Using these notions, quasi-sure analysis on the Wiener space can be well developed. We would refer the reader to Itô's very beautiful exposition [55] for details.

Bibliography of Kiyosi Itô[2]

(A) Papers

[1] On stochastic processes (infinitely divisible laws of probability) (Doctoral thesis). *Japan. Journ. Math.* **XVIII**, 261-301 (1942).
[2] Differential equations determining a Markoff process (original Japanese: Zenkoku Sizyo Sugaku Danwakai-si). *Journ. Pan-Japan Math. Coll.* No.1077 (1942).
[3] On the ergodicity of a certain stationary processes. In: *Proc. Imp. Acad. Tokyo* **20**, 54-55 (1944).
[4] A kinematic theory of turbulence. In: *Proc. Imp. Acad. Tokyo* **20**, 120-122 (1944).
[5] On the normal stationary process with no hysteresis. In: *Proc. Imp. Acad. Tokyo* **20**, 199-202 (1944).
[6] A screw line in Hilbert space and its application to the probability theory. *Proc. Imp. Acad. Tokyo* **20**, 203-209 (1944).
[7] Stochastic integral. In: *Proc. Imp. Acad. Tokyo* **20**, 519-524 (1944).
[8] On Student's test. *Proc. Imp. Acad. Tokyo* **20**, 694-700 (1944).
[9] On a stochastic integral equation. In: *Proc. Imp. Acad. Tokyo* **22**, 32-35 (1946).
[10] Stochastic differential equations in a differentiable manifold. *Nagoya Math. Journ.* **1**, 35-47 (1950).
[11] Brownian motions in a Lie group. In: *Proc. Imp. Acad. Tokyo* **26**, 4-10 (1950).
[12] On stochastic differential equations. *Mem. Amer. Math. Soc.* **4**, 1-51 (1951),
[13] On a formula concerning stochastic differentials. *Nagoya Math. Journ.* **3**, 55-65 (1951).
[14] Multiple Wiener integral. *Journ. Math. Soc. Japan* **3**, 157-169 (1951).
[15] Stochastic differential equations in a differentiable manifold (2). *Mem. Coll. Science, Univ. Kyoto, Ser. A.*, **28**, 81-85 (1953).
[16] Stationary random distributions. *Mem. Coll. Science, Univ. Kyoto, Ser. A.*, **28**, 209-223 (1953).
[17] Complex multiple Wiener integral. *Japan Journ. Math.* **22**, 63-86 (1952).
[18] Isotropic random current. In: *Proc. Third Berkeley Symp. Math. Statist. Prob.* **II**, 125-132 (1955).
[19] Spectral type of the shift transformation of differential processes with stationary increments. *Trans. Amer. Math. Soc.* **81**, 253-263 (1956).
[20] Potentials and random walk (with H. P. McKean, Jr.). *Illinois Journ. Math.* **4**, 119-132 (1960).
[21] Wiener integral and Feynman integral. In: *Proc. Fourth Berkeley Symp. Math. Statist. Prob.* **II**, 227-238 (1960).

[2] Those marked with * are not contained in the list in Kiyosi Itô Selected Papers

[22] Construction of diffusions. *Ann. Fac. Sci. Univ. Clermont* **2**, 23-32 (1962).
[23] The Brownian motion and tensor fields on Riemannian manifold. In: *Proc. Intern. Congr. Mathemat.* (Stockholm), 536-539 (1962).
[24] Brownian motion on a half line (with H. P. McKean, Jr.). *Illinois Journ. Math.* **7**, 181-231 (1963).
[25] The expected number of zeros of continuous stationary Gaussian processes. *Journ. Math. Kyoto Univ.* **3**, 207-216 (1964).
[26] On stationary solutions of a stochastic differential equation (with M. Nisio). *Journ. Math. Kyoto Univ.* **4**, 1-75 (1964).
[27] Transformation of Markov processes by multiplicative functionals (with S. Watanabe). *Ann. Inst. Fourier, Univ. Grenoble* **XV**, 13-30 (1965).
[28] The canonical modification of stochastic processes. *Journ. Math. Soc. Japan* **20**, 130-150 (1968).
[29] On the convergence of sums of independent Banach space valued random variables. (with M. Nisio). *Osaka Journ. Math.* **5**, 35-48 (1968).
[30] Generalized uniform complex measures in the Hilbertian metric space with their application to the Feynman integral. In: *Proc. Fifth Berkeley Symp. Math. Statist. Prob.* **II**, 145-161 (1965).
[31] On the oscillation functions of Gaussian processes (with M. Nisio). *Math. Scand.* **22**, 209-223 (1968).
[32] Canonical measurable random functions. In: *Proc. Int. Conf. Funct. Anal. Rel. Topics* (Tokyo), 369-377 (1969).
[33] The topological support of a Gaussian measure on Hilbert space. *Nagoya Math. Journ.* **38**, 181-183 (1970).
[34] Poisson point processes attached to Markov processes. In: *Proc. Sixth Berkeley Symp. Math. Statist. Prob.* **III**, 225-239 (1970).
[35] Stochastic differentials of continuous local martingales. In: *Stability of Stochastic Dynamical Systems* (Lecture Notes in Mathematics **294**), Springer-Verlag, Berlin, 1-7 (1972).
[36] Stochastic integration. In: *Vector and Operator Valued Measures and Applications*, Academic Press, New York, 141-148 (1973).
[37] Stochastic differentials, *Appl. Math. and Opt.* **1**, 374-381 (1974).
[38] Stochastic parallel displacement. In: *Probabilistic Methods in Differential Equations* (Lecture Notes in Mathematics **451**), Springer-Verlag, Berlin, 1-7 (1975).
[39] Stochastic calculus. In: *Mathematical Problems in Physics* (Lecture Notes in Physics **39**), Springer-Verlag, 218-223 (1975).
[40] Extension of stochastic integrals. In: *Proc. Int. Symp. Stochastic Differential Equations* (Kyoto), 95-109 (1976)
[41] Introduction to stochastic differential equations (with S. Watanabe). In: *Proc. Int. Symp. Stochastic Differential Equations* (Kyoto), i-xxx (1976).
[42] Continuous additive $\mathcal{S}'$-processes.

[43] Stochastic Analysis in infinite dimensions. In: *Stochastic Analysis* (A. Friedman and M. Pinsky, eds.), Academic Press, New York, 187-197 (1978).

[44] Infinite dimensional Ornstein-Uhlenbeck processes. In: *Taniguchi Symp. SA*, Katata, 197-224 (1982).

[45] Regularization of linear random functionals (with M. Nawata). In *Probability Theory and Mathematical Statistics, Fourth USSR-Japan Symposium Proceedings, 1982* (Lecture Notes in Mathematics **1021**), Springer-Verlag, Berlin, 257-267 (1983).

[46] Distribution-valued processes arising from independent Brownian motions. *Math. Zeit.* **182**, 17-33 (1983).

[47] A stochastic differential equation in infinite dimensions. In: *Contemporary Math.* **26**, 163-169 (1984).

[48]* Malliavin's C^∞-functionals of a centered Gaussian system. IMA preprint Series, Univ. Minnesota, No **327**(1987).

[49]* Malliavin calculus on a Segal space, in *Stochastic Analysis, Proc. of Japanese-French Seminar, Paris, 1987*, (eds. M. Métivier and S. Watanabe), Lecture Notes in Mathematics **1322**, Springer-Verlag, Berlin, 50-72 (1988).

[50]* Positive generalized functions on $(\mathbf{R}^\infty, \mathcal{B}^\infty, N^\infty)$, in *White Noise Analysis, Mathematics and Applications*, (eds. T. Hida, H.-H. Kuo, J. Potthoff and L. Streit), World Scientific, 166-179 (1990).

[51]* On Malliavin calculus, in *Proceedings of 1989 Singapore Probability Conference*, (eds. L. H. Y. Chen, K. P. Choi, K. Hu and J.-H. Lou), Walter de Gruyter, 47-72 (1992).

[52]* An elementary approach to Malliavin fields, in *Asymptotic problems in probability theory: Wiener functionals and asymptotics, Proceedings of Taniguchi Symp. Sanda and Kyoto, 1990*, (eds. K. D. Elworthy and N. Ikeda), Pitman Research Notes in Math. Series **284**, Longman, 35-89 (1993).

[53]* Semigroups in probability theory, in *Functional analysis and related topics, Proceedings of the International Conference in Memory of Professor Kôsaku Yosida, RIMS, Kyoto Univ. 1991*, (ed. H. Komatsu), Lecture Notes in Mathematics **1540**, Springer-Verlag, Berlin, 69-83 (1993).

[54]* On Malliavin tensor fields. *Communs. Pure and Appl. Math.* **XLVII**, 377-403 (1994). (The third of five special issues dedicated to Henry McKean.)

[55]* A measure-theoretic approach to Malliavin calculus, to appear in *'New Trends in Stochastic Analysis', Proc. Taniguchi Symposium, Sept. 1994, Charingworth, (eds. K. D. Elworthy, S. Kusuoka and I. Shigekawa)*, World Scientific, to appear.

(B) Books

[a] *Foundation of Probability Theory* (in Japanese), Iwanami, Tokyo, 1943.

[b] *Probability Theory* (in Japanese), Iwanami, Tokyo, 1952.

[c]* *Stochastic Processes* I, II (in Jpapanese), Iwanami Series of Modern Applied Mathematics A. 13, I, II, Iwanami, Tokyo, 1957.

[d] *Diffusion Processes and Their Sample Paths* (with H. P. McKean, Jr.), Springer, 1965; reprint of the 1974 Edition in the Springer Series of *Classics in Mathematics*, 1996.

[e] *Probability Theory* (in Japanese), Iwanami Series of Fundamental Mathematics, Analysis (I) vii, Iwanami, Tokyo, 1978, revised 1983.

[f] *Introduction to Probability Theory* (English translation of [e] Chapters I-IV), Cambridge Univ. Press, 1984.

[g] *Foundations of Stochastic Differential Equations in Infinite Dimensional Spaces,* (CBMS-NSF Reg. Conf. Ser. in Appl. Math. 47) SIAM, 1984.

(C) Lecture Notes

[i]* Lectures on Stochastic Processes, Tata Institute for Fundamental Research, Bombay 1961.

[ii]* Stochastic Processes, Lecture Notes Series, No. **16**, Aarhus University, 1969.

Kiyoshi Itô

Born: September 7, 1915 in Mie Prefecture, Japan
Currently Professor Emeritus at Kyoto University
Doctor of Science, the Imperial University of Tokyo, 1945

Awards

Asahi Prize, 1978
Japan Academy Prize and Imperial Prize, 1978
Fujiwara Prize, 1985
Wolf Foundation Prize, 1987

Memberships

Membre Associé Étranger, Académie des Science, France, 1989
Member of Japan Academy, 1991

Honorary degrees

Docteur Honoraris Causa, Université Paris VI, 1981
Honorable Doctor of Mathematics, ETH, Zürich, 1987
Degree of Doctor of Science Honoraris Causa, The University of Warwick, 1992

Lévy measure of superprocesses; absorption processes

E. B. Dynkin

Department of Mathematics, Cornell University, White Hall,
Ithaca, NY 14853-7901, USA

Summary. We prove that the (modified) Lévy measure of a superprocess X is the lifting of its discontinuous branching characteristic. We also investigate, for a wide class of functions f, the continuous martingale part of process $\langle f^t, X_t \rangle$. (For time-homogeneous processes with local branching, similar results have been obtained earlier by El Karoui and Roelly.)

We use the expression for the Lévy measure to establish the predictability of absorption processes. For a particle system, an absorption process can be obtained by freezing every particle at the first exit from a subset Q in the product of time interval and state space. The mass distribution X_t' of particles frozen during time interval $[0, t]$ is a measure on the complement of Q. The monotone increasing measure-valued process X_t' is called an absorption process. A process of this kind can be constructed for every superprocess X in E and every finely open Borel subset Q of $\mathbb{R}_+ \times E$. In a recent paper by Dynkin and Kuznetsov, absorption processes have been used to describe linear additive functionals of superprocesses.

1. Introduction

1.1. Superprocesses. Let E be a metrizable Luzin space and let $\mathcal{M} = \mathcal{M}(E)$ be the space of all finite measures on the Borel σ-algebra $\mathcal{B}_E$ of E. Equipped with the topology of weak convergence, $\mathcal{M}$ is also a metrizable Luzin space. A *superprocess* $X = (X_t, \mathcal{F}_X(I), P_{r,\mu})$ is an $\mathcal{M}$-valued right Markov process characterized by the following elements:[1] (a) a right Markov process $\xi = (\xi_t, \mathcal{F}(I), \Pi_{r,x})$ in E; (b) a (positive) continuous additive functional K of ξ; (c) a transformation ψ in the space of positive Borel functions on $S = \mathbb{R}_+ \times E$. The transition function of X is determined by the condition: for every $f \in p\mathcal{B}_E$[2] and every $\mu \in \mathcal{M}$,

$$P_{r,\mu} \exp\langle -f, X_t \rangle = \exp\langle -u^r, \mu \rangle \tag{1.1}$$

where[3]

$$u(r,x) + \Pi_{r,x} \int_r^t \psi(u)(s, \xi_s) K(ds) = \Pi_{r,x} f(\xi_t) \qquad \text{for all } r < t. \tag{1.2}$$

---

† Partially supported by National Science Foundation Grant DMS-9301315

[1] Basic facts on (nonhomogeneous) right Markov process and their additive functionals are collected in the Appendix.

[2] Notation $p\mathcal{B}$ ($b\mathcal{B}$) stands for the set of all positive (bounded) $\mathcal{B}$-measurable functions.

[3] We use notation $u^r(x)$ or $u(r,x)$ for a function u from S to $\mathbb{R}_+$.

In this paper we assume that

$$\psi(z)(s,x) = b^s(x)z(s,x)^2 + \int_{\mathcal{M}} \operatorname{Exp}\langle z^s,\nu\rangle \ell^s(x,d\nu) \tag{1.3}$$

where

$$\operatorname{Exp}(u) = e^{-u} - 1 + u, \tag{1.4}$$

$b \in p\mathcal{B}_S$ and ℓ is a kernel from S to $\mathcal{M}$.[4] Without changing ψ, we can set $\ell^s(x,\{0\}) = 0$.

It follows from (1.1)—(1.3) that, for all $r < t \in \mathbb{R}_+, f \in p\mathcal{B}_E, \mu \in \mathcal{M}$,

$$P_{r,\mu}\langle f, X_t\rangle = \Pi_{r,\mu} f(\xi_t). \tag{1.5}$$

Conditions on ξ, K, b, ℓ sufficient for the existence of a superprocess X are stated in Section 5.8 of the Appendix. If ξ is a Hunt process, then X can be chosen to be a Hunt process too. [In a time-homogeneous setting, this was proved in [10]. Only minor adjustments are needed for time-inhomogeneous processes.] *Through all this paper we assume that both ξ and X are Hunt processes.*

Equation (1.2) can be rewritten in the form

$$u^r(x) + \Pi_{r,x}[\frac{1}{2}\int_r^t u^s(\xi_s)^2 Q(ds) + \int_r^t \int_{\mathcal{M}} \operatorname{Exp}\langle u^s,\nu\rangle \ell(ds,d\nu)] = \Pi_{r,x} f(\xi_t) \tag{1.6}$$

where

$$Q(ds) = b^s(\xi_s)K(ds) \tag{1.7}$$

and

$$\ell(ds,d\nu) = \ell(\xi_s,d\nu)K(ds). \tag{1.8}$$

Additive functional Q is called the *continuous branching characteristic* of X and measure-valued additive functional $\ell(ds,d\nu)$ is called the *discontinuous branching characteristic* of X (cf. Section 1.3 in [3]).

1.2. Lifting and projection. Let C and c be continuous additive functionals of X and ξ respectively and let

$$P_{r,\mu}\int_r^\infty f(s)C(ds) = \Pi_{r,\mu}\int_r^\infty f(s)c(ds) \tag{1.9}$$

for every positive Borel function f and for all $r \in \mathbb{R}_+, \mu \in \mathcal{M}$. Then we say that C is the *lifting* of c and that c is the *projection* of C.

---

[4]We say that $\ell(x,B)$ is a kernel from a measurable space $(E,\mathcal{B})$ to a measurable space $(\tilde{E},\tilde{\mathcal{B}})$ if $\ell(x,\cdot)$ is, for every $x \in E$, a measure on $\tilde{\mathcal{B}}$ and if $\ell(\cdot,B)$ is, for every $B \in \tilde{\mathcal{B}}$, a $\mathcal{B}$-measurable function on E.

Suppose that ℓ and $\mathcal{L}$ are continuous $\mathcal{M}(Z)$-valued additive functionals of ξ and X and that

$$P_{r,\mu}\int_r^\infty\int_Z F(s,z)\mathcal{L}(ds,dz)=\Pi_{r,\mu}\int_r^\infty\int_Z F(s,z)\ell(ds,dz) \tag{1.10}$$

for every positive measurable function F on $\mathbb{R}_+\times Z$ and for all r,μ. Then we say that $\mathcal{L}$ is the *lifting* of ℓ and that ℓ is the *projection* of $\mathcal{L}$.

1.3.Modified Lévy measure of X. The definition of the Lévy measure is given in Section 5.7.

Theorem 1.1. *The modified Lévy measure L of a superprocess X is the lifting of its discontinuous branching characteristic ℓ.*

1.4. Continuous martingale part of process $\langle f^t, X_t\rangle$. The Markov semigroup of ξ is given by the formula

$$T_t^r f(x)=\Pi_{r,x}f(\xi_t). \tag{1.11}$$

We put $f\in D_A$ and $Af=\varphi$ if $f,\varphi\in b\mathcal{B}_S$ and if, for all $r<t\in\mathbb{R}_+$,

$$T_t^r f^t=f^r+\int_r^t T_s^r\varphi^s\, ds. \tag{1.12}$$

Note that functional

$$(Af)^s(\xi_s)ds \tag{1.13}$$

is the compensator of $f^t(\xi_t)-f^r(\xi_r)$ and functional

$$\langle (Af)^s, X_s\rangle\, ds \tag{1.14}$$

is the compensator of

$$Y_t^r=\langle f^t,X_t\rangle-\langle f^r,X_r\rangle. \tag{1.15}$$

Put

$$D_A^+=\{f\in D_A: f\ge 0, Af\le 0\}. \tag{1.16}$$

If $\varphi\in bp\mathcal{B}_S$ and if $\varphi^s=0$ for all sufficiently large s, then

$$f^r=\int_r^\infty T_s^r\varphi^s ds \tag{1.17}$$

belongs to D_A^+ and $Af(r,x)=\varphi^r(x)$.

It is easy to check that, if $f\in D_A$, then

$$M_t^r(f)=\langle f^t,X_t\rangle-\langle f^r,X_r\rangle-\int_r^t\langle (Af)^s,X_s\rangle\, ds. \tag{1.18}$$

is an additive X-martingale.

Theorem 1.2. *Let Q be the continuous branching characteristic of a superprocess X and let $f \in D_A^+$. Denote by $M^c(f)$ the continuous part of the additive X-martingale (1.18). The quadratic variation $\langle M^c(f)\rangle$ of $M^c(f)$ is the lifting of functional*

$$2[(Af)^s(\xi_s)]^2 Q(ds).$$

Remark. Theorem 1.2 holds if we replace the Lebesgue measure ds by an arbitrary diffuse measure $\sigma(ds)$ in formulae (1.12) and (1.18).

1.5. Exit measures and absorption process. To every finely open Borel set[5] Q in S, there corresponds a random measure X_Q called the exit measure from Q. It satisfies conditions similar to (1.1)—(1.2):

$$P_{r,\mu} \exp\langle -f, X_Q\rangle = \exp\langle -u^r, \mu\rangle, \tag{1.19}$$

$$u(r,x) + \Pi_{r,x} \int_r^\tau \psi(u)(s,\xi_s) K(ds) = \Pi_{r,x} f(\tau, \xi_\tau) \tag{1.20}$$

where $\tau = \inf\{s : (s, \xi_s) \notin Q\}$ and $f \in p\mathcal{B}_S$.

Put $Q_{<t} = \{(s,x) \in Q : s < t\}$. Measures $X_t^* = X_{Q_{<t}}$ satisfy equations

$$P_{r,\mu} \exp\langle -f, X_t^*\rangle = \exp\langle -u^r, \mu\rangle, \tag{1.21}$$

$$u(r,x) + \Pi_{r,x} \int_r^t \psi^*(u)(s,\xi_s^*) K(ds) = \Pi_{r,x} f(t, \xi_t^*) \tag{1.22}$$

where $\xi_t^* = \xi_{\tau\wedge t}$ is the process ξ stopped at the first exit from Q and $\psi^*(u)(s,x) = 1_Q(s,x)\psi(u)(s,x)$. By [6], process X_t^* can be chosen to be a right process. We see from (1.21)—(1.22) that this is a superprocess with parameters ξ^*, K^*, ψ^* where K^* is a functional of ξ^* induced by K.

We get the *absorption process* X_t' by restricting measures X_t^* to the complement Q^c of Q. It follows from [7, Lemma 3.1] that X_t' is monotone increasing.

Theorem 1.3. *The absorption process is predictable if measure $\ell^s(x,\cdot)$ is concentrated on set $\{\nu : \nu(Q^c) = 0\}$ for every $(s,x) \in Q$.*

We say that branching is *local* if each measure $\ell^s(x,\cdot)$ is concentrated on the ray $\{u\delta_x, u \geq 0\}$ [δ_x is the Dirac's measure at point x]. Clearly, the condition of Theorem 1.3 is satisfied in the case of local branching.

Remark. In [8] the absorption process is defined as the restriction of X_Q to $[0,t] \times E$. The difference between this process and the absorption process in the sense of the present paper is, $P_{r,\mu}$-a.s. for every pair (r,μ), a deterministic process and therefore Theorem 1.3 holds also for the absorption process in the sense of [8].

---

[5] A set Q is called finely open if, for every $(r,x) \in Q$, there exists, $\Pi_{r,x}$-a.s., $t > r$ such that $(s,\xi_s) \in Q$ for all $s \in (r,t)$.

1.6. Theorems 1.1 and 1.2 can be deduced from results of El Karoui and Roelly [9] (presented also in the exposition [1, Section 6.1]) under the following restrictions on X:

(a) ξ is a time-homogeneous Hunt process;

(b) $K(dt) = dt$;

(c) the branching is local.

In the general case, the statement of the results and their proofs are based on a relationship between additive functionals of ξ and X established in our previous work. In particular, the lifting operation is used to define an multiplicative martingale which plays a central role in [9] as well as in the proof of Theorems 1.1 and 1.2.

Acknowledgments. This investigation was motivated by our joint work with S. E. Kuznetsov on linear additive functionals of superdiffusions. Numerous discussions with him had a considerable impact on the contents of the present paper.

2. Functionals $\mathcal{Q}(f)$ and $\mathcal{L}$ and multiplicative martingale $\Phi(f)$

2.1. Existence and uniqueness of lifting. By [3, Theorem 6.2.3], the lifting exists and is unique for every continuous additive functional c with the property

$$\sup_{r\le t, x\in E} \Pi_{r,x} c(r,t] < \infty \qquad \text{for all } t. \tag{2.1}$$

The lifting also exists and is unique for every continuous $\mathcal{M}$-valued additive functional ℓ subject to condition: there exists a strictly positive measurable function g on $\mathbb{R}_+ \times \mathcal{M}$ such that

$$\sup_{r\le t, x\in E} \Pi_{r,x} \int_r^\infty \int_{\mathcal{M}} g(s,\nu)\ell(ds,d\nu) < \infty \qquad \text{for all } t. \tag{2.2}$$

2.2. The log-Laplace equation.

Theorem 2.1. *Suppose that a continuous additive functional c of ξ satisfies condition (2.1) and let C be the lifting of c. Then*

$$P_{r,\mu} \exp(-C(r,t]) = \exp\langle -u^r, \mu\rangle \tag{2.3}$$

where

$$u(r,x) + \Pi_{r,x} \int_r^t \psi(u)(s,\xi_s)K(ds) = \Pi_{r,x} c(r,t] \qquad \text{for all } r<t. \tag{2.4}$$

We call (2.4) the *log-Laplace equation for C.*

Proof. Put $h_t^r(x) = P_{r,x}C(r,t] = \Pi_{r,x}c(r,t]$. For every partition $\Lambda = \{r = t_0 < t_1 < \cdots < t_n = t\}$ of interval $[r,t]$, consider

$$C_\Lambda(r,t] = \sum_{i=0}^{n-1} \langle h_{t_{i+1}}^{t_i}, X_{t_i}\rangle. \tag{2.5}$$

Choose a monotone increasing sequence of Λ_n with the union everywhere dense in $[r,t]$ and set $C_n = C_{\Lambda_n}$. By [3, Theorem 2.4.1]

$$\lim_{n\to\infty} P_{r,\mu}|C_n(r,t] - C(r,t]| = 0 \tag{2.6}$$

for every $\mu \in \mathcal{M}$.

By [5, Theorem 1.2],

$$P_{r,\mu}\exp(-C_n(r,t]) = \exp\langle -u_n^r, \mu\rangle \tag{2.7}$$

where

$$u_n(r,x) + \Pi_{r,x}\int_r^t \psi(u_n)(s,\xi_s)K(ds) = P_{r,x}C_n(r,t]. \tag{2.8}$$

It follows from (2.6) and (2.7) that (2.3) holds with $u = \lim u_n$. Note that $P_{r,x}C_n(r,t] = h_t^r(x)$ and $u_n \le h_t^r$ by (2.8). Equation (2.4) follows from (2.8),(2.1) and the dominated convergence theorem.

2.3. Multiplicative X-martingale $\Phi(f)$. Under assumptions on parameters of a superprocess stated in Section 5.8, condition (2.1) holds for

$$c(ds) = [(Af)^s(\xi_s)]^2 Q(ds) \tag{2.9}$$

with an arbitrary $f \in D_A^+$. We denote by $\mathcal{Q}$ the lifting of this functional. Condition (2.2) holds with $g(s,\nu) = |\nu| \wedge |\nu|^2 + 1_{\nu=0}$. We denote the lifting of ℓ by $\mathcal{L}$.

Theorem 2.2. *For every $f \in D_A$, formula*

$$\Phi_t^r(f) = \exp\{-[\langle f^t, X_t\rangle - \langle f^r, X_r\rangle - \int_r^t \langle (Af)^s, X_s\rangle ds$$

$$+ \int_r^t \int_{\mathcal{M}} \operatorname{Exp}\langle f^s, \nu\rangle \mathcal{L}(ds, d\nu) + \mathcal{Q}_t^r(f)]\} \tag{2.10}$$

defines a multiplicative X-martingale.

Proof. By Theorem 2.1,

$$P_{r,\mu}\Phi_t^r(f) = e^{-\langle u^r - f^r, \mu\rangle} \tag{2.11}$$

where u satisfies equation (2.4) with

$$c(r,t] = f^t(\xi_t) - f^s(\xi_s) - \int_r^t (Af)^s(\xi_s)ds$$
$$+ \int_r^t \int_{\mathcal{M}} \operatorname{Exp}\langle f^s, \nu\rangle \ell(ds, d\nu) + \int_r^t (Af)^s(\xi_s)^2 Q(ds).$$

Note that $\Pi_{r,x} c(r,t] = f^r(x) + \Pi_{r,x} \int_r^t \psi(f)(s,\xi_s)K(ds)$ and therefore f is a solution of (2.4). By a generalized Gronwall's inequality [4, Lemma 3.2], (2.4) has a unique solution. Therefore $u = f$ and $P_{r,\mu}\Phi_t^r(f) = 1$ by (2.11).

3. Lévy measure of X

3.1. To prove Theorem 1.1 we need several lemmas.

Lemma 3.1. *Let $(E,\mathcal{B})$ be a measurable space and let $S = \mathbb{R}_+ \times E, \mathcal{N} = \mathbb{R}_+ \times \mathcal{M}(E)$. Suppose that $\mathcal{H}$ consists of bounded strictly positive $\mathcal{B}_S$-measurable functions and is closed under addition and under multiplication by constants $c > 0$. Let $\mathcal{B}(\mathcal{N})$ be the σ-algebra in $\mathcal{N}$ generated by functions $G(s,\nu) = \langle f^s, \nu\rangle$ with $f \in \mathcal{H}$. If N and P are two measures on $\mathcal{B}(\mathcal{N})$ and if*

$$\int_{\mathcal{N}} e^{-\langle f^s,\nu\rangle} N(ds,d\nu) = \int_{\mathcal{N}} e^{-\langle f^s,\nu\rangle} P(ds,d\nu) < \infty \qquad \text{for all } f \in \mathcal{H}, \tag{3.1}$$

then $N = P$.

Proof. Fix $f_0 \in \mathcal{H}$ and note that $\mathcal{B}(\mathcal{N})$ is generated by functions $\langle f^s + f_0^s, \nu\rangle$ with $f \in \mathcal{H}$. Finite measures $N_0(ds,d\nu) = e^{-\langle f_0^s,\nu\rangle} N(ds,d\nu)$ and $P_0(ds,d\nu) = e^{-\langle f_0^s,\nu\rangle} P(ds,d\nu)$ satisfy condition

$$\int_{\mathcal{N}} e^{-\langle f^s,\nu\rangle} N_0(ds,d\nu) = \int_{\mathcal{N}} e^{-\langle f^s,\nu\rangle} P_0(ds,d\nu) < \infty \tag{3.2}$$

for all $f \in \mathcal{H}$ and for $f = 0$. By the multiplicative system theorem $N_0 = P_0$ and therefore $N = P$.

Lemma 3.2. *Let $\mathcal{H}, \mathcal{N}$ and $\mathcal{B}(\mathcal{N})$ be the same as in Lemma 3.1. Suppose that F is a mapping from $\mathcal{H}$ to $\mathbb{R}_+$ given by the formula*

$$F(f) = C(f) + \int_{\mathcal{N}} \operatorname{Exp}\langle f^s, \nu\rangle N(ds,d\nu).$$

where $C(f) \geq 0$, $C(\lambda f) = \lambda^2 C(f)$ for all $\lambda > 0$ and N is a measure on $\mathcal{B}(\mathcal{N})$ such that $N\{\mathbb{R}_+ \times \{0\}\} = 0$. Then C and N are determined uniquely by F.

Proof. Put $\varphi(\lambda) = F(\lambda f)$ and denote by n_f the image of N under the mapping $\nu \to \langle f,\nu\rangle$ from $\mathcal{M}$ to $\mathbb{R}_+$. Note that

$$\varphi(\lambda) = \int_0^\lambda d\lambda_1 \int_0^{\lambda_1} g(\lambda_2)d\lambda_2$$

where

$$g(\lambda) = 2C(f) + \int_{\mathbb{R}_+} e^{-\lambda u} u^2 n_f(du).$$

Since $\varphi(\lambda) < \infty$ and since g is monotone decreasing, we conclude that $g(\lambda) < \infty$ for all λ and $\varphi'' = g$. Hence F determines uniquely $g(\lambda)$. By the dominated convergence theorem, $g(\lambda) \to 2C(f)$ as $\lambda \to \infty$. Hence F determines $C(f)$ and $\int_{\mathbb{R}_+} e^{-\lambda u} u^2 n_f(du)$. By Lemma 3.1, it determines measure $u^2 n_f(du)$. Since $n_f\{0\} = N\{\mathbb{R}_+ \times \{0\}\} = 0$, measure n_f is also determined by F. It remains to note that

$$\int_{\mathcal{N}} e^{-\langle f^s, \nu \rangle} N(ds, d\nu) = \int_{\mathbb{R}_+} e^{-u} n_f(du)$$

and to apply Lemma 3.1 once more.

Remark. If N and P are random measures and if relation (3.1) holds a.s., then $N = P$ a.s. Lemma 3.2 can be modified similarly.

Lemma 3.3. *Functions $G(s, \nu) = \langle f^s, \nu \rangle$ where $f \in D_A^+$ generate the Borel σ-algebra in $\mathcal{N}$.*

Proof. Since $\mathcal{N}$ is a metrizable Luzin space, its Borel σ-algebra is generated by every family of Borel functions which separate points (see, e.g. [2, Lemma 2.1]). Therefore it is sufficient to show that, for every $z_1 \neq z_2 \in S$ there exists f of the form (1.17) such that

$$G(r, \nu) = \int_r^\infty \langle T_s^r \varphi^s, \nu \rangle ds$$

separates z_1 and z_2. This is easy to get by using that $T_s^r \varphi^s(x) = \Pi_{r,x} \varphi^s(\xi_s) \to \varphi^r(x)$ as $s \downarrow r$ if φ is bounded and continuous.

3.2.

Lemma 3.4. *If $f \in D_A^+$ or if f is bounded and uniformly continuous, then $X_t^r = X_t - X_r$ and Y_t^r given by (1.15) are connected by the formula*

$$P_{r,\mu}\{Y\{s\} = \langle f^s, X\{s\} \rangle \quad \text{for all} \quad s > r\} = 1 \tag{3.3}$$

Proof. Put $F(s, \nu) = \langle f^s, \nu \rangle$. We need to show that

$$P_{r,\mu}\{\lim_{s \to t-} F(s, X_s) = F(t, X_{t-}) \quad \text{for all} \quad t > r\} = 1.$$

By Lemma 2.3.1 in [3], it is sufficient to prove that

$$P_{r,\mu} F(T_n, X_{T_n}) \to P_{r,\mu} F(T, X_T) \tag{3.4}$$

for every sequence of r-stopping times $T_n \uparrow T$.

If f is bounded and uniformly continuous, then F is continuous and (3.4) follows from the definition of a Hunt process.

If $f \in D_A^+$, then $f^t(\xi_t)$ is right continuous $\Pi_{r,\mu}$-a.s. (see [2, Theorem 5.1]. Equation (1.12) implies that, for every r-stopping time τ for ξ,

$$\Pi_{r,x} f^\tau(\xi_\tau) = f^r(x) + \Pi_{r,x} \int_r^\tau \varphi^s(\xi_s) ds.$$

Clearly, this implies

$$\Pi_{r,x} f^{\tau_n}(\xi_{\tau_n}) \to \Pi_{r,x} f^\tau(\xi_\tau) \tag{3.5}$$

for every sequence of r-stopping times $\tau_n \uparrow \tau$ for ξ. By Theorem 6.2.2 and Lemma 2.3.1 in [3], (3.5) implies (3.4).

Lemma 3.5. *Let*

$$Z = \Phi e^C \tag{3.6}$$

where Φ is a strictly positive multiplicative X-martingale and C is a continuous additive functional of X. Then C is the compensator of the local X-martingale

$$\tilde{Z}_t^r = \int_r^t (Z_{s-})^{-1} dZ_s. \tag{3.7}$$

Proof. We fix r and consider Z, C, Φ as functions in t. By Ito's formula (see, e.g., [3, p.122]), $dZ = d\Phi e^C + dCe^C\Phi$ and therefore $\tilde{Z}_t^r = C_t^r +$ local martingale.

3.3. Proof of Theorems 1.1–1.2. We apply Lemma 3.5 to $\Phi(f)$ defined by (2.10), $Z_t^r = \exp(-Y_t^r)$ with Y given by (1.15) and to

$$C_t^r = -\int_r^t \langle (Af)^s, X_s\rangle ds + \int_r^t \int_{\mathcal{M}} \operatorname{Exp}\langle f^s, \nu\rangle \mathcal{L}(ds, d\nu) + Q_t^r(f). \tag{3.8}$$

By (1.18),

$$Y_t^r = \int_r^t \langle (Af)^s, X_s\rangle ds + M_t^r(f)$$

and, by Ito's formula,[6]

$$Z_t^r - 1 = \int_r^t e^{-Y_{s-}}(-d_s Y_s^r + \frac{1}{2} d_s \langle M^c\rangle_s^r) + \sum_r^t e^{-Y_{s-}} \operatorname{Exp}(Y\{s\}).$$

This implies

$$\tilde{Z}_t^r = -Y_t^r + \frac{1}{2}\langle M^c\rangle_t^r + \sum_r^t \operatorname{Exp}(Y\{s\}).$$

[6] Writing $\sum_r^t$ means the sum over $s \in (r, t]$.

By Lemma 3.4, $Y\{s\}$ is indistinguishable from $\langle f^s, X\{s\}\rangle$ and therefore the compensator of the last term is equal to $\int_r^t \int_{\mathcal{M}} \langle f^s, \nu\rangle L(ds, d\nu)$ where L is the modified Lévy measure of X. The compensator of the first term is equal to $-\int_r^t \langle (Af)^s, X_s\rangle ds$. Since the compensator or $\tilde{Z}$ is given by (3.8), we get the equation

$$\int_r^t \int_{\mathcal{M}} \operatorname{Exp}\langle f^s, \nu\rangle \mathcal{L}(ds, d\nu) + \mathcal{Q}_t^r(f)$$

$$= \int_r^t \int_{\mathcal{M}} \operatorname{Exp}\langle f^s, \nu\rangle L(ds, d\nu) + \frac{1}{2}\langle M^c\rangle_t^r(f). \tag{3.9}$$

Theorems 1.1–1.2 follows from (3.9) and Lemmas 3.2–3.3.

4. Predictability of absorption process

4.1. We apply results on the Lévy measure to superprocess X^* described in Section 1.5. Let $Y_t = \langle f^t, X_t^*\rangle$ and let $Y\{s\}$ be the jump of Y_t at s. By Lemma 3.4, Theorem 1.1 and definition of the Lévy measure, if f is bounded and uniformly continuous, then

$$P_{r,\mu} \sum_r^t \operatorname{Exp}(Y\{s\}) = \Pi_{r,\mu} \int_r^t \psi_0^*(f)(s, \xi_s^*) K^*(ds) \tag{4.1}$$

where

$$\psi_0^*(f)(s,x) = \int_{\mathcal{M}} \operatorname{Exp}\langle f^s, \nu\rangle 1_Q(s,x) \ell^s(x, d\nu).$$

4.2. Proof of Theorem 1.3. Note that $Y_t = \langle f, X_t^*\rangle = \langle \varphi, X_t'\rangle$ for $f = \varphi 1_{Q^c}$. It is sufficient to prove that Y_t is predictable for bounded continuous φ. Consider a sequence of uniformly continuous functions $0 \le \varphi_n \le 1$ such that $\varphi_n \downarrow 1_{Q^c}$ and put $Y_t^n = \langle f^n, X_t^*\rangle = \langle \varphi_n, X_t'\rangle$ where $f_n = \varphi_n 1_{Q^c}$. By(4.1),

$$P_{r,\mu} \sum_r^t \operatorname{Exp}(Y^n\{s\}) = \Pi_{r,\mu} \int_r^t \psi_0(f_n)(s, \xi_s^*) 1_Q(s, \xi_s^*) K^*(ds). \tag{4.2}$$

By the condition of Theorem 1.3, $\psi_0(f) = 0$ on Q and therefore the right side in (4.2) tends to 0 as $n \to \infty$. Thus there exist Ω_0 and a sequence $n_k \to \infty$ such that $P_{r,\mu}(\Omega_0) = 1$ and

$$\sum_r^t \operatorname{Exp}(Y^{n_k}\{s\}) \to 0 \qquad \text{for all } \omega \in \Omega_0. \tag{4.3}$$

To simplify notation, we assume that $n_k = k$. Clearly, $Y_s^k \to \langle \varphi, X_s'\rangle$ for all ω, s. Since $\operatorname{Exp}(u) > 0$ for all $u \ne 0$, (4.3) implies that, for all $\omega \in \Omega_0$ and all s, $Y^k\{s\} \to 0$. By Theorem 3.1 in [6], there exists Ω_1 with $P_{r,\mu}(\Omega_1) = 1$ such that, for $\omega \in \Omega_1$, Y_s^k is right continuous in s and therefore $Y^k\{s\} = Y_s^k - Y_{s-}^k$ for all s. For $\omega \in \Omega_0 \cap \Omega_1$, $\langle \varphi, X_s' \rangle = \lim Y_{s-}^k$. Since Y_{s-}^k is predictable, $\langle \varphi, X_s' \rangle$ is also predictable.

5. Appendix

5.1 Markov processes. A Markov process ξ in a measurable space $(E, \mathcal{B})$, with a sample space Ω is described by a function $\xi : \mathbb{R}_+ \times \Omega \to E$, a family of σ-algebras $\mathcal{F}[r,t], 0 \le r \le t$ in Ω and a family of probability measures $\Pi_{r,x}$ on Ω subject to the conditions:

5.1.A. For every $B \in \mathcal{B}$ and every $s \in [r,t]$, $\{\omega : \xi_s(\omega) \in B\} \in \mathcal{F}[r,t]$.

5.1.B. For all $r < t \in \mathbb{R}_+, Y \in \mathcal{F}[r,t], Z \in \mathcal{F}[r,\infty)$,

$$\Pi_{r,x} YZ = \Pi_{r,x}(Y \Pi_{t,\xi_t} Z).$$

For an arbitrary interval I, we denote by $\mathcal{F}(I)$ the minimal σ-algebra which contains $\mathcal{F}[r,t]$ for all $r < t \in I$.

5.2. Right processes and Hunt processes. A Markov process ξ is called *right* if $(E, \mathcal{B})$ is a metrizable Luzin space and if :

5.2.A. For each ω, $\xi_t(\omega)$ is right continuous with left limits.

5.2.B. For all $r < t$, $f \in p\mathcal{B}$ and $\mu \in \mathcal{M}$, $\Pi_{s,\xi_s} f(\xi_t)$ is, $\Pi_{r,\mu}$-a.s., right continuous in $s \in [r,t)$.

5.2.C. For every $r \le t$, the σ-algebra $\mathcal{F}[r,t]$ is complete with respect to all measures $\Pi_{r,\mu}$ and it coincides with the intersection of $\mathcal{F}[r,u]$ over all $u > t$.

All processes in this paper are right. [By Theorem 3.1 in [2], every right process has the strong Markov property.]

We say that τ is an r-stopping time if it takes values in $[r,\infty)$ and if, for every $\mu \in \mathcal{M}$ and every t, the set $\{\tau \le t\} \in \mathcal{F}[r,t]$.

A right Markov process is called a *Hunt process* if ξ_t is quasi-left continuous which means: $\xi_{\tau_n} \to \xi_\tau$ $\Pi_{r,\mu}$-a.s. for every sequence of r-stopping times τ_n such that $\tau_n \uparrow \tau$.

5.3. Predictable processes. Denote by $\mathcal{P}_r$ the σ-algebra in $(r,\infty) \times \Omega$ generated by functions $F(t,\omega)$ which are left continuous in t and adapted to $\mathcal{F}(t,\infty)$.

We say that a stochastic process Y_t is ξ-predictable if, for every r, μ, the restriction of $Y(t,\omega)$ to $(r,\infty) \times \Omega$ is $\Pi_{r,\mu}$-indistinguishable from a $\mathcal{P}_r$-measurable function.

5.4. Additive functionals. A (positive) *additive functional* of ξ is a random measure $A(\omega, dt)$ on $(0,\infty)$ such that, for every r, μ and every open interval $I \subset (r,\infty)$, $A(\cdot, I)$ is $\mathcal{F}(I)$-measurable.

To every Borel function $\rho \ge 0$ on $S = \mathbb{R}_+ \times E$ and to every measure σ on $\mathbb{R}_+$ there corresponds an additive functional

$$A(ds) = \rho^s(\xi_s)\sigma(ds). \tag{5.1}$$

Two additive functionals A and B are called *equivalent* if, for all r, μ, the processes $A(r,t]$ and $B(r,t]$ are $\Pi_{r,\mu}$-indistinguishable. A is *continuous* if, for all r, μ, $\Pi_{r,\mu}\{A\{t\} = 0$ for all $t > r\} = 1$. The functional (5.1) is continuous if the measure σ is diffuse.

Let $(Z, \mathcal{Z})$ be a measurable Luzin space. An $\mathcal{M}(Z)$*-valued additive functional* of ξ is a random measure $\Gamma(\omega; dt, dz)$ on $\mathbb{R}_+ \times \mathcal{Z}$ such that $\Gamma(I \times B)$ is $\mathcal{F}(I)$-measurable for every open interval I and every $B \in \mathcal{Z}$. We say that Γ is continuous if $\Gamma([0,t] \times B)$ is continuous for every $B \in \mathcal{Z}$.

5.5. Additive adapted functions. Let $\xi = (\xi_t, \mathcal{F}(I), \Pi_{r,x})$ be a right Markov process. Suppose that a function $Y_t^r : \Omega \to \mathbb{R}$ is given for every pair $r < t \in \mathbb{R}_+$. We say that Y is ξ*-adapted* if, for every $r < t$, the function Y_t^r is measurable relative to $\mathcal{F}[r,t]$. A function Y is *additive* if

$$Y_s^r + Y_t^s = Y_t^r \qquad \text{for all } \; r < s < t \in \mathbb{R}_+.$$

The right continuity in r implies the right continuity in t and vice versa. We write

$$Y\{s\} = \lim_{r \uparrow s, t \downarrow s} Y_t^r$$

(assuming that the limit exists).

To every Borel function $f^t(x)$ there corresponds a ξ-adapted additive function $Y_t^r = f^t(\xi_t) - f^r(\xi_r)$.

If A is an additive functional of ξ such that $A(0,t] < \infty$ for all t, then $A_t^r = A(r,t]$ is a right continuous ξ-adapted function.

An additive ξ*-martingale* is an additive function Y_t^r with the property: for every $r \in \mathbb{R}_+$ and every $x \in E$, Y_t^r is a martingale relative to $(\mathcal{F}[r,t], \Pi_{r,x})$. Additive submartingales, supermartingales, local martingales and semimartingales over ξ are defined analogously. To every finite positive additive functional A there corresponds an additive supermartingales $A_t^r = A(r,t]$.

Note that a ξ-adapted additive function Y is an additive ξ-martingale if and only if

$$\Pi_{r,x} Y_t^r = 0 \qquad \text{for all } \; r < t \in \mathbb{R}_+. \tag{5.2}$$

Every additive ξ-martingale M admits a unique representation as the sum of a continuous and a purely discontinuous parts.

We deal also with multiplicative ξ-martingales. Note that a ξ-adapted multiplicive function Y is an multiplicative ξ-martingale if and only if

$$\Pi_{r,x} Y_t^r = 1 \qquad \text{for all } \; r \le t \in \mathbb{R}_+. \tag{5.3}$$

5.6. Compensator. Suppose that Y is an additive ξ-adapted function and c is a continuous additive functional of ξ. We say that c is the *compensator* of Y if $Y_t^r - c(r,t]$ is an additive ξ-martingale.

5.7. The Lévy measure. The Lévy measure of ξ is a random measure $N(ds, dy)$ on $S = \mathbb{R}_+ \times E$ concentrated on the set $\{(s, y) : \xi_{s-} \neq y\}$ and such that the additive functional

$$\tilde{F}_t^r = \int_r^t \int_E f^s(\xi_s, y) N(ds, dy)$$

is the compensator of the additive function

$$F_t^r = \sum_{r<s\le t} f^s(\xi_{s-}, \xi_s)$$

for every bounded positive Borel function f on $\mathbb{R}_+ \times E \times E$ with the property $f^s(x, x) = 0$ for all s, x.

A proof of the existence and uniqueness of the Lévy measure for every Hunt process can be found, for instance, in [3,p.41].

If E is a linear space, then we introduce the *modified Lévy measure* L by the formula

$$L(ds, d\nu) = N(ds, \xi_{s\_} + d\nu).$$

It does not charge set $\mathbb{R}_+ \times \{0\}$ and it is a $E \setminus \{0\}$-valued additive functional. Put $\xi_t^r = \xi_t - \xi_r$, $\xi\{s\} = \xi_s - \xi_{s-}$. If $g^s(x, z)$ is a positive Borel function such that $g^s(x, 0) = 0$ for all x, then

$$\int_r^t \int_E g^s(\xi_{s-}, z) L(ds, dz)$$

is the compensator of

$$\sum_{r<s\le t} g^s(\xi_{s-}, \xi\{s\}).$$

5.8. Suppose that $\xi = (\xi_t, \mathcal{F}(I), \Pi_{r,x})$ is a right process and K is a continuous additive functional with the property: $\Pi_{r,x} K(r, t) \to 0$ uniformly in x as $r, t \to s$ for every s. Let Q, ℓ be given by (1.7) and (1.8). For every $\nu \in \mathcal{M}(E)$, we set:

$$|\nu| = \langle 1, \nu \rangle = \nu(E), \quad \nu_x(B) = \nu(B \setminus \{x\}).$$

The existence of a superprocess with branching parameters Q, ℓ is proved in [3, Theorem 5.3.1] under the following conditions:

5.8.A. The functions

$$b^s(x), \int_{\{|\nu|\le 1\}} |\nu_x| \ell^s(x, d\nu), \int_{\mathcal{M}} |\nu| \wedge |\nu|^2 \ell^s(x, d\nu)$$

are bounded on every set $[0, T] \times E$.

5.8.B. The measure

$$\bar{\ell}(A,B) = \int_A K(ds) \int_{\mathcal{M}} \ell^s(\xi_{s-}, d\nu)\nu(B)$$

is dominated by the Lévy measure $N_\xi(ds, dy)$ of the process ξ in the following sense: for every positive Borel function $f^s(x,y)$ such that $f^s(x,x) = 0$ for all s, x,

$$\int f^s(\xi_{s-}, y)\bar{\ell}(ds, dy) \leq \int f^s(\xi_{s-}, y) N_\xi(ds, dy) \quad \text{a.s.}$$

References

1. D. A. Dawson, *Measure-valued Markov processes*, École d'Été de Probabilités de Saint Flour, 1991, Lecture Notes in Math. **1541** (1993), Springer.
2. E. B. Dynkin, *Regular Markov processes*, Uspekhi Mat. Nauk **28, 2** (1973), 35 - 64, Reprinted in Markov Processes and Related Problems of Analysis, London Mathematical Society Lecture Note Series, vol. 54, Cambridge Univ. Press, London, New York, 187 -218.
3. ———, *An Introduction to Branching Measure-Valued Processes*, American Mathematical Society, Providence, Rhode Island, 1994.
4. ———, *Branching particle systems and superprocesses*, Ann. Probab. **19** (1991), 1157-1194.
5. ———, *Path processes and historical superprocesses*, Probab. Th. Rel. Fields **90** (1991), 89-115.
6. ———, *On regularity of superprocesss*, Probab. Th. Rel. Fields **95** (1993), 263-281.
7. E. B. Dynkin and S.E. Kuznetsov, *Linear additive functionals of superdiffusions and related nonlinear P.D.E.*, preprint (1995).
8. ———, *Nonlinear parabolic P.D.E. and additive functionals of superdiffusions*, preprint (1995).
9. N. El Karoui and S. Roelly-Coppoletta, *Propriétés de martingales, explosion et représentation de Lévy-Khintchine d'une classe de processus de branchement à valeurs mesures*, Stochastic Processes Appl. **38** (1991), 239-266.
10. P. J. Fitzsimmons, *Construction and regularity of measure-valued Markov branching processes*, Israel J. Math. **64** (1988), 337-361.
11. ———, *On the martingale problem for measure-valued branching processes*, Seminar on Stochastic processes 1991 (E. Cinlar, K. L. Chung, M.J.Sharpe, eds.), Birkhauser, Boston, 1992.

A class of integration by parts formulae in stochastic analysis I

K. D. Elworthy and Xue-Mei Li

Mathematics Institute, University of Warwick, Coventry CV4 7AL, UK

1. Introduction

Consider a Stratonovich stochastic differential equation

$$dx_t = X(x_t) \circ dB_t + A(x_t)dt \tag{1.1}$$

with C^∞ coefficients on a compact Riemannian manifold M, with associated differential generator $\mathcal{A} = \frac{1}{2}\Delta_M + Z$ and solution flow $\{\xi_t : t \geq 0\}$ of random smooth diffeomorphisms of M. Let $T\xi_t : TM \to TM$ be the induced map on the tangent bundle of M obtained by differentiating ξ_t with respect to the initial point. Following an observation by A. Thalmaier we extend the basic formula of [EL94] to obtain

$$\mathbb{E}dF\left(T\xi_\cdot\left(h_\cdot\right)\right) = \mathbb{E}F\left(\xi_\cdot(x)\right)\int_0^T \left\langle T\xi_s\left(\dot{h}_s\right), X\left(\xi_s(x)\right)dB_s\right\rangle \tag{1.2}$$

where $F \in \mathcal{F}C_b^\infty(C_x(M))$, the space of smooth cylindrical functions on the space $C_x(M)$ of continuous paths $\gamma : [0,T] \to M$ with $\gamma(0) = x$, dF is its derivative, and $h_\cdot$ is a suitable adapted process with sample paths in the Cameron-Martin space $L_0^{2,1}([0,T]; T_xM)$. Set $\mathcal{F}_t^x = \sigma\{\xi_s(x) : 0 \leq s \leq t\}$. Taking conditional expectation with respect to $\mathcal{F}_T^x$, formula (1.2) yields integration by parts formulae on $C_x(M)$ of the form

$$\mathbb{E}dF(\gamma)(\bar{V}^h) = \mathbb{E}F(\gamma)\delta\overline{V}^h(\gamma) \tag{1.3}$$

where $\bar{V}^h$ is the vector field on $C_x(M)$

$$\bar{V}^h(\gamma)_t = \mathbb{E}\left\{T\xi_t(h_t)\,|\xi_\cdot(x) = \gamma\right\}$$

and $\delta\overline{V}^h : C_x(M) \to \mathbb{R}$ is given by

$$\delta\overline{V}^h(\gamma) = \mathbb{E}\left\{\int_0^T < T\xi_s(\dot{h}_s), X(\xi_s(x))dB_s > |\xi_\cdot(x) = \gamma\right\}.$$

When $h_\cdot$ is adapted to $\mathcal{F}_\cdot^x$ results from [ELJL95] extending [EY93] give explicit expressions for $\bar{V}^h$ and $\delta\bar{V}^h$ in terms of the Ricci curvature of the LeJan-Watanabe connection associated to (1.1). Equation (1.3) then reduces to a Driver's integration by parts formula, Theorem 3.3 below, but no hypothesis of torsion skew symmetry of the connection is required: the integration by parts formulae follow for the adjoint of any metric connection. In particular for any such connection there is a Hilbert "tangent space" of "good" directions obtained by parallel translation of the Cameron-Martin space of paths in T_xM. (In fact it is the "Ricci flow" or "Dohrn-Guerra parallel translation" (see Nelson [Nel84]), leading to the "damped gradient" ([FM93]) which occurs more naturally.) However, in Remark 2.4, we show that in this case $\bar{V}_h$ is in the class for which integration by parts formulae are known, so that the results of 2.3, 3.3, 3.5 are not claimed to be new in substance.

Although this filtering out of the extraneous noise gives intrinsic results comparable to those of Driver [Dri92], this viewpoint throws away a lot of the structure we have. Moreover integration by parts formulae such as (1.2) should have some connection with quasi-invariance properties of flows associated to the vector fields. Flows for the $\bar{V}^h$ on $C_x(M)$ do not appear to be easy to analyse in general. However in §3 we show that in the context of Diff M valued processes there are very natural flows associated and (1.2) has a rather natural geometric interpretation. This leads to another elementary proof of (1.2) and in Theorem 4.1 we use this method to obtain integration by parts formulae for the free path space.

There are at least 3 proofs of (1.2). The first given here is via Itô's formula and elementary martingale calculus (it requires F to be cylindrical), the second given here is based on the Girsanov-Maruyama theorem (and works for more general F), and a third method would be to deduce it from the standard integration by parts formula on Wiener space applied to the functional $F \circ \xi$, c.f. [Bis81]. Indeed this work was stimulated by D. Bell and D. Nualart pointing out that this third approach could be used to deduce the basic formula of [EL94]. The point made (and carried out) in [Elw92] and [EL94] that the first approach can be applied directly to 'Ricci flows' instead of derivative flows to give intrinsic formulae without stochastic flows, also needs to be emphasized: see also [SZ]. As such it gives the details of how 'Bismut's formula' (essentially integration by parts when F is a function of paths evaluated at just one time t) leads to the full integration by parts formula.

There are also now many proofs of Driver's results for $C_x(M)$ and for the free path space and their extensions. See [Hsu95], [ES95], [LN] (with a very concise proof), [AM], [Aid], and [CM].

Acknowledgment: This research was supported by SERC grant GR/H67263 and stimulated and helped by our contacts with A. Thalmaier.

2. The integration by parts formula from finite dimensional manifolds to path spaces

In this section we deduce by induction an integration by parts formula on the path space from a formula on the base manifold M. The key is to obtain formula (2.7) for M.

Let $h : \Omega\times[0,T] \to T_xM$ be an adapted process with $h(\omega) : [0,T] \to T_xM$ in $L^{2,1}$ for almost all ω.

Lemma 2.1. *If $h : \Omega \times [0,T] \to T_xM$ is adapted, $L^{2,1}$ for a.s. ω and $\left(\int_0^T |\dot{h}_s|^2 ds\right)^{1/2} \in L^{1+\epsilon}$ for some $\epsilon > 0$. Then for $t < T$,*

$$\begin{aligned}&\mathbb{E}\left\{\int_0^t < T\xi_s(\dot{h}_s), X(\xi_s(x))dB_s > |\xi_T(x)\right\}\\ &= \mathbb{E}\left\{\int_t^T < T\xi_s(-), X(\xi_s(x))dB_s > \frac{h_t - h_0}{T-t}\, |\xi_T(x)\right\}.\end{aligned} \tag{2.1}$$

If furthermore $h.$ is non-random then for $t \le T$,

$$\begin{aligned}&\mathbb{E}\left\{\int_0^t < T\xi_s(\dot{h}_s), X(\xi_s(x))dB_s > |\xi_T(x)\right\}\\ &= \mathbb{E}\left\{\int_0^t < T\xi_s(-), X(\xi_s(x))dB_s > \left(\frac{h_t - h_0}{t}\right) |\xi_T(x)\right\}.\end{aligned} \tag{2.2}$$

Proof. First by the Burkholder-Davis-Gundy inequality, for some constant c_1,

$$\mathbb{E}\left|\int_0^T < T\xi_s(\dot{h}_s), X(\xi_s(x))dB_s >\right| \le c_1\mathbb{E}\left(\int_0^T |T\xi_s(\dot{h}_s)|^2 ds\right)^{\frac{1}{2}}$$

$$\le c_1\left(\mathbb{E}\sup_{0\le s\le T} |T_x\xi_s|^{\frac{1+\epsilon}{\epsilon}}\right)^{\frac{\epsilon}{1+\epsilon}} \left[\mathbb{E}\left(\int_0^T |\dot{h}_s|^2 ds\right)^{\frac{1+\epsilon}{2}}\right]^{\frac{1}{1+\epsilon}}.$$

This is finite since $\sup_{0\le s\le t} |T_x\xi_s| \in L^q$ for all $1 \le q < \infty$, e.g. see [Li94]. Moreover, since the adapted processes in $L^\infty(\Omega, \mathcal{F}, \mathbb{P}; C^1([0,T]; T_xM))$ are dense in the subspace of adapted processes in $L^{1+\epsilon}(\Omega, \mathcal{F}, \mathbb{P}; L^{2,1}([0,T]; T_xM))$, this estimate allows us to assume that h belongs to the former space.

Set $M_t = \int_0^t < T_x\xi_s(-), X(\xi_s(x))dB_s >$. Then $\{M.\}$ is a T_x^*M valued local martingale. If $0 = t_0 < t_1 < \ldots < t_l = t$ is a partition of $[0,t]$, $\Delta_j t = t_{j+1} - t_j$, and $\Delta_j M = M_{t_{j+1}} - M_{t_j}$, then

$$\sum_{j=1}^{l-1} \Delta_j M(\dot{h}_{t_j}) \to \int_0^t \dot{h}_s dM_s = \int_0^t < T\xi_s(\dot{h}_s), X(\xi_s(x))dB_s > \tag{2.3}$$

and the convergence is in L^1.

On the other hand if $v_0 \in T_xM$ and P_t is the probabilistic semigroup associated to the S.D.E. and f a bounded measurable function then

$$d(P_Tf)(v_0) = \frac{1}{T}\mathbb{E}f(\xi_T(x)) \int_0^T \langle T\xi_s(v_0), X(\xi_s(x))dB_s\rangle . \tag{2.4}$$

See [EL94]. However by an observation of Thalmaier: the same proof shows that for any $r, h \in [0,T]$ with $h > 0$ and $r + h \leq T$

$$d(P_Tf)(v_0) = \frac{1}{h}\mathbb{E}f(\xi_T(x)) \int_r^{r+h} \langle T\xi_s(v_0), X(\xi_s(x))dB_s\rangle$$

c.f. [SZ]. From these two formulae we obtain:

$$\begin{aligned}&\mathbb{E}\left\{\tfrac{1}{T}\int_0^T < T\xi_s(v_0), X(\xi_s(x))dB_s > |\xi_T(x)\right\}\\ &= \mathbb{E}\left\{\tfrac{1}{h}\int_r^{r+h} < T\xi_s(v_0), X(\xi_s(x))dB_s > |\xi_T(x)\right\}.\end{aligned} \tag{2.5}$$

For any $0 \leq r \leq T$, let $\{\xi_s^r(x) : r \leq s \leq T, x \in M\}$ be the solution flow to (1.1) starting from x at time r. The flow $\xi_\cdot^r$ can be taken to be adapted to a filtration $\{\mathcal{F}_s^r : r \leq s \leq T\}$ independent of $\mathcal{F}_r$, and then we have $\xi_s^r\xi_r = \xi_s$, almost surely, $r \leq s \leq T$. From this, time homogeneity, and (2.5),

$$\begin{aligned}&\mathbb{E}\left\{\sum_{j=1}^{l-1} \Delta_j M(\dot{h}_{t_j})\,|\xi_T(x)\right\}\\
&= \mathbb{E}\left\{\sum_{j=1}^{l-1} \tfrac{\Delta_j t}{\Delta_j t} \int_{t_j}^{t_{j+1}} \left\langle T\xi_s^{t_j}\left(T\xi_{t_j}\left(\dot{h}_{t_j}\right)\right), X\left(\xi_s^{t_j}\left(\xi_{t_j}(x)\right)\right) dB_s\right\rangle \left|\xi_T^{t_j}(\xi_{t_j}(x))\right.\right\}\\
&= \mathbb{E}\left\{\sum_{j=1}^{l-1} \tfrac{\Delta_j t}{T-t} \int_t^T \left\langle T\xi_s^{t_j}\left(T\xi_{t_j}(\dot{h}_{t_j})\right), X\left(\xi_s^{t_j}\left(\xi_{t_j}(x)\right)\right) dB_s\right\rangle \left|\xi_T^{t_j}(\xi_{t_j}(x))\right.\right\}\\
&= \mathbb{E}\left\{\sum_{j=1}^{l-1} \tfrac{\Delta_j t}{T-t} \int_t^T < T\xi_s(\dot{h}_{t_j}), X(\xi_s(x))dB_s > |\xi_T(x)\right\}\\
&\to \mathbb{E}\left\{\int_t^T < T\xi_s(-), X(\xi_s(x))dB_s > \tfrac{h_t - h_0}{T-t}\,|\xi_T(x)\right\}.\end{aligned}$$

Comparing with (2.3) this gives the first required identity. When $h_\cdot$ is non-random the second follows immediately from (2.5). ∎

Remark:

As in [SZ] a further modification is possible replacing (2.5) by:

$$\begin{aligned}&\frac{1}{T}\mathbb{E}\left\{\int_0^T < T\xi_s(v_0), X(\xi_s(x))dB_s > |\xi_T(x)\right\}\\
&= \frac{1}{\int_0^T \Psi(r)dr}\mathbb{E}\left\{\int_0^T \Psi(s) < T\xi_s(v_0), X(\xi_s(x))dB_s > |\xi_T(x)\right\}\end{aligned}$$

for $\Psi : [0,T] \to \mathbb{R}$ integrable with $\int_0^T \Psi(r)dr \neq 0$. The argument leads to, for non-random h,

$$\begin{aligned}&\mathbb{E}\left\{\int_0^t < T\xi_s(\dot{h}_s), X(\xi_s(x))dB_s > |\xi_T(x)\right\}\\ &= \mathbb{E}\left\{\int_0^t \Psi(s) < T\xi_s(-), X(\xi_s(x))dB_s > \left(\frac{h_t - h_0}{\int_0^t \Psi(r)dr}\right) |\xi_T(x)\right\}.\end{aligned} \tag{2.6}$$

Corollary 2.2. *Under the conditions of the lemma, for any C^1 function $f : M \to \mathbb{R}$,*

$$\mathbb{E}f\left(\xi_T(x)\right) \int_0^T < T\xi_s(\dot{h}_s), X(\xi_s(x))dB_s >= \mathbb{E}df\left(T\xi_T\left(h_T - h_0\right)\right). \tag{2.7}$$

Proof. First by the composition property of solution flows,

$$\begin{aligned}&\mathbb{E}\left\{\int_t^T < T\xi_s(-), X(\xi_s(x))dB_s > \frac{h_t - h_0}{T-t} |\xi_T(x)\right\}\\ &= \mathbb{E}\left\{\int_t^T < T\xi_s^t(-), X(\xi_s^t(\xi_t(x)))dB_s > \frac{T\xi_t(h_t - h_0)}{T-t} |\xi_T^t(\xi_t(x))\right\}.\end{aligned}$$

As in the proof of the lemma, (2.1) yields

$$\begin{aligned}&\mathbb{E}f(\xi_T(x)) \int_0^t < T\xi_s(\dot{h}_s), X(\xi_s(x))dB_s >\\ &= \mathbb{E}f(\xi_T^t(\xi_t(x))) \int_t^T \langle T\xi_s^t(-), X(\xi_s^t(\xi_t(x))dB_s\rangle \frac{T\xi_t(h_t - h_0)}{T-t}\\ &= \mathbb{E}\left\{dP_{T-t}(f)\left(T\xi_t(h_t - h_0)\right)\right\}\end{aligned}$$

by [EL94], since $\mathcal{F}_\cdot^t$ is independent of $\mathcal{F}_t$. Now let t increase to T and the required result follows. ∎

Next consider a cylindrical function F on $C_x(M)$, the space of continuous paths with base point x. Write

$$F(\gamma_\cdot) = f(\gamma_{t_1}, \ldots, \gamma_{t_k}),$$

for $(t_1, \ldots, t_k) \in [0,T]^k$, $\gamma \in C_x(M)$ and f a smooth function on M^k. Suppose $h_0 = 0$ and consider the tangent vector field $V^h(\xi_\cdot(x))$ along $\{\xi_t(x) : 0 \le t \le T\}$ on $C_x(M)$ given by

$$V^h(\xi_\cdot)_t = T_x\xi_t(h_t).$$

Then

$$dF(V^h(\xi_\cdot)) = \sum_{j=1}^k d^j f_{\underline{\xi_t}}\left(V^h(\xi_\cdot)_{t_j}\right). \tag{2.8}$$

Here $\xi_{\underline{t}} = (\xi_{t_1}, \ldots, \xi_{t_k})$ and $d^j f$ is the partial derivative of f in the jth direction.

Let

$$\delta V^h(\xi.) = \int_0^T < T_x\xi_s(\dot{h}_s), X(\xi_s(x))dB_s > .$$

Theorem 2.3. *Let $h : [0,T] \times \Omega \to T_xM$ be an adapted stochastic process with almost surely all $h(\omega) \in L_0^{2,1}$ and $\mathbb{E}\left(\int_0^T |\dot{h}_s|^2 ds\right)^{\frac{1+\epsilon}{2}} < \infty$ for some $\epsilon > 0$. Then*

$$\mathbb{E}dF(V^h(\xi.)) = \mathbb{E}F(\xi.(x))\delta V^h(\xi.). \tag{2.9}$$

Proof. We prove by induction on k. When $k = 1$, this is just (2.7), the formula for functions. Let $\Omega = C_0([0,T];\mathbb{R}^n)$ be the canonical probability space. We set $\Omega_1 = C_0([0,t_1];\mathbb{R}^n)$ and $\Omega_2 = C_0([t_1,T];\mathbb{R}^n)$. There is then the standard decomposition of filtered spaces

$$\begin{aligned} &\{\Omega, \mathcal{F}, \mathcal{F}_t, 0 \le t \le T, \mathbb{P}\} \\ = \ &\{\Omega_1, \mathcal{F}, \mathcal{F}_t, 0 \le t \le t_1, \mathbb{P}_1\} \times \{\Omega_2, \mathcal{F}, \mathcal{F}_t^{t_1}, t_1 \le t \le T, \mathbb{P}_2\} \end{aligned}$$

in the sense that $\mathcal{F}_t = \mathcal{F}_t * \Omega_2$ if $t \le t_1$, and $\mathcal{F}_t = \mathcal{F}_{t_1} * \mathcal{F}_t^{t_1}$ if $t \ge t_1$. As before let $\xi_t^{t_1}(y_0), t_1 \le t \le T, y_0 \in M$ be the solution flow to (1.1) starting at time t_1, i.e. $\xi_{t_1}^{t_1}(y_0) = y_0$. We will consider it as a function of $\omega_2 \in \Omega_2$, adapted to $\mathcal{F}_\cdot^{t_1}$, while $\{\xi_t : 0 \le t \le t_1\}$ will be considered on Ω_1, and $\{\xi_t : t_1 \le t \le T\}$ on $\Omega_1 \times \Omega_2 = \Omega$. The composition property for flows gives

$$\xi_t^{t_1}\left(\xi_{t_1}(x, \omega_1), \omega_2\right) = \xi_t\left(x, (\omega_1, \omega_2)\right), \quad \text{each } t_1 \le t \le T, a.s.$$

Assume the required result holds for cylindrical functions depending on $k-1$ times, some $k \in \{2, 3 \ldots\}$. Take $y_0 \in M$ and define $f_1^{y_0} : M^{k-1} \to \mathbb{R}$ and $F_1^{y_0} : \Omega_2 \to \mathbb{R}$ by:

$$f_1^{y_0}(x_1, \ldots, x_{k-1}) = f(y_0, x_1, \ldots, x_{k-1})$$

and

$$F_1^{y_0}(\omega_2) = f(y_0, \xi_{t_2}^{t_1}(y_0, \omega_2), \ldots, \xi_{t_k}^{t_1}(y_0, \omega_2)).$$

Take $h_\cdot^1 : \Omega_2 \to L_0^{2,1}([t_1,T]; T_{y_0}M)$, adapted to $\mathcal{F}_\cdot^{t_1}$, and with $\mathbb{E}(\int_{t_1}^T |\dot{h}_s^1|^2 ds)^{\frac{1+\epsilon}{2}}$ finite. By time homogeneity our inductive hypothesis gives

$$\begin{aligned} &\sum_{j=2}^k \int_{\Omega_2} d^j f\left(y_0, \xi_{t_2}^{t_1}(y_0,\omega_2), \ldots, \xi_{t_k}^{t_1}(y_0,\omega_2)\right) \left(T\xi_{t_j}^{t_1}(h_{t_j}^1(\omega_2), \omega_2)\right) d\mathbb{P}_2(\omega_2) \\ &= \int_{\Omega_2} f\left(y_0, \xi_{t_2}^{t_1}(y_0,\omega_2), \ldots, \xi_{t_k}^{t_1}(y_0,\omega_2)\right) \times \\ &\quad \int_{t_1}^T \left\langle T\xi_r^{t_1}(\dot{h}_r^1(\omega_2), \omega_2), X(\xi_r^{t_1}(y_0,\omega_2))dB_r(\omega_2)\right\rangle d\mathbb{P}_2(\omega_2). \end{aligned} \tag{2.10}$$

Now for $\omega_1 \in \Omega_1$ (outside of a certain measure zero set) we can take $y_0 = \xi_{t_1}(x_0, \omega_1)$ and

$$h_t^1(\omega_2) = T\xi_{t_1}\left(h_t(\omega_1, \omega_2) - h_{t_1}(\omega_1), \omega_1\right).$$

Then, for almost all $\omega_1 \in \Omega_1$, we have $h^1_\cdot$ adapted to $\mathcal{F}_\cdot^{t_1}$. Substitute this in (2.10). Using the composition property, and then integrating over Ω_1 yields

$$\begin{aligned} &\textstyle\sum_{j=2}^k \mathbb{E} d^j f(\xi_{\underline{t}}) \left(T\xi_{t_j}(h_{t_j} - h_{t_1})\right) \\ &= \mathbb{E} f(\xi_{\underline{t}}(x)) \textstyle\int_{t_1}^T \left\langle T\xi_r(\dot{h}_r), X(\xi_r(x)) dB_r \right\rangle. \end{aligned} \tag{2.11}$$

On the other hand we can define $g : M \to \mathbb{R}^1$ by

$$g(x) = \int_{\Omega_2} f\left(x, \xi_{t_2}^{t_1}(x, \omega_2), \ldots, \xi_{t_k}^{t_1}(x, \omega_2)\right)$$

and apply formula (2.7) to g to obtain:

$$\begin{aligned} &\int_{\Omega_1} dg(T\xi_{t_1}(h_{t_1})) d\mathbb{P}_1(\omega_1) \\ &= \int_{\Omega_1} g(\xi_{t_1}(x)) \int_0^{t_1} \left\langle T\xi_r(\dot{h}_r)), X(\xi_r(x_0)) dB_r \right\rangle d\mathbb{P}_1(\omega_1). \end{aligned}$$

But note that

$$\int_{\Omega_1} dg(T\xi_{t_1}(h_{t_1})) d\mathbb{P}_1(\omega_1) = \sum_{j=1}^k \mathbb{E} d^k f_{\xi_{\underline{t}}}(T\xi_{t_j}(h_{t_1})),$$

and therefore

$$\sum_{j=1}^k \mathbb{E} d^j f_{\xi_{\underline{t}}}(T\xi_{t_j}(h_{t_1})) = \mathbb{E} f(\xi_{\underline{t}}) \int_0^{t_1} \left\langle T\xi_r(\dot{h}_r), X(\xi_r(x)) dB_r \right\rangle \tag{2.12}$$

Adding (2.11) we arrive at (2.9):

$$\sum_{j=1}^k \mathbb{E} d^j f_{\xi_{\underline{t}}}(T\xi_{t_j}(h_{t_j})) \quad = \quad \mathbb{E} f(\xi_{\underline{t}}(x)) \int_0^T \left\langle T\xi_r(\dot{h}_r), X(\xi_r(x)) dB_r \right\rangle.$$

∎

B. Let $\tilde{\nabla}$ be a metric connection for the manifold M with torsion T, and $\tilde{\nabla}'$ its adjoint connection defined by

$$\tilde{\nabla}'_{V_1} V_2 = \tilde{\nabla}_{V_1} V_2 - T(V_1, V_2).$$

Here V_1, V_2 are vector fields. Let $\tilde{R}$ be the curvature tensor of $\tilde{\nabla}$ and define $\tilde{Ric}^{\#} : TM \to TM$ by $\tilde{Ric}^{\#}(v) = \text{trace}\, \tilde{R}(v, -)-$. If $\{x_s\}$ is a diffusion on M with generator $\frac{1}{2}\text{trace}\tilde{\nabla}\text{grad} + L_Z$ denote by $\tilde{//}_s$ the parallel transport along $\{x_s\}$, and $\{\tilde{B}_s : 0 \le s \le t\}$ the martingale part of the anti-development of $\{x_s : 0 \le s \le t\}$ using $\tilde{//}_s$, a Brownian motion on $T_{x_0}M$. Let $v_s = \tilde{W}_s^Z(v_0)$ be the solution to

$$\frac{\tilde{D}'}{\partial s} v_s = -\frac{1}{2}\tilde{\text{Ric}}^{\#}(v_s) + \tilde{\nabla} Z(v_s)$$

starting from $v_0 \in T_{x_0}M$. Here $\tilde{D}'$ denotes the covariant differentiation along the paths of $\{x_t\}$ using the adjoint connection. We will show that (2.9) implies Driver's integration by parts formula. However we do not need to assume $\tilde{\nabla}'$ (or equivalently $\tilde{\nabla}$) is torsion skew symmetric.

Corollary 2.4. *Let F be a cylindrical function on $C_{x_0}(M)$. Suppose $h : [0,T] \times \Omega \to T_{x_0}M$ is adapted to the filtration of $\{x_s : 0 \le s < \infty\}$ and such that $h(\omega)$ is in $L_0^{2,1}$ for almost all ω and $h \in L^{1+\epsilon}\left(\Omega, \mathcal{F}, \mathbb{P}; L_0^{2,1}([0,T]; T_{x_0}M)\right)$ for some $\epsilon > 0$. Then*

$$\mathbb{E} dF(\tilde{W}_{\cdot}^Z(h_{\cdot})) = \mathbb{E} F(\xi_{\cdot}(x_0)) \int_0^T < \tilde{W}_s^Z(\dot{h}_s), \tilde{//}_s d\tilde{B}_s > . \tag{2.13}$$

When $\tilde{\nabla}'$ is metric for some Riemannian metric on M, it suffices to have $h \in L^1\left(\Omega, \mathcal{F}, \mathbb{P}; L_0^{2,1}([0,T])\right)$.

Proof. By a result of [ELJL95] we can choose X such that $\tilde{\nabla}$ equals the Le Jan-Watanabe connection induced from the stochastic differential equation

$$dx_t = X(x_t) \circ dB_t + Z(x_t)dt$$

and the solution flow $\{\xi_{\cdot}(x)\}$ has generator $\frac{1}{2}\text{trace}\tilde{\nabla}\text{grad} + L_Z$ (c.f. Corollary 3.4 of [ELJL95]). Moreover the conditioned process of the derivative flow $T\xi_t(v_0)$ with respect to the natural filtration of $\{\xi_{\cdot}(x_0)\}$ is given by $\{\tilde{W}_{\cdot}^Z(v_0)\}$:

$$\mathbb{E}\{T\xi_t(v_0) \mid \mathcal{F}_T^{x_0}\} = \tilde{W}_t^Z(v_0),$$

by Theorem 3.2 of [ELJL95] extending [EY93]. The result follows since $\tilde{B}_t$ equals $\int_0^t \tilde{//}_s^{-1} X(\xi_s(x_0)) dB_s$.

If $\tilde{\nabla}'$ is metric for some Riemannian metric then $\sup_{0 \le s \le t} |\tilde{W}_s^Z|$ is in $L^\infty(\Omega, \mathcal{F}, \mathbb{P})$ and so the Burkholder-Davis-Gundy inequality used as in the proof of Lemma 2.1 allows us to take $\epsilon = 0$. ∎

Remarks 2.5. (i). Let $S : TM \times TM \to TM$ be a tensor fields of type (1,2), and let ∇ refer to the Levi-Civita connection of M. Then, by [KN69] p.146, a connection $\tilde{\nabla}$ can be defined by

$$\tilde{\nabla}_{V_1}(V_2) = \nabla_{V_1}(V_2) + S(V_1, V_2)$$

for vector fields V_1, V_2. and all linear connections on M can be obtained this way. It is easy to see that $\tilde{\nabla}$ is metric if and only if

$$< S(W,U), V >= - < U, S(W,V) >$$

for all vector fields U, V, W, i.e. if and only if $S(W,-)$ is skew symmetric. On the other hand the adjoint connection is given by

$$\tilde{\nabla}'_{V_1}(V_2) = \nabla_{V_1}(V_2) + S(V_2, V_1)$$

so that it is torsion skew symmetric if also $S(-,W)$ is skew symmetric. In terms of the Levi-Civita connection our vector fields $\bar{V}^h$ for which the integration by parts formula hold therefore satisfy an equation of the form

$$D\bar{V}_t^h = -S(\bar{V}_t^h, \circ dx_t) + A_t(\bar{V}_t^h)dt + W_t^h(\dot{h}_t)dt + \nabla A(\bar{v}_t^h)dt$$

where A_t is linear (also depending on S). In particular they are "tangent processes" in the sense proposed by Driver, for which integration by parts formulae are known: see [Dri95b], [CM], [AM], and [Aid], [Dri95a].

(ii) For cylinder functions depending on one time only such integration by parts formulae go back to Bismut [Bis84].

3. Geometric intepretation and a shorter proof

A. The processes $T_x\xi_t(h_t)$ cannot strictly speaking be considered as tangent vectors or vector fields on $C_x(M)$. In some sense they form tangent vectors at $\xi_\cdot(x,-)$ to the space of processes (or semi-martingales)

$$[0,T] \times \Omega \to M$$

since $T_x\xi_t(h_t(\omega),\omega) \in T_{\xi_t(x,\omega)}M$ for $(t,\omega) \in [0,T] \times \Omega$ or equivalently as 'tangent vectors' to the space of random variables

$$\Omega \to C_x(M)$$

at $\omega \mapsto \xi_\cdot(x,\omega)$. However c.f. [Dri92] there is still no natural associated flow. In fact the most natural interpretation takes into account the variable x and replaces $C_x(M)$ by $P_{id}\mathrm{Diff}M$ the space of paths on the diffeomorphism group of M, as we now describe.

Let Diff$M$ be the space of C^∞ diffeomorphisms of M. We can consider it with a rather formal differential structure or if the reader prefers it can be replaced by a suitable Sobolev space of diffeomorphisms, to give a Hilbert manifold (as in [Elw82] following [EM70]). In any case the tangent space $T_\alpha(\mathrm{Diff}M)$ will be identified with all vector fields on M over α i.e. smooth

$v : M \to TM$ such that $v(x) \in T_{\alpha(x)}M$ for all $x \in M$. If $P\mathrm{Diff}M$ refers to continuous paths $\phi : [0,T] \to \mathrm{Diff}M$ with $\phi(0) = id_M$ then $T_\phi P\mathrm{Diff}M$ will be identified with continuous $v : [0,T] \to T\mathrm{Diff}M$ vanishing at $t = 0$, such that $v(t) \in T_{\phi(t)}\mathrm{Diff}M$, or equivalently $v : [0,T] \times M \to TM$ with $v(t)(x) \in T_{\phi(t)(x)}M$.

B. Given our S.D.E. (1.1) now take $h \in L_0^{2,1}([0,T];\mathbb{R}^n)$. There is $X^{h_\cdot}$, the time dependent vector field $X(\cdot)(h_t)$ on M. From this we obtain a field U^h on $P\mathrm{Diff}M$ by

$$U^h(\phi)_t(x) = T_x\phi_t(X(x)h_t). \tag{3.1}$$

This is just the left invariant vector field on $P\mathrm{Diff}M$ corresponding to $X^{h_\cdot} \in T_eP\mathrm{Diff}M$ for $e(t) = id_M$, $0 \le t \le T$.

For each $0 \le t \le T$ let $H_t^\tau : M \to M$, $\tau \in \mathbb{R}$ be the solution flow to the vector field $X(\cdot)(h_t)$ so

$$\begin{cases} \frac{\partial}{\partial\tau}H_t^\tau(x) & = & X(H_t^\tau(x))h_t \\ H_t^0(x) & = & x. \end{cases} \tag{3.2}$$

Lemma 3.1. *The vector field U^h on PDiffM has solution flow $\Phi_\tau : PDiffM \to PDiffM$, $\tau \in \mathbb{R}$ given by $\Phi_\tau(\phi)_t(x) = \phi_t(H_t^\tau(x))$.*

Proof. By left invariance we can suppose $\phi = e$. We then need only to observe that

$$\frac{\partial}{\partial\tau}H_t^\tau(x) = TH_t^\tau\,(X(x)h_t)$$

for each $0 \le t \le T$: a standard property of ordinary, time-independent dynamical systems which is seen by differentiating the identity

$$H_t^{\tau+\sigma} = H_t^\tau \circ H_t^\sigma(x)$$

with respect to σ at $\sigma = 0$. ∎

C. In the case where h is random, with $h : \Omega \to L_0^{2,1}([0,T];\mathbb{R}^d)$ adapted, we can use the same notation to obtain a variation of our stochastic flow $\{\xi_t : 0 \le t \le T\}$ on M generated by the vector field V^h, and given explicitly by

$$\xi_\cdot^\tau = \Phi_\tau(\xi_\cdot),$$

i.e.

$$\xi_t^\tau(x) = \xi_t(H_t^\tau(x)). \tag{3.3}$$

In particular

$$\frac{\partial}{\partial\tau}\xi_t^\tau(x)\,|_{\tau=0} = T\xi_t\,(X(x)h_t)\,. \tag{3.4}$$

Using the structure of $C_x(M)$ as a C^∞ Banach manifold let $BC^1(C_x(M))$ be the space of C^1 maps $F : C_x(M) \to \mathbb{R}$ such that there is a constant $|dF|_\infty$ with

$$|dF(v)| \leq |dF|_\infty \sup_{0\leq t\leq T} |v_t| \tag{3.5}$$

for all tangent vectors $v : [0,T] \to TM$ to $C_x(M)$. Set $V_t^{X(h)}(x) = T\xi_t\left(X(x)(h_t)\right)$, which gives rise to a vector field along $\{\xi_\cdot(x)\}$ on $C_x(M)$.

Proposition 3.2. *Suppose $h : [0,T] \times \Omega \to T_x M$ is adapted, belongs to $L_0^{2,1}$ a.s. and such that $\mathbb{E}\left(\int_0^T |\dot{h}_s|^2 ds\right)^{\frac{1+\epsilon}{2}} < \infty$ for some $\epsilon > 0$. Then for each $x \in M$ the processes $\xi_\cdot^\tau(x)$, $\tau \in \mathbb{R}$ have mutually equivalent laws $\mathbb{P}_\tau^x$, $\tau \in \mathbb{R}$ on $C_x(M)$ with*

$$\frac{d\mathbb{P}_\tau^x}{d\mathbb{P}_0^x} =$$

$$exp\left\{\int_0^T \langle X(\xi_s^\tau(x))^* T\xi_s\left(\frac{\partial}{\partial s}H_s^\tau(x)\right), dB_s\rangle - \frac{1}{2}\int_0^T |T\xi_s\left(\frac{\partial}{\partial s}H_s^\tau(x)\right)|^2 ds\right\}.$$

Moreover, for any $F \in BC^1(C_x(M))$,

$$\mathbb{E}dF(V_\cdot^{X(h)}) = \mathbb{E}F(\xi_\cdot)\int_0^T \left\langle X(\xi_s(x))dB_s, V_s^{X(\dot{h})}(x)\right\rangle.$$

Proof. For the equivalent part note that $\{\xi_t^\tau : 0 \leq t \leq T\}$ satisfies the equation:

$$d\xi_t^\tau(x) = X\left(\xi_t^\tau(x)\right) \circ dB_t + A(\xi_t^\tau(x))dt + T\xi_t\left(\frac{\partial}{\partial t}H_t^\tau(x)\right) dt.$$

A straightforward argument shows that

$$\int_0^T \left|X(\xi_s^\tau(x))^* T\xi_s\left(\frac{\partial}{\partial s}H_s^\tau(x)\right)\right|^2 ds < \infty, \quad a.s.$$

Therefore if we set

$$M_t^\tau = \int_0^t \left\langle X(\xi_s^\tau(x))^* T\xi_s\left(\frac{\partial}{\partial s}H_s^\tau(x)\right), dB_s\right\rangle,$$

then by the Girsanov-Maruyama theorem, $\mathbb{P}_\tau^x$ is equivalent to $\mathbb{P}_0^x$ and

$$\frac{d\mathbb{P}_\tau^x}{d\mathbb{P}_0^x} = e^{M_T^\tau - \frac{1}{2}<M^\tau>_T}. \tag{3.6}$$

Consequently,

$$\mathbb{E}F(\xi_\cdot^\tau(x)) = \mathbb{E}F(\xi_\cdot(x))\frac{d\mathbb{P}_\tau^x}{d\mathbb{P}_0^x}.$$

Now suppose $h.$ and $\int_0^{\cdot} |\dot{h}_s|^2 ds$ are bounded on $[0,T] \times \Omega$. Differentiating with respect to τ at $\tau = 0$ and using (3.2) gives

$$\mathbb{E} dF(T\xi.(X(x)h.)) = \mathbb{E} F(\xi.(x)) \frac{\partial}{\partial \tau} \left(\frac{d\mathbb{P}^x_\tau}{d\mathbb{P}^x_0} \right)_{\tau=0},$$

since $|dF|$ is bounded and $\sup_{0 \le s \le T} |T\xi_s| \in \cap_{1 \le p < \infty} L^p$.

The second statement follows from differentiation of (3.6), using the fact that $\left(\frac{d\mathbb{P}^x_\tau}{d\mathbb{P}^x_0} \right)_{\tau=0} = 1$ and $\frac{\partial}{\partial t} H^\tau_t(x) |_{\tau=0} = 0$:

$$\begin{aligned}
\frac{\partial}{\partial \tau} \left(\frac{d\mathbb{P}^x_\tau}{d\mathbb{P}^x_0} \right)_{\tau=0} &= \left(\frac{d\mathbb{P}^x_\tau}{d\mathbb{P}^x_0} \right)_{\tau=0} \cdot \left[\left(\frac{\partial}{\partial \tau} M^\tau_T \right)_{\tau=0} - \frac{1}{2} \left(\frac{\partial}{\partial \tau} \langle M^\tau_T \rangle \right)_{\tau=0} \right] \\
&= \int_0^T \left\langle X(\xi^\tau_s(x)) dB_s, \frac{D}{\partial \tau} \left[T\xi_s \left(\frac{\partial}{\partial s} H^\tau_s(x) \right) \right] \right\rangle_{\tau=0} \\
&= \int_0^T \left\langle X(\xi_s(x)) dB_s, T\xi_s \left(\frac{D}{\partial s} X(H^\tau_s(x)) h_s \right) \Big|_{\tau=0} \right\rangle \\
&= \int_0^T \left\langle X(\xi_s(x)) dB_s, T\xi_s (X(x) \dot{h}_s) \right\rangle .
\end{aligned}$$

For general h take a sequence of bounded h_n which converges to h in $L^{\frac{1+\epsilon}{2}}(\Omega, L^{2,1}_0([0,T]))$ to finish the proof. See the proof of theorem 4.1. ∎

The following is an analogue of Corollary 2.4: here $\tilde{\nabla}$ is any metric connection and $\tilde{W}^Z_\cdot$ is as in Corollary 2.4,

Theorem 3.3. *Let $F \in BC^1(C_x(M))$ and $h(\omega) \in L^{2,1}_0([0,T]; \mathbb{R}^n)$ a.s.. Suppose $h.$ is adapted to the filtration of $\{\mathcal{F}^x_\cdot\}$ and such that $\mathbb{E} \left(\int_0^T |\dot{h}_s|^2 ds \right)^{\frac{1+\epsilon}{2}} < \infty$ for some $\epsilon > 0$. Then*

$$\mathbb{E} dF(\tilde{W}^Z_\cdot(h.)) = \mathbb{E} F(\xi.(x)) \int_0^T < \tilde{W}^Z_s(\dot{h}_s), \tilde{//}_s d\tilde{B}_s > . \tag{3.7}$$

If $\tilde{\nabla}'$ is metric for some Riemannian metric, we can take $\epsilon = 0$.

4. Integration by parts for the free path space

It is easy to modify the proof of Proposition 3.2 to the case where $h(0) \neq 0$ and so obtain an integration by parts formula for the free path space $PM = \cup_{x \in M} P_x M$ with uniform topology and measure given by the Riemannian measure of M together with the laws of $\{\xi.(x) : x \in M\}$. In fact it is straightforward to generalize to the case of an x-dependent $h.$. For this let $C^1(TM)$ be the space of C^1 vector fields on M with its usual topology:

Theorem 4.1. *Let* $h : [0,T] \times \Omega \to C^1(TM)$ *be a cadlag adapted process such that the* T_xM *valued process* $h.(x)$ *has sample paths in* $L^{2,1}([0,T];T_xM)$ *for each* $x \in M$ *with* $|h_0(\cdot)| + \sqrt{\int_0^t |\dot{h}_s(\cdot)|^2 ds}$ *in* $L^{1+\epsilon}(\Omega \times M;\mathbb{R})$ *for some* $\epsilon > 0$. *Let* F *be in* $BC^1(PM;\mathbb{R})$. *Then*

$$\begin{aligned}&\mathbb{E}\int_M dF\left(T_x\xi.(h.(\omega)(x))\right)dx\\ = \;&\mathbb{E}\int_M F(\xi.(x))\left\{-divh_0(x) + \int_0^T \left\langle T\xi_s(\dot{h}_s(x)), X(\xi_s(x))dB_s\right\rangle\right\}dx.\end{aligned} \tag{4.1}$$

Proof. Proceed as for Proposition 3.2 but with $X(x)h_t$ replaced by $h_t(x)$. In particular the definition (2.3) of H_t^τ becomes

$$\begin{aligned}\frac{\partial}{\partial\tau}H_t^\tau(x) &= h_t\left(H_t^\tau(x)\right)\\ H_t^0(x) &= x.\end{aligned}$$

while ξ_t^τ is defined by (3.3). However now $\xi_0^\tau(x) = \xi_0\left(H_0^\tau(x)\right)$: the starting point is transported by the flow of $h_0(x)$.

We first assume $h.$ and $\int_0^\cdot |\dot{h}_s|^2 ds$ are bounded on $\Omega \times M$. Then the Girsanov-Maruyama theorem gives us equivalence between the measures $\mathbb{P}_\tau^x$ and $\mathbb{P}_0^{H_0^\tau(x)}$ with

$$\int_M \mathbb{E}F\left(\xi.^\tau(x)\right)dx = \int_M \mathbb{E}F\left(\xi.(H_0^\tau(x))\right)\frac{d\mathbb{P}_\tau^x}{d\mathbb{P}_0^{H_0^\tau(x)}}dx.$$

On differentiating this there is the extra term

$$\begin{aligned}&\int_M dF\left(T\xi.\left(\frac{\partial}{\partial\tau}H_0^\tau(x)\Big|_{\tau=0}\right)\right)dx\\ = \;&\int_M dF\left(T_x\xi.\left(h_0(x)\right)\right)dx\\ = \;&\int_M d_x\left(F\circ\xi.\right)\left(h_0(x)\right)dx\end{aligned}$$

where $d_x\left(F\circ\xi.\right)$ refers to the derivative in M of $F\circ\xi. : M\times\Omega\to\mathbb{R}$. Now apply the classical Stokes theorem on M to get:

$$\begin{aligned}&\mathbb{E}\int_M dF(T_x\xi.(h.(\omega)(x)))dx\\ = \;&\mathbb{E}\int_M F(\xi.(x))\left\{-divh_0(x) + \int_0^T < T_x\xi_s(\dot{h}_s(x)), X(\xi_s(x))dB_s >\right\}dx.\end{aligned}$$

For general h let τ_R be the first exit time of $||h.||_{C^1} + \int_0^\cdot |\dot{h}_s(x)|^2 ds$ from $[0,R)$. Set $h_t^R(x) = h_{t\wedge\tau_R}(x)\chi_{\{||h_0||_{C^1}<R\}}$. We have:

$$\mathbb{E}\int_M dF(T_x\xi.(h^R_.(\omega)(x)))dx$$
$$= \mathbb{E}\chi_{\{||h_0||_{C^1}<R\}}\int_M F(\xi.(x))\Big\{-div h_0(x) +$$
$$\int_0^{T\wedge\tau_R}\langle T_x\xi_s(\dot{h}_s(x)), X(\xi_s(x))dB_s\rangle\Big\}dx.$$

Now let $R \to \infty$. The left hand side converges to $\mathbb{E}\int_M dF(T\xi.(h.(\omega)(x)))dx$ since

$$|dF(T\xi.(h^R_.(\omega)(-)))| \le \tilde{c}\sup_t |T\xi_t(\omega)| \sup_t |h_t(-,\omega)|$$

and $\sup_x \mathbb{E}\left(\sup_t |T\xi_t| \int_M \sup_t |h_t(x,\omega)|dx\right) < \infty$ from

$$\sup_t |h_t(x)| \le |h_0(\omega)| + \int_0^T |\dot{h}_s(\omega)|ds$$
$$\le |h_0(\omega)| + \sqrt{T}\sqrt{\int_0^T |\dot{h}_s(\omega)|^2 ds} \in L^{1+\epsilon}(\Omega \times M)$$

Using Burkholder-Davis-Gundy inequality to justify the integration on the right hand side we see that it converges to the right hand side of (4.1). ∎

Just as before the intrinsic formulae can be deduced using [ELJL95]:

Theorem 4.2. *Let F be in $BC^1(PM;\mathbb{R})$ and h be as in Theorem 4.1 but with $h.(x)$ adapted to the filtration of $\{\mathcal{F}^x_.\}$, and $div h_0 \in L^1(\Omega \times M, \mathbb{R})$. Then for any metric connection $\tilde{\nabla}$ on M,*

$$\begin{aligned} &\mathbb{E}\int_M dF\left(\tilde{W}^Z_.(h.(\omega)(x))\right)dx \\ = &\ \mathbb{E}\int_M F(\xi.(x))\left\{-div h_0(x) + \int_0^T \left\langle \tilde{W}^Z_s(\dot{h}_s(x)), \tilde{//}_s d\tilde{B}_s\right\rangle\right\}dx. \end{aligned} \tag{4.2}$$

If furthermore $\tilde{\nabla}'$ is metric with respect to a Riemannian metric, we can take $\epsilon = 0$.

Proof. The proof is just as that of Theorem 3.3. ∎

References

[Aid] S. Aida. On the irreducibility of certain Dirichlet forms on loop spaces over compact homogeneous spaces. To appear in 'New Trends in stochastic Analysis', Proc. Taniguchi Symposium, Sept. 1995, Charingworth, ed. K. D. Elworthy and S. Kusuoka, I. Shigekawa, World Scientific Press.

[AM] H. Airault and P. Malliavin. Integration by parts formulas and dilation vector fields on elliptic probability spaces. Institut Mittag-Leffler preprints No. 24, 1994/95.

[Bis81] J. M. Bismut. Martingales, the Malliavin calculus and harmonic theorems. In D. Williams, editor, *Stochastic Integrals, Lecture Notes in Maths. 851*, pp. 85–109. Springer-Verlag, 1981.

[Bis84] J. M. Bismut. *Large deviations and the Malliavin calculus. Progress in Math. 45.* Birkhaüser, 1984.

[CM] A.-B. Cruzeiro and P. Malliavin. Curvatures of path spaces and stochastic analysis. Institut Mittag-Leffler preprints No. 16, 1994/95.

[Dri92] B. Driver. A Cameron-Martin type quasi-invariance theorem for Brownian motion on a compact Riemannian manifold. *J. Funct. Anal.*, Vol. 100, pp. 272–377, 1992.

[Dri95a] B. Driver. The Lie bracket of adapted vector fields on Wiener spaces. Preprint, 1995.

[Dri95b] Bruce K. Driver. Towards calculus and geometry on path spaces. In *Stochastic Analysis: AMS Proceedings of symposium in pure Math. Series*, pp. 423–426. AMS. Providence, Rhode Island, 1995.

[EL94] K.D. Elworthy and Xue-Mei Li. Formulae for the derivatives of heat semigroups. *J. Funct. Anal.*, Vol. 125, No. 1, pp. 252–286, 1994.

[ELJL95] K. D. Elworthy, Yves Le Jan, and Xue-Mei Li. Concerning the geometry of stochastic differential equations and stochastic flows. To appear in 'New Trends in stochastic Analysis', Proc. Taniguchi Symposium, Sept. 1995, Charingworth, ed. K. D. Elworthy and S. Kusuoka, I. Shigekawa, World Scientific Press, 1995.

[Elw82] K.D. Elworthy. *Stochastic Differential Equations on Manifolds.* Lecture Notes Series 70, Cambridge University Press, 1982.

[Elw92] K. D. Elworthy. Stochastic flows on Riemannian manifolds. In M. A. Pinsky and V. Wihstutz, editors, *Diffusion processes and related problems in analysis, volume II. Birkhauser Progress in Probability*, pp. 37–72. Birkhauser, Boston, 1992.

[EM70] D. G. Ebin and J. Marsden. Groups of diffeomorphisms and the motion of an incompressible fluid. *Ann. of Math.*, Vol. 92, No. 1, pp. 102–163, 1970.

[ES95] O. Enchev and D.W. Stroock. Towards a Riemannian geometry on the path space over a Riemannian manifold. *J. Funct. Anal.*, Vol. 134, No. 2, pp. 392–416, 1995.

[EY93] K. D. Elworthy and M. Yor. Conditional expectations for derivatives of certain stochastic flows. In J. Azéma, P.A. Meyer, and M. Yor, editors, *Sem. de Prob. XXVII. Lecture Notes in Maths. 1557*, pp. 159–172. Springer-Verlag, 1993.

[FM93] S. Fang and P. Malliavin. Stochastic analysis on the path spaces of a Riemannian manifold. *J. Funct. Anal.*, Vol. 118, pp. 249–274, 1993.

[Hsu95] E. Hsu. Inégalités de sobolev logarithmiques sur un espace de chemins. *C. R. Acad. Sci. Paris, t. 320. Série I.*, pp. 1009–1012, 1995.

[KN69] S. Kobayashi and K. Nomizu. *Foundations of differential geometry, Vol. II.* Interscience Publishers, 1969.

[Li94] Xue-Mei Li. Stochastic differential equations on noncompact manifolds: moment stability and its topological consequences. *Probab. Theory Relat. Fields*, Vol. 100, No. 4, pp. 417–428, 1994.

[LN] R. Leandre and J. Norris. Integration by parts and Cameron-Martin formulas for the free-path space of a compact Riemannian manifold. Warwick Preprints: 6/1995.

[Nel84] E. Nelson. *Quantum Flucatuations*. Princeton University Press, Princeton, 1984.

[SZ] D. W. Stroock and O. Zeitouni. Variations on a theme by Bismut. Preprint.

Present address of Xue-Mei Li
Mathematics Department, U-9, MSB 111, University of Connecticut,
196 Auditorium Road, Storrs, Connecticut 06269, USA

Smooth measures and continuous additive functionals of right Markov processes

P. J. Fitzsimmons and R. K. Getoor

Department of Mathematics, University of California, San Diego, 9500 Gilman Drive, La Jolla, CA 92093-0112, USA

Summary. The Revuz correspondence sets up a bijection between the class of positive continuous additive functionals of a Markov process and a certain class of "smooth" measures on the state space of the process. We consider the correspondence in the context of a Borel right process with a distinguished excessive measure. A "nest" type characterization of smooth measures is provided, as well as a capacitary characterization of nests. Our results extend work of Revuz, Fukushima, and others.

1. Introduction

Consider a one-dimensional Brownian motion $B = (B_t)$. As is well-known, the Brownian local times $(L_t^x)_{t\geq 0, x\in \mathbb{R}}$ serve as occupation densities with respect to Lebesgue measure: if $f : \mathbb{R} \to [0, \infty[$ is a locally integrable function, then

$$(1.1) \qquad \int_0^t f(B_s)\, ds = \int_{\mathbb{R}} L_t^y\, f(y)\, dy, \qquad \forall t \geq 0$$

almost surely. The left side of (1.1) is the simplest type of positive continuous additive functional (PCAF) of Brownian motion. According to a result of Itô and McKean [IMcK65], the most general (finite) PCAF of Brownian motion is obtained by substituting a general positive Radon measure $\mu(dy)$ for $f(y)dy$ on the right side of (1.1). This correspondence between positive Radon measures and PCAFs is a bijection.

Although local times are found only in one-dimensional situations, the correspondence noted above persists quite generally even though (1.1) fails to have meaning. A clue to the general case may be found by Laplace-transforming in t and then taking expectations in (1.1):

$$(1.2) \qquad P^x \int_0^\infty e^{-qt} f(B_t)\, dt = \int_{\mathbb{R}} u^q(x,y) f(y)\, dy,$$

[2] The research of this author was supported, in part, by NSF Grant DMS 92-24990.

Key words and phrases. Continuous additive functional, smooth measure, nest, homogeneous random measure, capacity.

1990 AMS Subject classification. Primary: 60J55; secondary 60J45, 60J40.

where P^x is the law of B started from x, $P^x(F)$ denotes the integral of the functional F with respect to the measure P^x, and $u^q(x,y) = P^x \int_0^\infty e^{-qt} d_t L_t^y$. More generally, if A is a PCAF of B admitting the representation

$$A_t = \int_{\mathbb{R}} L_t^x \, \mu(dx),$$

then

$$P^x \int_0^\infty e^{-qt} g(B_s) \, dA_t = \int_{\mathbb{R}} u^q(x,y) g(y) \, \mu(dy).$$

Now suppose that $X = (X_t)$ is a strong Markov process with state space E and q-potential density $u^q(x,y)$ relative to some excessive reference measure m on E; that is,

$$P^x \int_0^\infty e^{-qt} g(X_t) \, dt = \int_E u^q(x,y) g(y) \, m(dy),$$

where now P^x denotes the law of X started at x. Under suitable conditions, if $A = (A_t)$ is any finite PCAF of X, then there is a σ-finite measure ν_A such that

$$P^x \int_0^\infty e^{-qt} g(X_t) \, dA_t = \int_E u^q(x,y) g(y) \, \nu_A(dy) \tag{1.3}$$

for all positive measurable g. The measure ν_A is called the Revuz measure of A in honor of D. Revuz, who proved (1.3) and established the bijectivity of the correspondence $A \leftrightarrow \nu_A$ in the context of standard processes in duality. Revuz' work [**Re70**] contains the definitive resolution (in this context) of a problem studied by a number of earlier authors, including Meyer [**M62**], Volkonskii[**V60**], Wentzell [**W61**], McKean-Tanaka [**McKT61**], and Blumenthal-Getoor [**BG64**].

In studying the PCAFs of the symmetric Hunt process associated with a regular Dirichlet form, Fukushima [**Fu79**] was led to broaden the notion of continuous additive functional to allow for an exceptional set of starting points. This is quite natural in the Dirichlet space setting, since the law of the associated Hunt process is itself determined modulo an exceptional set of starting points. (In fact, a similar notion of PCAF appears already in Sharpe's study [**Sh71**] of dual multiplicative functionals of standard processes in duality. For recent work on the connection between (not necessarily finite) PCAFs admitting exceptional sets and multiplicative functionals, see [**BDMM87**], [**St92**], [**G95**], [**K95**].) Both Revuz and Fukushima, in their different contexts, were able to completely characterize the measures (termed *smooth measures*, following McKean-Tanaka [**McKT61**]) that can occur as Revuz measures of PCAFs. Roughly speaking, such a measure can charge no set that the process visits only countably often (an obvious consequence of a representation such as (1.3)) and such a measure must be locally "quasi-Radon."

Our aim in this paper is to establish the Revuz correspondence in the general context of Borel right processes, the role of reference measure being filled by a distinguished excessive measure. (This excessive measure need not be a reference measure, nor are any duality hypotheses imposed.) We are able to completely characterize the class of smooth measures in this context, in a manner analogous to the characterization of Fukushima.

To streamline the exposition we shall impose a transience hypothesis, but this entails no loss of generality: All of our results can be applied to the (transient!) 1-subprocess of a non-transient process to obtain for the non-transient process the analogues of the results presented here. See the beginning of section 5 for more on this point.

After introducing notation and proving a preliminary result on strongly supermedian functions in section 2, we proceed in section 3 to characterize Revuz measures in terms of "nests," a concept introduced by Fukushima [**Fu79**]. In section 4 we provide a characterization of nests in terms of the natural capacity associated with a right process and an excessive measure. This result shows that our characterization of smooth measures generalizes that of Fukushima. (In: Tue, 12 Dec 1995 13:23:18 -0800 nests is via a capacity condition, which is then shown to be equivalent to an exit time condition. We find it convenient to proceed in the opposite direction.) Section 5 contains some complementary remarks.

We close this introduction with a few words on notation. If $(F, \mathcal{F}, \mu)$ is a measure space, then $b\mathcal{F}$ (resp. $p\mathcal{F}$) denotes the class of bounded real-valued (resp. $[0, \infty]$-valued) $\mathcal{F}$-measurable functions on F. For $f \in p\mathcal{F}$ we shall use $\mu(f)$ to denote the integral $\int_F f\, d\mu$; similarly, if $D \in \mathcal{F}$ then $\mu(f; D)$ denotes $\int_D f\, d\mu$. We write $\mathcal{F}^*$ for the universal completion of $\mathcal{F}$; that is, $\mathcal{F}^* = \cap_\nu \mathcal{F}^\nu$, where $\mathcal{F}^\nu$ is the ν-completion of $\mathcal{F}$ and the intersection runs over all finite measures on $(F, \mathcal{F})$. If $(E, \mathcal{E})$ is a second measurable space and $K = K(x, dy)$ is a kernel from $(F, \mathcal{F})$ to $(E, \mathcal{E})$ (*i.e.*, $F \ni x \mapsto K(x, A)$ is $\mathcal{F}$-measurable for each $A \in \mathcal{E}$ and $K(x, \cdot)$ is a measure on $(E, \mathcal{E})$ for each $x \in F$), then we write μK for the measure $A \mapsto \int_F \mu(dx) K(x, A)$ and Kf for the function $x \mapsto \int_E K(x, dy) f(y)$.

2. Preliminaries

Throughout the paper $X = (\Omega, \mathcal{F}, \mathcal{F}_t, \theta_t, X_t, P^x)$ will denote the canonical realization of a transient Borel right Markov process with state space $(E, \mathcal{E})$. We shall use the standard notation for Markov processes as found, for example, in [**BG68**], [**G90**], [**DM87**] and [**Sh88**]. Briefly, X is a strong Markov process with right continuous sample paths, the state space E (with Borel sets

$\mathcal{E}$) is homeomorphic to a Borel subset of a compact metric space, and the transition semigroup $(P_t)_{t\geq 0}$ of X preserves the class $b\mathcal{E}$ of bounded $\mathcal{E}$-measurable functions. It follows that the resolvent operators $U^q := \int_0^\infty e^{-qt} P_t\, dt$, $q \geq 0$, also preserve Borel measurability. We shall write U for U^0; the transience hypothesis we have imposed amounts to the assumption that there is a bounded, strictly positive function $b \in \mathcal{E}^*$ such that Ub is bounded. Substituting $U^1 b$ for b, we can always take b to be measurable over the σ-field $\mathcal{E}^e$ generated by the functions in $\mathcal{E}^*$ that are q-excessive for some $q \geq 0$. In the sequel, all named subsets of E are taken to be $\mathcal{E}^e$-measurable unless explicit mention is made to the contrary; likewise, all named functions are $\mathcal{E}^e$-measurable and positive (*i.e.*, taking values in $[0, \infty]$).

As usual, to allow for the possibility $P_t 1_E(x) < 1$, a cemetery state Δ is adjoined to E as an isolated point, and the process is sent to Δ at its lifetime ζ. We adhere to the convention that a function (resp. measure) defined on E (resp. $\mathcal{E}^*$) is extended to the cemetery state Δ by declaring its value at Δ (resp. $\{\Delta\}$) to be 0.

We fix once for all an excessive measure m. Thus, m is a σ-finite measure on $(E, \mathcal{E}^e)$ and $mP_t \leq m$ for all $t > 0$. Since X is a right process, we then have $\lim_{t\to 0} mP_t = m$, setwise.

Recall that a set B is m-polar provided $P^m(T_B < \infty) = 0$, where $T_B := \inf\{t > 0 : X_t \in B\}$ denotes the hitting time of B. A property or statement $P(x)$ will be said to hold quasi-everywhere (q.e.), or for quasi-every $x \in E$, provided it holds for all x outside some m-polar subset of E. It would be more proper to use the term "m-quasi-everywhere," but since the measure m will remain fixed the abbreviation to "q.e." will cause no confusion. Similarly, the qualifier "a.e. m" will be abbreviated to "a.e." On the other hand, certain terms (*e.g.*, polar) have a longstanding meaning without reference to a background measure, and so we shall use the more precise term "m-polar" to maintain the distinction. Notice that any finely open m-null set is m-polar. Consequently, any excessive function vanishing a.e. vanishes q.e.

If T is a stopping time, then the associated stopping operator P_T is defined by

$$P_T f(x) := P^x(f(X_T); T < \zeta), \qquad f \in p\mathcal{E}^*.$$

In case $T = T_B$, we write P_B instead of P_{T_B}. A function $f \in p\mathcal{E}^e$ is *strongly supermedian* provided $P_T f \leq f$ for every stopping time T. We collect in the next lemma some facts about strongly supermedian functions to be used in the sequel. The proof of part (a) is adapted from [**BG**, VI(3.2)].

(2.1) Lemma. *(a) If f is strongly supermedian and $f = 0$ a.e., then $f = 0$ q.e.*

(b) If (f_n) is a sequence of strongly supermedian functions, then $\liminf_n f_n$ *is also strongly supermedian.*

(c) Let a stopping time T be the limit of a sequence (T_n) of perfect exact terminal times. Then $f(x) := P^x(T < \zeta)$ and $g(x) := P^x(T < \infty)$ are strongly supermedian functions.

*Proof.*In view of [**G75**, (12.15)], to prove (a) it suffices to show that each compact subset of $\{f \geq \epsilon\}$ is m-polar, for each $\epsilon > 0$. To this end fix $\epsilon > 0$ and let K be a compact subset of $\{f \geq \epsilon\}$. Then $X_{T_K} \in K$ on $\{T_K < \infty\}$, so for $x \in E$ we have

$$f(x) \geq P_K f(x) \geq \epsilon P^x(T_K < \infty).$$

Thus, the excessive function $x \mapsto P^x(T_K < \infty)$ vanishes a.e., hence q.e., so that K is m-polar. Point (b) follows immediately from Fatou's lemma. Clearly the stopping time T in point (c) is a perfect terminal time, which implies that $P_S f \leq f$ and $P_S g \leq g$ for all stopping times S. On the other hand, T need not be exact, so f and g need not be excessive. Nonetheless, they are $\mathcal{E}^e$-measurable. To see this define $\mathcal{F}^e := \sigma\{h(X_t) : h \in b\mathcal{E}^e, t \geq 0\}$ Then by [**Sh88**, (60.2)], for each n there is an $\mathcal{F}^e$-measurable time T_n^* such that $T_n = T_n^*$ almost surely. Thus $T = \lim_n T_n^*$ almost surely, and so $f(x) = P^x(T < \zeta) = P^x(\lim_n T_n^* < \zeta)$ is $\mathcal{E}^e$-measurable because of [**Sh88**, (8.6)]. In the same way, $g \in \mathcal{E}^e$. □

3. Continuous additive functionals

In this paper we shall adopt the following extended definition of the notion of positive continuous additive functional, due to Fukushima [**Fu79**], allowing for an exceptional set of starting points. Following Getoor and Sharpe [**GSh84**, (6.8)] we call a Borel m-polar set whose complement is absorbing an *m-inessential set.*

(3.1) Definition. A *positive continuous additive functional* (abbreviated PCAF) is an $(\mathcal{F}_t)$-adapted process $A = (A_t)_{t\geq 0}$, with values in $[0,\infty]$, for which there exists a *defining set* $\Lambda \in \mathcal{F}$ and an m-inessential set N (called an *exceptional set* for A) such that

(i) $P^x(\Lambda) = 1$ for all $x \notin N$;
(ii) $\theta_t\Lambda \subset \Lambda$ for all $t \geq 0$;
(iii) For all $\omega \in \Lambda$, the map $t \mapsto A_t(\omega)$ is continuous on $[0,\infty[$, and finite-valued on $[0,\zeta(\omega)[$;
(iv) For all $\omega \in \Lambda$ and all $s,t \geq 0$, $A_{t+s}(\omega) = A_t(\omega) + A_s(\theta_t\omega)$;
(v) For all $t \geq 0$, $A_t([\Delta]) = 0$, where $[\Delta]$ denotes the constant path $t \mapsto \Delta$. In particular, $A_t(\omega) = A_\zeta(\omega)$ for all $t \geq \zeta(\omega)$ and all $\omega \in \Lambda$.

We say that two PCAFs A and B are *m-equivalent* provided they have a common defining set Λ and a common exceptional set N such that $A_t(\omega) = B_t(\omega)$ for all $t \geq 0$ and all $\omega \in \Lambda$. Two PCAFs A and B are m-equivalent if and only if $P^m(A_t \neq B_t) = 0$ for all $t > 0$. To see this write N_A (resp. Λ_A) for an exceptional set (resp. a defining set) for A, and similarly for B. Define $N_0 := N_A \cup N_B$, $\Lambda_0 := \Lambda_A \cap \Lambda_B$, and $\Lambda_1 := \{A_t = B_t, \forall t\}$. Clearly

N_0 is m-inessential, and it is easy to check that $\theta_t(\Lambda_0 \cap \Lambda_1) \subset \Lambda_0 \cap \Lambda_1$. If $x \in E \setminus N_0$, then

$$P^x([\Lambda_0 \cap \Lambda_1]^c) \leq P^x(\Lambda_0^c) + P^x(\Lambda_0 \setminus \Lambda_1) = P^x(\Lambda_0 \setminus \Lambda_1)$$

since $P^x(\Lambda_0) = 1$ for all $x \in E\setminus N_0$. The function $g : x \mapsto P^x(\Lambda_0\setminus\Lambda_1)$ vanishes a.e. by hypothesis; moreover, $g|_{E\setminus N_0}$ is readily seen to be excessive for X restricted to the absorbing set $E \setminus N_0$. Thus g vanishes q.e. on $E \setminus N_0$, hence q.e. since N_0 is m-inessential. By [**GSh84**, (6.12)], there is an m-inessential set N containing $N_0 \cup \{g > 0\}$. It follows that N (resp. $\Lambda := \Lambda_0 \cap \Lambda_1$) is a common exceptional set (resp. defining set) for A and B.

A PCAF may be viewed as the distribution function of a (diffuse) homogeneous random measure (defined below). In many respects, homogeneous random measures are easier to handle, and they will play a crucial role in our treatment of PCAFs. In what follows we shall write $\mathbb{R}^{++}$ for the open half-line $]0,\infty[$ and $\mathcal{R}^{++}$ for the class of Borel subsets of $\mathbb{R}^{++}$.

(3.2) Definition. A homogeneous random measure (HRM) is a kernel κ from $(\Omega, \mathcal{F}^*)$ to $(\mathbb{R}^{++}, \mathcal{R}^{++})$ such that

(i) For each ω, the measure $\kappa(\omega, \cdot)$ is carried by $]0, \zeta(\omega)]$;

(ii) There is a sequence $(\kappa_n)_{n\geq 1}$ of bounded kernels from $(\Omega, \mathcal{F}^*)$ to $(\mathbb{R}^{++}, \mathcal{R}^{++})$, such that $\kappa = \sum_n \kappa_n$;

(iii) For all $\omega \in \Omega$, $t > 0$, and $B \in \mathcal{R}^{++}$,

$$\kappa(\theta_t\omega, B) = \kappa(\omega, B + t).$$

An HRM is *optional* provided each of the kernels κ_n can be chosen so that $t \mapsto \kappa_n(\cdot,]0,t])$ is $\mathcal{F}_t$-optional (*i.e.*, $\mathcal{F}_t$-adapted, since $t \mapsto \kappa_n(\cdot,]0,t])$ is right continuous). An HRM is *diffuse* provided the measure $\kappa(\omega, \cdot)$ is diffuse for each $\omega \in \Omega$.

As we shall see, every diffuse optional HRM which puts finite mass on compact subsets of $[0,\zeta[$ gives rise to a PCAF. Before stating this precisely we need to recall the definition of Revuz measure. Recall that a set $B \in \mathcal{E}^e$ is *m-semipolar* provided

$$P^m(X_t \in B \text{ for uncountably many } t) = 0.$$

(3.3) Definition. The *Revuz measure* associated with an HRM κ is the measure ν_κ defined on $(E, \mathcal{E}^e)$ by the formula

$$\nu_\kappa(f) :=\uparrow \lim_{t\to 0} \frac{1}{t} P^m \int_0^t f(X_s)\, \kappa(ds), \quad f \in p\mathcal{E}^e. \tag{3.4}$$

The limit in (3.4) coincides with

$$\uparrow \lim_{q\to\infty} q \cdot mU^q_\kappa(f),$$

where

$$U^q_\kappa f(x) := P^x \int_0^\zeta e^{-qt} f(X_t)\, \kappa(dt)$$

is the q-potential operator associated with κ. It is clear that the Revuz measure of a diffuse optional HRM charges no m-semipolar set. Similarly, if A is a PCAF with exceptional set N, then we have the q-potential operator

$$(3.5) \qquad U^q_A f(x) := P^x \int_0^\zeta e^{-qt} f(X_t)\, dA_t, \qquad x \in E \setminus N,$$

and the Revuz measure

$$(3.6) \quad \nu_A(f) :=\uparrow \lim_{t\to 0} \frac{1}{t} P^m \int_0^t f(X_s)\, dA_s =\uparrow \lim_{q\to\infty} q \cdot mU^q_A(f), \quad f \in p\mathcal{E}^e.$$

The Revuz measure of a PCAF is σ-finite [**Re70**, III.1], and (as for HRMs) charges no m-semipolar set. See [**FG88**] and [**G90**], in addition to [**Re70**], for further details on Revuz measures.

The following basic existence and uniqueness result is an immediate consequence of [**Fi87**, (5.21),(5.23),(5.27)].

(3.7) Theorem. *Let μ be a σ-finite measure on $(E, \mathcal{E}^e)$ charging no m-semipolar set. Then there exists a diffuse optional HRM κ with Revuz measure μ. If κ' is a second diffuse optional HRM with Revuz measure μ, then $\kappa(\omega,\cdot) = \kappa'(\omega,\cdot)$ for P^m a.e. $\omega \in \Omega$.*

For the proof of the next proposition, the reader is referred to [**G95**, (4.3)].

(3.8) Proposition. *Let κ be a diffuse optional HRM. For $t \geq 0$ define*

$$A_t := \kappa(]0, t+]) := \lim_n \kappa(]0, t + 1/n])$$

$$S := \inf\{t : A_t = \infty\} = \inf\{t : \kappa(]0,t]) = \infty\}.$$

Then S is a perfect terminal time of X, A is $(\mathcal{F}_t)$-adapted and right continuous, and the only possible discontinuity of $t \mapsto A_t$ is a jump of infinite magnitude at S (and only on $\{S < \zeta\}$). If $t < S$ then $A_t = \kappa(]0,t]) < \infty$. Moreover, A satisfies the additivity condition (3.1)(iv) on all of Ω, and the "exactness" condition $\lim_{s\to 0} A_{t-s}(\theta_s\omega) = A_t(\omega)$ for all $t > 0$ and $\omega \in \Omega$.

(3.9) Remark. With S as in (3.8), the function $x \mapsto P^x(S < \zeta)$ is $\mathcal{E}^e$-measurable, hence strongly supermedian; see the top of p. 91 of [**G95**].

Before connecting the notions PCAF and HRM, we require a bit more terminology

(3.10) Definitions.

(i) $\tau(B) := T_{B^c} \wedge \zeta = \inf\{t > 0 : X_t \notin B\}$ denotes the *exit time* of X from B.
(ii) A *nest* is an increasing sequence $(B_n) \subset \mathcal{E}^e$ such that $P^m(\lim_n \tau(B_n) < \zeta) = 0$.
(iii) $\mathcal{S}^*$ denotes the class of all measures on $(E, \mathcal{E}^e)$ that charge no m-semipolar set.
(iv) A measure $\mu \in \mathcal{S}^*$ is *smooth* provided there is a nest (G_n) of finely open sets with $\mu(G_n) < \infty$ for all n. We say that such a sequence (G_n) *reduces* μ. We write $\mathcal{S}$ for the class of smooth measures.

Our use of the term nest is consistent with [**MR92**], but it is analogous to what is called a generalized nest in [**FOT94**, (2.2.17)]. Since the hitting time of a set coincides almost surely with the hitting time of its fine closure, the exit time from a set coincides with the exit time from the fine interior of the set. Thus, there is no loss of generality in assuming the G_n in (3.10)(iv) to be finely open, and this assumption simplifies certain arguments.

(3.11) Theorem. *(a) The Revuz measure ν_A of any PCAF A is an element of $\mathcal{S}$. Two PCAFs with identical Revuz measures are m-equivalent.*

(b) Conversely, given $\mu \in \mathcal{S}$ there exists a (unique, up to m-equivalence) PCAF with Revuz measure μ.

Proof.(a) Fix a strictly positive function b in $L^1(m)$ such that $Ub \leq 1$ and define

$$\varphi(x) = P^x \int_0^\zeta \exp(-A_t) b(X_t)\, dt, \qquad x \in E \setminus N,$$

where N is an exceptional set for A. Clearly $\varphi(x) \leq Ub(x) \leq 1$ and $\varphi(x) > 0$ for all $x \in E \setminus N$. Also, it is easy to check that

$$\varphi(x) = Ub(x) - U_A\varphi(x), \qquad x \in E \setminus N.$$

Consequently, φ is finely continuous on $E \setminus N$, being the difference of functions excessive for the restriction of X to the absorbing set $E \setminus N$. Thus, the sets

$$G_n := \{\varphi > 1/n\} \setminus N, \qquad n = 1, 2, \ldots,$$

form an increasing sequence of finely open sets. Moreover, $U_A 1_{G_n} \leq n \cdot U_A \varphi \leq n \cdot Ub$ on $E \setminus N$, which implies that $\nu_A(G_n) \leq n \cdot m(b) < \infty$; *cf.* [**Re70**, p. 508]. Writing τ_n for the exit time $\tau(G_n)$ we therefore have $\varphi(X_{\tau_n}) \leq 1/n$ a.s. P^x, and so $P_{\tau_n}\varphi(x) \leq 1/n$, for all $x \in E \setminus N$. Thus, for $x \in E \setminus N$

$$\begin{aligned} 1/n \geq P_{\tau_n}\varphi(x) &= P^x \left(\int_{\tau_n}^\zeta \exp(-A_{t-\tau_n}(\theta_{\tau_n}))\, b(X_t)\, dt \right) \\ &\geq P^x \int_{\tau_n}^\zeta \exp(-A_t) b(X_t)\, dt. \end{aligned}$$

Since $b > 0$ and $\exp(-A_t) > 0$ a.s. P^x on $\{t < \zeta\}$, we must have $\lim_n \tau_n = \zeta$, a.s. P^x for all $x \in E \setminus N$. Thus, (G_n) is a nest, and ν_A is a smooth measure. We defer the proof of the uniqueness assertion until after the proof of part (b)

(b) Let us now fix a smooth measure μ and a nest (G_n) reducing μ. Since the set $E \setminus \cup_n G_n$ is m-polar, it is μ-null. Thus μ is σ-finite, being carried by $\cup_n G_n$. Consequently, (3.7) applies and there is a diffuse optional HRM κ with Revuz measure μ. Let the increasing process A and the terminal time S be as promised by (3.8), and let us check that $P^m(S < \zeta) = 0$. First notice that because the limit in (3.4) is monotone increasing, $P^m \int_0^t 1_{G_n}(X_s)\, \kappa(ds) \le t\mu(G_n) < \infty$. It follows that for each $t > 0$,

$$P^m(\kappa(]0,t]); t < \tau(G_n)) \le P^m \int_0^t 1_{G_n}(X_s)\, \kappa(ds) < \infty.$$

Thus, $t \mapsto A_t = \kappa]0, t+]$ is finite on $[0, \tau(G_n)[$ a.s. P^m for each n. But $P^m(\lim_n \tau(G_n) < \zeta) = 0$ since (G_n) is a nest. This implies that $P^m(S < \zeta) = 0$, and so $P^x(S < \zeta) = 0$ for q.e. x by (2.1)(a) and Remark (3.9). By [**GSh84**, (6.12)], there is an m-inessential set N such that $N \supset \{x \in E : P^x(S < \zeta) > 0\}$. It is now a simple matter to check that A is a PCAF with defining set $\{S \ge \zeta\}$ and exceptional set N.

Finally, let us prove the uniqueness assertion made in part (a) of the theorem. Let A and B be PCAFs with common Revuz measure μ, defining sets Λ_A and Λ_B, and exceptional sets N_A and N_B. Notice that $N := N_A \cup N_B$ is m-inessential, and serves as a common exceptional set for A and B. Let (G_n^A) (resp. (G_n^B)) be a nest of finely open sets defined as in the proof of part (a), with respect to A (resp. B), and define $G_n := G_n^A \cap G_n^B$. Clearly (G_n) is a nest and $\mu(G_n) < \infty$ for each n. Moreover, $U_A 1_{G_n} \le n$ and $U_B 1_{G_n} \le n$ on $E \setminus N$. We now apply [**G95**, (6.11)] (to the restriction of X to the absorbing set $E \setminus N$) and find that

$$\text{(3.12)} \qquad m(g \cdot U_A 1_{G_n}) = \mu(1_{G_n} \hat{U} g) = m(g \cdot U_B 1_{G_n}), \qquad \forall g \in p\mathcal{E}.$$

Here $\hat{U}$ is the 0-potential operator for the moderate Markov process $\hat{X}$ in weak duality with X (with respect to m), as constructed in [**Fi87**]. Now (3.12) implies that $U_A 1_{G_n} = U_B 1_{G_n}$, first a.e. and then q.e., since both of these functions are excessive for X restricted to $E \setminus N$. Let ν be a probability measure equivalent to m, and notice that by the preceding discussion, the processes $U_A 1_{G_n}(X_t)$ and $U_B 1_{G_n}(X_t)$ are indistinguishable, bounded, right continuous, positive supermartingales over the filtered probability space $(\Omega, \mathcal{F}, \mathcal{F}_t, P^\nu)$. By the uniqueness of the Doob-Meyer decomposition, their associated increasing processes, namely $A^n := \int_0^\cdot 1_{G_n}(X_s)\, dA_s$ and $B^n := \int_0^\cdot 1_{G_n}(X_s)\, dB_s$, are P^ν-indistinguishable. Since $\nu \sim m$, we conclude that $P^m(A_t \ne B_t, t < \tau(G_n)) = 0$. But (G_n) is a nest, so $P^m(A_t \ne B_t) = 0$ for all $t > 0$. □

Combining (3.8) with the proof of (3.11) we obtain the following

(3.13) Corollary. *Any PCAF A is m-equivalent to a PCAF A' for which the additivity property $A'_{t+s}(\omega) = A'_t(\omega) + A'_s(\theta_s\omega)$ holds for all $\omega \in \Omega$ and all $s, t \geq 0$, and for which the exactness condition $\lim_{s\to 0} A'_{t-s}(\theta_s\omega) = A'_t(\omega)$ holds for all $\omega \in \Omega$ and all $t > 0$.*

The above exactness condition plays an important role in the work of Baxter, Dal Maso and Mosco **[BDMM87]**, Sturm **[St92]**, Getoor **[G95]**, and Kuwae **[K95]** concerning generalized PCAFs (positive but not necessarily finite continuous additive functionals).

4. Capacity and nests

In this section we shall give a capacitary characterization of nests, generalizing work found in Fukushima **[Fu79]**, **[FOT94]** and Ma-Röckner **[MR92]**. We discuss two capacities associated with X and m. The first of these, denoted C, is easier to work with and provides a simpler characterization of nests. But it is the second which is in many respects more natural; this capacity, denoted Γ, is the direct generalization to our context of the classical Newtonian capacity.

In order to define C we require the following

(4.1) Lemma. *There is a finite measure π such that (i) π and m are mutually absolutely continuous and (ii) $\pi U \leq m$. Things being so, the excessive measures πU and m are mutually absolutely continuous.*

Proof. Because X is transient, we can appeal to **[G90**, (2.10)**]** to find potentials $\pi_n U$ increasing setwise to m, such that $\pi_n(E) < \infty$ and $\pi_n \ll m$ for all n. The measure $\pi := \sum_{n=1}^{\infty} 2^{-n}[1 + \pi_n U^1(E)]^{-1}\pi_n U^1$ then does the job. □

Fix $b > 0$ such that $Ub \leq 1$, and define

$$(4.2) \qquad C(B) := \pi P_B U b = P^\pi \int_{T_B \wedge \zeta}^{\zeta} b(X_t)\, dt, \qquad B \in \mathcal{E}^e.$$

Notice that $C(E) \leq \pi(E) < \infty$. It is well-known **[GSt87**, (7.3)**]** that C is monotone increasing, strongly subadditive, countably subadditive and ascending. Of course, C depends on the choice of the measure π and the function b, but regardless of this choice a set B is m-polar if and only if $C(B) = 0$. The following characterization of nests should be compared to **[MR92**, Thm. III.2.11, Prop. IV.5.30]

(4.3) Proposition. *An increasing sequence $(B_n) \subset \mathcal{E}^e$ is a nest if and only if $C(E \setminus B_n) \to 0$ as $n \to \infty$.*

Proof. This follows immediately from the definitions once we rewrite (4.2) as

$$C(E \setminus B_n) = P^\pi \int_{\tau(B_n)}^{\zeta} b(X_t)\, dt.$$

□

We now recall from Getoor and Steffens [GSt87] the definition of the capacity Γ. Let L be the energy functional associated with X ([GSt87] or [G90, Sect. 3])

$$L(\xi, h) := \sup\{\nu(h) : \nu U \leq \xi\},$$

where ξ is an excessive measure and h is an excessive function. The capacity $\Gamma = \Gamma_m$ is now defined by

$$\Gamma(B) := L(m, P_B 1), \qquad B \in \mathcal{E}^e. \tag{4.4}$$

Like C, Γ is monotone increasing, strongly subadditive, countably subadditive, and ascending. Moreover, a set B is m-polar if and only if $\Gamma(B) = 0$. We shall have need of the balayage operator R_B as well; R_B operates on the convex cone of excessive measures, and is the L-dual of the hitting operator P_B. That is, defining

$$R_B\xi(f) := L(\xi, P_B U f), \qquad B \in \mathcal{E}^e,$$

we have the identity

$$L(R_B\xi, h) = L(\xi, P_B h), \qquad B \in \mathcal{E}^e,$$

for any excessive measure ξ and any excessive function h. In particular, $\Gamma(B) = L(R_B m, 1)$.

Our characterization of nests in terms of Γ relies on a notion of "small" set, which we now introduce. In the context of regular Dirichlet forms, compact sets are strong equilibrium sets as defined below.

(4.5) Definitions.

(i) A set $B \in \mathcal{E}^e$ is an *equilibrium set* provided $R_B m = \gamma_B U$ for some finite measure γ_B. In this case $\Gamma(B) = \gamma_B(E) < \infty$.

(ii) An equilibrium set B is a *strong equilibrium set* provided γ_B charges no m-polar set and $P_{T_n} P_B 1 \to 0$ q.e. for any increasing sequence (T_n) of stopping times satisfying $P^x(\lim_n T_n < \zeta) = 0$ for q.e. x.

Because of [GSh84, (3.3)], one can restrict attention to increasing sequences of *hitting* times in checking that an equilibrium set B is a strong equilibrium set, provided $P_t P_B 1 \to 0$ a.e. as $t \to \infty$. Since an excessive measure dominated by a potential is itself a potential [G90, (5.23)], it is clear that any subset of an equilibrium set is an equilibrium set. The same is true for strong equilibrium sets. To see this, suppose that B is a strong equilibrium set and $D \subset B$. Then D is an equilibrium set and $\gamma_D U \leq \gamma_B U$. Moreover, $P_{T_n} P_D 1 \leq P_{T_n} P_B 1$ so the final condition for strong equilibrium sets is met by D. Also, because $\gamma_D U \leq \gamma_B U$, a theorem of Rost ([Ro71] or [Fi91]) tells us that there is a (randomized) stopping time T such that $\gamma_D = \gamma_B P_T$. Now let N be an m-polar set, and appeal once more to [GSh84, (6.12)] to find an m-inessential set $N^* \supset N$. Using the evident fact that 1_{N^*} is a strongly supermedian function, we compute

$$\gamma_D(N) \leq \gamma_D(N^*) = \gamma_B P_T 1_{N^*} \leq \gamma_B(N^*) = 0.$$

Thus, γ_D charges no m-polar set.

In many concrete situations, the condition in (4.5)(ii) that γ_B charge no m-polar set is automatically satisfied. Indeed, we have the following

(4.6) Proposition. *Let ρU be the potential part of m and let B be an equilibrium set. If ρ charges no m-polar set, then neither does γ_B.*

*Proof.*This results on combining [**G90**, (10.34)] with [**FG91**, p. 142]. □

Suppose, for example, that X is in weak duality (with respect to m) with a second Borel right process $\hat{X}$, and that the sector condition is satisfied by X and m, as in [**Fi89**]. Then (3.9) and (4.11) of [**Fi89**] imply that ρ charges no m-polar set. For a second example, suppose that X and $\hat{X}$ are m-standard processes in weak duality with respect to m, and that $\hat{X}$ is even m-special standard. Furthermore assume that $x \mapsto \hat{U}^q 1(x)$ is lower semi-continuous for some $q > 0$. Under these conditions, an argument of Revuz [**Re71**] may be adapted to show that if μU is *any* potential dominated by m, then μ charges no m-polar set.

We can now state the main result of the paper, which should be compared to [**FOT94**, Eq. (2.2.17), Lem. 5.1.6]

(4.7) Theorem. *An increasing sequence $(G_n) \subset \mathcal{E}^e$ of finely open sets is a nest if and only if*

$$\Gamma(H \setminus G_n) \to 0, \qquad \text{as } n \to \infty \tag{4.8}$$

for each finely closed strong equilibrium set H.

The proof of (4.7) requires some preparation. We begin with a simple criterion for strong equilibrium sets, and then proceed to show that there exists a nest of strong equilibrium sets. The conditions imposed in the following result can be sharpened with more effort, but the form we present is adequate for our purposes.

(4.9) Proposition. *Let $G \in \mathcal{E}^e$ be finely open and suppose that (i) there is a strictly positive function f with $Uf < \infty$ q.e. such that $G \subset \{Uf \geq 1\}$, and (ii) there is a finite measure μ charging no m-polar set such that $m(B) \leq \mu U(B)$ for all Borel sets $B \subset G$. Then G is a strong equilibrium set.*

*Proof.*By condition (ii) and [**G90**, (4.26),(4.29)], $R_G m \leq \mu U$. Thus, $R_G m$ is a potential, say $R_G m = \gamma_G U$. Because $\gamma_G U \leq \mu U$, the measure γ_G charges no m-polar set, by the argument given below (4.5). Moreover, $\Gamma(G) = \gamma_G(E) \leq$

$\mu(E) < \infty$. Finally, if (T_n) is an increasing sequence of stopping times with limit at least ζ, a.s. P^x for q.e. x, then

$$P_{T_n} P_G 1(x) \leq P_{T_n} U f(x) = P^x \int_{T_n \wedge \zeta}^{\zeta} f(X_t)\, dt,$$

which tends to 0 provided $Uf(x) < \infty$ and $P^x(\lim_n T_n < \zeta) = 0$, hence for q.e. $x \in E$. Thus, G is a strong equilibrium set. □

The following construction is the key to Theorem (4.7), and may be of independent interest.

(4.10) Proposition. *There exists a nest comprised of finely closed strong equilibrium sets.*

*Proof.*Recall from (4.1) the finite measure $\pi \sim m$ with $\pi U \leq m$ (and then $\pi U \sim m$). Let u be an m-fine version of the Radon-Nikodym derivative $d(\pi U)/dm$. More precisely, by results in **[FG91]** there is an m-inessential set N_u, with $\rho(N_u) = 0$ where ρU is the potential part of m, such that u is finely continuous on $E \setminus N_u$. Moreover, we can (and do) assume that u is bounded above by 1 and Borel measurable. Notice that N_u, being the complement of an absorbing set, is finely closed. Define an increasing sequence of finely closed sets by

$$H_n := [\{u \geq 1/n\} \cap \{Ub \geq 1/n\} \setminus N_u] \cup N_u, \qquad n = 1, 2, \ldots \tag{4.11}$$

where the function $b > 0$ is such that $Ub \leq 1$. The finely open set $\{u > 1/n\} \cap \{Ub > 1/n\} \setminus N_u$ contains $H_n \setminus N_u$ and is a strong equilibrium set by (4.9). Thus $H_n \setminus N_u$ is a strong equilibrium set, hence so it H_n, as is easily verified. It remains to show that (H_n) is a nest. Notice that

$$\tau(H_n) = \tau(G_n) \wedge \tau(D_n),$$

where $G_n := \{u \geq 1/n\} \cup N_u$ and $D_n := \{Ub \geq 1/n\}$. The sequence (D_n) is a nest, since for all $x \in E$,

$$1/n \geq P_{\tau(D_n)} Ub(x) = P^x \int_{\tau(D_n)}^{\zeta} b(X_t)\, dt.$$

Thus, it suffices to show that (G_n) is also a nest.

From **[FG91]** we know that $t \mapsto u(X_t)$, in addition to being right continuous, has left limits on $]0, \zeta[$ a.s. P^m. Moreover, when u is composed with the Kuznetsov process associated with X and m, we obtain a supermartingale in reversed time. It follows that

$$P^m(\exists\, 0 < s < t \text{ such that } u(X_s) > 0, u(X_t) \wedge u(X)_{t-} = 0) = 0. \tag{4.12}$$

Write τ_n for $\tau(G_n)$ and notice that $\tau_n = T_{\{u<1/n\}} \wedge \zeta$, a.s. P^m, since N_u is m-polar. Thus

(4.13) $$T := \uparrow \lim_n \tau_n \le T_{\{u=0\}} \wedge \zeta, \qquad \text{a.s. } P^m,$$

and $u(X_{\tau_n}) \le 1/n$ on $\{\tau_n < \zeta\}$. Hence, $u(X_T) \wedge u(X)_{T-} = 0$ a.s. P^m on $\{0 < T < \zeta\}$. In view of (4.12), $u(X_s) = 0$ for all $s \in [0, T[$, a.s. P^m. Therefore

$$P^m(T; T < \zeta) = P^m\left(\int_0^T 1_{\{u=0\}}(X_s)\, ds\,; T < \zeta\right)$$
$$\le P^m \int_0^\infty 1_{\{u=0\}}(X_t)\, dt = mU(u = 0) = 0,$$

the final equality following because $\{u = 0\}$ is m-null since $m \ll \pi U = u \cdot m$. Thus, $P^m(0 < T < \zeta) = 0$. Now on the event $\{T = 0\}$ we have $\tau_n = 0$ for all n. However $P^x(\tau_n = 0) = 0$ if $x \in \{u > 1/n\} \setminus N_u$, and $\cap_n[\{u > 1/n\} \setminus N_u]^c = \{u = 0\} \cup N_u$ which is m-null. It follows that $P^m(T = 0) = 0$. We have now established that $P^m(T < \zeta) = 0$. That is, (G_n) is a nest, as was to be shown. □

(4.14) Remark. If X has left limits on $]0, \zeta[$, then one may take each H_n constructed in the proof of (4.10) to be relatively compact. To see this, use [**MR92**, Thm. IV.1.15] to choose a nest of compacts (K_n). Then $(H_n \cap K_n)$ is a nest of relatively compact, finely closed, strong equilibrium sets.

We now record several lemmas, which taken together will prove (4.7).

(4.15) Lemma. *Let (B_n) be a decreasing sequence from $\mathcal{E}^e$. If $\Gamma(B_n) \to 0$, then $\lim_n P^x(T_{B_n} < \infty) = 0$ for q.e. x.*

Proof. Since $\pi U \le m$, we have $\pi P_{B_n} 1 = L(\pi U, P_{B_n} 1) \le \Gamma(B_n) \to 0$. Thus the decreasing sequence of excessive functions $(P_{B_n} 1)$ converges to 0 a.e. π, hence a.e. m. By (2.1), the limit vanishes q.e., as claimed. □

(4.16) Lemma. *Let (G_n) be a nest of finely open sets and let H be a finely closed, strong equilibrium set. Then $\lim_n P^x(T_{H\setminus G_n} < \infty) = 0$ for q.e. $x \in E$.*

Proof. First note that $T_{H\setminus G_n} < \infty$ if and only if $T_{H\setminus G_n} < \zeta$. Thus,

$$\lim_n P^x(T_{H\setminus G_n} < \infty) = P^x(\Lambda),$$

where $\Lambda := \cap_n\{T_{H\setminus G_n} < \zeta\}$. Now $\lim_n T_{H\setminus G_n} \ge \lim_n \tau(G_n) = \zeta$ a.s. P^x for q.e. x, and so $P_{H\setminus G_n} P_H 1 \to 0$ q.e. since H is a strong equilibrium set; that is

$$0 = \lim_n P_{H\setminus G_n} P_H 1(x) = P^x(T_{H\setminus G_n} + T_H \circ \theta_{T_{H\setminus G_n}} < \zeta, \forall n), \text{ for q.e. } x.$$

It follows that for P^m-a.e. $\omega \in \Lambda$, the last exit time of X from H is strictly smaller than ζ. But this is absurd, unless $P^m(\Lambda) = 0$, because $X_{T(H\setminus G_n)} \in H$ (since H is finely closed) and $T_{H\setminus G_n} \to \zeta$, a.s. P^m on Λ. Consequently, $P^m(\Lambda) = 0$, and an appeal to (2.1)(c) finishes the proof. □

(4.17) Lemma. *Let (F_n) be a decreasing sequence of strong equilibrium sets. If $P_{F_n}1 \to 0$ q.e., then $\Gamma(F_n) \to 0$.*

Proof. We make use of the representation

$$\Gamma(B) = Q_m(T_B < \infty; 0 < S \leq 1)$$

where S is any finite stationary time over the Kuznetsov process (Y_t, Q_m) associated with X and m. See [**G90**, (10.14)]. We shall also use the following form of the strong Markov property. Let F be an equilibrium set (so that $R_F m = \gamma_F U$). Then upon combining (6.19) and (7.5) of [**G90**] one obtains $Q_m(\alpha = T_F, W(h)^c) = 0$, the set $W(h)$ being as defined on page 57 of [**G90**]. In particular, $Q_m(T_F = -\infty) = 0$. Now using the strong Markov property [**G90**, (6.15)] together with (6.20), (7.5), (8.23) of [**G90**] yields

$$(4.18) \qquad Q_m(\phi(T_F)G(\theta_{T_F}); T_F < \infty) = \int_{\mathbb{R}} \phi(t)\, dt \cdot \int_E P^x(G)\, \gamma_F(dx),$$

for any positive Borel function ϕ on $\mathbb{R}$ and any $G \in p\mathcal{F}^*$. See also [**FG91**, (2.3)].

Now by hypothesis, each F_n is a strong equilibrium set, so $Q_m(T_{F_n} = -\infty) = 0$. Thus,

$$(4.19) \qquad \lim_n \Gamma(F_n) = Q_m(-\infty < T_{F_n} < \infty, \forall n; 0 < S \leq 1).$$

If the left side of (4.19) is strictly positive, then

$$Q_m(-\infty < T_{F_n} < \infty, \forall n) > 0.$$

Define events $\Lambda_0 := \{-\infty < \alpha = T_{F_n} < \infty, \forall n\}$ and $\Lambda_k := \{\alpha < T_{F_k}, T_{F_n} < \infty, \forall n\}$, for $k = 1, 2, \ldots$. Clearly

$$\{-\infty < T_{F_n} < \infty, \forall n\} \subset \cup_{k=0}^{\infty} \Lambda_k,$$

so to finish the proof it will suffice to show that $Q_m(\Lambda_k) = 0$ for $k = 0, 1, 2, \ldots$.

Fix $k \geq 1$. If $Q_m(\Lambda_k) > 0$, then there is a rational t such that

$$(4.20) \qquad Q_m(\alpha < t < T_{F_k}, T_{F_n} < \infty, \forall n) > 0.$$

Now choose a strictly positive function $f \in L^1(m)$. Then (4.20) implies

$$(4.21) \qquad Q_m(f(Y_t); \alpha < t < T_{F_k}, T_{F_n} < \infty, \forall n) > 0.$$

But using the simple Markov property of Q_m at time t and the terminal time property of the T_{F_n},

$$(4.22) \qquad \begin{aligned} Q_m(f(Y_t); &\alpha < t < T_{F_k}, T_{F_n} < \infty, \forall n) \\ &\leq Q_m(f(Y_t) P^{Y_t}(T_{F_n} < \infty, \forall n)) \\ &= \int_E f(x) \lim_n P^x(T_{F_n} < \infty)\, m(dx) \end{aligned}$$

and the last term in (4.22) vanishes because of the hypothesis. Thus, $Q_m(\Lambda_k) = 0$ for $k \geq 1$.

To handle the case $k = 0$ we shall use (4.18) in combination with the identity $T_{F_n} = \alpha + T_{F_n} \circ \theta_\alpha$ on $\{\alpha > -\infty\}$. Notice that if $Q_m(\Lambda_0) > 0$, then $Q_m(\phi(\alpha); \Lambda_0) > 0$, where ϕ is any strictly positive Borel function on $\mathbb{R}$ with finite Lebesgue integral. Thus, using (4.18) for the third equality below,

$$\begin{aligned} Q_m(\phi(\alpha); \Lambda_0) &= Q_m(\phi(\alpha); -\infty < \alpha = T_{F_n} < \infty, \forall n) \\ &= Q_m(\phi(T_{F_1}); -\infty < \alpha = T_{F_n} < \infty, \forall n) \\ &\leq Q_m(\phi(T_{F_1}); -\infty < T_{F_1} < \infty, T_{F_n} \circ \theta_{T(F_1)} < \infty, \forall n) \\ &= \int_{\mathbb{R}} \phi(s)\, ds \cdot \int_E P^x(T_{F_n} < \infty, \forall n)\, \gamma_{F_1}(dx) \end{aligned}$$

But $P^x(T_{F_n} < \infty, \forall n) = \lim_n P_{F_n} 1(x)$ vanishes for q.e. x, and γ_{F_1} charges no m-polar set, by hypothesis. It follows that $Q_m(\phi(\alpha); \Lambda_0) = 0$, and the lemma is proved. □

Proof of (4.7). Let (G_n) be a nest of finely open sets, and let H be a finely closed, strong equilibrium set. Then $P_{H \setminus G_n} 1 \to 0$ q.e., by (4.16). Now (4.17) implies that $\Gamma(H \setminus G_n) \to 0$, as desired.

Conversely, suppose that (G_n) is an increasing sequence of finely open sets for which (4.8) holds for each finely closed strong equilibrium set H. Let (H_k) be the nest of finely closed, strong equilibrium sets promised by (4.10). Evidently,

$$\tau(G_n) \geq T_{H_k \setminus G_n} \wedge \tau(H_k).$$

But $\lim_n T_{H_k \setminus G_n} = \infty$ a.s. P^m by (4.15). Thus,

$$\lim_n \tau(G_n) \geq \lim_k \tau(H_k) = \zeta, \qquad \text{a.s. } P^m,$$

since (H_k) is a nest. It follows that (G_n) is a nest. □

5. Concluding Remarks

All of the results presented in earlier sections are valid for a general (non-transient) Borel right process X. Indeed, one merely has to apply our results to the 1-subprocess of X (namely, the process $X^{(1)}$ obtained by killing X at an independent unit mean exponential time), and then interpret these results in terms of X. In particular, all of the results of section 3 remain valid as stated, without any alteration of the definitions used in that section. In section 4 one must replace the capacities C and Γ by the analogous 1-capacities, and one must use (strong) 1-equilibrium sets (*i.e.*, (strong) equilibrium sets of $X^{(1)}$) in the appropriate places. These changes being made, the main result Theorem (4.7) is valid in the general case.

We now briefly discuss an alternative characterization of the class of smooth measures. Let us call a function $g \in bp\mathcal{E}^e$ *m-regular* provided (i) $t \mapsto g(X_t)$ is right continuous a.s. P^m, and (ii) $\lim_n P^x(g(X_{T_n})) = P^x(g(X_T))$ for a.e. x, whenever (T_n) is an increasing sequence of stopping times with limit T. (Equivalently, the predictable projection of $g(X.)$ is left continuous a.s. P^m. When X is a Hunt process, a function is m-regular if and only if it is m-quasi-continuous; see Le Jan **[LJ82]**.) It is not hard to check that if g is m-regular and $g > 0$ q.e., then $\inf_{0 \le s \le t} g(X_s) > 0$ on $\{t < \zeta\}$ a.s. P^m.

Now, given $\mu \in \mathcal{S}^*$, let κ be the associated diffuse optional HRM. Suppose that there is an m-regular function g with $g > 0$ q.e. and $\mu(g) < \infty$. Then $P^m(\int_0^t g(X_s)\kappa(ds)) \le t\mu(g) < \infty$, whence

$$(5.1) \quad \infty > \int_0^t g(X_s)\kappa(ds) \ge \inf_{0 \le s \le t} g(X_s) \cdot \kappa]0,t], \qquad \text{a.s. } P^m \text{ on } \{t < \zeta\},$$

from which it follows that $\kappa]0,t] < \infty$ a.s. P^m on $\{t < \zeta\}$, and then that μ is the smooth measure associated with the PCAF $A_t := \kappa]0,t+]$ as in (3.8).

Conversely, let A be a PCAFwith Revuz measure ν_A, and recall the function φ used in the proof of (3.11):

$$(5.2) \qquad \varphi(x) = P^x \int_0^\zeta \exp(-A_t) b(X_t)\,dt, \qquad x \in E \setminus N,$$

where $0 < b \in \mathcal{E}^e$ with $Ub \le 1$ and N is an exceptional set for A. If T is a stopping time then

(5.3)

$$P^x(\varphi(X_T)) = P^x\left(\int_{T\wedge\zeta}^\zeta \exp(-(A_t - A_T)) b(X_t)\,dt\,; T < \zeta\right), \qquad x \in E \setminus N,$$

from which it follows easily that φ is m-regular. Clearly $\varphi > 0$ on $E \setminus N$ (hence q.e.) and $\nu_A(\varphi) < \infty$ as in the proof of (3.11). We have thus proved the following characterization of $\mathcal{S}$.

(5.4) Proposition. *A measure $\mu \in \mathcal{S}^*$ is smooth if and only if there exists an m-regular function g with $g > 0$ q.e. and $\mu(g) < \infty$.*

Finally, let μ be an element of $\mathcal{S}^*$, let κ be the associated diffuse optional HRM, and let the explosion time $S = \inf\{t : \kappa]0,t] = \infty\}$ be as in (3.8). Mimicking (5.2), lets us define

$$(5.5) \qquad \psi(x) := P^x \int_0^\zeta \exp(-\kappa]0,t])\, b(X_t)\,dt, \qquad x \in E,$$

where $Ub \le 1$ as before. One can show that φ is finely continuous, that $t \mapsto \varphi(X_t)$ admits left limits almost surely, and that

$$(5.6) \qquad S = \lim_n T_{\{\varphi \le 1/n\}} = \inf\{t \ge 0 : \varphi(X_t) \wedge \varphi(X)_{t-} = 0\},$$

where $\varphi(X)_{0-} := \varphi(X_0)$. This representation of S has been proved by Sturm [**St92**, (2.5)] in case X is Brownian motion, and by Getoor [**G95**, p. 93] for standard processes in weak duality under the assumption that semipolar sets are m-polar.

References

[BDMM87] Baxter, J., Dal Maso, G., Mosco, U. (1987): Stopping times and Γ-convergence. Trans. Amer. Math. Soc. **303** 1–38.

[BG64] Blumenthal, R.M., Getoor, R.K. (1964): Additive functionals of Markov processes in duality. Trans. Amer. Math. Soc. **112** 131–163.

[BG68] Blumenthal, R.M., Getoor, R.K. (1968): Markov Processes and Potential Theory. Academic Press, New York.

[DM87] Dellacherie, C., Meyer, P.-A. (1987): Probabilités et Potentiel, Ch. XII–XVI. Hermann, Paris.

[Fi87] Fitzsimmons, P.J. (1987): Homogeneous random measures and a weak order for the excessive measures of a Markov process. Trans. Amer. Math. Soc. **303** 431–478.

[Fi89] Fitzsimmons, P.J. (1989): Markov processes and nonsymmetric Dirichlet forms without regularity. J. Funct. Anal. **85** 287–306.

[Fi91] Fitzsimmons, P.J. (1991): Skorokhod embedding by hitting times. Seminar on Stochastic Processes 1990, pp. 183–191, Birkhäuser, Boston-Basel-Berlin, .

[FG88] Fitzsimmons, P.J., Getoor, R.K. (1988): Revuz measures and time changes. Math. Z. **199** 233–256.

[FG91] Fitzsimmons, P.J., Getoor, R.K. (1991): A fine domination principle for excessive measures. Math. Z. **207** 137–151.

[Fu79] Fukushima, M. (1979): On additive functionals admitting exceptional sets. J. Math. Kyoto Univ. **19** 191–202.

[FOT94] Fukushima, M., Oshima, Y., Takeda, M. (1994): Dirichlet Forms and Markov Processes. De Gruyter, Berlin-New York.

[G75] Getoor, R.K. (1975): Markov Processes: Ray Processes and Right Processes. Lecture Notes in Math. **440**. Springer-Verlag, Berlin-Heidelberg-New York.

[G90] Getoor, R.K. (1990): Excessive Measures. Birkhäuser, Boston-Basel-Berlin.

[G95] Getoor, R.K. (1995): Measures not charging semipolars and equations of Schrödinger type. Potential Analysis **4** 79–100.

[GSh84] Getoor, R.K., Sharpe, M.J. (1984): Naturality, standardness, and weak duality for Markov processes. Z. fur Warscheinlichkeitstheorie verw. Gebiete **64** 1–62.

[GSt87] Getoor, R.K., Steffens, J. (1987): The energy functional, balayage, and capacity. Ann. Inst. H. Poincaré **23** 321–357.

[IMcK65] Itô, K., McKean, H.P. (1965): Diffusion processes and their Sample Paths. Springer-Verlag, Berlin-Heidelberg-New York.

[K95] Kuwae, K. (1995): Permanent sets of measures charging no exceptional sets and the Feynman-Kac formula. To appear in Forum Math.

[LJ82] Le Jan, Y. (1982): Quasi-continuous functions associated with a Hunt process. Proc. Amer. Math. Soc. **86** 133–138.

[MR92] Ma, Z.-M. and Röckner, M. (1992): Introduction to the Theory of (Non-Symmetric) Dirichlet Forms. Springer-Verlag, Berlin-Heidelberg-New York.

[McKT61] McKean, H.P., Tanaka, H. (1961): Additive functionals of the Brownian path. Mem. Coll. Sci. Univ. Kyoto, Ser. A, **33** 479–506.

[Me62] Meyer, P.-A. (1962): Fonctionelles multiplicatives et additives de Markov. Ann. Inst. Fourier Grenoble **12** 125–230.

[Re70] Revuz, D. (1970): Mesures associees aux fonctionelles additives de Markov, I. Trans. Amer. Math. Soc. **148** 501–531.

[Re71] Revuz, D. (1971): Remarques sur les potentiels de mesure. Séminaire de Probabilité V, Lecture Notes in Math. **191**, pp. 275–277. Springer-Verlag, Berlin-Heidelberg-New York.

[Ro71] Rost, H. (1971): The stopping distributions of a Markov process. Z. Warscheinlichkeitstheorie verw. Gebiete **14** 1–16.

[Sh71] Sharpe, M.J. (1971): Exact multiplicative functionals in duality. Indiana Univ. Math. J. **21** 27–61.

[Sh88] Sharpe, M.J. (1988): General Theory of Markov Processes. Academic Press, San Diego.

[St92] Sturm, K.Th. (1992): Measures charging no polar sets and additive functionals of Brownian motion. Forum Math. **4** 257–297.

[V60] Volkonskii, V.A. (1960): Additive functionals of Markov processes. Trudy Moscov. Math. Obsc. **9** 143–189. [English transl. in Selected Math. Trans. Stat. and Prob. **5**, Amer. Math. Soc., Providence, 1965.]

[W61] Wentzell, A.D. (1961): Non-negative functionals of Markov processes. Soviet Math. Dokl. **2** 218–221.

On decomposition of additive functionals of reflecting Brownian motions

Masatoshi Fukushima[1] and Matsuyo Tomisaki[2]

[1] Department of Mathematical Science, Faculty of Engineering Science,Osaka University, Toyonaka, Osaka 560, Japan
[2] Department of Mathematics, Faculty of Education, Yamaguchi University, Yamaguchi 753, Japan

1. Reflecting Brownian motions on non-smooth domains

Let X be a locally compact separable Hausdorff metric space and m be a positive Radon measure on X with full support. For an m-symmetric Hunt process $\mathbf{M} = (X_t, P_x)$ on X with the associated Dirichlet form $(\mathcal{E}, \mathcal{F})$ being regular on $L^2(X; m)$, the following decomposition of additive functionals (AF's in abbreviaton) is known ([11]):

$$u(X_t) - u(X_0) = M_t^{[u]} + N_t^{[u]}, \quad P_x - \text{almost surely}, \tag{1.1}$$

which holds for quasi every (q.e. in abbreviation) $x \in X$. Here u is a quasi-continuous function in the space $\mathcal{F}$, $M_t^{[u]}$ is a martingale AF with quadratic variation being associated with the energy measure of u, $N_t^{[u]}$ is a continuous AF of zero energy and 'for q.e. $x \in X$' means 'for every $x \in X$ outside a set of zero capacity'. (1.1) is beyond a semimartingale decomposition in that $N_t^{[u]}$ is of zero quadratic variation P_m-a.s. but not necessarily of bounded variation P_x-a.s. on each finite time interval. Nevertheless both $M_t^{[u]}$ and $N_t^{[u]}$ are well computable from u through the Dirichlet form $\mathcal{E}$ and accordingly the decomposition (1.1) has proved to be a useful substitute of Ito's formula for symmetric Markov processes.

Since the decomposition (1.1) is formulated based entirely upon the potential theory of the regular Dirichlet space, we have to admit exceptional sets of zero capacity for the identity (1.1) and for the definition of AF's involved in (1.1) as well. However it has been known that the identity (1.1) and the involved AF's can be formulated without any exceptional set of zero capacity if we assume the absolute continuity of the transition function $p_t(x, B)$ of the process $\mathbf{M}$([11, 9, 10]):

$$p_t(x, \cdot) \prec m. \tag{1.2}$$

It is this strict version of the decomposition (1.1) that we are going to utilize in this paper.

(Normally) reflecting Brownian motions provide us with good examples to which decomposition (1.1) is applicable. Let D be a domain in the Euclidean

d-space R^d for $d \geq 2$ and $\overline{D} = D \cup \partial D$ be its closure. We are specifically concerned with a normally reflecting Brownian motion on $\overline{D}$ when the boundary ∂D is continuous but non- smooth, e.g. convex, Lipschitz, Hölder etc. There have been at least three different approaches to it;
by *Dirichlet form* (Bass-Hsu[1, 2] for a bounded Lipschitz domain),
by *Skorohod equation* (Tanaka[16] for a convex domain) and
by *submartingale problem* (Varadhan-R.Williams[17] for a two dimensional wegde and DeBlassie-Toby[6, 7] for a two dimensional outward cusp domain).

For a domain $D \subset R^d$, we denote the Lebesgue measure on D by $m(dx)$ or simply by dx and we consider a Dirichlet form $(\mathcal{E}, \mathcal{F})$ on $L^2(D) = L^2(D; m)$ defined by

$$\mathcal{F} = H^1(D), \quad \mathcal{E}(u,v) \doteq \frac{1}{2}\int_D \nabla u(x) \cdot \nabla v(x) dx, \tag{1.3}$$

where $H^1(D) = \{u \in L^2(D) : \partial_i u \in L^2(D), 1 \leq i \leq d\}$ the Sobolev space of order 1. Let $\{T_t, t > 0\}$ be the corresponding strongly continuous semigroup of Markovian symmetric operators on $L^2(D)$. Suppose D is of class C in the sense that its boundary ∂D is locally expressible as a graph of continuous function of $d-1$ variables, then (1.3) is regular as a Dirichlet form on $L^2(\overline{D})$ rather than on $L^2(D)$ because $C_0^\infty(\overline{D})$ is dense in $H^1(D)$ ([14]). Further the Dirichlet form (1.3) is strongly local. Hence, according to general theorems (see Theorem 7.2.2 and Problem 5.7.1 of [11]), there exists uniquely up to a set of zero capacity a conservative diffusion process (a conservative Hunt process with continuous sample paths) $\mathbf{M} = (X_t, P_x)$ on $\overline{D}$ which is associated with the form (1.3) in the sense that its transition function $p_t(x, B) = P_x(X_t \in B)$ satisfies that

$$p_t f \text{ is a version of } T_t f \text{ for any } f \in B_0(D), \tag{1.4}$$

where $B_0(D)$ denotes the set of bounded functions on D vanishing outside a bounded set. We may call this diffusion $\mathbf{M}$ a (normally) reflecting Brownian motion on $\overline{D}$.

Now let D be a bounded Lipschitz domain and $\mathbf{M} = (X_t, P_x)$ be a reflecting Brownian motion on $\overline{D}$ in the above sense. Let $\phi_i(x) = x_i, 1 \leq i \leq d$, be the coordinate functions. By the divergence theorem, we then have

$$\mathcal{E}(\phi_i, v) = -\frac{1}{2}\int_{\partial D} n_i(x) v(x) \sigma(dx), \quad v \in C_0^\infty(\overline{D}), \tag{1.5}$$

where σ is the surface measure on ∂D and $\mathbf{n} = (n_1, n_2, \cdots, n_d)$ is the inward unit normal vector defined σ−a.e. on ∂D. We can easily check (see the proof of Lemma 1 in §3) that σ is of finite 1-energy integral in the sense that

$$\int_{\partial D} |v| d\sigma \leq C\sqrt{\mathcal{E}_1(v,v)} \quad v \in C_0^\infty(\overline{D}),$$

and consequently there is a positive continuous AF L_t of **M** with Revuz measure σ. Applying the decomposition (1.1) to functions $\phi_i \in H^1(D)$ and combining the identity (1.5) with a characterization of $N^{[\phi_i]}$ given in [11], Bass-Hsu[1] obtained the following Skorohod type expression of the sample path X_t of **M**:

$$X_t = x + B_t + \frac{1}{2}\int_0^t \mathbf{n}(X_s)dL_s \quad P_x - a.s. \tag{1.6}$$

holding for q.e. $x \in \overline{D}$, where B_t is a d-dimensional Brownian motion starting at the origin.

Up to this point, we need not use any result in another work of Bass-Hsu[2], in which they constructed a conservative diffusion process **M** on $\overline{D}$ satisfying not only the condition (1.4) but also the strong Feller property of the resolvent G_λ:

$$G_\lambda(B_0(D)) \subset C(\overline{D}). \tag{1.7}$$

This property implies the absolute continuity of the resolvent and of the transition function as well([11;Theorem 4.2.4]) and accordingly the strict version of the decomposition (1.1) is applicable. Besides Bass-Hsu[2] proved that the surface measure σ is of bounded 1-potential by making use of a Gauss upper bound of the transition density derived from a Sobolev inequality and Carlen-Kusuoka-Stroock's theorem [3]. Hence the associated positive continuous AF L_t can be taken in the ordinary strict sense and the strict version of the Skorohod type decomposition (1.6) can be established holding for every $x \in \overline{D}$ rather than q.e. x ([11;Example 5.2.2]).

One can look at the strict version of (1.6) as a Skorohod equation; given a Brownian path B_t starting at the origin and for each fixed $x \in \overline{D}$, we look for the pair of functionals (X_t, L_t) satisfying (1.6) under the requirement that L_t increases only when $X_t \in \partial D$. When D is a convex domain in R^d, Tanaka[16] proved that the solution of this problem can be uniquely and deterministically constructed from B_t. Since a bounded convex domain is Lipschitz, the strict decomposition (1.6) enables us to conclude that, for a bounded convex domain D, the Bass-Hsu reflecting Brownian motion on $\overline{D}$ is identical in law with Tanaka's one.

On the other hand, by extending a work of Varadhan-R.Williams on an infinite two-dimensional wedge[17], DeBlassie-Toby[6] formulated as a unique solution of a submartingale problem a normally reflecting Brownian motion on a two-dimensional standard outward cusp domain

$$C = \{(x, y) \in R^2 : y \geq |x|^\gamma\} \quad 0 < \gamma < 1, \tag{1.8}$$

and constructed it from the normally reflecting Brownian motion on the upper half plane by means of a conformal map and a random time change. A subsequent paper by DeBlassie-Toby[7] further demonstrated that the constructed process admits the Skorohod representation if $\gamma > \frac{1}{2}$ but otherwise the process starting at the origin fails even to be a semimartingale.

It is therefore tempting to extend Bass-Hsu's construction of the normally reflecting Brownian motion with strong Feller resolvent to a more general domain in R^d admitting outward and inward cusp boundary points and check if the constructed process admits the strict semimartingale representation (1.6) whenever the Hölder exponent γ at each outward cusp boundary point is greater than $\frac{1}{2}$ regardless the dimension d. This has been carried out affirmatively in [12] but under a rather strong restriction that $\gamma > (d-1)/d$ due to a technical difficulty. In [13], this restriction is completely removed and furthermore, by using the strict version of the decomposition (1.1) again, the constructed process is linked to a submartingale problem and accordingly identified in law with DeBlassie-Toby's one in the case of the two dimensional standard outward cusp domain. Actually [12, 13] deal with a normally reflecting diffusion process associated with a general second order uniformly elliptic differential operator of divergence form with measurable coefficients a_{ij}. The normally reflecting Brownian motion is a special case that $a_{ij} = \frac{1}{2}\delta_{ij}$.

In the next section, we summarize those results in [13] about construction and decomposition but we shall do so only in the Brownian motion case for the sake of simplicity. In the final section, we present some detailed proof of the strict smoothness of the surface measure under the stated condition on the outward cusp boundary points - a key step leading to the strict Skorohod representation (1.6).

In this paper, we are only concerned with a domain of class C and a normally reflecting Brownian on its Euclidean closure. For an arbitrary bounded domain D however, one can conceive a normally reflecting Brownian motion on its Martin-Kuramochi type compactification D^*([8]) and a stationally diffusion on the Euclidean closure $\overline{D}$ induced from it. The decomposition (1.1) of AF's has been also effectively utilized in the study of these processes. See the works by Chen[4], Chen-Fitzsimmons-Williams[5], Williams-Zheng[18] and Pardoux-Williams[15] in this connection.

2. Construction and decomposition on a Lipschitz domain with outward and inward cusps

Let F be a real valued function defined on a set $E \subset R^k$ including the origin such that $F(x) = \alpha|x|^\gamma + f(x)$, where $0 < \gamma < 1$, $\alpha \in R$, and f is a k-dimensional Lipschitz continuous function vanishing at the origin. Here $|\cdot|$ denotes the Euclidean norm. We call such F a Hölder function and we let

$$\mathrm{Exp}(F) = \gamma, \quad \mathrm{Höl}(F) = \alpha,$$

$$\mathrm{Lip}(F) = \mathrm{Lip}(f) = min\{K > 0 : |f(x) - f(y)| \le K|x-y|,\ x, y \in E\}.$$

For $x = (x_1, \cdots, x_d) \in R^d$, we let $x' = (x_1, \cdots, x_{d-1})$ so that $x = (x', x_d)$.

From now on, we consider a domain $D \subset R^d$ with $d \ge 2$ satisfying the next condition (H):

(H) There are four constants $\gamma \in (0,1), \delta > 0, A \geq 1, M > 0$ and a locally finite covering $\{U_j\}_{j \in J}$ of ∂D with the following properties:

(i) For each $j \in J$, there are a Hölder function F_j of $d-1$variables and a constant $r_j > \delta$ such that
F_j is defined on the $d-1$-dimensional ball centered at the origin with radius r_j,
$\text{Exp}(F_j) \geq \gamma$,
$\text{Höl}(F_j) = 0$, or $1/A \leq \text{Höl}(F_j) \leq A$, or $-A \leq \text{Höl}(F_j) \leq -1/A$,
$\text{Lip}(F_j) \leq M$,
$U_j \cap D = \{\zeta = (\zeta', \zeta_d) : |\zeta| < r_j, F_j(\zeta') < \zeta_d\}$, for some Cartesian coordinate system $\zeta = (\zeta', \zeta_d)$.

(ii) $\partial D \subset \cup_{j \in J} \tilde{U}_{j,\delta}$, where $\tilde{U}_{j,\delta} = \{x \in U_j : \text{dist}(x, \partial U_j) > \delta\}$.

We denote by J_+, J_0 and J_- the set of $j \in J$ for which $\text{Höl}(F_j)$ is positive, vanishing and negative respectively. For $j \in J$, denote by $a_j (\in \partial D)$ the origin in U_j with respect to the coordinate system ζ. a_j is called an outward (resp. inward) cusp boundary point of D if $j \in J_+$ (resp.$j \in J_-$). If D is bounded, condition (H) reduces to a simple one that every point x of ∂D has a neighbourhood U_x such that $\partial D \cap U_x$ is the graph of a Hölder function of $d-1$ variables. If further $J_+ \cup J_- = \emptyset$, then D is just a bounded Lipschitz domain.

Let $(\mathcal{E}, \mathcal{F})$ be the Dirichlet form on $L^2(D)$ defined by (1.3) and $\{G_\lambda, \lambda > 0\}$ be the associated resolvent on $L^2(D)$. It is Markovian in the sense that $0 \leq \lambda G_\lambda f \leq 1$ whenever $0 \leq f \leq 1$ and it is well defined as a bounded linear operator on $L^p(D)$ for any $p \in [1, \infty]$. Denote by $C_\infty(\overline{D})$ the space of those functions in $C(\overline{D})$ vanishing at infinity.

Theorem 1. (i) $G_\lambda(L^2(D) \cap L^p(D)) \subset C(\overline{D}), \quad p > 1 + (d-1)/\gamma.$
(ii) $G_\lambda \left(C_\infty(\overline{D})\right)$ *is dense subspace of* $C_\infty(\overline{D})$.
(iii) *There is a function* $G_\lambda(x,y)$ *continuous on* $\overline{D} \times \overline{D}$ *off diagonal such that*

$$G_\lambda f(x) = \int_{\overline{D}} G_\lambda(x,y) f(y) dy, \quad x \in \overline{D}, \ f \in C_\infty(\overline{D}).$$

The first assertion means that G_λ has the strong Feller property (1.7). This theorem is proved in [13] following a method of Stampacchia and Moser for PDE. A key step in the proof is a modified Sobolev inequality of Moser's type, which has been proved in [12] under the restriction stated in the preceding section but the restriction is removed in [13] thanks to a specific transformation of the standard cusp domain onto a rectangular set. We will not go into further details of the proof but we note the following Sobolev inequaities which are derived in the course of the proof and will be utilized in the next section.

Proposition 1. (i) *There is a positive constant* C *such that*

$$\|u\|_{L^q(D)} \leq C\sqrt{\mathcal{E}_1(u,u)}, \quad u \in H^1(D), \tag{2.1}$$

for $2 \le q \le 2(d-1+\gamma)/(d-1-\gamma)$.
(ii) *Assume the absence of outward cusp boundary point:* $J_+ = \emptyset$. *Then the above inequality is valid for* $2 \le q \le 2d/(d-2)$ *in case* $d \ge 3$ *and for* $2 \le q < \infty$ *in case* $d = 2$.

Theorem 1 readily implies the next theorem:

Theorem 2. *There exists a conservative diffusion process* $\mathbf{M} = (X_t, P_x)$ *on* $\overline{D}$ *with resolvent* G_λ *of Thoerem 1.* $\mathbf{M}$ *is associated with the Dirichlet form (1.3) and the transition function of* $\mathbf{M}$ *satisfies (1.2) and (1.4).*

Proof By Theorem 1 (ii) and the Hille-Yosida theorem, there exists a strongly continuous Markovian semigroup $\{T_t, t > 0\}$ on $C_\infty(\overline{D})$ such that $G_\lambda f$ is the Laplace transform $T_t f$ for $f \in C_\infty(\overline{D})$. We have then a Feller transition function by $T_t f(x) = \int_{\overline{D}} p_t(x, dy) f(y)$, which gives rise to a Hunt process $\mathbf{M}$ on $\overline{D}$. $\mathbf{M}$ is associated with the regular Dirichlet form (1.3) on $L^2(\overline{D})$ since G_λ is. Therefore we can apply general theorems in [11] to the associated pair $\mathcal{E}$ and $\mathbf{M}$. In particular, $p_t(x, \cdot)$ is absolutely continuous because $G_\lambda(x, \cdot)$ is ([11;Theorem 4.2.4]). Since $\mathcal{E}$ has the strong local property, we can invoke Theorem 4.5.4 and Problem 5.7.1 of [11] to conclude that $\mathbf{M}$ is a conservative diffusion process on $\overline{D}$.

$\mathbf{M}$ of Theorem 2 is called the reflecting Brownian motion on $\overline{D}$. We next formulate a decomposition of the sample path of $\mathbf{M}$ and its Skorohod representation. Notice that, under the present condition (H) for the domain D, the surface measure σ on ∂D is well defined with a local expression

$$\sigma(E) = \int_{E_*} \sqrt{1 + |\nabla F_j(\zeta')|^2} d\zeta', \quad E \subset U_j \cap \partial D, \tag{2.2}$$

where $E_* = \{\zeta' : (\zeta', F_j(\zeta')) \in E\}$. Further, the unit inward normal vector $\mathbf{n}(\zeta) = (n_1(\zeta), \cdots, n_d(\zeta))$ makes sense σ-a.e. on ∂D by

$$\mathbf{n}(\zeta) = (-\nabla F_j(\zeta'), 1)/\sqrt{(1 + |\nabla F_j(\zeta')|^2}, \zeta \in U_j \cap \partial D.$$

Theorem 3. *Let* $\mathbf{M} = (X_t, P_x)$ *be the diffusion process of Theorem 2.*
(i) *The sample path* $X_t = (X_t^1, X_t^2, \cdots, X_t^d)$ *of* $\mathbf{M}$ *admits a unique decomposition*

$$X_t^i - X_0^i = B_t^i + N_t^i, \quad 1 \le i \le d, P_x - \text{a.s. for any } x \in \overline{D}, \tag{2.3}$$

where $B_t = (B_t^1, B_t^2, \cdots, B_t^d)$ *is a d-dimensional Brownian motion starting at the origin* P_x*-a.s. and* N_t^i *are CAF's in the strict sense locally of zero energy.*
(ii) *Suppose that*

$$\text{Exp}(F_j) > 1/2, \quad j \in J_+, \tag{2.4}$$

then the surface measure σ *on* ∂D *is smooth in the strict sense.*
(iii) *Under the condition (2.4), the Skorohod representation (1.6) of* X_t *holds for every* $x \in \overline{D}$ *with a PCAF* L_t *in the strict sense with Revuz measure* σ.

The second assertion (ii) will be proved in the next section.

Proof of (i). The coordinate functions $\phi_i, 1 \le i \le d$, are locally in $H^1(D)$ and their coenergy measures $d\mu_{\langle\phi_i,\phi_j\rangle}$ equal $\delta_{ij}dx$, which are obviously smooth in the strict sense with associated CAF in the strict sense being the constant functionals $\delta_{ij}\cdot t$. Hence the strict and local version of (1.1) for $u = \phi_i$ holds by virtue of [10;Theorem 2] and we get the strict decomposition (2.3) for $X_t^i = \phi_i(X_t)$, local continuous martingale AF's $B_t^i = M_t^{[\phi_i]}$ with quadratic covariations $\delta_{ij}\cdot t$ and CAF's N_t^i locally of zero energy.

Proof of (ii) $\Longrightarrow$ (iii).

The divergence theorem (1.5) still holds for the present domain with the surface measure σ and the unit inward normal vector $\mathbf{n}$ given above. Hence, if σ is smooth in the strict sense, then N_t^i admits the expression as in the last term of (1.6) with L_t being as in the statement (iii) on account of [9;Corollary 3.1].

3. Strict smoothness of the surface measure

In this section, we give a proof of the second assertion of Theorem 3 the strict smoothness of the surface measure σ under the condition (2.4). As for the definition of the strict smoothness of a Borel measure, we refer to [11]. We also use the notation S_0 in [11] to denote the set of all positive Radon measure on $\overline{D}$ with finite energy integral. For a positive Borel measure μ, its (λ-)potential is defined by $G_\lambda\mu(x) = \int_{\overline{D}} G_\lambda(x,y)\mu(dy),\ x \in \overline{D}$, where $G_\lambda(x,y)$ is the resolvent appearing in Theorem 1. It is known ([12]) that a positive Radon measure μ on $\overline{D}$ is smooth in the strict sense if the following condition is satisfied:

$$\text{For any ball } B \subset R^d, \quad I_B\cdot\mu \in S_0 \ \text{ and } G_\lambda I_B\cdot\mu(x) < \infty \quad \forall x \in \overline{D}. \tag{3.1}$$

Note that this condition is automatically satisfied if $I_B\cdot\mu$ is of bounded potential for any ball B.

For any ball B, the compact set $\overline{B}\cap\partial D$ can be covered by a finite number of open sets $\tilde{U}_{j,\delta}$ appearing in the condition (H) (ii) for the domain D. Besides $G_\lambda(x,y)$ is jointly continuous off diagonal by Theorem 2.1 (iii). For the proof of (3.1), it is therefore sufficient to show

$$I_\Gamma\cdot\sigma \in S_0 \quad \text{and} \quad G_\lambda I_\Gamma\sigma(x) < \infty, \qquad x \in \Gamma, \tag{3.2}$$

or even a stronger condition

$$\sup_{x\in\Gamma} G_\lambda I_\Gamma\sigma(x) < \infty, \tag{3.3}$$

where

$$\Gamma = \{\zeta = (\zeta', \zeta_d) : |\zeta| < \rho,\ \zeta_d = F_j(\zeta')\} \subset U_j \cap \partial D$$

for each fixed $j \in J$ and $\rho < r_j$. $\mathrm{Exp}(F_j)$ will be denoted by γ_j. $c_1, c_2, \cdots$ will denote some positive constants. We further let

$$\Gamma_* = \{\zeta' : (\zeta', F_j(\zeta')) \in \Gamma\} (\subset \{\zeta', |\zeta'| < \rho\}).$$

We shall prove (3.2) when $j \in J_+$ and $\gamma_j > 1/2$ and prove (3.3) when $j \in J_- \cup J_0$.

Lemma 1. *Let $j \in J_+$ and suppose $\gamma_j > 1/2$. Then $I_\Gamma \cdot \sigma \in S_0$.*

Proof We have to show

$$\int_{\Gamma_*} |u(\zeta', F_j(\zeta'))| \sigma(d\zeta') \leq c_1 \sqrt{\mathcal{E}_1(u,u)}, \quad u \in C_0^\infty(\overline{D}). \tag{3.4}$$

By (2.2), the surface measure σ has a density $\sigma(\zeta')$ with respect to $d\zeta'$ satisfying

$$\sigma(\zeta') \leq c_2 |\zeta'|^{\gamma_j - 1}. \tag{3.5}$$

Hence the square of the left hand side of (3.4) is dominated by

$$c_2^2 \int_{\Gamma_*} u(\zeta', F_j(\zeta'))^2 d\zeta' \cdot \int_0^\rho r^{2\gamma_j + d - 4} dr.$$

The second factor is finite under the stated condition. Consider a function $\psi \in C_0^\infty(U)$ taking value 1 on the set Γ. Then from the expression

$$u(\zeta', F_j(\zeta')) = - \int_{F_j(\zeta')}^{\sqrt{r^2 - |\zeta'|^2}} \frac{\partial}{\partial \zeta_d} \{\psi(\zeta', \zeta_d) u(\zeta', \zeta_d)\} d\zeta_d, \quad \zeta' \in \Gamma_*,$$

we see that the first factor is dominated by $c_3 \mathcal{E}_1(u,u)$.

To proceed further,we need a comparison lemma for the resolvent:

Lemma 2. *Let K be a compact subset of $\overline{D}$ and U be a bounded domain containing K such that the domain $D_1 = D \cap U$ possesses the property (H). Denote by $G_\lambda^1(x,y), x, y \in \overline{D}_1$, the resolvent of the reflecting Brownian motion (denoted by $\mathbf{M}_1$) on $\overline{D}_1$. Then*

$$G_\lambda(x,y) \leq G_\lambda^1(x,y) + c_4, \quad x, y \in K, x \neq y$$

for some positive constant c_4 depending on the set K.

Proof Let $F = \overline{D} \cap U$. On account of [11;Theorem 4.4.3], the parts of $\mathbf{M}$ and $\mathbf{M}_1$ on the set F have a common Dirichlet form on $L^2(F)$ and hence a common resolvent density. Then the lemma follows from the Dynkin formula and off diagonal continuity of $G_\lambda(x,y)$.

Lemma 3. *For each $j \in J$ and each compact subset K of $U_j \cap \overline{D}$, $G_\lambda(x,y)$ for $x,y \in K, x \neq y$, is dominated by the following quantity, where c_5 is a positive constant depending on j and K:*

$$c_5|x-y|^{-\frac{d-1-\gamma_j}{\gamma_j}} \quad \text{if } j \in J_+, d \geq 2. \tag{3.6}$$

$$c_5|x-y|^{-d+2} \quad \text{if } j \in J_0 \cup J_-, d \geq 3. \tag{3.7}$$

$$c_5|x-y|^{-\epsilon} \quad \text{for any } \epsilon > 0 \quad \text{if } j \in J_0 \cup J_-, d = 2. \tag{3.8}$$

Proof Notice that the Sobolev inequality in Proposition 1 (i) holds with D and γ being replaced by $D_j = U_j \cap D$ and γ_j respectively. Since the domain D_j is bounded, we can invoke Carlen-Kusuoka-Stroock[3] to conclude in the same way as in [2;Section 2] that the resolvent density $G^1_\lambda(x,y)$ of the reflecting Brownian motion on $\overline{D}_j$ admits the estimate

$$G^1_\lambda(x,y) \leq c_6|x-y|^{-\beta}, \quad x,y \in \overline{D}_j, \tag{3.9}$$

for $\beta = 4/(q-2)$. In particular, by taking $q = 2(d-1+\gamma_j)/(d-1-\gamma_j)$, we see that (3.9) is valid for $\beta = (d-1-\gamma_j)/\gamma_j$. We can then use Lemma 2 to get (3.6).

In the case that $j \in J_0 \cup J_-$, Proposition 1 (ii) is applicable to the domain D_j and we see the validity of (3.9) for $\beta = d-2$ [resp. $\beta = \epsilon > 0$] by taking $q = 2d/(d-2)$ [resp. $q = 4/\epsilon + 2$]. We again use Lemma 2 to get (3.7)[resp.(3.8)].

Lemma 4. *Let $j \in J_+$ and suppose $\gamma_j > 1/2$. Then $G_\lambda I_\Gamma \cdot \sigma(\zeta) < \infty, \zeta \in \Gamma$.*

Proof Keeping the expression

$$G_\lambda I_\Gamma \cdot \sigma(\zeta) = \int_{\Gamma_*} G_\lambda(\zeta, (\eta', F(\eta')))\sigma(\eta')d\eta' \tag{3.10}$$

and the bound (3.5) of σ in mind, we first prove the finiteness of the potential for $\zeta = 0$. From (3.6),

$$G_\lambda I_\Gamma \cdot \sigma(0) \leq c_6 \int_{\Gamma_*} g(\eta')^{-1}|\eta'|^{\gamma_j - 1}d\eta',$$

where $g(\eta') = (|\eta'|^2 + |F_j(\eta')|^2)^{\frac{d-1-\gamma_j}{\gamma_j}}$. Denote Höl$(F_j)$ by α_j. Since $g(\eta')$ is dominating $(\alpha_j/2)^{\frac{d-1-\gamma_j}{\gamma_j}}|\eta'|^{d-1-\gamma_j}$, $|\eta'| < \delta$, for some $\delta > 0$, we obtain

$$G_\lambda I_\Gamma \cdot \sigma(0) \leq c_7 \int_0^\delta r^{d+\gamma_j-3-(d-1-\gamma_j)}dr + c_7\delta^{-\frac{d-1-\gamma_j}{\gamma_j}} \int_\delta^\rho r^{d+\gamma_j-3}dr$$

which is finite under the present assumption on γ_j.

Next take a $\zeta \in \Gamma, \zeta \neq 0$. We can choose a neighbourhood V_1 and V_2 of ζ such that $0 \notin V_1$, $\overline{V}_1 \subset V_2 \subset U_j$ and $D_2 = V_2 \cap D$ is a Lipschitz domain.

Let $\tilde{\Gamma} = \Gamma \cap V_1$. Then, by the same reasoning as in the proof of Lemma 3, we see that $G_\lambda(\zeta,\eta), \eta \in \tilde{\Gamma}$, is dominated by $c_8|\zeta-\eta|^{-d+2}$ in case that $d \geq 3$ and by $c_9|\zeta-\eta|^{-\epsilon}, \epsilon > 0$, in case that $d = 2$. Since $\sigma(\eta)$ is bounded on $\tilde{\Gamma}$, we get the finiteness of $G_\lambda I_{\tilde{\Gamma}} \cdot \sigma(\zeta)$ and hence of $G_\lambda I_\Gamma \cdot \sigma(\zeta)$.

Lemma 5. *(3.3) is valid when $j \in J_0 \cup J_-$.*

Proof When $j \in J_0 \cup J_-$, we have the bound (3.7) and (3.8) of the resolvent. If $j \in J_0$, then the density function σ of the surface measure in uniformly bounded on Γ and hence (3.3) is immediate from the expression (3.10).

Suppose $j \in J_-$ and $d \geq 3$. (3.5),(3.7) and (3.10) lead us to

$$G_\lambda I_\Gamma \cdot \sigma(\zeta) \leq c_{10} \int_{|\eta'|<\rho} r(\zeta',\eta')d\eta', \quad \zeta \in \Gamma,$$

where $r(\zeta',\eta') = |\zeta'-\eta'|^{-d+2}|\eta'|^{-1+\gamma}$. For $\zeta \neq 0,\ \zeta \in \Gamma$, we put $\xi' = \zeta'/|\zeta'|$ so that $|\xi'| = 1$. Then

$$G_\lambda I_\Gamma \cdot \sigma(\zeta) \leq c_{11}|\zeta'|^\gamma \int_{|\eta'|<\frac{\rho}{|\zeta'|}} r(\xi',\eta')d\eta'. \tag{3.11}$$

The last integral is dominated by a constant $c_{12} = \int_{|\eta'|<2} r(\xi',\eta')d\eta' < \infty$ if $\rho/2 \leq |\zeta'| < \rho$ and by $c_{12} + c_{13}|\zeta'|^{-\gamma}$ if $|\zeta'| < \rho/2$. Thus we arrive at (3.3).

Suppose finally $j \in J_-$ and $d = 2$. Then, from (3.8), we obtain the bound (3.11) with γ and $r(\xi',\eta')$ being replaced by $\gamma - \epsilon$ and $|\xi'-\eta'|^{-\epsilon}|\eta'|^{-1+\gamma}$ respectively. Therefore we attain (3.3) again by choosing ϵ smaller than γ.

The proof of the strict smoothness of the surface measure under condition (2.4) is complete by Lemma 1, Lemma 4 and Lemma 5.

Finally we note the following lemma. Denote by Ξ_+ the totality of outward cusp boundary points.

Lemma 6. *For any neighbourhood W of Ξ_+, $I_{\partial D \setminus W} \cdot \sigma$ is smooth in the strict sense.*

Indeed $I_{\Gamma\setminus W} \cdot \sigma$ is of bounded potential in view of Lemma 5 and the last part of the proof of Lemma 4. Let us consider the process **M** of Theorem 2 and let $\tilde{L}_t$ be a PCAF in the strict sense with Revuz measure $I_{\partial D\setminus W} \cdot \sigma$. For any function $f \in C_b^2(\overline{D})(\subset H^1_{loc}(D))$ which is constant on the set W, the Gauss-Green formula reads

$$\mathcal{E}(f,v) + \frac{1}{2}\int_D \Delta f \cdot v dx = -\frac{1}{2}\int_{\partial D} \nabla f \cdot \mathbf{n} I_{\partial D\setminus W} d\sigma.$$

By [9;Corollary 3.1] again, we can conclude from this that

$$f(X_t) - f(X_0) - \frac{1}{2}\int_0^t \Delta f(X_s)ds$$

is a sum of a martingale AF (in the strict sense) and a continuous AF (in the strict sense) $\frac{1}{2}\int_0^t \nabla f \cdot \mathbf{n}(X_s)d\tilde{L}_s$, which is increasing whenever $\nabla f \cdot \mathbf{n} \geq 0$ σ-a.e. Thus M is linked to a submartingale problem as was mentioned in §1.

References

1. R.F.Bass and P.Hsu: The semimartingale structure of reflecting Brownian motion, Proc.A.M.S.,**108**(1990),1007-1010
2. R.F.Bass and P.Hsu: Some potential theory for reflecting Brownian motion in Hölder and Lipschitz domains, Ann.Probab.,**19** (1991),486-506
3. E.A.Carlen, S.Kusuoka and D.W.Stroock: Upper bounds for symmetric Markov transition functions, Ann.Inst.H.Poincaré Probab.Statist.,**23**(1987),245-287
4. S.Q.Chen: On reflecting diffusion processes and Skorohod decompositions, Probab.Theory Relat.Fields **94**(1993),281-315
5. Z.Q.Chen, P.J.Fitzsimmons and R.J.Williams: Quasimartingales and strong Caccioppoli set, Potential Analysis,**2**(1993),219-243
6. R.D.DeBlassie and E.H.Toby: Reflecting Brownian motion in a cusp, Trans.Am.Math.Soc.,**339**(1993),297-321
7. R.D.DeBlassie and E.H.Toby: On the semimatingale representation of reflecting Brownian motion in a cusp, Prob.Theory Relat.Fields, **94**(1993),505-524
8. M.Fukushima: A construction of reflecting barrier Brownian motions for bounded domains, Osaka J.Math.,**4**(1967),183-215
9. M.Fukushima: On a strict decomposition of additive functionals for symmetric diffusion processes, Proc.Japan Acad.,**70** Ser. A(1994),277-281
10. M.Fukushima: On a decomposition of additive functionals in the strict sense for a symmetric Markov process, in "*Dirichlet Forms and Stochastic Processes*',eds. Z.Ma, M.Röckner, J.Yan, pp155-169,Walter de Gruyter 1995
11. M.Fukushima,Y.Oshima and M.Takeda: *Dirichlet forms and symmetric Markov processes*, Walter de Gruyter, Berlin-New York 1994
12. M.Fukushima and M.Tomisaki:Reflecting diffusions on Lipschitz domains with cusps-Analytic construction and Skorohod representation, Potential Analysis **4**(1995),377-408
13. M.Fukushima and M.Tomisaki: Construction and decomposition of reflecting diffusions on Lipschitz domains with Hölder cusps, Probab. Theory Relat. Fields, to appear
14. V.G.Maz'ja: *Sobolev spaces*, Springer-Verlag, Berlin-Heidelberg-New York, 1985
15. E.Pardoux and R.J.Williams: Symmetric reflected diffusions, Ann.Inst.Henri Poincaré **30**(1994),13-62
16. H.Tanaka: Stochastic differential equations with reflecting boundary condition in convex regions, Hiroshima Math.J.,**9**(1979),163-177
17. S.R.S.Varadhan and R.J.Williams: Brownian motion in a wedge with oblique reflection, Comm.Pure.Appl.Math.,**38**(1985),405-443
18. R.J.Williams and W.A.Zheng: On reflecting Brownian motion-a waek convergence approach, Ann.Inst.Henri Poincaré **26**(3)(1990),461-488

Equilibrium fluctuations for lattice gas

T. Funaki

Graduate School of Mathematical Sciences, University of Tokyo, Komaba, Meguro-ku, Tokyo 153, Japan

1. Introduction

Professor K. Itô initiated the study of both equilibrium and non-equilibrium fluctuations for a class of systems consisting of a large number of particles [5,6,7]. He especially took a system of independent Brownian particles as a model and derived an infinite-dimensional ($\mathcal{D}'$-valued) Ornstein-Uhlenbeck process in the scaling limit of central limit theorem's type for the counting measures associated with the position of particles. This result was afterward generalized to an interacting case by Spohn [11] in an equilibrium situation. The corresponding law of large numbers, equivalently, the hydrodynamic limit for interacting Brownian particles was established by Varadhan [13].

In this paper we shall discuss the equilibrium fluctuations for the lattice gas. The lattice gas is a Markovian system of particles on a cubic lattice jumping randomly to neighboring sites under the restriction that at most one particle can occupy each site. For the sake of simplicity we disregard boundary conditions and consider the system on the d-dimensional periodic lattice $\Gamma_N = (\mathbb{Z}/N\mathbb{Z})^d (= \{1, 2, \dots, N\}^d)$. Our system is reversible under Bernoulli measures, but does not fulfill the so-called gradient condition. The hydrodynamic limit for this model was investigated by [4]. The equilibrium fluctuations for non-gradient models were discussed by Lu [9] for Ginzburg-Landau lattice model and then by Chang [3] for generalized symmetric exclusion process. We shall essentially use the same methods developed by them. The main idea lies in the employment of the perturbation technique combined with estimates on the central limit theorem variances originally due to Varadhan [14] to treat an outward diverging factor caused by the non-gradientness of the model. Non-equilibrium fluctuations were established by Chang and Yau [1] for Ginzburg-Landau model of gradient type; however, no rigorous proof is known so far for the non-gradient models, see [12] for a conjecture.

2. Model and result

The model and notation of the present article are the same as those in [4]. The state space for the lattice gas on Γ_N is denoted by $\mathcal{X}_N = \{0,1\}^{\Gamma_N}$, the set of all configurations $\eta = \{\eta_x; x \in \Gamma_N\}$ with $\eta_x = 0$ or 1 indicating that the site x is vacant or occupied, respectively. For $x, y \in \Gamma_N$ and $\eta \in \mathcal{X}_N$,

$\eta^{x,y}$ denotes an element of $\mathcal{X}_N$, obtained from η by exchanging the values of η_x and η_y; thus $(\eta^{x,y})_x = \eta_y$, $(\eta^{x,y})_y = \eta_x$ and $(\eta^{x,y})_z = \eta_z$ if $z \neq x, y$. Let τ_x, $x \in \Gamma_N$, be the shift operators acting on $\mathcal{X}_N$ by $(\tau_x\eta)_y = \eta_{y+x}, y \in \Gamma_N$, with addition being modulo N. They also act on functions f on $\mathcal{X}_N$ by $(\tau_x f)(\eta) = f(\tau_x\eta)$. The notations $\eta^{x,y}$ and τ_x also indicate corresponding ones for $\mathcal{X} = \{0,1\}^{\mathbb{Z}^d}$, the configuration space on the whole lattice. For $\Lambda = \Gamma_N$ or $\mathbb{Z}^d$, Λ^* (or Λ^{**}) denote the set of all unoriented (or oriented) bonds $b = \{x, y\}$ inside Λ (i.e., $x, y \in \Lambda$ and $|x - y| = 1$). We sometimes write η^b instead of $\eta^{x,y}$ for bonds $b = \{x, y\}$.

The generator L_N of our lattice gas on Γ_N is given by

$$L_N = \sum_{b \in \Gamma_N^*} c_b(\eta)\pi_b,$$

where π_b is the operator acting on functions f on $\mathcal{X}_N$ by

$$\pi_b f(\eta) = f(\eta^b) - f(\eta).$$

The family of functions $\{c_b(\eta) \equiv c_{x,y}(\eta); b = \{x, y\} \in (\mathbb{Z}^d)^*\}$ on $\mathcal{X}$ which determine the jump rate of particles between two neighboring sites x and y is supposed to satisfy the following three conditions (a) - (c):

(a) Positive and local: $c_{x,y}(\eta) > 0$ and depends only on $\{\eta_z; |z - x| \leq r\}$ for some $r > 0$.
(b) Spatially homogeneous: $c_{x,y} = \tau_x c_{0,y-x}$, $\{x, y\} \in (\mathbb{Z}^d)^*$.
(c) Detailed balance under Bernoulli measures: $c_{x,y}(\eta)$ is independent of $\{\eta_x, \eta_y\}$.

In view of (a), the jump rate $c_b(\eta)$ is naturally regarded as a function on $\mathcal{X}_N$ for $b \in \Gamma_N^*$, at least if N is large enough that $N > 2r$. The third condition (c) is equivalent to the symmetry of L_N relative to the Bernoulli measures $\nu_\rho, \rho \in [0,1]$, on $\mathcal{X}_N$ such that $\nu_\rho\{\eta_x = 1\} = \rho$ for every x; ν_ρ sometimes denote those on $\mathcal{X}$.

Let $\eta^N(t) = \{\eta_x^N(t), x \in \Gamma_N\}$ be the Markov process on $\mathcal{X}_N$ governed by the infinitesimal generator $\mathcal{L}_N = N^2 L_N$. The factor N^2 comes from the time change. Its macroscopic empirical-density field, i.e., the measure-valued process is defined by

$$\rho^N(t, d\theta) = N^{-d} \sum_{x \in \Gamma_N} \eta_x^N(t)\delta_{x/N}(d\theta), \quad \theta \in \mathbb{T}^d,$$

where $\mathbb{T}^d = \mathbb{R}^d/\mathbb{Z}^d$ is a d-dimensional torus identified with $[0,1)^d$ and δ_θ is the δ-measure at θ. It was shown in [4] under certain additional assumptions on the initial distributions of $\eta^N(t)$ and on the diffusion coefficient that $\rho^N(t, d\theta)$ converges in probability to $\rho(t,\theta)d\theta$ as $N \to \infty$ for every $t \geq 0$ and $\rho(t,\theta)$ is a solution of a nonlinear diffusion equation with diffusion coefficient $D(\rho) = \{D_{ij}(\rho)\}$ defined by (2.1) below. See the equation (1.6) in [4].

Our problem is to discuss the limit of the fluctuation of $\rho^N(t, d\theta)$ arround its mean. Let us explain $D(\rho)$ first, since it appears also in the limit of the density fluctuation fields. We introduce a quadratic form $\ell \cdot \widehat{c}(\rho)\ell$, $\ell \in \mathbb{R}^d$, for each $\rho \in [0,1]$ via the variational formula

$$\ell \cdot \widehat{c}(\rho)\ell = \inf_{F \in \mathcal{F}_0^d} \ell \cdot \widehat{c}(\rho; F)\ell,$$

where $\cdot$ means the inner product of $\mathbb{R}^d$, $\mathcal{F}_0^d$ denotes the class of all $\mathbb{R}^d$-valued local functions on $\mathcal{X}$ and

$$\ell \cdot \widehat{c}(\rho; F)\ell = \frac{1}{2} \sum_{|x|=1} E^{\nu_\rho} \left[c_{0,x} \left(\ell \cdot \left\{ x(\eta_x - \eta_0) - \pi_{0,x}\left(\sum\nolimits_{y \in \mathbb{Z}^d} \tau_y F\right) \right\} \right)^2 \right].$$

A $d \times d$ symmetric matrix $\widehat{c}(\rho) = \{\widehat{c}_{ij}(\rho)\}_{1 \le i,j \le d}$, which is denoted in the same notation, corresponds to the quadratic form introduced above. We also introduce the compressibility:

$$\chi(\rho) = \rho - \rho^2.$$

Then the diffusion coefficient $D(\rho) = \{D_{ij}(\rho)\}_{1 \le i,j \le d}$ is defined as

$$D(\rho) = \frac{\widehat{c}(\rho)}{2\chi(\rho)}, \quad \rho \in [0,1]. \tag{2.1}$$

Now we are at the position to describe our problem and a main result in this paper. We fix $\rho \in (0,1)$ and consider $\eta^N(t)$ having an initial distribution $\nu = \nu_\rho$. Notice that the distribution of $\eta^N(t)$ is ν for all $t \ge 0$. Define the fluctuation $\xi^N(t, d\theta)$ of the macroscopic density field $\rho^N(t, d\theta)$ arround its mean as

$$\begin{aligned} \xi^N(t, d\theta) :&= N^{d/2}\{\rho^N(t, d\theta) - E[\rho^N(t, d\theta)]\} \\ &= N^{-d/2} \sum_{x \in \Gamma_N} \{\eta_x^N(t) - \rho\} \delta_{x/N}(d\theta). \end{aligned}$$

For $J \in C(\mathbb{T}^d)$, the integral $\int_{\mathbb{T}^d} J(\theta)\, \xi^N(t, d\theta)$ is simply denoted by $\xi^N(t, J)$. We are interested in the asymptotic behavior of ξ^N as $N \to \infty$. Let P^N be the distribution of ξ^N on the space $D([0,T], \mathcal{M}(\mathbb{T}^d))$ where $\mathcal{M}(\mathbb{T}^d)$ stands for the class of all signed measures on $\mathbb{T}^d$. The main result is formulated as follows.

Theorem 1. *The sequence P^N converges to P as $N \to \infty$ weakly on the space $D([0,T], H^{-\alpha}(\mathbb{T}^d)), \alpha > d/2 + 1$, where $H^{-\alpha}(\mathbb{T}^d) = \{\xi \mid \|\xi\|_{-\alpha}^2 = ((-\Delta + 1)^{-\alpha}\xi, \xi)_{L^2(\mathbb{T}^d)} < \infty\}$ is the Sobolev space of order $-\alpha$ and P is* *the distribution of the stationary Ornstein-Uhlenbeck process* $\xi(t)$ *which is a solution of an infinite-dimensional SDE:*

$$d\xi(t) = \mathrm{Tr}(D(\rho)\partial^2\xi(t))\,dt + \mathrm{Tr}(\sqrt{\hat{c}(\rho)}d\nabla \mathbf{w}_t). \tag{2.2}$$

In this equation, $\partial^2\xi = \{\partial_i\partial_j\xi\}_{1\le i,j\le d}$ $(\partial_i = \partial/\partial\theta_i)$, $\nabla\mathbf{w}_t = \{\partial_i w_t^j\}_{1\le i,j\le d}$ *and* $\mathbf{w}_t = \{w_t^j(\theta)\}_{1\le j\le d}$ *is a system of independent cylindrical Brownian motions on* $L^2(\mathbb{T}^d)$, *in other words, each* $dw_t^j(\theta)/dt$ *(formal derivative in* t*) is a space-time white noise. The initial data* $\xi(0)$ *for (2.2) is a (spatial) white noise with intensity* $\chi(\rho)$, *namely, it is a mean zero generalized Gaussian field on* $\mathbb{T}^d$ *with covariance structure* $E[(\xi(0,J))^2] = \chi(\rho)\|J\|^2_{L^2(\mathbb{T}^d)}$ *for all* $J \in C^\infty(\mathbb{T}^d)$.

The proof consists of two steps: showing the tightness of $\{P^N\}$ (Section 5) and identifying the Ornstein-Uhlenbeck process in the limit due to Holley and Stroock's martingale approach (Section 3).

3. Martingale approach

Since our model is of non-gradient type, the term of bounded variation of the semimartingale $\xi^N(t,J)$ contains a diverging factor. To overcome this difficulty, we employ the perturbation method. Let us define $\tilde{\xi}_K^N(t,J)$, $J \in C^\infty(\mathbb{T}^d)$, $K = 1,2,\dots$, by

$$\tilde{\xi}_K^N(t,J) = \xi^N(t,J) + \zeta_K^N(t,J),$$

where

$$\zeta_K^N(t,J) = N^{-d/2-1}\sum_{x\in\Gamma_N}\nabla^{(N)}J(x/N)\cdot\tau_x F_K(\eta^N(t)),$$

$\nabla^{(N)}J(x/N) \equiv \{\nabla_i^{(N)}J(x/N)\}_{1\le i\le d} := \{N(J((x+e_i)/N) - J(x/N))\}_{1\le i\le d}$, $e_i \in \mathbb{Z}^d$ denotes the i-th unit vector and $F_K = \{F_K^i\}_{1\le i\le d} \in \mathcal{F}_0^d$ are chosen in such a manner that $\lim_{K\to\infty}\hat{c}(\rho;F_K) = \hat{c}(\rho)$. Recall that $\eta^N(t)$ is a stationary process with distribution ν for each $t \ge 0$. Since F_K are local functions, it is easily seen that

$$\lim_{N\to\infty} E[(\zeta_K^N(t,J))^2] = 0, \quad t \ge 0.$$

Therefore, the term $\zeta_K^N(t,J)$ is negligible, but it plays an important role when we compute the differential of $\tilde{\xi}_K^N(t,J)$ in t. Applying Itô's formula, we have

$$\tilde{\xi}_K^N(t,J) = \tilde{\xi}_K^N(0,J) + m_K^N(t,J) + \int_0^t a_K^N(\eta^N(s),J)\,ds \tag{3.1}$$

where

$$a_K^N(\eta, J) = -N^{-d/2+1} \sum_{i=1}^{d} \sum_{x \in \Gamma_N} \{W_{x,x+e_i} - L_N(\tau_x F_K^i)\} \nabla_i^{(N)} J(x/N),$$

$W_{x,y}(\eta) = c_{x,y}(\eta)(\eta_y - \eta_x)$ is a current from y to x, $m_K^N(t, J)$ is a martingale with predictable quadratic variational process

$$\langle m_K^N(t, J) \rangle = \int_0^t b_K^N(\eta^N(s), J)\, ds$$

and

(3.2)

$$b_K^N(\eta, J) = N^{-d} \sum_{i=1}^{d} \sum_{x \in \Gamma_N} c_{x,x+e_i} \Big[(\eta_{x+e_i} - \eta_x) \nabla_i^{(N)} J(x/N)$$
$$- \pi_{x,x+e_i} \{ \textstyle\sum_{y \in \Gamma_N} \tau_y F_K \cdot \nabla^{(N)} J(y/N) \} \Big]^2 .$$

The asymptotic behavior of the martingale term can be easily computed. Indeed, since F_K is local and J is smooth, one can replace $\nabla^{(N)} J(y/N)$ with $\nabla^{(N)} J(x/N)$ in the right hand side of (3.2) with an error $O(s(F_K)/N)$ where $s(F)$ denotes the size of the support of $F \in \mathcal{F}_0^d$ and, when η is ν-distributed, the law of large number holds for $b_K^N(\eta, J)$ and hence we have

$$\lim_{N \to \infty} b_K^N(\eta, J) = \int_{\mathbb{T}^d} \nabla J(\theta) \cdot \hat{c}(\rho; F_K) \nabla J(\theta)\, d\theta.$$

Therefore,

$$\lim_{K \to \infty} \lim_{N \to \infty} b_K^N(\eta, J) = \int_{\mathbb{T}^d} \nabla J(\theta) \cdot \hat{c}(\rho) \nabla J(\theta)\, d\theta.$$

This coincides with the variance of the martingale term $(\mathrm{Tr}(\sqrt{\hat{c}(\rho)} \nabla \mathbf{w}_t), J)$ appearing in the limit equation (2.2).

The treatment of the drift term of (3.1) is more difficult. The essentials lie in the following proposition which asserts that the so-called gradient replacement holds in an equilibrium situation even by multiplting a large factor $N^{d/2}$ by that shown in the non-equilibrium situation [4, Theorem 3.2], i.e., a much faster convergence can be proven in the equilibrium.

Proposition 2. *For every $G \in C(\mathbb{T}^d)$ and $1 \le i \le d$,*

$$\lim_{K \to \infty} \overline{\lim_{N \to \infty}} N^{-d/2+1} E\Big[\Big| \int_0^t \sum_{x \in \Gamma_N} G(x/N) X_{x,i}^{K,N}(\eta^N(s))\, ds \Big|\Big] = 0, \tag{3.3}$$

where

$$X_{x,i}^{K,N}(\eta) = W_{x,x+e_i} - L_N(\tau_x F_K^i) - \sum_{j=1}^{d} D_{ij}(\rho)(\eta_{x+e_j} - \eta_x).$$

Once this proposition is shown, the drift term $a_K^N(\eta^N(s), J)$ (more precisely, its integration in s) can be asymptotically replaced with

$$\begin{aligned}
&-N^{-d/2+1} \sum_{i,j=1}^{d} D_{ij}(\rho) \sum_{x\in\Gamma_N} (\eta_{x+e_j}^N(s) - \eta_x^N(s))\nabla_i^{(N)} J(x/N) \\
&= -N^{-d/2+1} \sum_{i,j=1}^{d} D_{ij}(\rho) \sum_{x\in\Gamma_N} (\eta_x^N(s) - \rho) \\
&\qquad\qquad \times \{\nabla_i^{(N)} J((x-e_j)/N) - \nabla_i^{(N)} J(x/N)\} \\
&= \sum_{i,j=1}^{d} D_{ij}(\rho)\xi^N(s, \partial_i\partial_j J) + o(1).
\end{aligned}$$

Here, $o(1)$ means an error term such that $\lim_{N\to\infty} E[\{o(1)\}^2] = 0$. Hence, we obtain the drift term in the limiting Ornstein-Uhlenbeck process, see (2.2).

4. Proof of Proposition 2

We follow the strategy outlined in [3]. First, we show that $G(x/N)$ can be locally replaced with a constant function. More precisely, we take $\ell > 0$ and let $\Lambda(x,\ell) = \{y \in \Gamma_N | -\ell \le y_i - x_i < \ell, 1 \le i \le d\}$ be the box of side length 2ℓ with center x. For simplicity, we assume $M = N/2\ell \in \mathbb{Z}_+$ (otherwise, a small care is required for boxes at the end). For $x \in \Gamma_N$, we denote $[x]_\ell := q\ell$ if $x \in \Lambda(q\ell, \ell)$ for some $q \in (\mathbb{Z}/M\mathbb{Z})^d$. Then we have

Lemma 3. *For every $G \in C(\mathbb{T}^d)$ and for $A_x = W_{x,x+e_i}, L_N(\tau_x F_K^i)$ or $\eta_{x+e_j} - \eta_x$, we have*

$$\lim_{K\to\infty}\lim_{\ell\to\infty}\overline{\lim_{N\to\infty}} N^{-d/2+1} E[|\int_0^t \sum_{x\in\Gamma_N} \{G(x/N) - G([x]_\ell/N)\}A_x(\eta^N(s))\, ds|] = 0.$$

Here, the limit in K is necessary only for $A_x = L_N(\tau_x F_K^i)$. A similar assertion holds if $G([x]_\ell/N)$ is replaced with $G([x]_\ell/N) \times 1_{\Lambda(q\ell,\ell-s(F_K)-r)}(x)$.

For the proof of this lemma we need the following exponential estimates:

Lemma 4. *Let $\{H_b \in \mathbb{R}\}_{b\in\Gamma_N^{**}}$ be given. Then we have*

$$E[\exp(|\int_0^t \sum_{b=\{x,y\}\in\Gamma_N^{**}} H_b A_b(\eta^N(s))\,ds|)]$$
$$\leq \begin{cases} 2\exp(2tN^{-2}\|c\|_\infty \sum_{b\in\Gamma_N^{**}} H_b^2), & \text{when } A_b = W_b, \\ 2\exp(tN^{-2}\|c\|_\infty C(F_K^i) \sum_{b\in\Gamma_N^{**}} H_b^2), & \text{when } A_b = L_N(\tau_x F_K^i), \\ 2\exp(2tN^{-2}\|c^{-1}\|_\infty \sum_{b\in\Gamma_N^{**}} H_b^2), & \text{when } A_b = \eta_y - \eta_x, \end{cases}$$

where $C(F)$ denotes a constant depending only on F.

Lemma 4 can be proved by combining Feynman-Kac formula, spectral theorem and an integration by parts formula relative to ν, cf. [3]. The assertion in Lemma 3 can be immediately shown by using Jensen's inequality and the estimates given in Lemma 4. Note that the ratio of the volumes of two boxes $\Lambda(q\ell, \ell - s(F_K) - r)$ and $\Lambda(q\ell, \ell)$ goes to 1 as $\ell \to \infty$.

In view of Lemma 3, (3.3) follows if one can show

(4.1)
$$\lim_{K\to\infty} \lim_{\ell\to\infty} \overline{\lim_{N\to\infty}} N^{-d/2+1} E[|\int_0^t \sum_{q\in(\mathbb{Z}/M\mathbb{Z})^d} G(q\ell/N) X_{q,i}^{K,\ell}(\eta^N(s))\,ds|] = 0,$$

where

$$X_{q,i}^{K,\ell}(\eta) = \sum_{x\in\Lambda(q\ell,\ell)} W_{x,x+e_i} - \sum_{x\in\Lambda(q\ell,\ell-s(F_K)-r)} L_N(\tau_x F_K^i)$$
$$- \sum_{x\in\Lambda(q\ell,\ell)} \sum_{j=1}^d D_{ij}(\rho)(\eta_{x+e_j} - \eta_x).$$

However, we have

$$N^{-d+2} E[|\int_0^t \sum_{q\in(\mathbb{Z}/M\mathbb{Z})^d} G(q\ell/N) X_{q,i}^{K,\ell}(\eta^N(s))\,ds|^2]$$
$$\leq 2t\|G\|_\infty^2 \ell^{-d} E^\nu[\Delta_{\ell,m,\zeta}(X_{0,i}^{K,\ell})],$$

cf. [3]. Here,

$$\Delta_{\ell,m,\zeta}(X) := E^{\nu_{\ell,m}}[X(-L_\zeta)^{-1}X]$$

is the so-called *central limit theorem variance* defined for such X that $E^{\nu_{\ell,m}}[X] = 0$, where $\nu_{\ell,m}$ is the canonical Gibbs measure in $\Lambda_\ell = \{y \in \mathbb{Z}^d \mid -\ell \leq y_i \leq \ell, 1 \leq i \leq d\}$ given particles' number m (in our case, it is simply a uniform measure on $\{\xi \in \mathcal{X}_{\Lambda_\ell} | \sum_{x\in\Lambda_\ell} \xi_x = m\}$, $\mathcal{X}_{\Lambda_\ell} = \{0,1\}^{\Lambda_\ell}$) and $L_\zeta \equiv L_{\Lambda_\ell,\zeta}$ is a generator of the lattice gas on Λ_ℓ having an external condition $\zeta \in \mathcal{X}$, see (5.1) in [4]. Since Corollary 5.1 in [4] asserts that

$$\lim_{\substack{\ell,m\to\infty \\ \ell^{-d}m\to\rho}} \ell^{-d}\Delta_{\ell,m,\zeta}(X_{0,i}^{K,\ell}) = \frac{1}{2}[e_i\cdot\hat{c}(\rho;F_K)e_i - e_i\cdot\hat{c}(\rho)e_i]$$

holds uniformly in $\rho \in [0,1]$ and $\zeta \in \mathcal{X}$, we obtain (4.1).

5. Tightness

To show the tightness of the family of distributions of $\xi^N(t)$ on the space $D([0,T], H^{-\alpha}(\mathbb{T}^d))$, $\alpha > d/2+1$, we decompose it as

$$\xi^N(t,J) = \xi^N(0,J) + m^N(t,J) + \int_0^t a^N(\eta^N(s),J)\,ds$$

where m^N and a^N are defined as in (3.1) by taking $F_K = 0$. We may prove the tightness of martingale and drift terms separately. Although the drift term contains an outward diverging factor, its tightness can be shown similarly to [9,2] based on the first estimate of Lemma 4 and Garsia-Rodemich-Ramsey's lemma. On the other hand, to show the tightness of $m^N(t)$ which has jumps, we use infinite-dimensional version of Rebolledo's theorem, [8]. Namely, it is sufficient to show the following two assertions:
(i) For all $t \in [0,T]$, $\{m^N(t)\}_N$ is tight as a family of $H^{-\alpha}(\mathbb{T}^d)$-valued random variables and
(ii) a family of quadratic variational processes $\{\langle m^N\rangle(t)\}_N$ is tight in the space $C([0,T],\mathbb{R})$, where

$$\begin{aligned}\langle m^N\rangle(t) &\equiv \mathrm{Tr}\langle\langle m^N\rangle\rangle(t)\\ &:= \sum_{j=1}^\infty \langle (m^N(t),\psi_j)_{-\alpha}\rangle = \sum_{j=1}^\infty \lambda_j^{-\alpha}\langle m^N(t,\varphi_j)\rangle,\end{aligned}$$

$(\cdot,\cdot)_{-\alpha}$ denotes the inner product of $H^{-\alpha}(\mathbb{T}^d)$, $\{\psi_j = \lambda_j^{\alpha/2}\varphi_j\}_{j=1}^\infty$ is a complete orthonormal system (CONS) of $H^{-\alpha}(\mathbb{T}^d)$ and $\{\varphi_j\}_{j=1}^\infty$ is a CONS of $L^2(\mathbb{T}^d)$ such that $(-\Delta+1)\varphi_j = \lambda_j\varphi_j$ with eigenvalues λ_j arranged in small order. Since $\langle m^N\rangle(t)$ is continuous in t, the Aldous condition [A] (see [8]) for $\{\langle m^N\rangle(t)\}_N$ follows from the assertion (ii) above, cf. [10].

To see (ii), using (3.2), we have

$$\begin{aligned}\langle m^N\rangle(t) = \sum_{j=1}^\infty \lambda_j^{-\alpha}\int_0^t N^{-d}\sum_{i=1}^d\sum_{x\in\Gamma_N} c_{x,x+e_i}(\eta^N(s))\\ \times(\eta_{x+e_i}^N(s)-\eta_x^N(s))^2(\nabla_i^{(N)}\varphi_j(x/N))^2\,ds,\end{aligned}$$

and therefore

$$\left|\frac{d}{dt}\langle m^N\rangle(t)\right| \le \text{const} \sum_{j=1}^{\infty} \lambda_j^{-\alpha} \|\nabla \varphi_j\|^2_{L^2(\mathbb{T}^d)}$$
$$\le \text{const} \sum_{j=1}^{\infty} \lambda_j^{-\alpha+1}.$$

Accordingly, we have $\sup_N |d\langle m^N\rangle(t)/dt| < \infty$ if $\alpha > d/2+1$ noting that λ_j behaves like $\lambda_j \sim \text{const } j^{2/d}$ as $j \to \infty$, and (ii) is shown. To see (i),

$$E[\|m^N(t)\|^2_{-\alpha}] = \sum_{j=1}^{\infty} \lambda_j^{-\alpha} E[\langle m^N(t, \varphi_j)\rangle].$$

This becomes finite and bounded in N if $\alpha > d/2+1$. Noting Rellich's theorem (i.e., the imbedding $H^{-\alpha}(\mathbb{T}^d) \subset H^{-\alpha'}(\mathbb{T}^d)$ is compact if $\alpha < \alpha'$), we have (i).

References

[1] C.C. Chang and H.T. Yau: Fluctuations of one-dimensional Ginzburg-Landau models in non-equilibrium, Commun. Math. Phys., **145** (1992), 209-234.

[2] C.C. Chang: Equilibrium fluctuations of gradient reversible particle systems, Probab. Theory Relat. Fields, **100** (1994), 269-283.

[3] C.C. Chang: Equilibrium fluctuations of nongradient reversible particle systems, to appear in: Nonlinear Stochastic PDE's: Hydrodynamic Limit and Burgers' Turbulence (eds. Funaki and Woyczynski), IMA volume **77**, Springer, 1995, pp. 41-51.

[4] T. Funaki, K. Uchiyama and H.T. Yau: Hydrodynamic limit for lattice gas reversible under Bernoulli measures, to appear in: Nonlinear Stochastic PDE's: Hydrodynamic Limit and Burgers' Turbulence (eds. Funaki and Woyczynski), IMA volume **77**, Springer, 1995, pp. 1-40.

[5] K. Itô: Motion of infinitely many particles (in Japanese), Sûri-kaiseki Kenkyûsho Kôkyûroku, **367** (1979), 1-33.

[6] K. Itô: Stochastic analysis in infinite dimensions, in: Stochastic Analysis (eds. Friedman and Pinsky), Proceedings (Evanston 1978), Academic Press, 1980, pp. 187-197.

[7] K. Itô: Distribution-valued processes arising from independent Brownian motions, Math. Z., **182** (1983), 17-33.

[8] A. Joffe and M. Metivier: Weak convergence of sequences of semimartingales with applications to multitype branching processes, Adv. Appl. Prob., **18** (1986),

[9] S. Lu: Equilibrium fluctuations of a one-dimensional nongradient Ginzburg-Landau model, Ann. Probab., **22** (1994), 1252-1272.
[10] M. Metivier and S. Nakao: Equivalent conditions for the tightness of a sequence of continuous Hilbert valued martingales, Nagoya Math. J., **106** (1987), 113-119.
[11] H. Spohn: Equilibrium fluctuations for interacting Brownian particles, Commun. Math. Phys., **103** (1986), 1-33.
[12] H. Spohn: Large Scale Dynamics of Interacting Particles, Springer, 1991.
[13] S.R.S. Varadhan: Scaling limit for interacting diffusions, Commun. Math. Phys., **135** (1991), 313-353.
[14] S.R.S. Varadhan: Nonlinear diffusion limit for a system with nearest neighbor interactions - II, In: Asymptotic problems in probability theory: stochastic models and diffusions on fractals (eds. Elworthy and Ikeda), Longman, 1993, pp. 75-128.

Hall's transform and the Segal–Bargmann map

Leonard Gross[1] and Paul Malliavin[2]

[1] Department of Mathematics, Cornell University, White Hall, Ithaca, NY 14853-7901, USA
[2] 10 rue Saint Louis en L'Ile, 75004, Paris, France

Summary. It is shown how Hall's transform for a compact Lie group can be derived from the infinite dimensional Segal-Bargmann transform by means of stochastic analysis.

1. Introduction and Statement of Results

Consider a connected, simply connected Lie group K whose Lie algebra $\mathfrak{k}$, which will be identified with $T_e(K)$, possesses an Ad K invariant inner product $\langle\ ,\ \rangle$. This class of Lie groups includes all connected, simply connected compact Lie groups, as well as $\mathbb{R}^n$, and also products of these two types of groups (and that's all of them). Let $t > 0$. Associated to such a group, a choice of Ad K invariant inner product, and time t, there are three Hilbert spaces, which are very different in their construction and appearance, yet are naturally isomorphic. The three Hilbert spaces are as follows.

For any element ξ in $\mathfrak{k}$ let us write $\widetilde{\xi}$ for the corresponding left invariant vector field on K. If $e_1, \ldots, e_n$ is an orthonormal basis of $\mathfrak{k}$ then

$$\Delta := \sum_{j=1}^{n} (\widetilde{e}_j)^2 \quad \text{on} \quad C_c^\infty(K) \tag{1.1}$$

is independent of the basis and, as an operator in $L^2(K, dx)$, has a closure which is a negative self–adjoint operator. There is a strictly positive, smooth probability density ρ_t on K, the heat kernel [Da, R], such that the heat semigroup $e^{t\Delta/2}$ is given by convolution by ρ_t:

$$e^{t\Delta/2} f = \rho_t * f \quad f \in L^2(K, dx). \tag{1.2}$$

The first of the three Hilbert spaces which we wish to discuss in this paper is $L^2(K, \rho_t(x)dx)$.

The second Hilbert space will be a space of holomorphic functions over the complexification, G, of K. Denote by $\mathfrak{g} = \mathfrak{k} \otimes \mathbb{C}$ the complexification of the Lie algebra of K. Then G may be defined as the connected, simply connected Lie group whose Lie algebra is $\mathfrak{g}$. Since $\mathfrak{k} \subset \mathfrak{g}$ there is a homomorphism of

Partial support was provided to the first author by NSF grant 9501238-DMS

K into G and it is well known (see e.g. [Hal]) that this is an injection with closed range. Thus we may and will regard K as a closed subgroup of G. Write $i = \sqrt{-1}$ and let

$$\Delta_c = \sum_{j=1}^{n} \{(\tilde{e}_j)^2 + (\widetilde{ie_j})^2\} \quad \text{on} \quad C_c^\infty(G) \tag{1.3}$$

where $e_1, \dots, e_n$ is an orthonormal basis of $\mathfrak{k}$. As before, this operator is independent of the choice of orthonormal basis of $\mathfrak{k}$, and, as before, there is a heat kernel μ_t determined by the equation

$$e^{t\Delta_c/4} f = \mu_t * f \quad f \in L^2(G, dy). \tag{1.4}$$

G is naturally a complex Lie group. Let us write $\mathcal{H}(G)$ for the space of holomorphic functions on G. Then the second Hilbert space which will be of interest to us is

$$\mathcal{H}L^2(G, \mu_t) := \mathcal{H}(G) \cap L^2(G, \mu_t(y)dy). \tag{1.5}$$

For example if $K = \mathbb{R}^n$ then $G = \mathbb{C}^n$ and ρ_t and μ_t are Gaussian measures. If $K = SU(2)$ then we may take $G = SL(2, \mathbb{C})$.

The third Hilbert space has a more algebraic character. Denote by $T(\mathfrak{g})$ the tensor algebra over $\mathfrak{g}$ and by J the two-sided ideal in $T(\mathfrak{g})$ generated by

$$\{\xi \otimes \eta - \eta \otimes \xi - [\xi, \eta]; \xi, \eta \in \mathfrak{g}\}.$$

Whereas $T(\mathfrak{g})$ is the weak direct sum, $\sum_{j=0}^\infty \mathfrak{g}^{\otimes n}$, the algebraic dual space $T'(\mathfrak{g})$ may be identified with the strong direct sum (direct product), $\sum_{j=0}^\infty (\mathfrak{g}^*)^{\otimes j}$, under the pairing

$$\langle \alpha, \beta \rangle = \sum_{k=0}^{\infty} \langle \alpha_k, \beta_k \rangle \quad \alpha \in T'(\mathfrak{g}), \ \beta \in T(\mathfrak{g}) \tag{1.6}$$

where

$$\alpha = \sum_{k=0}^{\infty} \alpha_k \quad \alpha_k \in (\mathfrak{g}^*)^{\otimes k} \tag{1.7}$$

and $\beta = \sum_{k=0}^\infty \beta_k$ (finite sum) with $\beta_k \in \mathfrak{g}^{\otimes k}$. J^0 will denote the annihilator of J in the dual space $T'(\mathfrak{g})$. Define

$$\|\alpha\|_t^2 = \sum_{k=0}^{\infty} (t^k/k!) |\alpha_k|^2_{(\mathfrak{g}^*)^{\otimes k}} \tag{1.8}$$

when α is given by (1.7). The third Hilbert space may now be defined as

$$J_t^0 = \{\alpha \in J^0 : \|\alpha\|_t^2 < \infty\}. \tag{1.9}$$

Theorem 1.1. *(Hall's transform)[Hal, D] Let f be a function in $L^2(K, \rho_t(x)dx)$. Then the convolution $(\rho_t * f)(x)$ exists for all x in K and $\rho_t * f$ has a unique analytic continuation $\mathcal{H}_t f$ to G. Moreover $\mathcal{H}_t f$ is in $L^2(G, \mu_t)$ and the map*

$$\mathcal{H}_t : L^2(K, \rho_t) \to \mathcal{H}L^2(G, \mu_t)$$

is unitary.

Theorem 1.2. *(Driver)[D] Suppose that u is in $\mathcal{H}(G)$. Define the kth Taylor coefficient of u at e to be the element $D^k u(e)$ in $(\mathfrak{g}^*)^{\otimes k}$ determined by*

$$(1.10)\qquad \langle D^k u(e), \xi_1 \otimes \cdots \otimes \xi_k \rangle = \widetilde{\xi}_1 \cdots \widetilde{\xi}_k u(e) \quad \xi_j \in \mathfrak{g},\ j = 1, \dots, k.$$

Write $D^0 u(e) = u(e)$. Then the map

$$(1.11)\qquad (1 - D)_e^{-1} : u \to \sum_{k=0}^{\infty} D^k u(e)$$

from $\mathcal{H}(G)$ into $T'(\mathfrak{g})$ is a unitary operator from $\mathcal{H}L^2(G, \mu_t)$ onto J_t^0.

Note that the right side of (1.10) is complex linear in each ξ_j because u is holomorphic.

We will discuss now the historical evolution of the proofs of these two theorems, in part because it is the objective of this paper to reverse some of the steps in this evolution and in part because the techniques of the present work rely on the stochastic analysis that led to the ideas embodied in Theorems 1.1 and 1.2. Consider the K valued Brownian motion $k(\cdot)$ on $[0,1]$ which starts at e and whose transition semigroup is $e^{t\Delta/2}$. Denote by P the corresponding path space measure on $\{k \in C([0,1]; K) : k(0) = e\}$. The conditional measure $P_0 := P(\quad | \, k(1) = e)$ is a probability measure on the loop space $\mathcal{L} := \{k \in C([0,1]; K) : k(0) = k(1) = e\}$. $\mathcal{L}$ is clearly a group under pointwise multiplication. The Sobolev space

$$H_0(K) = \{k \in \mathcal{L} : k \text{ is absolutely continuous and } \int_0^1 |k(s)^{-1}\dot{k}(s)|^2 ds < \infty\}$$

is a subgroup of $\mathcal{L}$. Moreover the right action $k \to k k_0$ $(k \in \mathcal{L}, k_0 \in H_0(K))$ preserves the measure class of P_0 [MM]. It makes sense therefore to ask whether the action of $H_0(K)$ on $\mathcal{L}$ is ergodic. That is, if $f : \mathcal{L} \to \mathbb{R}$ satisfies $f(kk_0) = f(k)$ with P_0 probability one for each $k_0 \in H_0(K)$ is f equal a.e. to a constant? An affirmative answer has interesting ramifications for certain Schrödinger operators over $\mathcal{L}$ [G4]. For this reason such an ergodicity theorem was proved in [G4]. We will state more precisely in Section 7 (Lemma 7.5) the form of this theorem that will be needed in this paper. This ergodicity theorem has recently been reproven by G. Sadasue [Sad] by use of fundamental methods from quasi–sure analysis, [AM]. The proof of ergodicity given in [G4] was in the spirit of one of the standard proofs of ergodicity of the irrational flow on the torus: expand a given flow invariant function in a double Fourier series and show that all the Fourier coefficients are zero except for the constant term. In the context of a loop group one transfers the ergodicity problem to the $\mathfrak{k}$ valued Wiener process $X := k^{-1} \circ dk$ and uses a multiple Ito integral expansion for square integrable functions of the process

X instead of Fourier series. One obtains as an incidental byproduct of this method a natural unitary operator

$$U_t : L^2(K, \rho_t) \to J_t^0.$$

The construction of this unitary operator U_t, given in [G4], involves the entire Wiener process X in spite of the fact that neither the domain nor range of U_t involves X. It will not be necessary to describe here in more detail exactly how the indirectly defined map U_t arises from the ergodicity proof given in [G4]: the nature of U_t has been greatly clarified by several subsequent papers [Ha1,2 Hij1,2, D, D-G]. B. Hall, [Ha1], discovered the transformation $\mathcal{H}_t$ (for a compact group), motivated by the existence of the map U_t . O. Hijab independently [Hij1] gave an explicit formula for the map U_t, which we will come back to in a moment. B. Driver [D] then proved that the power series map $(1-D)_e^{-1}$ of Theorem 1.2 is unitary and, by combining this with Hijab's formula, arrived at a second derivation of Hall's transform in the form $\mathcal{H}_t = ((1-D)_e^{-1})^{-1}U_t$. It is most illuminating however to write this identity in the form

$$U_t = (1-D)_e^{-1}\mathcal{H}_t : L^2(K, \rho_t) \to J_t^0.$$

In fact Hijab's formula is the real version of this: $U_t f = (1-D)_e^{-1} e^{t\Delta/2} f$. That is, the analytic continuation of $e^{t\Delta/2} f$ can be omitted if one merely wishes to go all the way to J_t^0 from $L^2(K, \rho_t)$ (because Taylor coefficients can be obtained by differentiating in "real" directions.)

It is the nominal objective of this paper to give a third proof of Theorem 1.1. Our proof is longer than the combined proof of Driver and Hijab. It is also longer than the original proof of Hall, although it does not make use of structure theory of complex semi–simple Lie groups as Hall's proof does. Furthermore we will use a recent extension [DG] of Driver's theorem (Theorem 1.2) in a serious way in Section 6. Our real motivation is to show how, by reintroducing the preceding K valued Brownian motion — and the corresponding G valued Brownian motion — one can derive Hall's transform for a finite dimensional group K from the infinite dimensional linear version of this transform known as the Segal–Bargmann transform.

Sections two, three and four are entirely expository. We will give a self contained account of the Segal–Bargmann transform and its inverse, the Fourier–Wiener transform, as it was developed in the work of Cameron, Martin, Bargmann, Segal and Krée. We will present it in a form which is aimed at the applications we will make. For an explanation of these topics which emphasizes their role in quantum field theory see the recent book [BSZ,Ch.I] and the work [Ni]. For connections with other parts of classical analysis see [Car] and the book [Fol]. For recent applications of the Segal-Bargmann transform to quantum field theory see [Pan, Ped, Zh]. For work containing some of the original physical motivation see [Se1,2,3,4,5]. For extensions of some of these isomorphisms to quotients of compact groups see [Hal, G6]. For a "particle like" decomposition of J_t^0 see [G5].Just as the Fourier transform over $\mathbb{R}^n$ has an extension to tempered distributions, so also the Segal-Bargmann transform has an extension to various classes of generalized functions. For reference to the vast literature on the extension (known as the S-transform) to Hida distributions see [HKPS, KLPS, Ku, Lee1,2, Ob, Pot] and their bibliographies.

It is a pleasure to acknowledge illuminating discussions with Bruce Driver.

2. Two Transforms Over $\mathbb{R}^n$ and $\mathbb{C}^n$

The content of this section is entirely expository. The material is taken from the 1945 and 1947 papers of Cameron and Martin [Ca, CM1,2] and from the 1961 paper of V. Bargmann [B1]. We have arranged the material in such a way as to show the relation between the two transforms and to emphasize aspects that will be important to us later. We have avoided use of Hermite polynomials.Our intention is to emphasize the role of the heat semigroup $e^{t\Delta/2}$. It is the heat semigroup, rather than special functions, which plays the key role in extending these transforms to groups. However in Remark 2.13 we will discuss the well known connection of these transforms with Hermite polynomials and with Wick ordering.

Notation 2.1. For two vectors a and b in $\mathbb{C}^n$ we will write

$$a \cdot b = \sum_{j=1}^{n} a_j b_j \qquad a, b \in \mathbb{C}^n. \tag{2.1}$$

If $z = x + iy$ with x and $y \in \mathbb{R}^n$ we will write $\overline{z} = x - iy$. Then $a \cdot b$ is bilinear in a and b while the usual inner product on $\mathbb{C}^n$ is $a \cdot \overline{b}$ and the usual norm on $\mathbb{C}^n$ is $|u|^2 = u \cdot \overline{u}$.

Notation 2.2. Define

$$p_t(u) = (2\pi t)^{-n/2} \exp\{-|u|^2/2t\} \qquad u \in \mathbb{R}^n \tag{2.2}$$

$$m_t(z) = (\pi t)^{-n} \exp\{-|z|^2/t\} \qquad z \in \mathbb{C}^n. \tag{2.3}$$

p_t and m_t are the fundamental solutions for Laplacians on $\mathbb{R}^n$ and $\mathbb{C}^n$ respectively. Specifically, if $z = (z_1, \ldots, z_n) \in \mathbb{C}^n$ with $z_j = x_j + iy_j$ define

$$\Delta = \sum_{j=1}^{n} \partial^2/\partial x_j^2 \tag{2.4}$$

and

$$\Delta_c = \sum_{j=1}^{n} (\partial^2/\partial x_j^2 + \partial^2/\partial y_j^2). \tag{2.5}$$

Then, denoting convolution by an asterisk, we have

$$e^{(t/2)\Delta} = p_t * \tag{2.6}$$

and

$$e^{(t/4)\Delta_c} = m_t * . \tag{2.7}$$

The operators in these equations should be interpreted first as operators over $L^2(\mathbb{R}^n, dx)$ and $L^2(\mathbb{C}^n, dz)$ respectively where dx and $dz = dxdy$ refer to Lebesgue measure. But we will want to interpret them as operators on larger classes of functions in the theorems below.

The measures $p_t(x)dx$ and $m_t(z)dz$ have well known Laplace transforms given by

$$\int_{\mathbb{R}^n} e^{x\cdot a} p_t(x)dx = e^{(t/2)a\cdot a}, \qquad a \in \mathbb{C}^n \tag{2.8}$$

and

$$\int_{\mathbb{C}^n} e^{x\cdot a + y\cdot b} m_t(z)dxdy = e^{(t/4)(a\cdot a + b\cdot b)}, \qquad a \text{ and } b \in \mathbb{C}^n. \tag{2.9}$$

Indeed these equations may be used to define p_t and m_t. For this purpose it suffices to use only a and b in $i\mathbb{R}^n$, which then gives the Fourier transform. Almost all of our computations in this section will be based only on the identities (2.8) and (2.9).

The following two theorems will be proved in this section.

Theorem 2.3. *(Bargmann) [B1] Let $t > 0$ and let $1 < q < \infty$. For any function f in $L^q(\mathbb{R}^n, p_t(x)dx)$ the convolution $p_t * f$ has a unique analytic continuation, $S_t f$, to $\mathbb{C}^n$ given by*

(2.10)

$$(S_t f)(z) = e^{-(2t)^{-1} z\cdot z} \int_{\mathbb{R}^n} f(u) e^{t^{-1} z\cdot u} p_t(u) du \qquad z \in \mathbb{C}^n, \ f \in L^q(\mathbb{R}^n, p_t).$$

Moreover S_t is a unitary operator from $L^2(\mathbb{R}^n, p_t)$ onto $\mathcal{H}L^2(\mathbb{C}^n, m_t)$.

Theorem 2.4. *(Cameron and Martin) [CM2]. Define, for $t > 0$,*

$$(C_t f)(y) = (S_{2t} f)(iy) \qquad y \in \mathbb{R}^n, \ f \in L^2(\mathbb{R}^n, p_{2t}). \tag{2.11}$$

Then C_t extends uniquely to a unitary operator (which we also denote by C_t) from $L^2(\mathbb{R}^n, p_t)$ onto itself. Moreover

$$(C_t^{-1} f)(y) = (C_t f)(-y) \qquad f \in L^2(\mathbb{R}^n, p_t). \tag{2.12}$$

Lemma 2.5. *Let $1 < q < \infty$. The equation (2.10) defines a holomorphic function on $\mathbb{C}^n$. Its restriction to $\mathbb{R}^n$ is exactly $p_t * f$.*

Proof: Here and in the following q' will always denote the conjugate index to q. For each $z \in \mathbb{C}^n$ the function $u \to e^{z\cdot u/t}$ is in $L^{q'}(\mathbb{R}^n, p_t)$. So the integral in (2.10) exists for each $z \in \mathbb{C}^n$. Moreover on the set $\{z \in \mathbb{C}^n : |z| \le a\}$ we have $|f(u)||e^{z\cdot u/t}| \le |f(u)|e^{a|u|/t}$ which is integrable. It follows that $S_t f$ is continuous on $\mathbb{C}^n$ and that for any closed contour C in the z_j plane we have

$$\int_C dz_j \int_{\mathbb{R}^n} f(u)e^{z\cdot u/t}p_t(u)du = \int_{\mathbb{R}^n} f(u)\int_C e^{z\cdot u/t}dz_j p_t(u)du = 0.$$

(Only in this equation will dz_j mean a 1–form instead of Lebesgue measure.) Hence by Morera's theorem the integral in (2.10) is holomorphic in each z_j. Thus $(S_t f)$ is in $\mathcal{H}(\mathbb{C}^n)$. Since $e^{-(2t)^{-1}|x|^2+t^{-1}x\cdot u}p_t(u) = p_t(x-u)$ we see that $(S_t f)(x) = (p_t * f)(x)$ for x in $\mathbb{R}^n$. Q.E.D.

Notation 2.6. A multiindex, $\alpha = (k_1, \dots, k_n)$, is an n–tuple of nonnegative integers. For such a multiindex we write $z^\alpha = \prod_{j=1}^n z^{k_j}$ and $|\alpha| = \sum_{j=1}^n k_j$ and $\alpha! = \prod_{j=1}^n k_j!$. In the following lemma α will run over all multiindices.

Lemma 2.7. *(Polynomial basis over $\mathbb{C}^n$) [B1] The set $\{z^\alpha\}$ forms an orthogonal basis of $\mathcal{H}L^2(\mathbb{C}^n, m_t)$. If $f \in \mathcal{H}(\mathbb{C}^n)$ and has the pointwise convergent power series*

$$f(z) = \sum_\alpha a_\alpha z^\alpha \qquad a_\alpha \in \mathbb{C} \tag{2.13}$$

then

$$\int_{\mathbb{C}^n} |f(z)|^2 m_t(z)dz = \sum_\alpha |a_\alpha|^2 t^{|\alpha|}\alpha!. \tag{2.14}$$

The series (2.13) is also convergent in the $L^2(\mathbb{C}^n, m_t)$ sense if either side (hence both sides) of (2.14) is finite.

Proof: Let $D(\sigma)$ be the polydisc $\{z \in \mathbb{C}^n : \sup_j |z_j| \le \sigma\}$. Consider first the case $n = 1$. Let $M_\sigma(j,k) = \int_{|z|<\sigma} z^j \bar{z}^k m_t(z)dxdy$. Put $z = re^{i\theta}$. Then $M_\sigma(j,k) = \int_0^{2\pi} e^{i\theta(j-k)}d\theta \int_0^\sigma r^{j+k+1}e^{-r^2/t}(\pi t)^{-1}dr$ from which follows

(i) $M_\sigma(j,k) = 0$ if $j \ne k$.

(ii) $M_\sigma(k,k) \uparrow k!t^k$ as $\sigma \uparrow \infty$.

(iii) $\int_{\mathbb{C}} z^j \bar{z}^k m_t(z)dxdy = \delta_{jk} k!t^k$.

Thus in n dimensions, since the power series (2.13) converges uniformly on the product set $D(\sigma)$ and since $z^\alpha \bar{z}^\beta m_t(z)$ is itself a product of functions of $z_1, \ldots, z_n$ respectively, we have

$$\int_{D(\sigma)} |f(z)|^2 m_t(z)dz = \sum_{\alpha,\beta} \int_{D(\sigma)} a_\alpha \bar{a}_\beta z^\alpha \bar{z}^\beta m_t(z)dz$$
$$= \sum_\alpha |a_\alpha|^2 \prod_{j=1}^n M_\sigma(k_j, k_j)$$

where, in the last sum, we have written $\alpha = (k_1, \ldots, k_n)$, and have used (i). Now let $\sigma \uparrow \infty$ and use the monotone convergence theorem on both sides of the last equality. Then (2.14) follows from (ii). The set $\{z^\alpha\}$ is an orthogonal set by (iii) and the functions $(t^{|\alpha|}\alpha!)^{-1/2} z^\alpha$ are orthonormal. Therefore if the right side of (2.14) is finite then the sequence $f_N(z) := \sum_{|\alpha| \le N} a_\alpha z^\alpha$ converges in the $L^2(m_t)$ sense to some function g. Since a subsequence converges a.e. to g we have $f = g$ a.e.. So the series in (2.13) converges to f in the $L^2(m_t)$ sense. In particular if f is in $\mathcal{H}L^2(m_t)$ then (2.13) converges to f in the $L^2(m_t)$ sense and the set $\{z^\alpha\}$ is an orthogonal basis.

Q.E.D.

Corollary 2.8. *(Bargmann's Pointwise bounds) [B1] If f is in $\mathcal{H}L^2(\mathbb{C}^n, m_t)$ then*

$$|f(z)|^2 \le e^{|z|^2/t} \|f\|^2_{L^2(\mathbb{C}^n, m_t)} \qquad z \in \mathbb{C}^n. \tag{2.15}$$

Proof: Using (2.13) and (2.14) we have

$$|f(z)|^2 = \Big| \sum_\alpha (t^{|\alpha|}\alpha!)^{1/2} a_\alpha (t^{|\alpha|}\alpha!)^{-1/2} z^\alpha \Big|^2$$
$$\le \Big(\sum_\alpha t^{|\alpha|}\alpha! |a_\alpha|^2 \Big) \Big(\sum_\alpha (t^{|\alpha|}\alpha!)^{-1} |z^\alpha|^2 \Big)$$
$$= \|f\|^2_{L^2} \prod_{j=1}^n \sum_{k=0}^\infty |z_j|^{2k}/(t^k k!) \tag{2.16}$$

which is (2.15).

Q.E.D.

Corollary 2.9. *(Monotonicity of L^2 norm.) Let $0 \le k < n$. Denote by $m_t^{(k)}$ the density (2.3) for dimension k instead of n. Then*

$$\int_{\mathbb{C}^k} |f(z)|^2 m_t^{(k)}(z)dz \le \int_{\mathbb{C}^n} |f(z)|^2 m_t^{(n)}(z)dz. \qquad f \in \mathcal{H}(\mathbb{C}^n) \tag{2.17}$$

Proof: If f is given by (2.13) on $\mathbb{C}^n$ then f is given by the same series on $\mathbb{C}^k$ but with those coefficients a_α set equal to zero for which $max(\alpha_{k+1},...,\alpha_n) > 0$. (2.16) now follows from (2.14). Q.E.D.

Remark 2.10. If $f \in \mathcal{H}L^2(\mathbb{C}^n, m_t)$ then the conditional expectation of f given $\mathbb{C}^k$ is exactly the restriction of f to $\mathbb{C}^k$. (To see this use $\int_{\mathbb{C}^{n-k}} z^\alpha m_t^{(n-k)}(z)dz = 0$ when $\alpha \neq 0$.) Of course the inequality (2.17) also follows from this.

In Section 3 we are going to reformulate Lemma 2.7 in a way that will allow us, in Section 6, to replace $\mathbb{C}^n$ by an arbitrary connected, complex Lie group. (See Lemmas 3.7 and 6.3.)

Lemma 2.11. *(Fundamental sets) Let*

$$f_a(u) = e^{a \cdot u} \qquad a \in \mathbb{C}^n, \qquad u \in \mathbb{R}^n \tag{2.18}$$

and $g_a(z) = e^{a \cdot z}$ for a and $z \in \mathbb{C}^n$. Let $1 < q < \infty$. The sets $\{f_a : a \in \mathbb{R}^n\}$ and $\{f_a : a \in i\mathbb{R}^n\}$ are each fundamental in $L^q(\mathbb{R}^n, p_t)$. The set $\{g_a : a \in \mathbb{R}^n\}$ is fundamental in $\mathcal{H}L^2(\mathbb{C}^n, m_t)$.

Proof: If $h \in L^{q'}(\mathbb{R}^n, p_t)$ and $(f_a, h)_{L^2(\mathbb{R}^n, p_t)} = 0$ for all a in $i\mathbb{R}^n$ then $h(u)p_t(u)$, which is in $L^1(\mathbb{R}^n, du)$, has Fourier transform equal to zero. Hence $h = 0$ a.e. Thus $\{f_a : a \in i\mathbb{R}^n\}$ is fundamental in $L^q(\mathbb{R}^n, p_t)$. But by the proof of Lemma 2.5 the function $\mathbb{C}^n \ni a \to (f_a, h)$ is holomorphic. Therefore if it is zero for all $a \in \mathbb{R}^n$ it is zero for all a in $i\mathbb{R}^n$. Hence $\{f_a : a \in \mathbb{R}^n\}$ is also fundamental in $L^q(\mathbb{R}^n, p_t)$. Now note that $g_a \in \mathcal{H}L^2(\mathbb{C}^n, m_t)$. Suppose that $(g_a, \varphi)_{L^2(\mathbb{C}^n, m_t)} = 0$ for all $a \in \mathbb{R}^n$ and some φ in $\mathcal{H}L^2(\mathbb{C}^n, m_t)$. For $|a| \le 1$ we have $|g_a(z)| \le e^{|z|}$, which is in $L^2(\mathbb{C}^n, m_t)$. Using this estimate one justifies easily repeated differentiations with respect to $a_1, \ldots, a_n$ at $a = 0$ in the identity $\int_{\mathbb{C}^n} e^{z \cdot a}\overline{\varphi(z)} m_t(z)dz = 0$, by which means one obtains $(z^\alpha, \varphi)_{L^2(\mathbb{C}^n, m_t)} = 0$ for all multiindices α. It now follows from Lemma 2.7 that $\varphi = 0$. Q.E.D.

Lemma 2.12. *(Transforms of exponentials) Continuing the notation of Lemma 2.11 we have*

$$(S_t f_a)(z) = e^{z \cdot a + (t/2) a \cdot a} \qquad z \text{ and } a \in \mathbb{C}^n \tag{2.19}$$

$$(C_t f_a)(y) = e^{y \cdot ia + ta \cdot a} \qquad y \in \mathbb{R}^n, \ a \in \mathbb{C}^n \tag{2.20}$$

$$(S_t f_a, S_t f_b)_{L^2(\mathbb{C}^n, m_t)} = (f_a, f_b)_{L^2(\mathbb{R}^n, p_t)} \qquad a, b \in \mathbb{C}^n \tag{2.21}$$

$$(C_t f_a, C_t f_b)_{L^2(\mathbb{R}^n, p_t)} = (f_a, f_b)_{L^2(\mathbb{R}^n, p_t)} \qquad a, b \in \mathbb{C}^n. \tag{2.22}$$

Proof: By (2.10) and the Laplace transform formula (2.8) for p_t

$$(S_t f_a)(z) = e^{-(2t)^{-1}z\cdot z} \int_{\mathbb{R}^n} e^{a\cdot u} e^{t^{-1}z\cdot u} p_t(u)du$$
$$= e^{-(2t)^{-1}z\cdot z} e^{(t/2)(a+t^{-1}z)\cdot(a+t^{-1}z)}$$

which is (2.19). Replace t by $2t$ in (2.19) and z by iy to get (2.20). The remaining two identities are also immediate consequences of the Laplace transform formulas (2.8) and (2.9). We have

$$(2.23) \qquad (f_a, f_b)_{L^2(\mathbb{R}^n, p_t)} = \int_{\mathbb{R}^n} e^{(a+\bar{b})\cdot u} p_t(u)du = e^{(t/2)(a+\bar{b})\cdot(a+\bar{b})}$$

while, by (2.19) and (2.9)

$$(S_t f_a, S_t f_b)_{L^2(\mathbb{C}^n, m_t)} = \int_{\mathbb{C}^n} e^{x\cdot a + y\cdot ia + (t/2)a\cdot a} e^{x\cdot\bar{b} - y\cdot i\bar{b} + (t/2)\bar{b}\cdot\bar{b}} m_t(z)dz$$
$$= \exp\{(t/2)(a\cdot a + \bar{b}\cdot\bar{b}) + (t/4)(a+\bar{b})\cdot(a+\bar{b}) - (t/4)(a-\bar{b})\cdot(a-\bar{b})\}$$
$$= \exp\{(t/2)(a+\bar{b})\cdot(a+\bar{b})\}$$

which proves (2.21). Finally (2.22) follows from (2.20) and (2.23) via the identities

$$(C_t f_a, C_t f_b)_{L^2(\mathbb{R}^n, p_t)} = \int_{\mathbb{R}^n} e^{y\cdot ia + ta\cdot a} e^{-y\cdot i\bar{b} + t\bar{b}\cdot\bar{b}} p_t(y)dy$$
$$= \exp\{t(a\cdot a + \bar{b}\cdot\bar{b}) - (t/2)(a-\bar{b})\cdot(a-\bar{b})\}$$
$$= \exp\{(t/2)(a+\bar{b})\cdot(a+\bar{b})\}.$$

Q.E.D.

Proof of Theorems 2.3 and 2.4. Denote by $\mathcal{E}$ the set of finite linear combinations of the exponential functions f_a for $a \in \mathbb{R}^n$. If $f = \Sigma\alpha_j f_{a_j}$ is in $\mathcal{E}$ then applying (2.21), we have

$$\|S_t f\|^2_{L^2(\mathbb{C}^n, m_t)} = \Sigma\alpha_j\bar{\alpha}_k(S_t f_{a_j}, S_t f_{a_k})$$
$$= \Sigma\alpha_j\bar{\alpha}_k(f_{a_j}, f_{a_k})_{L^2(\mathbb{R}^n, p_t)} = \|f\|^2_{L^2(\mathbb{R}^n, p_t)}.$$

Thus, since, by Lemma 2.11, $\mathcal{E}$ is dense in $L^2(\mathbb{R}^n, p_t)$, the restriction $S_t \mid \mathcal{E}$ extends to an isometry $\widehat{S}_t$, from $L^2(\mathbb{R}^n, p_t)$ into $\mathcal{H}L^2(\mathbb{C}^n, m_t)$. By (2.19) the range of $\widehat{S}_t$ contains the exponentials g_a for $a \in \mathbb{R}^n$, which constitute a fundamental set by Lemma 2.11. Hence $\widehat{S}_t$ is unitary. A similar argument based on (2.22) shows that $C_t \mid \mathcal{E}$ extends uniquely to a unitary operator $\widehat{C}_t$ from $L^2(\mathbb{R}^n, p_t)$ onto itself. We must show that these two unitaries are actually given by the original definitions (2.10) and (2.11). But if $f \in L^2(\mathbb{R}^n, p_t)$ and $f_k \in \mathcal{E}$ with $\|f_k - f\|_{L^2(\mathbb{R}^n, p_t)} \to 0$ then by (2.10) $(S_t f_k)(z)$ converges to $(S_t f)(z)$ for each z in $\mathbb{C}^n$, while by Bargmann's pointwise bounds (2.15) and the unitarity of $\widehat{S}_t$, $(S_t f_k)(z) - (\widehat{S}_t f)(z)$ also converges to zero for each z. So $S_t f = \widehat{S}_t f$ everywhere. In the case of C_t we must show that $\widehat{C}_t$ is given by (2.11) for f in the dense subspace $L^2(\mathbb{R}^n, p_{2t})$. For such an f we can find a sequence f_k in $\mathcal{E}$ which converges to f in the $L^2(\mathbb{R}^n, p_{2t})$ norm. By what has already been proved for S_{2t} it follows that $(C_t f_k)(y)$ converges to $(C_t f)(y)$ for all y in $\mathbb{R}^n$. On the other hand f_k also converges to f in $L^2(\mathbb{R}^n, p_t)$ norm because $p_t(u) \le$ const. $p_{2t}(u)$ for all u in $\mathbb{R}^n$. So $(C_t f_k) \equiv \widehat{C}_t f_k$ converges to $\widehat{C}_t f$ in $L^2(\mathbb{R}^n, p_t)$. Hence $C_t f = \widehat{C}_t f$ a.e., which is what has to be shown.

Finally to prove (2.12) it suffices to show that $(C_t^2 f)(-y) = f(y)$ for a fundamental set of functions f, such as the exponentials f_a with $a \in \mathbb{C}^n$. But (2.20) shows that $C_t f_a = e^{ta\cdot a} f_{ia}$. Applying C_t again gives $C_t^2 f_a = e^{ta\cdot a} e^{t(ia)\cdot(ia)} f_{-a} = f_{-a}$ which, in view of the definition (2.18), completes the proof of Theorem 2.4.

Remark 2.13. (Connection with Hermite polynomials and Wick ordering) The Hermite polynomials, $H_k^t(u)$ [He], associated with time t, may be defined by the identity

$$\sum_{k=0}^{\infty} (a^k/k!) H_k^t(u) = e^{au-(t/2)a^2} \quad u \in \mathbb{R},\ a \in \mathbb{C}. \tag{2.24}$$

$H_k^t(u)$ is a polynomial in u whose highest order term is u^k. Moreover they are mutually orthogonal in $L^2(\mathbb{R}, p_t)$. $S_t H_k^t$ and $C_t H_k^t$ can be computed from (2.19) and (2.20) (in one dimension) as follows. Multiply (2.19) by $e^{-(t/2)a^2}$ and use (2.24) on the left of (2.19) to find $\sum_{k=0}^{\infty} (a^k/k!)(S_t H_k^t)(z) = e^{za}$. The interchange of the sum with S_t follows from the fact that the series (2.24) is convergent in $L^2(\mathbb{R}, p_t)$. Now expand e^{za} in a power series and compare the coefficients of a^k to find

$$(S_t H_n^t)(z) = z^n \quad z \in \mathbb{C},\ n = 0, 1, 2, \ldots. \tag{2.25}$$

Thus $S_t^{-1} z^n = H_n$. So S_t^{-1} represents "Wick ordering" as a unitary map from $\mathcal{H}L^2(\mathbb{C}, m_t)$ onto $L^2(\mathbb{R}, p_t)$. Now multiply (2.20) by $e^{-ta^2/2}$ and expand the argument of C_t by (2.24) again to find

$$\begin{aligned}\sum_{k=0}^{\infty} (a^k/k!)(C_t H_k^t)(y) &= e^{y(ia)-(t/2)(ia)^2} \\ &= \sum_{k=0}^{\infty} ((ia)^k/k!) H_k^t(y)\end{aligned}$$

for all $a \in \mathbb{C}$ and $y \in \mathbb{R}$. Hence

$$C_t H_n^t = i^n H_n^t \tag{2.26}$$

The equality (2.26) and its multivariable version have been discussed frequently in the mathematics and physics literature for a long time. See e.g. [CM2 , page 105].

Remark 2.14. While the holomorphic functions z^α form a natural orthogonal basis for the space $\mathcal{H}L^2(\mathbb{C}^n, m_t)$ there is also a natural extension of this basis to an orthogonal basis of the full space $L^2(\mathbb{C}^n, m_t)$. This basis has been described by K. Ito [Ito2] in infinite dimensions and recently applied by I. Shigekawa [Shi2].

Remark 2.15. The equations (2.11) and (2.12) show that S_t^{-1} can be expressed in terms of $C_{t/2}$. Indeed for $f \in L^2(\mathbb{R}^n, p_t)$ these equations give

$$C_{t/2}((S_t f)(-i\cdot)) = f. \tag{2.27}$$

The equation (2.10) expresses $(S_t f)(z)$ as a multiple, by $e^{-(2t)^{-1}z\cdot z}$, of the ordinary Laplace transform of $f(u)p_t(u)$. The transform C_t will similarly be recognized as convertible into one of the standard inversion formulas for the Laplace transform. See [CM2] for more discussion of this. B. Hall [Ha2] has recently shown how to describe the inverse of S_t in the compact group case by a formula analogous to (2.11).

3. Power Series, Fock Space, and the Bargmann-Krée-Segal skeleton theorem

We are going to amplify on Lemma 2.7, which expresses a unitary equivalence between holomorphic functions in $\mathcal{H}L^2(\mathbb{C}^n, m_t)$ and their power series expansion coefficients. The content of this section is entirely expository. The notation is taken from [D] and [DG]. The slightly unorthodox choice of norms in Fock space in Equation (3.4) is dictated by their good relation to the natural norms on universal enveloping algebras that we will use in Section 6 and also their convenient relation with Ito multiple integral expansions (cf. [G4]).

Notation 3.1. Let V be a separable complex Hilbert space. Denote by V^* its dual space. $T(V)$ will denote the algebra of algebraic tensors over V. Specifically, we will write V^k for the set of finite sums of tensor products of k elements of V and will denote by $T(V)$ the weak direct sum $\sum_{k=0}^{\infty} V^k$. $T'(V)$ will denote its algebraic dual space. $V^{\otimes k}$ will denote the Hilbert space tensor product while $V^{\odot k}$ will denote the closed subspace of $V^{\otimes k}$ consisting of symmetric tensors. $(V^*)^{\otimes k}$ and $V^{\otimes k}$ are naturally topologically dual to each other.

Let $t > 0$. Define Hilbert spaces $\mathcal{F}_t(V)$ and $\mathcal{F}^t(V)$ as follows. The **contravariant Fock space** $\mathcal{F}_t(V)$ consists of those elements

$$\beta = \sum_{k=0}^{\infty} \beta_k \quad \beta_k \in V^{\odot k} \quad k = 0, 1, 2, \ldots \tag{3.1}$$

of the strong direct sum $\sum_{k=0}^{\infty} V^{\odot k}$ satisfying

$$\|\beta\|_t^2 = \sum_{k=0}^{\infty} (k!/t^k)|\beta_k|^2 < \infty. \tag{3.2}$$

The **covariant Fock space** $\mathcal{F}^t(V)$ consists of those elements

$$\alpha = \sum_{k=0}^{\infty} \alpha_k \quad \alpha_k \in (V^*)^{\odot k} \quad k = 0, 1, 2, \ldots \tag{3.3}$$

of the strong direct sum $\sum_{k=0}^{\infty}(V^*)^{\odot k}$ satisfying

$$\|\alpha\|_t^2 = \sum_{k=0}^{\infty} (t^k/k!)|\alpha_k|^2 < \infty. \tag{3.4}$$

Then $\mathcal{F}_t(V)$ and $\mathcal{F}^t(V)$ are complex Hilbert spaces in the inner products which are the polarizations of the norms given in (3.2) and (3.4) respectively. Moreover $\mathcal{F}^t(V)$ and $\mathcal{F}_t(V)$ are dual in the bilinear pairing given by

$$\langle \alpha, \beta \rangle = \sum_{k=0}^{\infty} {}_{(V^*)^{\otimes k}}\langle \alpha_k, \beta_k \rangle_{V^{\otimes k}}. \tag{3.5}$$

Notation 3.2. For each z in V we write

$$\exp z = \sum_{k=0}^{\infty} (1/k!) z^{\otimes k} \quad z \in V. \tag{3.6}$$

Then $\exp z$ is in $\mathcal{F}_t$ and a straightforward computation starting from (3.2) gives

$$\|\exp z\|_t^2 = e^{|z|^2/t} \quad z \in V. \tag{3.7}$$

Definition 3.3. A function $f : V \to \mathbb{C}$ is holomorphic if it is Frechet differentiable at each point z of V and its Frechet derivative $Df(z)$ is complex linear for each point z in V. We will write $\mathcal{H}(V)$ for the space of holomorphic functions on V.

It will be useful to note the well known fact that f is holomorphic on V if and only if for each z and w in V $f(z+\lambda w)$ is a holomorphic function of λ on $\mathbb{C}$ while f itself is bounded on some neighborhood of z. See e.g. [HP].

Notation 3.4. Let $f: V \to \mathbb{C}$ be a holomorphic function. Define the directional derivative by $D_h f(z) = (d/ds)\big|_{s=0} f(z+sh)$. Then for each z in V, $D_{h_1} \cdots D_{h_k} f(z)$ is a symmetric k–linear form on V and is in fact continuous in each variable separately. It extends to a linear form on V^k which we denote by $D^k f(z)$. Thus

$$\langle D^k f(z), h_1 \otimes \cdots \otimes h_k \rangle = D_{h_1} \cdots D_{h_k} f(z), \quad z \in V, \ j = 1, \ldots, k. \tag{3.8}$$

With $D^0 f(z)$ defined as $f(z)$ the infinite sum $\sum_{k=0}^{\infty} D^k f(z)$ is then a well defined linear form on $T(V)$. We will follow [D] in using the suggestive notation $(1-D)_z^{-1} f$ for this sum. Thus we will write

$$(1-D)_z^{-1} f = \sum_{k=0}^{\infty} D^k f(z). \tag{3.9}$$

Then $(1-D)_z^{-1} f$ always exists as an element of the dual space $T'(V)$.

If $w\langle h_1, \ldots, h_k\rangle$ is a k–linear form on V which is continuous in each variable h_j (separately) then the Hilbert–Schmidt norm of w is defined to be

$$|w|^2 = \sum_{j_1, \ldots, j_k = 1}^{\infty} |w\langle e_{j_1}, f_{j_2}, \ldots, h_{j_k}\rangle|^2 \tag{3.10}$$

where $\{e_1, e_2, \ldots\}, \ldots, \{h_1, h_2, \ldots\}$ are each orthonormal bases of V (which can be taken to be the same orthonormal basis). The independence of $|w|^2$ from the choice of orthonormal bases can be seen by changing one orthonormal basis at a time and using the fact that the dual space norm of a continuous linear functional on V can be computed as the sum of the absolute squares of its components on any orthonormal basis. w is said to be of Hilbert–Schmidt type if $|w| < \infty$. In particular $D^k f(z)$ is of Hilbert–Schmidt type if its Hilbert–Schmidt norm

$$|D^k f(z)|^2 = \sum_{j_1, \ldots, j_k = 1}^{\infty} |\langle D^k f(z), e_{j_1} \otimes \cdots \otimes e_{j_k}\rangle|^2 \tag{3.11}$$

is finite for some orthonormal basis $e_1, e_2, \ldots$ of V.

Suppose that w is a k–linear, separately continuous form on V and we regard it as a linear form on V^k. If β is in V^k then $|\langle w, \beta\rangle| \le |w| \ |\beta|_{V^{\otimes k}}$ where $|w|$ denotes the norm given by (3.10) and $|\beta|_{V^{\otimes k}}$ denotes the norm of β in the Hilbert space $V^{\otimes k}$. In fact

$$|w| = \sup\{|\langle w, \beta\rangle|; |\beta|_{V^{\otimes k}} = 1\}. \tag{3.12}$$

In short we may and will identify a k–linear form whose Hilbert–Schmidt norm (3.10) is finite as an element of $(V^*)^{\otimes k}$.

If $D^k f(z)$ is of Hilbert–Schmidt type for each k, and further

$$(3.13) \qquad \|(1-D)_z^{-1} f\|_t^2 := \sum_{k=0}^{\infty} (t^k/k!)|D^k f(z)|^2 < \infty$$

then, by (3.4), $(1-D)_z^{-1} f$ is in the covariant Fock space $\mathcal{F}^t$.

Notation 3.5. Fix $t > 0$. For any finite dimensional inner product space M of dimension n denote by $m_t^M(dz)$ the measure on M given by the density (2.3) when M is identified with $\mathbb{C}^n$ by any orthonormal basis of M.Let

$$(3.14.) \qquad \|f\|_t^2 = \sup_M \int_M |f(z)|^2 m_t^M(dz) \qquad f \in \mathcal{H}(V).$$

wherein the supremum is taken over all finite dimensional complex subspaces of V. Now we define

$$(3.15.) \qquad \mathcal{H}^t = \{f \in \mathcal{H}(V) : \|f\|_t^2 < \infty\}.$$

A *cylinder* function on V is a function that depends on only finitely many "coordinates" in the sense that for some finite dimensional projection P on V, as a real Hilbert space, $f(Pz) = f(z)$ for all $z \in V$. One says that f is based on the range of P. Define

$$(3.16.) \qquad \mathcal{H}_0^t = \{f \in \mathcal{H}^t : f \text{ is a cylinder function}\}.$$

If f is in $\mathcal{H}(V)$ and is based on a finite dimensional complex subspace M_0, and if M is another finite dimensional complex subspace containing M_0, then by (2.3) m_t^M is the product measure $m_t^{M_0} \times m_t^{M-M_0}$. Since $f|M$ does not depend on the $M - M_0$ coordinates it follows that $\int_M |f(z)|^2 m_t^M(dz) = \int_{M_0} |f(z)|^2 m_t^{M_0}(dz)$. Because of the monotonicity of norms expressed in Corollary 2.9 it therefore follows that

$$\|f\|_t^2 = \int_{M_0} |f(z)|^2 m_t^{M_0}(dz)$$

when f is in $\mathcal{H}(V)$ and is based on M_0. Hence $\mathcal{H}_0^t$ is an inner product space in the norm (3.14). It will be proven in this section that $\mathcal{H}_0^t$ is dense in $\mathcal{H}^t$, that $\mathcal{H}^t$ is complete in the norm (3.14), and that $\mathcal{H}^t$ is therefore a Hilbert space.

Theorem 3.6. *(Bargmann-Krée)[B2,3, Kr1,2,3] Let $t > 0$ and let f be in $\mathcal{H}^t$. Then $(1-D)_0^{-1} f$ is in $\mathcal{F}^t$. Moreover the map*

$$(1-D)_0^{-1} : \mathcal{H}^t \to \mathcal{F}^t(V)$$

is a surjective isometry. The inverse is given by the map $w \to \langle w, \exp z\rangle$

Lemma 3.7. *Let M be a finite dimensional complex inner product space. For any function f in $\mathcal{H}(M)$ there holds*

$$\|(1-D)_0^{-1} f\|_t^2 = \int_M |f(z)|^2 m_t^M(dz) \quad f \in \mathcal{H}(M). \tag{3.17}$$

In particular if either side is finite so is the other.

Proof: Choose an orthonormal basis $e_1, \ldots, e_n$ of M and use it to identify M with $\mathbb{C}^n$. Then the given holomorphic function f has a power series representation of the form (2.13). We assert that

$$\langle D^k f(0), e_{j_1} \otimes \cdots \otimes e_{j_k} \rangle = \alpha! a_\alpha \quad f \in \mathcal{H}(\mathbb{C}^n) \tag{3.18}$$

where α is the multi-index such that α_r is the number of times that r occurs in the ordered set $\{j_1, \ldots, j_k\}$. In fact, writing $\partial_j = \partial/\partial z_j$, (3.18) merely asserts the obvious identity $\partial_{j_1} \cdots \partial_{j_k} f(0) = \alpha! a_\alpha$. Next we will show that

$$|D^k f(0)|^2 = k! \sum_{|\alpha|=k} \alpha! |a_\alpha|^2. \tag{3.19}$$

Given a multi-index α with $|\alpha| = k$ the number of ordered k-tuples $j_1, \ldots, j_k$ with α_r occurrences of r is $k!/\alpha!$. Hence, using (3.18) we find

$$\begin{aligned} |D^k f(0)|^2 &= \sum_{j_1, \ldots, j_k = 1}^{n} |\langle D^k f(0), e_{j_1} \otimes \cdots \otimes e_{j_k} \rangle|^2 \\ &= \sum_{|\alpha|=k} (k!/\alpha!) |\alpha! a_\alpha|^2 \end{aligned}$$

which is (3.19). Now, multiplying (3.19) by $t^k/k!$, summing over k, and using (2.14), we get

$$\begin{aligned} \|(1-D)_0^{-1} f\|_t^2 &= \sum_{k=0}^{\infty} (t^k/k!) |D^k f(0)|^2 \\ &= \sum_{k=0}^{\infty} t^k \sum_{|\alpha|=k} \alpha! |a_\alpha|^2 \\ &= \int_{\mathbb{C}^n} |f(z)|^2 m_t(z) dz. \end{aligned}$$

Q.E.D.

Corollary 3.8. *If $f \in \mathcal{H}(V)$ then*

$$\|(1-D)_0^{-1} f\|_t^2 = \|f\|_t^2. \tag{3.20}$$

Proof: Let $f \in \mathcal{H}(V)$ and suppose that M is a finite dimensional complex subspace of V. Let $F = f|M$. In view of the definition (3.8) we may write

$$\langle D^k F(0), h_1 \otimes \cdots \otimes h_k\rangle = \langle D^k f(0), h_1 \otimes \cdots \otimes h_k\rangle \qquad \text{for } h_j \in M, j = 1, \ldots, k$$

Therefore for any orthonormal basis $e_1, \ldots, e_n$ of M we have, by applying Lemma 3.7 to F,

$$\sum_{k=0}^{\infty} (t^k/k!) \sum_{j_1,\ldots,j_k=1}^{n} |\langle D^k f(0), e_{j_1} \otimes \cdots \otimes e_{j_k}\rangle|^2 = \int_M |f(z)|^2 m_t^M(dz). \tag{3.21}$$

Since, by Corollary 2.9, $\int_M |f(z)|^2 m_t^M(dz)$ increases with M, we may find an increasing sequence M_i of finite dimensional subspaces of V such that a) $\cup M_i$ is dense in V and b) $\|f\|_t^2 = \lim_{i\to\infty} \int_{M_i} |f(z)|^2 m_t^{M_i}(dz)$ (which may be infinite). Choose an orthonormal basis $e_1, e_2, \ldots$ of V such that $e_1, \ldots, e_{n_i}$ is an orthonormal basis of M_i. Replace M in (3.21) by M_i and n by n_i. Then for each k the summand on the left increases to $(t^k/k!)|(D^k f)(0)|^2$. By the monotone convergence theorem we may therefore take the limit as $i \to \infty$ to get (3.20). Q.E.D.

Proof of Theorem 3.6. If f is in $\mathcal{H}^t$ then (3.20) shows that $(1-D)_0^{-1} f$ is in $\mathcal{F}^t(V)$ and that the map $f \to (1-D)_0^{-1} f$ is isometric.

To prove surjectivity let w be in $\mathcal{F}^t(V)$. Define

$$f(z) = \langle w, \exp z\rangle \quad z \in V \tag{3.22}$$

where $\exp z$ is given by (3.6). In view of (3.7) we have, by Schwarz' inequality,

$$|f(z)|^2 \le \|w\|_t^2 e^{|z|^2/t} \quad z \in V. \tag{3.23}$$

Let w_k denote the component of w in $(V^*)^{\odot k}$. So $w = \sum_{k=0}^{\infty} w_k$. Since w_k is symmetric one can compute that $D_h\langle w_k, z^{\otimes k}\rangle = k\langle w_k, h \otimes z^{\otimes(k-1)}\rangle$,and similarly

$$D_{h_1} \cdots D_{h_k}\langle w_k, z^{\otimes k}\rangle = k!\langle w_k, h_1 \otimes h_1 \otimes \cdots \otimes h_k\rangle.$$

Write $f_k(z) = (1/k!)\langle w_k, z^{\otimes k}\rangle$. Then f_k is in $\mathcal{H}(V)$ and the definition (3.8) shows that $(D^k f_k)(0) = w_k$. But by Schwarz's inequality we have

$$\left|f(z) - \sum_{k=0}^{N} f_k(z)\right| = \left|\left\langle w - \sum_{k=0}^{N} w_k, \exp z\right\rangle\right|$$
$$\le \left\|w - \sum_{k=0}^{N} w_k\right\|_t \exp(|z|^2/2t),$$

which converges to zero uniformly on bounded sets. Hence $f \in \mathcal{H}(V)$. Moreover we can compute the derivatives of f term by term in the power series $f = \sum_{k=0}^{\infty} f_k$. At $z = 0$ only the term f_k contributes to the kth derivative. Hence by a previous computation we have $(D^k f)(0) = w_k$. Adding these we find

$$(1 - D)_0^{-1} f = w \quad \text{if} \quad f(z) = \langle w, \exp z\rangle, \; w \in \mathcal{F}^t(V). \tag{3.24}$$

Hence by Corollary 3.8 f is in $\mathcal{H}^t$. Q.E.D.

Theorem 3.9. *(Bargmann-Krée–Segal skeleton theorem) [B2,3, Kr1,2,3, Se5] $\mathcal{H}^t$ is a Hilbert space in the norm defined in (3.14). Moreover $\mathcal{H}_0^t$ is dense in $\mathcal{H}^t$.*

Proof. Since $\mathcal{H}^t$ is, by Theorem 3.6, isometrically isomorphic to $\mathcal{F}^t(V)$, $\mathcal{H}^t$ is a Hilbert space in the norm given by (3.14). If $p : \mathbb{C}^n \to \mathbb{C}$ is a holomorphic polynomial and $e_1, \dots, e_n$ is any orthonormal set in V^* then the function $f(z) = p(\langle e_1, z\rangle, \dots, \langle e_n, z\rangle)$ is a cylinder function on V, is also in $\mathcal{H}^t$, and $w := (1 - D)_0^{-1} f$ is an algebraic symmetric tensor over V^*. Every algebraic symmetric tensor w arises in this way (e.g. from $f(z) := \langle w, \exp z\rangle$.) Since the algebraic symmetric tensors are dense in $\mathcal{F}^t(V)$, $\mathcal{H}_0^t$ is dense in $\mathcal{H}^t$. Q.E.D.

4. The Segal–Bargmann Transform and the Fourier–Wiener Transform

This section is expository, as were the previous two sections. It is based on the early papers of Cameron and Martin [Ca, CM1,2], of Krée [Kr1,2,3] and of Segal [Se1,2,5].

We need to extend the two transforms of Section 2 to infinite dimensions in order to derive Hall's transform for a finite dimensional compact group. A feature of the transforms that we will need to use in our application is the invariance under orthogonal transformations and translations. In order to avoid technical irrelevancies related to orthogonal invariance we are going to develop the infinite dimensional theory in this section in terms of cylinder set measures on a real Hilbert space H_r rather than in terms of Wiener measure on some completion of H_r. The latter will be carried out in Section 7 in the concrete case needed for our application. Here we will focus on the minimum amount of structure needed to formulate the main theorems. To this end let us review the notion of a cylinder set measure on a real Hilbert space H_r and at the same time establish notation. Denote by $\mathcal{F}$ the set of finite dimensional subspaces F of H_r and by P_F the orthogonal projection of H_r onto F. A cylinder set in H_r is a set C of the form $C = P_F^{-1} B$ where B is a Borel set in the finite dimensional subspace F. C is said to be based on F.

The collection $\mathcal{S}_F$ of all cylinder sets based on F is clearly a σ–field and the union

$$\mathcal{R} = \bigcup_{F \in \mathcal{F}} \mathcal{S}_F$$

is a field. A set function $m : \mathcal{R} \to [0,1]$ is called a <u>cylinder set measure</u> if $m \mid \mathcal{S}_F$ is countably additive and $m(H_r) = 1$. m is automatically additive on $\mathcal{R}$. A function $f : H_r \to \mathbb{C}$ is a <u>cylinder function</u> if f is $\mathcal{S}_F$ measurable for some $F \in \mathcal{F}$. Such a function f is said to be <u>based on</u> F. The cylinder functions based on F are exactly those functions on H_r of the form $f(x) = \varphi(P_F x)$ for some Borel function $\varphi : F \to \mathbb{C}$. If f is any cylinder function based on a subspace $F \in \mathcal{F}$ then $\int_{H_r} |f(x)| m(dx)$ is well defined because $m \mid \mathcal{S}_F$ is countably additive. Moreover if the last integral is finite then $\int_{H_r} f(x) m(dx)$ is also well defined. In particular if $F_1 \in \mathcal{F}$ and $F_1 \supset F$ then f is also based on F_1. But it is easy to see that $\int_{H_r} f(x) m(dx)$ is not altered by viewing f as based on F_1. Finally if $m^F(B) := m(P_F^{-1} B)$ is the countably additive probability measure on F induced by m then

$$\int_{H_r} f(x) m(dx) = \int_F f(x) m^F(dx). \tag{4.1}$$

The map $\varphi \to \varphi \circ P_F$ is therefore a unitary map from $L^2(F, m^F)$ onto $L^2(H_r, \mathcal{S}_F, m)$, the space of square integrable $\mathcal{S}_F$ measurable cylinder functions on H_r. In this section we will use these unitary maps to transfer the two finite dimensional transforms of Section 2 to infinite dimensions by letting F vary over all finite dimensional subspaces. Consistency will be an issue. Let us note immediately that if also $G \in \mathcal{F}$ and $F \subset G$ then $L^2(H_r, \mathcal{S}_F, m)$ is a closed subspace of $L^2(H_r, \mathcal{S}_G, m)$.

If $\langle\ ,\ \rangle$ denotes the inner product on H_r then the function $x \to \langle x, a\rangle$ is a cylinder function on H_r based on the one dimensional subspace span$\{a\}$ and so is $e^{i\langle x,a\rangle}$. The Fourier transform $\widehat{m}(a) := \int_{H_r} e^{i\langle x,a\rangle} m(dx)$ determines m uniquely. For further discussion of these concepts see [Kuo].

<u>Notation 4.1</u>. We want to consider now a complex Hilbert space H_c with Hermitian inner product $(\ ,\ )$. We will want to use the set $\mathcal{F}_c$ of finite dimensional complex subspaces M of H_c and the corresponding complex linear orthogonal projection P_M. H_c is also a real Hilbert space with respect to the real inner product $\langle z, w\rangle = Re(z,w)$. Of course $\mathcal{F}_c \subsetneqq \mathcal{F}$. There is a unique cylinder set measure m_t on H_c, as a real Hilbert space, whose Fourier transform is

$$\int_{H_c} e^{iRe(w,z)} m_t(dz) = e^{-(t/4)|w|^2}. \tag{4.2}$$

If $H_c = \mathbb{C}^n$ then Equation (4.2) agrees with Equation (2.9) for $a, b \in i\mathbb{R}^n$. and m_t is then the measure with density given by (2.3). If H_c is infinite dimensional and $M \in \mathcal{F}_c$ is of complex dimension n then the induced measure $m_t^M := m_t \circ P_M^{-1}$ on M is also given by the density (2.3) when M is identified with $\mathbb{C}^n$ by any choice of an orthonormal (w.r.t. $(\ ,\ )$) basis. We may therefore take over results from Sections 2 and 3.

Notation 4.2. Let H_r be a real Hilbert space with inner product $\langle\ ,\ \rangle$ and let p_t denote the centered isotropic Gaussian cylinder set measure on H_r whose Fourier transform is given by

$$\int_{H_r} e^{i\langle x,u\rangle} p_t(du) = \exp\{-(t/2)|x|^2\} \qquad x \in H_r. \tag{4.3}$$

Note that for any n–dimensional subspace F of H_r the measure p_t^F is given by the density (2.2) on $\mathbb{R}^n$ when F is identified with $\mathbb{R}^n$ by any choice of orthonormal basis of F. We will write for $1 < q < \infty$

$$L_0^q(H_r, p_t) = \Big\{\text{cylinder functions } f : H_r \to \mathbb{C} \mid \|f\|_{L^q(p_t)}^q := \int_{H_r} |f(u)|^q p_t(du) < \infty\Big\}. \tag{4.4}$$

Two cylinder functions which differ on a cylinder set of p_t measure zero will be identified. In the notation introduced earlier we may write

$$L_0^q(H_r, p_t) = \bigcup_{F \in \mathcal{F}} L^q(H_r, \mathcal{S}_F, p_t). \tag{4.5}$$

Notation 4.3. $L^q(H_r, p_t)$ will denote the completion of $L_0^q(H_r, p_t)$ in the norm (4.4).

Remark 4.4. As is well known H_r can be embedded in some (non–unique) Banach space B carrying a Gaussian measure w_t (Wiener measure) in such a way that $L_0^q(H_r, p_t)$ maps naturally to a dense subspace of $L^q(B, w_t)$. Once one has made such a choice of B and of an embedding $H_r \hookrightarrow B$ one can then identify the abstract completion $L^q(H_r, p_t)$ with $L^q(B, w_t)$. We will do this in the next section. B will be a classical Wiener space and H_r its Cameron–Martin space. But we will not make a choice of B in this section because the important issue of orthogonal invariance that we need to address would then be unnecessarily obscured. However in this section we will need to use one consequence of this identification: if $1 < q < \infty$ and q' is the conjugate index then $L^{q'}(H_r, p_t)$ is the dual space of $L^q(H_r, p_t)$. The natural pairing is given by

$$\phi_g(f) = \int_{H_r} f(u) g(u) p_t(du) \quad f \in L_0^q(H_r, p_t), g \in L_0^{q'}(H_r, p_t) \tag{4.6}$$

ϕ_g extends to a continuous linear functional (also denoted by ϕ_g) on $L^q(H_r, p_t)$.Since $L_0^{q'}$ is dense in $L^{q'}$ we may conclude that $\{\phi_g : g \in L_0^{q'}(H_r, p_t)\}$ is a separating family in $(L^q)^*$. We will need to use this observation in Proposition 4.7.

Remark 4.5. We have taken the terminology "skeleton theorem" in the statement of Theorem 3.9 from [Sug1]. H. Sugita shows in this paper that the Lévy stochastic area functional on Wiener space is very poorly related to its restriction to the Cameron–Martin subspace, in spite of the fact that the restriction is just a quadratic function. The Bargmann-Krée–Segal theorem shows, in contrast, that any function in the $L^2(H_c, m_t)$ completion of the holomorphic polynomials has a well defined "restriction" ($\approx$ skeleton) to the Cameron–Martin space H_c, a fact which is notoriously false in the absence of holomorphy. There have been several studies recently of L^p spaces of holomorphic functions on complex abstract Wiener spaces [Shi2, Sug2,3, Tan, FR] describing several senses in which a holomorphic function in L^p has a determining skeleton.

Notation 4.6. Let $H_c := H_r \otimes_{\mathbb{R}} C$ denote the complexification of H_r. We may write any element z of H_c uniquely in the form $z = x + iy$ with x and y in H_r. We will write $\overline{z} = x - iy$ in this case. H_c naturally carries two quadratic forms, the Hermitian inner product, which we denote by (,), and a bilinear form $z \cdot w$ given by

$$z \cdot w = (z, \overline{w}) \qquad z, w \in H_c \tag{4.7}$$

which reduces to $\langle z, w\rangle$ when z and w are both in H_r.

Proposition 4.7. *Let $t > 0$. Let $1 < q < \infty$ and let $z \in H_c$. Define*

$$(S_t f)(z) = e^{-(2t)^{-1} z \cdot z} \int_{H_r} f(u) e^{t^{-1} z \cdot u} p_t(du) \qquad f \in L_0^q(H_r, p_t) \tag{4.8}$$

a) For each point z in H_c the map $f \to (S_t f)(z)$ has a unique continuous linear extension to $L^q(H_r, p_t)$. We will denote the extension also by $(S_t f)(z)$.

b) If f is a cylinder function based on a finite dimensional subspace M of H_r then $S_t f$ is based on $M + iM$.

c) For each element f in $L^q(H_r, p_t)$ $S_t f$ is in $\mathcal{H}(H_c)$.

d) If f is in $L^q(H_r, p_t)$ and $S_t f = 0$ on H_c then $f = 0$.

e) If f is in $L_0^q(H_r, p_t)$ and x is in H_r then

$$(S_t f)(x) = \int_{H_r} f(x - u) p_t(du) \qquad x \in H_r \tag{4.9}$$

Proof: For fixed $z := x + iy$ in H_c the function $u \to e^{t^{-1} z \cdot u}$ is a cylinder function based on the subspace spanned by x and y. Moreover since $|e^{t^{-1} z \cdot u}| = e^{t^{-1} x \cdot u}$ (2.8) gives

$$\|e^{t^{-1} z \cdot u}\|_{L^{q'}(H_t, p_t)} = e^{(q'/2t)|x|^2} \le e^{(q'/2t)|z|^2}.$$

Choosing q' to be the conjugate index to q we find

$$(4.10)\qquad |(S_t f)(z)| \le |e^{-(2t)^{-1}z\cdot z}|e^{(q'/2t)|z|^2}\|f\|_{L^q(H_r,p_t)},\ f\in L^q_0(H_r,p_0).$$

Thus for each point z in H_c the map $f \to (S_t f)(z)$ has a unique continuous extension to $L^q(H_r,p_t)$. Of course (4.10) continues to hold for the extended map S_t. To prove b) it suffices to show that if $z := x+iy$ is in $M+iM$ while $w\in(M+iM)^\perp$ then $(S_t f)(z+w) = (S_t f)(z)$. But if w is orthogonal to $M+iM$ then $(w,x)=(w,y)=0$, so $w\cdot z=0$ and moreover $e^{t^{-1}w\cdot u}$ is independent (relative to p_t) of $f(u)e^{t^{-1}z\cdot u}$. Since $\int_{H_r} e^{t^{-1}w\cdot u}p_t(du) = e^{(2t)^{-1}w\cdot w}$ by (2.8) and since $(z+w)\cdot(z+w) = z\cdot z + w\cdot w$, we find

$$\begin{aligned}(S_t f)(z+w) &= e^{-(2t)^{-1}(z\cdot z+w\cdot w)}\int_{H_r} f(u)e^{t^{-1}z\cdot u}p_t(du)\int_{H_r} e^{t^{-1}w\cdot u}p_t(du)\\ &= (S_t f)(z).\end{aligned}$$

This establishes property b). But in addition, if f is based on M and z is in $M+iM$ we may write (4.8) as

$$(S_t f)(z) = e^{-(2t)^{-1}z\cdot z}\int_M f(u)e^{t^{-1}z\cdot u}p_t^M(u)du \qquad f\in L^q(H_r,\mathcal{S}_M,p_t).$$

By Lemma 2.5 $S_t f$ is holomorphic on $M+iM$. In view of property b) and Definition 3.3 it follows that $S_t f$ is in $\mathcal{H}(H_c)$. Now apply (4.10) to differences to conclude that for any element f in $L^q(H_r,p_t)$ there is a sequence of holomorphic functions on H_c which converge uniformly on bounded sets in H_c to $S_t f$. This proves c).

To prove d) suppose that f is in $L^q(H_r,p_t)$ and that $S_t f = 0$. If M is a finite dimensional subspace of H_r and g is a finite linear combination of the functions $\{e^{t^{-1}iy\cdot u} : y\in M\}$ then by assumption $\varphi_g(f)=0$, where φ_g is the extension of (4.6) to L^q. But by Lemma 2.11 such functions g are dense in $L^{q'}(H_r,\mathcal{S}_M,p_t)$. So $\varphi_g(f)=0$ for all g in $L^{q'}(H_r,\mathcal{S}_M,p_t)$. By Remark 4.4 the union over M of these spaces is a separating family in $L^q(H_r,p_t)^*$. Hence $f=0$.

Part e) may be proven as in Lemma 2.5 by using Part b). Thus if f is based on a finite dimensional subspace M then both $(S_t f)(x)$ and $\int_{H_r} f(x-u)p_t(du)$ are based on M as functions of x. Therefore it suffices to prove (4.9) in case x is in M. But if x is in M then we may write $\int_{H_r} f(x-u)p_t(du) = \int_M f(x-u)p_t^M(u)du$ and we may similarly write $(S_t f)(x)$ as the integral (2.10), wherein we identify M with R^n by some choice of orthonormal basis of M. Lemma 2.5 can now be applied. Q.E.D.

Theorem 4.8. *(Segal–Bargmann transform) [Se1,2,5, B1]) Let $t > 0$. Define S_t as in Proposition 4.7 with $q = 2$. Then S_t is a unitary operator from $L^2(H_r, p_t)$ onto $\mathcal{H}^t$. Furthermore*

$$S_t(L^2_0(H_r, p_t)) = \mathcal{H}^t_0 \tag{4.11}$$

Proof: By Proposition 4.7 $S_t f$ is in $\mathcal{H}(H_c)$ for any element f in $L^2(H_r, p_t)$. Let M be a finite dimensional subspace of H_r and write $M_c := M + iM \subset H_c$. We will take over the results of Theorem 2.3 by identifying M with $\mathbb{R}^n$ for some choice of orthonormal basis of M. Thus if $f \in L^2(H_r, \mathcal{S}_M, p_t)$ then

$$\int_{M_c} |(S_t f)(z)|^2 m_t^{M_c}(dz) = \int_M |f(u)|^2 p_t^M(du)$$

by Theorem 2.3. But by Part b) of Proposition 4.7, $S_t f$ is based on M_c. Thus the last equation may be written

$$\|S_t f\|^2_{L^2(H_c, m_t)} = \|f\|^2_{L^2(H_r, p_t)}, \qquad f \in L^2_0(H_r, p_t). \tag{4.12}$$

So $S_t(L^2_0(H_r, p_t) \subset \mathcal{H}^t_0$. To prove the equality (4.11) observe that any function g in $\mathcal{H}^t_0$ is based on some finite dimensional subspace $K \subset H_c$. But the two real linear maps from K to H_r given by $x + iy \to x$ and $x + iy \to y$ have finite dimensional ranges in H_r. If M is the span of these ranges then $K \subset M_c$. So g is also based on M_c. By Theorem 2.3, $g = S_t f$ for some function $f \in L^2(H_r, \mathcal{S}_M, p_t)$. In view of (4.5) this proves (4.11). We see that S_t, on cylinder functions, is isometric from a dense subspace of $L^2(H_r, p_t)$ onto a dense subspace of $\mathcal{H}^t$. S_t therefore extends to a unitary operator $\hat{S}_t$ between these spaces. It remains only to show that the extension $\hat{S}_t$ is actually given by the pointwise extension described in Proposition 4.7, Part a). But if $f \in L^2(H_r, p_t)$ and if $\{f_n\}_{n=1}^\infty \subset L^2_0(H_r, p_t)$ with $\|f - f_n\|_{L^2(H_r, p_t)} \to 0$ then by definition of $\hat{S}_t$ we have $\|\hat{S}_t f - S_t f_n\|_t \to 0$. In view of Theorem 3.6 the inequality (3.23) may be written

$$|g(z)|^2 \leq \|g\|_t^2 e^{|z|^2/t} \tag{4.13}$$

for g in $\mathcal{H}(H_c)$. It follows that $(\hat{S}_t f)(z) - (S_t f_n)(z) \to 0$ for each $z \in H_c$. Hence $(\hat{S}_t f)(z) = (S_t f)(z)$. Q.E.D.

The following theorem of Cameron, Martin [CM2] and Segal [Se1] is the infinite dimensional version of Theorem 2.4. As in the preceding theorem we need only apply Theorem 2.4 to one finite dimensional subspace at a time in a consistent way.

Theorem 4.9. *(The Fourier–Wiener transform of Cameron, Martin and Segal.) Let H_r be a real Hilbert space and define p_t as in (4.3). Define*

$$(C_t f)(y) = (S_{2t} f)(iy) \quad \text{for} \quad f \in L_0^2(H_r, p_{2t}). \tag{4.14}$$

Then C_t extends uniquely to an isometry from $L_0^2(H_r, p_t)$ onto itself and to a unitary operator from $L^2(H_r, p_t)$ onto itself. Moreover

$$(C_t^{-1} f)(y) = (C_t f)(-y) \quad \text{for} \quad f \in L_0^2(H_r, p_t). \tag{4.15}$$

Proof: If $f \in L_0^2(H_r, p_{2t})$ and is based on the finite dimensional subpsace F of H_r then by Proposition 4.7 $S_{2t} f$ is based on $F + iF$ and consequently $C_t f$ is based on F. By Theorem 2.4 C_t therefore extends uniquely to a unitary operator C_t^F from $L^2(H_r, \mathcal{S}_F, p_t)$ onto itself. If G is a finite dimensional subspace of H_r containing F then $L^2(H_R, \mathcal{S}_F, p_t)$ is a closed subspace of $L^2(H_r, \mathcal{S}_G, p_t)$. Moreover $C_t^F f$ and $C_t^G f$ are both given by (4.14) for f in $L^2(H_r, \mathcal{S}_F, p_{2t})$, which is dense in $L^2(H_r, \mathcal{S}_F, p_t)$. Hence C_t^F and C_t^G agree on $L^2(H_r, \mathcal{S}_F, p_t)$. It follows that the collection of extensions C_t^F give a well defined isometry of $L_0^2(H_r, p_t)$ onto itself. The identity (4.15) follows from (2.12).

Q.E.D.

Notation 4.10. Let H_r be a real Hilbert space and H_c its complexification. If O is an orthogonal transformation on H_r we denote by $O_c := O \otimes_{\mathbb{R}} 1$ its complexification. So $O_c(x + iy) = Ox + iOy$ for x and y in H_r.

Since O_c commutes with the conjugation $z \to \overline{z}$, O_c is both unitary with respect to the H_c inner product $(\ ,\ )$ and also orthogonal with respect to the bilinear form $z \cdot w = (z, \overline{w})$. If $f : H_r \to \mathbb{C}$ is a cylinder function based on a finite dimensional subspace M of H_r then $f \circ O$ is a cylinder function based on $O^{-1}M$. The measure p_t is orthogonally invariant in the sense that $p_t(O^{-1}B) = p_t(B)$ for any cylinder set B. The map $f \to f \circ O$ is therefore isometric from $L_0^q(H_r, p_t)$ onto itself and extends to an isometry $\widehat{O}$: $L^q(H_r, p_t) \to L^q(H_r, p_t)$. On the complex side note that $f \circ O_c$ is holomorphic when f is holomorphic because O_c is complex linear and bounded. Moreover one sees from (4.2) that m_t is invariant under O_c. Therefore the map $f \to f \circ O_c$ defines a unitary operator on $\mathcal{H}^t$. But we will not use this unitary operator here because we will want to go outside L^2 in our application.

Theorem 4.11. *(Orthogonal invariance of S_t and C_t) Let $t > 0$ and let O be an orthogonal transformation on H_r. Then, for $1 < q < \infty$,*

$$(S_t f)(O_c z) = (S_t \widehat{O} f)(z) \qquad f \in L^q(H_r, p_t), z \in H_c \tag{4.16}$$

and

$$\widehat{O} C_t = C_t \widehat{O}. \tag{4.17}$$

Proof: Since $\widehat{O}$ is isometric on L^q, it suffices, by Proposition 4.7, Part a), to prove (4.16) for $f \in L_0^q(H_r, p_t)$. But for such a function f we have, by (4.8),

$$\begin{aligned}(S_t f)(O_c z) &= e^{-(2t)^{-1}(O_c z)\cdot(O_c z)} \int_{H_r} f(u) e^{t^{-1}(O_c z)\cdot u} p_t(du) \\ &= e^{-(2t)^{-1} z\cdot z} \int_{H_r} f(u) e^{t^{-1} z\cdot O^{-1} u} p_t(du) \\ &= e^{-(2t)^{-1} z\cdot z} \int_{H_r} f(Ou) e^{t^{-1} z\cdot u} p_t(du) \\ &= (S_t \widehat{O} f)(z) \quad \text{for all} \quad z \in H_c.\end{aligned}$$

This proves (4.16). In particular for $q = 2$ and $f \in L_0^2(H_r, p_{2t})$ we also have $(S_{2t} f)(O_c iy) = (S_{2t}\widehat{O} f)(iy)$. Thus $(C_t f)(Oy) = (C_t \widehat{O} f)(y)$ for f in $L_0^2(H_r, p_{2t})$. Since $\widehat{O}$ and C_t are both unitary (4.17) follows. Q.E.D.

Remark 4.12. (Translation invariance of S_t) If $a \in H_r$ and $f : H_R \to \mathbb{C}$ is a cylinder function then the translated function $f^a(x) := f(x + a)$ is also a cylinder function. Since convolution commutes with translation we have $(S_t f)(x + a) = (S_t f^a)(x)$, and by analytic continuation

$$(4.18) \qquad (S_t f)(z + a) = (S_t f^a)(z) \qquad z \in H_c,\ a \in H_r$$

provided of course that f and f^a are in the domain of S_t. Even if $f \in L_0^2(H_r, p_t)$, f^a need not be in $L_0^2(H_r, p_t)$ because p_t is not translation invariant. On the other hand if $1 < q < \infty$ and $f \in L_0^q(H_r, p_t)$ then for any $s \in (1, q)$ we have

$$\begin{aligned}\int_{H_r} |f(x + a)|^s p_t(du) &= \int_{H_r} |f(v)|^s e^{-(2t)^{-1}|a|^2 + t^{-1}\langle a, u\rangle} p_t(du) \\ &\le C(s, q) \|f\|_q^s\end{aligned}$$

for some constant $C(s, q)$ by Hölders inequality. So $f^a \in L_0^s(H_r, p_t)$ for all $s \in (1, q)$. Therefore both sides of (4.18) are well defined and the equality holds.Moreover the last inequality shows that the map $f \to f(\cdot + a)$ extends to a bounded operator from L^q into L^s. Hence, using Proposition 4.7, Part a), we see that (4.18) holds for all $f \in L^q(H_r, p_t)$ when f^a is interpreted as an element of $L^s(H_r, p_t)$ in this way.

Remark 4.13. (S_t intertwines derivatives.) The identity (4.18) can be differentiated with respect to a in H_r at $a = 0$ for "smooth" functions f, showing that S_t commutes with differentiation. To make this precise one need only choose a suitable domain in L^q on which to allow a directional derivative to act. For example polynomials on H_r can form a useful domain for some purposes. But we will not need to pursue this in this work.

Remark 4.14. (Segal's duality transform) If H_r is a real separable Hilbert space and H_c is its complexification we can compose some of the maps studied so far to obtain Segal's duality transform [Se1]. Thus in the notation of Section 3 we see that the map

$$(1-D)_0^{-1}S_t : L^2(H_r, p_t) \to \mathcal{F}^t(H_c)$$

is unitary and factors through the intermediate space $\mathcal{H}^t$. Segal actually defined the duality transform as the inverse of this map. Taking into account the different normalization used in the present paper, an inspection of his proof of Theorem 3 of [Se1] shows that his construction of the duality transform, $\mathcal{D}$, may be described as follows. Let $w \in \mathcal{F}^t(H_c)$. Define $f_w(z) = \langle w, \exp z\rangle$. We have seen in Theorem 3.6 that the map $w \to f_w$ from $\mathcal{F}^t(H_c)$ onto $\mathcal{H}^t(H_c)$ is the inverse of $(1-D)_0^{-1}$. Moreover we have seen in Remark 2.15 and Theorem 4.9 that $S_t^{-1} : \mathcal{H}^t(H_c) \to L^2(H_r, p_t)$ can be expressed in terms of $C_{t/2}$. It is this expression of S_t^{-1} that Segal uses in his construction of $\mathcal{D} : \mathcal{F}^t(H_c) \to L^2(H_r, p_t)$, which he defines by $\mathcal{D} : w \to f_w \to S_t^{-1} f_w$.

Remark 4.15. (History.) Some of the identities derived in this section were derived originally by the physicist V. Fock in his early work [Fock1,2] on the emission and absorption of radiation. Bargmann, for example, cites one such identity of Fock at the end of [B2]. Cook [Co] was the first to give Fock's formalism a precise mathematical meaning and develop the functional analytic properties of the naturally associated quantized operators.

The Fourier-Wiener transform of Cameron and Martin can be expressed in terms of conditional expectations of Brownian motions. This was done by Hitsuda, [Hit1,2], along with a derivation of other properties of this map.

There have been many ways of describing the dualiy transform. The form that underlies most of the present paper is due to K. Ito [Ito1]. For a derivation that emphasizes diagonalization of field operators see [GJ]. For a connection with Markov processes see [Dyn]. For another view of the connection between the duality transform and the Segal-Bargmann transform see [Ko].

The equation (4.9) shows that even in infinite dimensions the Segal-Bargmann transform is (informally) constructed from the heat semigroup associated to a Laplacian, namely the infinite dimensional Laplacian discussed in [G1]. But there are serious analytic problems in giving meaning to the infinite dimensional heat semigroup in function spaces defined by L^p norms. In [G1] this semigroup was investigated in spaces of bounded continuous functions (on an abstract Wiener space.) The heat semigroup, $e^{t\Delta/2}$, as well as the associated group $e^{it\Delta/2}$, have also been investigated in the context of white noise analysis, [Ku, Ob]. In this context the Segal-Bargmann transform has been extended from L^2 to a space of generalized functions. The extension is referred to as the S-transform.

5. Holomorphic Functions on the Finite Energy Path Group

Definition 5.1. Let G be a connected complex Lie group with complex Lie algebra $\mathfrak{g} := T_e(G)$ and let $(\ ,\ )$ be a given Hermitian inner product on $\mathfrak{g}$. The finite energy path group of G, denoted $H(G)$, is

$$H(G) = \Big\{g \in C([0,1],G); g \text{ is absolutely continuous,} \quad g(0) = e, \text{ and } \int_0^1 |g(s)^{-1}\dot{g}(s)|^2 ds < \infty\Big\}. \tag{5.1}$$

Here we are using $|\ |$ to denote the norm in $\mathfrak{g}$ associated to the given inner product, $\dot{g}(s) = dg(s)/ds$, and $g(s)^{-1}\dot{g}(s)$ is the matrix group notation that we will use to denote the left translate of the tangent vector $\dot{g}(s)$ back to the identity e. Similarly we will write

$$H(\mathfrak{g}) = \Big\{z \in C([0,1];\mathfrak{g}) : z \text{ is absolutely continuous,} \quad z(0) = 0, \text{ and } |z|^2 = \int_0^1 |\dot{z}(s)|^2 ds < \infty\Big\}. \tag{5.2}$$

In this section we will define a notion of holomorphic function on $H(G)$ and show its equivalence with the usual notion of holomorphic function on the complex Hilbert space $H(\mathfrak{g})$. G will not be assumed to be simply connected in this section except in Corollary 5.13.

Let us note first that $H(G)$ is a group with respect to the pointwise product

$$(g_1g_2)(s) = g_1(s)g_2(s). \tag{5.3}$$

For an element h in $H(\mathfrak{g})$ we will write

$$(e^h)(s) = \exp h(s) \qquad 0 \le s \le 1. \tag{5.4}$$

Then e^h is in $H(G)$.

Notation 5.2. There is a canonical map θ from $H(\mathfrak{g})$ onto $H(G)$ given by the solution $g = \theta(z)$ to the differential equation

$$g(s)^{-1}\dot{g}(s) = \dot{z}(s) \qquad g(0) = e. \tag{5.5}$$

The inverse of θ is simply given by

$$(5.6) \qquad (\theta^{-1}g)(s) = \int_0^s g(\sigma)^{-1}\dot{g}(\sigma)d\sigma.$$

We will regard θ^{-1} as a global coordinate system on the manifold $H(G)$ and use it to define the topology of $H(G)$.

Definition 5.3. A function $\varphi : H(G) \to \mathbb{C}$ is differentiable at a point $g \in H(G)$ if the function $h \to \varphi(ge^h)$ is Fréchet differentiable on the complex Hilbert space $H(\mathfrak{g})$ at $h = 0$ with a complex linear derivative. We denote the derivative by $\varphi'(g)$. Thus $\varphi'(g) \in H(\mathfrak{g})^*$, the space of continuous, complex linear, functionals on $H(\mathfrak{g})$. A function $\varphi : H(G) \to \mathbb{C}$ is holomorphic on $H(G)$ if $\varphi'(g)$ exists for each $g \in H(G)$ and is a continuous function of g into $H(\mathfrak{g})^*$.

Remark 5.4. Of course if θ^{-1} is to be regarded as a global coordinate system on $H(G)$ in the usual way then to say that φ is holomorphic on $H(G)$ should be interpreted to mean that $\varphi \circ \theta$ is a holomorphic function on the complex Hilbert space $H(\mathfrak{g})$. We will first prove that this notion of holomorphy is equivalent to that given in Definition 5.3.

Notation 5.5 An element k in $H(G)$ defines a right action on $H(\mathfrak{g})$ given by

$$(5.7) \qquad (z \cdot k)(s) = \int_0^s (Ad\, k(\sigma)^{-1})\dot{z}(\sigma)d\sigma \quad z \in H(\mathfrak{g}),\ k \in H(G).$$

One verifies readily that $z \cdot k$ is again in $H(\mathfrak{g})$ and that $z \cdot (k_1 k_2) = (z \cdot k_1) \cdot k_2$. In addition to the linear map $z \to z \cdot k$ we will need the affine map $T_k : H(\mathfrak{g}) \to H(\mathfrak{g})$ given by

$$(5.8) \qquad T_k z = \theta^{-1}(k) + z \cdot k, \quad k \in H(G), z \in H(\mathfrak{g})$$

We further define

$$(5.9) \qquad ((ad\, h)z)(s) = \int_0^s ((ad\, h(\sigma))\dot{z}(\sigma)d\sigma$$

These maps relate the group operation in $H(G)$ to the vector space operations in $H(\mathfrak{g})$ in accordance with the following elementary proposition, which is a variant of identities used frequently elsewhere.

Proposition 5.6. *Let $k \in H(G)$ and let $z \in H(\mathfrak{g})$. Then*

$$(5.10.) \qquad \theta(z)k = \theta(T_k z) \qquad k \in H(G), z \in H(\mathfrak{g})$$

and

$$\theta(z+h) = \theta(h \cdot \theta(z)^{-1})\theta(z) \qquad \text{for } h \text{ and } z \in H(\mathfrak{g}) \tag{5.11}$$

Proof: Let $g = \theta(z)$ and $h = \theta^{-1}(k)$. Then

$$\begin{aligned}(g(s)k(s))^{-1}\frac{d}{ds}(g(s)k(s)) &= k(s)^{-1}g(s)^{-1}(\dot{g}(s)k(s) + g(s)\dot{k}(s)) \\ &= (z \cdot k)^{\cdot}(s) + \dot{h}(s)\end{aligned}$$

which is (5.10). Now replace z by h in (5.10) and replace k by $\theta(z)$ to get $\theta(h)\theta(z) = \theta(z + h \cdot \theta(z))$.Then replace h by $h \cdot \theta(z)^{-1}$ to get (5.11). Q.E.D.

Theorem 5.7. *Let $\varphi : H(G) \to \mathbb{C}$ be a function and let $f = \varphi \circ \theta$. Then φ is holomorphic in the sense of Definition 5.3 if and only if f is holomorphic on $H(\mathfrak{g})$. When either is holomorphic we have*

$$\langle \varphi'(g), h \rangle = \left\langle Df(z), h - (ad\, h)z \right\rangle \tag{5.12}$$

if z and h are in $H(\mathfrak{g})$ and $g = \theta(z)$.

We will need the following elementary lemmas, which have been used frequently in the past. We will collect them here for the readers convenience.

Lemma 5.8. *For each element z in $H(\mathfrak{g})$ the map $h \to T_{e^h} z$ from $H(\mathfrak{g})$ into itself is Fréchet differentiable at $h = 0$ with derivative given by*

$$(d/dt)T_{e^{th}} z|_{t=0} = h - (ad\, h)z. \tag{5.13}$$

Moreover the right side of (5.13) is jointly continuous in z and h.

Proof: In view of (5.8) it suffices to show that at $t = 0$

$$d/dt(e^{-th(s)}(d/ds)e^{th(s)})\big|_{t=0} = \dot{h}(s)$$

and

$$(d/dt)Ad\, e^{-th(s)}\dot{z}(s) = -[h(s), \dot{z}(s)]$$

with both t derivatives existing in the $L^2([0,1], \mathfrak{g})$ sense and uniformly for $|h| \le 1$. But this is the content of [G4, Lemma 3.4] and its proof. The second term in (5.13) is jointly continuous in h and z because its $H(\mathfrak{g})$ norm is denominated by $M|h|_\infty|z|$ where $M = \sup\{\|ad\,\xi\|_{\mathfrak{g}\to\mathfrak{g}} : |\xi|_{\mathfrak{g}} = 1\}$. The sup norm $|h|_\infty$ satisfies $|h|_\infty \le |h|$. Q.E.D.

Lemma 5.9. *Let $g \in H(G)$. There exists $\epsilon > 0$ such that $\theta(h \cdot g)(s)$ lies in a canonical coordinate neighborhood of the identity of G for all $s \in [0,1]$ when $|h| < \epsilon$. For such h the function $s \to log(\theta(h \cdot g)(s))$ from $\{h; |h| < \epsilon\}$ into $H(\mathfrak{g})$ is Fréchet differentiable at $h = 0$ with directional derivative $h \cdot g$, which is itself jointly continuous in h and g.*

Proof: This lemma is essentially Lemma 4.12 of [G3]. The map $h \to h \cdot g$ is a bounded, complex linear map from $H(\mathfrak{g})$ into itself. So we have $|h \cdot g| \to 0$ as $|h| \to 0$. Hence we may apply [G3, Lemma 4.12] to conclude that $\theta(h \cdot g)(s)$ lies in a canonical coordinate neighorhood of e in G for all s in $[0,1]$ if $|h|$ is sufficiently small. [G3, Lemma 4.12] also shows that the function $s \to \log(\theta(k \cdot g)(s))$ is in $H(\mathfrak{g})$ and is Fréchet differentiable at $h = 0$ with directional derivative given by $h \to h \cdot g$. The joint continuity of $h \cdot g$ is a straightforward computation that uses, in part, the fact that if g_n is a sequence that converges in the $H(G)$ topology then g_n converges uniformly on [0,1]. Q.E.D.

Proof of Theorem 5.7. If f is holomorphic, $g = \theta(z)$, and $f = \varphi \circ \theta$ then

$$\varphi(gk) = \varphi(\theta(z)k) = \varphi(\theta(T_k z)) = f(T_k z)$$

So by Lemma 5.8 $\varphi(ge^h)$, which equals $f(T_{e^h} z)$, is Fréchet differentiable in h at $h = 0$ and (5.12) holds. Since the right side of (5.12) is jointly continuous in z and h, $\varphi'(g)$ is continuous in g.

Conversely suppose that φ is holomorphic.For any element $g \in H(G)$ define $((Ad\, g)h)(s) = (Ad\, g(s))h(s)$. For each $g \in H(G)$ $Ad\, g$ is a bounded linear operator on $H(\mathfrak{g})$. Therefore the map $h \to \varphi(e^h g) = \varphi(ge^{(Ad\, g^{-1})h})$ is differentiable at $h = 0$. But if $g = \theta(z)$ then, by Lemma 5.9 and Equation (5.11), we have, for small $|h|$,

$$\begin{aligned} f(z+h) &= \varphi(\theta(z+h)) = \varphi(\theta(h \cdot g^{-1})g) \\ &= \varphi(e^{log\theta(h \cdot g^{-1})}g) = \varphi(ge^{(Adg^{-1})log\theta(h\cdot g^{-1})}) \end{aligned}$$

which is a composition of three differentiable functions for each fixed z. Hence $f(z+h)$ is Fréchet differentiable in h at $h = 0$ with derivative $\langle Df(z), h\rangle = \langle \varphi'(g), (Ad\, g^{-1})(h \cdot g^{-1})\rangle$. The right side is easily seen to be continuous in g and h. Hence f is holomorphic. Q.E.D.

Definition 5.10. A function $\varphi : H(G) \to \mathbb{C}$ is an endpoint function if there exists a function $v : G \to \mathbb{C}$ such that

$$\varphi(g) = v(g(1)) \qquad \forall\, g \in H(G). \tag{5.14}$$

Notation 5.11. Denote by $H_0(\mathfrak{g})$ the space

$$H_0(\mathfrak{g}) = \{h \in H(\mathfrak{g}) : h(1) = 0\} \tag{5.15}$$

and by $H_0(G)$ the subgroup

$$H_0(G) = \{g \in H(G) : g(1) = e\}. \tag{5.16}$$

Furthermore for any function $\varphi : H(G) \to \mathbb{C}$ and any element $h \in H(\mathfrak{g})$ we define the directional derivative

$$(\partial_h \varphi)(g) = (d/dt)\varphi(ge^{th})\big|_{t=0} \tag{5.17}$$

whenever the derivative exists.

Proposition 5.12. *Let G be a connected Lie group with a given inner product on its Lie algebra. A function $\varphi : H(G) \to \mathbb{C}$ is an endpoint function if and only if*

$$\varphi(gg_0) = \varphi(g) \ \forall\ g \in H(G) \text{ and } \forall\ g_0 \in H_0(G). \tag{5.18}$$

Proof: If $\varphi(g)$ depends only on $g(1)$ then since $(gg_0)(1) = g(1)$ when $g_0 \in H_0(G)$ the necessity of (5.18) follows. Conversely, if (5.18) holds and g_1 and g_2 are in $H(G)$ and satisfy $g_1(1) = g_2(1)$ then $g_0 := g_1^{-1}g_2$ is in $H_0(G)$ and $\varphi(g_2) = \varphi(g_1 g_0) = \varphi(g_1)$. So $\varphi(g)$ depends only on $g(1)$. Q.E.D.

Theorem 5.13. *Let G be a connected complex Lie group. Suppose that $(\ ,\ )$ is a Hermitian inner product on its Lie algebra $\mathfrak{g}$. Regarding $\mathfrak{g}$ as a real Lie algebra let $\mathfrak{k}$ be a subalgebra of $\mathfrak{g}$. Assume*

a) $\mathfrak{k} + i\mathfrak{k} = \mathfrak{g}$

b) The connected subgroup K of G such that $T_eK = \mathfrak{k}$ is closed in G.

c) Every element in $H_0(G)$ is homotopic to an element of $H_0(K)$.

Let $\varphi \in \mathcal{H}(H(G))$. Then φ is an endpoint function if and only if

$$\varphi(gk) = \varphi(g) \quad \forall\ g \in H(G) \text{ and } \forall\ k \in H_0(K). \tag{5.19}$$

Proof: Since $H_0(K) \subset H_0(G)$ (5.19) is necessary by Proposition 5.12. Assume now that (5.19) holds. Let $h \in H_0(\mathfrak{k})$. Then $e^{th} \in H_0(K)$. So $\varphi(ge^{th}) = \varphi(g)$ for all real t. Therefore $(\partial_h\varphi)(g) = 0$ for all h in $H_0(\mathfrak{k})$. But since φ is holomorphic $(\partial_h\varphi)(g)$ is complex linear in h. In view of hypothesis a) we can therefore assert that $(\partial_h\varphi)(g) = 0$ for all h in $H_0(\mathfrak{g})$ and for all g in $H(G)$. Hence for any element h in $H_0(\mathfrak{g})$ we have

$$d\varphi(ge^{sh})/ds = d\varphi(ge^{sh}e^{th})/dt\big|_{t=0} = (\partial_h\varphi)(ge^{sh}) = 0.$$

Therefore $\varphi(ge^{sh}) = \varphi(g)$ for all real s and in particular for $s = 1$. By induction it now follows that

$$\varphi(ge^{h_1}\cdots e^{h_n}) = \varphi(g) \quad \forall\ g \in H(G) \text{ and } h_j \in H_0(\mathfrak{g})\ j = 1, \ldots, n. \tag{5.20}$$

Now $H_0(G)$ is a closed subgroup of $H(G)$ and a neighborhood of the identity is covered by elements of the form e^h, $h \in H_0(\mathfrak{g})$. Thus if $H_0^0(G)$ denotes the connected component of the identity in $H_0(G)$ then any element of $H_0^0(G)$ is a finite product $e^{h_1} \cdots e^{h_n}$ as in (5.20). Thus

$$\varphi(gg_0) = \varphi(g) \quad \forall\, g \in H(G) \text{ and } \forall\, g_0 \in H_0^0(G). \tag{5.21}$$

By Proposition 5.12 it suffices to prove that $\varphi(gg_1) = \varphi(g)$ for all g in $H(G)$ and for all g_1 in $H_0(G)$. But if $g_1 \in H_0(G)$ then by assumption c) g_1 is homotopic to some element k in $H_0(K)$. That is, there is a continuous map $u : [0,1] \times [0,1] \to G$ such that $u(0,s) = k(s)$, $u(1,s) = g_1(s)$ and $u(t,0) = u(t,1) = e$. By standard smoothing arguments we can assume further that $u(t,\cdot) \in H_0(G)$ for each t. Then the loop $g_0(s) := g_1(s)k(s)^{-1}$ is homotopic to the constant map $e(\cdot)$ via the homotopy $u(t,s)k(s)^{-1}$. Hence g_0 is in $H_0^0(G)$. Since $g_1 = g_0 k$ (5.19) and (5.21) give $\varphi(gg_1) = \varphi(gg_0k) = \varphi(gg_0) = \varphi(g)$. Q.E.D.

Corollary 5.14. *Let K be a connected, simply connected Lie group with an Ad K invariant inner product $\langle\ ,\ \rangle$ on its Lie algebra. Let $G \supset K$ be its complexification. Extend $\langle\ ,\ \rangle$ to $\mathfrak{g}$ as a Hermitian inner product $(\ ,\ )$ and define $H(G)$ and $H(K)$ as above. Let $\varphi : H(G) \to \mathbb{C}$ be holomorphic. Then φ is an endpoint function if and only if condition (5.19) holds. Equivalently, define $f = \varphi \circ \theta$. Then φ is an endpoint function if and only if*

$$f(T_k z) = f(z)\ \forall\, z \in H(\mathfrak{g}) \text{ and } \forall\, k \in H_0(K). \tag{5.22}$$

Proof: In view of Equation (5.10) the conditions (5.19) and (5.22) are equivalent. The hypotheses of Theorem 5.13 are all satisfied. In fact Condition c) of Theorem 5.13 holds because every element of $H_0(G)$ is homotopic to the constant function $e(\cdot)$. So Theorem 5.13 is applicable. Q.E.D.

Let us remark that if K is compact and connected, but not simply connected, and G is its complexification, then Theorem 5.13 is still applicable and Corollary 5.14 still holds.

6. Isometric Embedding of Holomorphic Endpoint Functions

We continue the notation of Section 5. In this section the connected complex Lie group G that we will consider need not be simply connected nor need it be the complexification of some other group. The given inner product $(\ ,\ )$ on its Lie algebra need not be Ad G invariant. If $\mathfrak{g}$ has complex dimension d then it is a real inner product space of real dimension $2d$ with respect to the inner product $\mathrm{Re}(\xi, \eta)$. Let $e_1, \ldots, e_{2d}$ be any orthonormal basis of

$(\mathfrak{g}, \mathrm{Re}(\cdot,\cdot))$. For an element $\xi \in \mathfrak{g}$ denote by $\tilde{\xi}$ the left invariant vector field on G which extends ξ. Define a Laplacian, Δ, on G by the equation

$$\Delta\varphi = \sum_{j=1}^{2d} (\tilde{e}_j)^2\varphi, \quad \varphi \in C^\infty(G). \tag{6.1}$$

dx will denote a right invariant Haar measure. The restriction $\Delta \mid C_c^\infty(G)$ defines an essentially self–adjoint operator, which we again denote by Δ. This operator is negative and generates a Hermitian semigroup $e^{t\Delta/4}$ of contraction operators on $L^2(G, dx)$. It is given by convolution by a heat kernel μ_t : $(e^{t\Delta/4}\varphi)(x) = \int_G \varphi(xy^{-1})\mu_t(y)dy$. For each $t > 0$, μ_t is in $C^\infty(G)$, is strictly positive, is a probability density on G , and satisfies the heat equation

$$\partial\mu_t(x)/\partial t = (\Delta/4)\mu_t(x), \quad t > 0 \tag{6.2}$$

with initial condition

$$\int_G \varphi(x)\mu_t(x)dx \to \varphi(e) \quad \text{as} \quad t \downarrow 0 \quad \text{for} \quad \varphi \in C_c(G). \tag{6.3}$$

These properties are discussed further in [DG]. We will write $\mathcal{H}L^2(G, \mu_t)$ for the Hilbert space of holomorphic functions on G which are in L^2 with respect to the probability measure $\mu_t(x)dx$.

Theorem 6.1. *Let $u \in \mathcal{H}(G)$. Define θ as in Notation 5.2 and define*

$$f(z) = u(\theta(z)(1)) \quad \textit{for} \quad z \in H(\mathfrak{g}). \tag{6.4}$$

Then

$$\|f\|_{\mathcal{H}^t} = \|u\|_{L^2(G,\mu_t)} \qquad u \in \mathcal{H}(G),\ t > 0. \tag{6.5}$$

Remark 6.2. Although Equation (6.5) does not involve the "power series" of either f or u we are going to give a proof of that equation by computing both sides in terms of their "power series expansion coefficients." Other, more direct, avenues of proof that we have pursued seem to require as much work and yield less generality. We will rely heavily on the following result in [DG], which expresses $\|u\|_t^2$ in terms of the "expansion coefficients" (i.e., all derivatives) of u at the identity. A more direct proof of Theorem 6.1 would be desirable.

Lemma 6.3. *[DG] Let $u \in \mathcal{H}(G)$ and let α_k be the unique tensor in $(\mathfrak{g}^*)^{\otimes k}$ satisfying*

$$\langle \alpha_k, \xi_1 \otimes \cdots \otimes \xi_k \rangle = (\tilde{\xi}_1 \cdots \tilde{\xi}_k u)(e),\ \forall\, \xi_j \in \mathfrak{g},\ j = 1, \ldots, k,\ k = 1, 2, \ldots \tag{6.6}$$

and $\alpha_0 = u(e)$. *Then*

$$(6.7) \qquad \int_G |u(x)|^2 \mu_t(x)dx = \sum_{k=0}^{\infty} (t^k/k!)|\alpha_k|^2_{(\mathfrak{g}^*)^{\otimes k}} \quad u \in \mathcal{H}(G),\ t > 0.$$

Proof: This is the assertion of Theorem 2.5 in [DG]. (See also its proof in Section 5 of [DG].) Note that we have not made any statement about the finiteness of either side of (6.7). This means, of course, that one side is finite if and only if the other side is finite. Q.E.D.

Lemma 6.4. *Let* $u \in \mathcal{H}(G)$ *and define* α_k *as in the preceding lemma. Let* $f(z) = u(\theta(z)(1))$ *for* $z \in H(\mathfrak{g})$. *Then the power series for* f *is given by*

$$(6.8) \qquad f(z) = \sum_{k=0}^{\infty} \langle \alpha_k, \psi_k(z) \rangle$$

where

$$(6.9) \qquad \psi_k(z) = \int_{\Delta_k} \dot{z}(s_1) \otimes \cdots \otimes \dot{z}(s_k) ds_1 \cdots ds_k \in \mathfrak{g}^{\otimes k}$$

and

$$(6.10) \qquad \Delta_k = \{(s_1, \ldots, s_k) : 0 < s_1 < s_2 < \cdots < s_k < 1\}.$$

Proof: For a self-contained and short proof of this lemma the reader is referred to Proposition 5.1 in [D]. Although the proposition is stated there for smooth z the proof is equally valid for $z \in H(\mathfrak{g})$. A stochastic, L^2 version of this lemma was derived in [G4] (see Theorem 5.1). But the proof is longer and more combinatoric than the insightful proof given in [D]. Q.E.D.

Proof of Theorem 6.1: We are going to prove (6.5) by computing the norm of $(1 - D)_0^{-1} f$ in $\mathcal{F}^t(H(\mathfrak{g}))$ and then applying Theorem 3.6. The structure of $H(\mathfrak{g})$ as a function space allows us to describe the symmetric tensors over $H(\mathfrak{g})$ and over $H(\mathfrak{g})^*$ in an explicit manner — one which is common in the treatment of multiple Ito integrals. We will follow the discussion in [G4], Section 4.

In view of the definition (5.2) of $H(\mathfrak{g})$ we may and will identify $H(\mathfrak{g})$ with $L^2([0,1];\mathfrak{g})$. Then we can identify $L^2([0,1];\mathfrak{g})^{\otimes k}$ with $L^2([0,1]^k;\mathfrak{g}^{\otimes k})$. The symmetric tensors can then be characterized as follows. If τ is in the permutation group S_k and $\mathbf{s} = (s_1, \ldots, s_k)$ is in $[0,1]^k$ define $\tau\mathbf{s} = (s_{\tau^{-1}(1)}, \ldots, s_{\tau^{-1}(k)})$ and define an action of τ on $\mathfrak{g}^{\otimes k}$ by $\tau(\xi_1 \otimes \cdots \otimes \xi_k) = \xi_{\tau^{-1}(1)} \otimes \cdots \otimes \xi_{\tau^{-1}(k)}$. The natural action of τ on a function φ in $L^2([0,1]^k;\mathfrak{g}^{\otimes k})$ is given by

$$(6.11) \qquad (P_\tau \varphi)(\mathbf{s}) = \tau\varphi(\tau^{-1}\mathbf{s}).$$

A symmetric k–tensor over $L^2([0,1];\mathfrak{g})$ corresponds then to a function φ in $L^2([0,1]^k;\mathfrak{g}^{\otimes k})$ satisfying

$$P_\tau \varphi = \varphi \quad \text{for all} \quad \tau \quad \text{in} \quad S_k. \tag{6.12}$$

Taking Δ_k to be the open simplex defined in Equation (6.10) it is clear that if φ is symmetric, i.e. satisfies (6.12), then it is determined up to a set of measure zero in $[0,1]^k$ by its restriction to Δ_k. Moreover any function ψ in $L^2(\Delta_k, \mathfrak{g}^{\otimes k})$ has a unique (up to a set of Lebesgue measure zero) extension to a symmetric function φ in $L^2([0,1]^k;\mathfrak{g}^{\otimes k})$ and their norms are related by

$$\|\varphi\|^2_{L^2([0,1]^k;\mathfrak{g}^{\otimes k})} = k! \int_{\Delta_k} |\psi(s_1, \dots, s_k)|^2 ds_1 \cdots ds_k \tag{6.13}$$

because $[0,1]^k$ is (up to a set of measure zero) the disjoint union $\bigcup_{\tau \in S_k} \tau\Delta_k$, while τ acts unitarily on $\mathfrak{g}^{\otimes k}$. These two descriptions of the symmetric k–tensors, in terms of φ or its restriction $\psi = \varphi \mid \Delta_k$, are both useful. The same discussion applies of course to symmetric tensors over the dual space $H(\mathfrak{g})^*$, which we will identify with $L^2([0,1];\mathfrak{g}^*)$. Similarly the bilinear pairing between a symmetric function w in $L^2([0,1]^k;(\mathfrak{g}^*)^{\otimes k})$ and the symmetric function φ reduces to an integral over Δ_k by the identity

$$\int_{[0,1]^k} \langle w(\mathbf{s}), \varphi(\mathbf{s})\rangle d\mathbf{s} = k! \int_{\Delta_k} \langle w(\mathbf{s}), \varphi(\mathbf{s})\rangle d\mathbf{s}.$$

Now suppose that f is given by (6.4) and that $w = (1-D)_0^{-1} f$. Then $w_k = D^k f(0)$. To identify the coefficient w_k we need only look at the homogeneous terms of degree k in (6.9). To this end define v_k to be the symmetric (cf. (6.12)) function in $L^2([0,1]^k;(\mathfrak{g}^*)^{\otimes k})$ such that

$$v_k(\mathbf{s}) = \alpha_k \quad \text{for} \quad \mathbf{s} \in \Delta_k, \quad k = 0, 1, 2, \dots. \tag{6.14}$$

(Note that although v_k is constant on each simplex $\tau\Delta_k$ it may have different values on different simplices.) Now $z^{\otimes k}$ is given by the symmetric function $(s_1, \dots, s_k) \to \dot{z}(s_1) \otimes \cdots \otimes \dot{z}(s_k)$, which takes values in $\mathfrak{g}^{\otimes k}$.

Using symmetry, as before, to reduce to an integral over Δ_k we find

(6.15)

$$\begin{aligned}(1/k!)\langle v_k, z^{\otimes k}\rangle &= (1/k!) \int_{[0,1]^k} \langle v_k(\mathbf{s}), \dot{z}(s_1) \otimes \cdots \otimes \dot{z}(s_k)\rangle d\mathbf{s} \\ &= \int_{\Delta_k} \langle \alpha_k, \dot{z}(s_1) \otimes \cdots \otimes \dot{z}(s_k)\rangle d\mathbf{s} \\ &= \langle \alpha_k, \psi_k(z)\rangle\end{aligned}$$

where ψ_k is given by (6.9). (The reader should observe that the three angular brackets on the right refer to the pairing between $(\mathfrak{g}^*)^{\otimes k}$ and $\mathfrak{g}^{\otimes k}$.) Comparing (6.15) with Equation (6.8) we find that

$$f(z) = \sum_{k=0}^{\infty} (1/k!)\langle v_k, z^{\otimes k}\rangle. \tag{6.16}$$

Hence

$$w_k := D^k f(0) = v_k. \tag{6.17}$$

Since the action of τ on $(\mathfrak{g}^*)^{\otimes k}$ is unitary for each τ in S_k we have $|w_k(\mathbf{s})|^2 = |\alpha_k|^2$ a.e. on $[0,1]^k$. Hence $|w_k|^2 = \int_{[0,1]^k} |w_k(\mathbf{s})]^2 ds = |\alpha_k|^2$ for $k = 0, 1, 2, \ldots$. Therefore

$$\begin{aligned} \|w\|_t^2 &= \sum_{k=0}^{\infty} (t^k/k!)|w_k|^2 \\ &= \sum_{k=0}^{\infty} (t^k/k!)|\alpha_k|^2 \\ &= \|u\|^2_{L^2(G,\mu_t)} \end{aligned}$$

wherein we have used Lemma 6.3 in the last step. Applying Theorem 3.6 now yields (6.5) for any function u in $\mathcal{H}(G)$. Q.E.D.

Corollary 6.5. . *Denote by $\mathcal{H}_e^t$ the subspace of $\mathcal{H}^t$ consisting of endpoint functions. Then $\mathcal{H}_e^t$ is a closed subspace of $\mathcal{H}^t$ and the endpoint map defined in (6.4) is a unitary operator from $\mathcal{H}L^2(G, \mu_t)$ onto $\mathcal{H}_e^t$.*

Proof. The image of the endpoint map is a closed subspace of $\mathcal{H}^t$ by (6.5). Moreover if f is a holomorphic endpoint function then by Theorem 5.7 it has the form (6.4) for some function $u \in \mathcal{H}(G)$. If in addition f is in $\mathcal{H}_e^t$ then by (6.5) $\|u\|_{L^2(G,\mu_t)} < \infty$. So f is in the range of the endpoint map. Q.E.D.

7. Transfer of Endpoint Functions

In this section K will denote a connected, simply connected Lie group of compact type. $\langle\ ,\ \rangle$ will denote an $Ad\,K$ invariant inner product on its Lie algebra $\mathfrak{k}$. We denote by G the complexification of K and by $\mathfrak{g}$ the Lie algebra of G. Then $\mathfrak{g}$ is the complexification of $\mathfrak{k}$. We adopt the notation given in the Introduction. Denote by $(\ ,\ )$ and $\langle\ ,\ \rangle$ respectively the Hermitian inner product on $\mathfrak{g}$ and bilinear form on $\mathfrak{g}$ that extends the inner product on $\mathfrak{k}$. Then $H(\mathfrak{g})$ is naturally the complexification of $H(\mathfrak{k})$: every element of $H(\mathfrak{g})$ is uniquely of the form $h_1 + \sqrt{-1}h_2$ where h_1 and h_2 are in $H(\mathfrak{k})$.

We are going to apply the Segal–Bargmann transform S_t of Section 4 with $H_r = H(\mathfrak{k})$, $H_c = H(\mathfrak{g})$ and $t = 1$. Our goal in this section is to show that S_1 takes endpoint functions of K valued Brownian motion to holomorphic endpoint functions of G valued Brownian motion, thereby establishing a map from functions on K to holomorphic functions on G. We will then show that this map is exactly the Hall transform.

We have already discussed holomorphic endpoint functions of G valued finite energy paths in the previous two sections. In the present section we will first give the corresponding discussion for the K valued Brownian motion. There are significant technical differences arising from the fact that for the G valued process we have available the Bargmann-Krée–Segal skeleton theorem, which enables us to use finite energy paths and the rather simple characterization theorem of endpoint functions given in Theorem 5.13 and Corollary 5.14 . For the K valued process we will need a similar characterization of endpoint functions. But this time the characterization theorem is equivalent to a deep ergodicity theorem (cf. Lemma 7.5 and its proof.) The main theorem of this section is Theorem 7.13, which is Hall's theorem [Hal] as extended to the compact type case by B. Driver [D]. Let p denote Wiener measure on $\mathfrak{k}$ valued Wiener space, $W(\mathfrak{k})$, over $[0,1]$. The coordinate process X is then a $\mathfrak{k}$ valued Brownion motion. The Stratonovich stochastic differential equation

$$dk(s) = k(s) \circ dX(s) \qquad k(0) = e \tag{7.1}$$

determines the Ito map $\widetilde{\theta} : W(\mathfrak{k}) \to W(K) := \{k \in C([0,1]; K) : k(0) = e\}$ as an almost everywhere defined measurable function. The induced measure $P = \widetilde{\theta}_* p$ on $W(K)$ is the path space measure for the K valued Markov process with transition semigroup $\exp(t\Delta/2)$ where Δ is the Laplacian on K associated with the given $Ad\, K$ invariant inner product on $\mathfrak{k}$ as described in the Introduction.

We are going now to extend some of the results of Section 5 from the smooth category to the measure theoretic category.

<u>Notation 7.1</u>. Let $k \in H(K)$ and define

$$(T_k X)(s) = k(s)^{-1}\dot{k}(s) + \int_0^s (Ad\, k(\sigma)^{-1}) dX(\sigma) \qquad X \in W(\mathfrak{k}). \tag{7.2}$$

The integral can be interpreted as a stochastic integral, in which case $T_k X$ is defined for almost every X in $W(\mathfrak{k})$. But actually the integrand $Ad\, k(\sigma)^{-1}$ is absolutely continuous in σ with derivative in L^2. So by integration by parts one sees that the second term in (7.2) can be defined for all X in $W(\mathfrak{k})$ and is linear in X. Clearly $T_k X$ is again in $W(\mathfrak{k})$. The map T_k has been studied extensively [AHK, G2, MM, Shi1]. In the next two lemmas we will summarize its properties.

Lemma 7.2. *[AHK, Shi1, MM]. For any k in $H(K)$ T_k preserves the Wiener measure class.*

Lemma 7.3. *Let k be in $H(K)$. Then*

$$\widetilde{\theta}(T_k X) = \widetilde{\theta}(X)k \quad \text{a.s.} \quad [p]. \tag{7.3}$$

For further discussion of these two lemmas see e.g. [G3, pp.285,286].

Definition 7.4. A function ψ in $L^2(W(K), P)$ is an endpoint function if there exists a measurable function $u : K \to \mathbb{C}$ such that

$$\psi(g) = u(g(1)) \text{ for a.e. } g \text{ in } W(K). \tag{7.4}$$

Note that if k is in $H(K)$ then (by Lemmas 7.2 and 7.3) the map $g \to gk$ preserves the P measure class. Hence $\psi(gk)$ is a well defined random variable on $(W(K), P)$. If moreover ψ is an endpoint function, e.g. given by (7.4), and $k \in H_0(K)$ then

$$\psi(gk) = u(g(1)k(1)) = u(g(1)) = \psi(g) \text{ a.s. } P. \tag{7.5}$$

The converse of this observation is an ergodicity theorem, which is the principle input to the present paper. We have the following lemma.

Lemma 7.5. *(Ergodicity lemma [G4]) Let $\psi : W(K) \to \mathbb{C}$ be a measurable function in $L^2(W(K), P)$. Then ψ is an endpoint function if and only if for each function $k \in H_0(K)$*

$$\psi(gk) = \psi(g) \text{ a.s. } P. \tag{7.6}$$

Proof: This lemma is the main result of [G4]. The proof given in [G4] involves multiple Ito expansions of $\psi \circ \widetilde{\theta}$ in a very direct way and requires, and yields,a proof also of the isomorphism $U_t : L^2(K, \rho_t) \to J_t^0$ described in the Introduction of the present paper. This isomorphism is very close in spirit to Hall's transform — which we wish to prove in this paper. It is therefore fortuitous that recently a new proof of the ergodicity theorem, and therefore of the present lemma, has been found. The new proof, by Gaku Sadasue [Sad], is much shorter and has a more fundamental character. It is based on quasi–sure analysis [AM] rather than on the multiple Ito expansions used in [G4].

It will be convenient to restate the preceeding lemma in terms of the pullback of ψ to $W(\mathfrak{k})$.

Lemma 7.6. *Let $\psi \in L^2(W(K), P)$ and let $F = \psi \circ \widetilde{\theta}$. Then ψ is an endpoint function if and only if*

$$F \circ T_k = F \text{ a.e. on } W(\mathfrak{k}) \text{ for each } k \text{ in } H_0(K). \tag{7.7}$$

Proof: If k is in $H_0(K)$ then $(F \circ T_k)(X) = \psi(\widetilde{\theta}(T_k X)) = \psi(\widetilde{\theta}(X)k)$ a.s. $[p]$. So the condition (7.7) on F is equivalent to the condition (7.6) on ψ. Q.E.D.

Next, we will explain how the Segal–Bargmann transform, defined in Section 4 over an arbitrary real Hilbert space H_r, is realized in our present setting of $\mathfrak{k}$ valued Brownian motion. Write p_1 for the Gaussian cylinder set measure on $H(\mathfrak{k})$ with time parameter $t = 1$. If $e_1, \ldots, e_n$ is an orthonormal set in $H(\mathfrak{k})$ and $\varphi : \mathbb{R}^n \to \mathbb{C}$ is Borel measurable then the function $f(x) := \varphi(\langle x, e_1\rangle, \ldots, \langle x, e_n\rangle)$, defined for x in $H(\mathfrak{k})$, is a typical cylinder function based on span$(e_1, \ldots, e_n)$. Since the linear functionals $\{\langle x, e_j\rangle\}_{j=1}^n$ have the same joint (Gaussian) distribution with respect to p_1 as the stochastic integrals, $\left\{\int_0^1 \langle \dot{e}_j(s), dX(s)\rangle\right\}_{j=1}^n$, have with respect to p, the map $f \to \widetilde{f}$, where

$$(7.8) \qquad \widetilde{f}(X) = \varphi\Big(\int_0^1 \langle \dot{e}_1(s), dX(s)\rangle, \ldots, \int_0^1 \langle \dot{e}_n(s), dX(s)\rangle\Big),$$

determines an isometry of $L_0^q(H(\mathfrak{k}), p_1)$ into $L^q(W(\mathfrak{k}), p)$ for $1 \le q \le \infty$. In particular the completion of $L_0^2(H(\mathfrak{k}), p_1)$ can be naturally identified with $L^2(W(\mathfrak{k}), p)$. The definition (4.8) of S_t for $t = 1$ can now be written for any function f in $L^2(W(\mathfrak{k}), p)$ as

$$(7.9) \qquad (S_1 f)(z) = e^{-\langle z,z\rangle/2} \int_{W(\mathfrak{k})} f(X) e^{\int_0^1 \langle \dot{z}(s), dX(s)\rangle} p(dX), \;\; z \in H(\mathfrak{g}).$$

Note that the exponential factor in the integrand is in $L^q(W(\mathfrak{k}), p)$ for all $q < \infty$. Theorem 4.8 of Segal and Bargmann can be restated in the present context thus:

Lemma 7.7. *(Segal-Bargmann Theorem.) The map S_1 defined by (7.9) is a unitary operator from $L^2(W(\mathfrak{k}), p)$ onto $\mathcal{H}^1(H(\mathfrak{g}))$.*

For the next theorem we define

$$(7.10) \qquad T_k^c z = \theta^{-1} k + z \cdot k \qquad z \in H(\mathfrak{g}), \;\; k \in H(K).$$

This is the same map denoted T_k in (5.8). But we will emphasize the complex extension here to be consistent with the notation of Theorem 4.11 and to avoid confusion with the map T_k of Notation 7.1.

Theorem 7.8. *Let f be in $L^2(W(\mathfrak{k}), p)$ and let k be in $H(K)$. Then $f \circ T_k$ is in $L^q(W(\mathfrak{k}), p)$ for all $q < 2$ and*

$$(7.11) \qquad (S_1(f \circ T_k))(z) = (S_1 f)(T_k^c z) \text{ for all } z \in H(\mathfrak{g}), \;\; k \in H(K).$$

Proof: First note that the map $x \to x \cdot k$ is an orthogonal transformation on $H(\mathfrak{k})$. So the map $x \to h + x \cdot k$ is a rotation followed by a translation. By Theorem 4.11 and Remark 4.12 S_1 commutes with both translations and rotations and so (7.11) ought to hold. There is a minor technical issue arising from the fact that T_k is, and must be, defined on (almost all of) $W(\mathfrak{k})$, cf. (7.2), while the orthogonal transformations discussed in Theorem 4.11 are defined on $H(\mathfrak{k})$ — which is a set of measure zero in $W(\mathfrak{k})$. To see that the preceding heuristic argument is nevertheless correct let us consider first the rotation part of T_k. Let $(T_k^0 X)(s) = \int_0^s (Ad\, k(\sigma)^{-1}) dX(\sigma)$, which as we noted after Notation 7.1 is defined for all X in $W(\mathfrak{k})$. Let $(Ox)(s) = \int_0^s Ad\, k(\sigma)^{-1} \dot{x}(\sigma) d\sigma$ for $x \in H(\mathfrak{k})$. For y in $H(\mathfrak{k})$ we can compute readily that $\langle y, Ox \rangle = \langle y \cdot k^{-1}, x \rangle$. So the extended linear functional $x \to \langle y \cdot k^{-1}, x \rangle$ is, by (7.8), $\langle y \cdot k^{-1}, \cdot \rangle^{\sim}(X) = \int_0^1 (y \cdot k^{-1})^{\cdot}(s) dX(s) = \int_0^1 \langle \dot{y}(\sigma), Ad\, k(\sigma)^{-1} dX(\sigma) \rangle$ (stochastic integral) $= \int_0^1 \langle \dot{y}(\sigma), d(T_k^0 X)(\sigma) \rangle$. Applying this to each of the arguments of $\tilde{f}$ in (7.8) we have $(f \circ O)^{\sim}(X) = \tilde{f}(T_k^0 X)$ a.e. In this sense the orthogonal invariance of S_1 of Theorem 4.11 goes over to invariance under the transformation T_k^0. A similar (but simpler) argument applies to translations giving (7.11) for square integrable cylinder functions. But for fixed $z \in H(\mathfrak{g})$ both sides are continuous in f in $L^2(W(\mathfrak{k}), p)$ norm — the left side because $f \to f \circ T_k$ is continuous from L^2 to L^q for any $q \in (1,2)$ while $(S_1 g)(z)$ is continuous in g in L^q norm for any $q \in (1,2)$. Q.E.D.

Corollary 7.9. *Let ψ be in $L^2(W(K), P)$ and let $F = \psi \circ \tilde{\theta}$. Then $S_1 F$ is an endpoint function for G if and only if ψ is an endpoint function for K.*

Proof: If ψ is an endpoint function then $F \circ T_k = F$ for all k in $H_0(K)$ by Lemma 7.6 . Hence, by Theorem 7.8 $(S_1 F)(T_k^c z) = (S_1(F \circ T_k))(z) = (S_1 F)(z)$ for all $k \in H_0(K)$ and all z in $H(\mathfrak{g})$. Since $S_1 F$ is holomorphic, $S_1 F$ is therefore an endpoint function for G by Corollary 5.14 . Conversely, if $S_1 F$ is an endpoint function for G then $(S_1 F) \circ T_k^c = S_1 F$ for all $k \in H_0(K)$. By Theorem 7.8 we therefore have $S_1(F \circ T_k - F)(z) = 0$ for all z in $H(\mathfrak{g})$. Now $F \circ T_k - F$ is in $L^q(W(\mathfrak{k}), p)$ for some $q \in (1,2)$. Hence by the uniqueness theorem (Proposition 4.7, Part d)) $F \circ T_k - F = 0$. By the ergodicity lemma, Lemma 7.6, ψ is an endpoint function. Q.E.D.

Notation 7.10. Let u be in $L^2(K, \rho_1)$. Denote the corresponding endpoint function on $W(K)$ by u^e. Thus

$$u^e(g) = u(g(1)), \quad g \in W(K), \quad u \in L^2(K, \rho_1).$$

Since the distribution of $g(1)$ in K with respect to P has density ρ_1 the map $u \to u^e$ is isometric from $L^2(K, \rho_1)$ into $L^2(W(K), P)$. By the previous corollary $S_1 u^e$ is an endpoint function for G. That is, there is a unique holomorphic function $v : G \to \mathbb{C}$ such that

$$(S_1 u^e)(z) = v(\theta(z)(1)). \tag{7.12}$$

Define

$$Vu = v. \tag{7.13}$$

Since $u \to u^e \to S_1 u^e$ are isometric and $v \to v(\theta(z)(1))$ is, by Corollary 6.5, unitary from $\mathcal{H}L^2(G, \mu_1)$ onto $\mathcal{H}^1_e(H(\mathfrak{g}))$, V is an isometry from $L^2(K, \rho_1)$ into $\mathcal{H}L^2(G, \mu_1)$.

Corollary 7.11. *The map V defined in (7.13) is unitary.*

Proof: Only the surjectivity has not been established so far. If v is in $\mathcal{H}L^2(G, \mu_1)$ and v^e is the corresponding endpoint function then $v^e \circ \theta$ is in $\mathcal{H}^1_e(H(\mathfrak{g}))$ by Corollary 6.5. Hence $F := S_1^{-1} v^e$ is the pullback, $F = \psi \circ \widetilde{\theta}$, of an endpoint function ψ, by Corollary 7.9. So $\psi = u^e$ for some function u in $L^2(K, \rho_1)$ and therefore $v = Vu$. Q.E.D.

Corollary 7.12. *Let u be in $L^2(K, \rho_1)$ and let $v = Vu$. Then*

$$v(a) = (e^{\Delta/2} u)(a) \qquad a \in K. \tag{7.14}$$

Proof: Choose k in $H(K)$ with $k(1) = a$. Let $h = \theta^{-1} k$. Then $v(a) = v(\theta(h)(1)) = (S_1(u^e \circ \widetilde{\theta}))(h)$ since v depends only on the endpoint of $\theta(h)$. Since h is "real", i.e., is in $H(\mathfrak{k})$, equation (7.9) may be applied in the form

$$v(a) = \int_{W(\mathfrak{k})} u(\widetilde{\theta}(h - X)(1)) p(dX). \tag{7.15}$$

Now the map $X \to -X \cdot \theta(h)$ is orthogonal as a map from $H(\mathfrak{k})$ to itself, hence its extension to $W(\mathfrak{k})$ is measure preserving. (See the discussion in the proof of Theorem 7.8.) Thus in the last displayed equation we may replace $h - X$ by $h + X \cdot \theta(h)$. So

$$\begin{aligned} v(a) &= \int_{W(\mathfrak{k})} u(\widetilde{\theta}(T_k X)(1)) p(dX) \\ &= \int_{W(\mathfrak{k})} u(\widetilde{\theta}(X)(1) k(1)) p(dX) \\ &= \int_{W(\mathfrak{k})} u(\widetilde{\theta}(X) a) p(dX) \\ &= (e^{\Delta/2} u(\cdot a))(e) \\ &= (e^{\Delta/2} u)(a) \end{aligned}$$

since $e^{\Delta/2}$ is a translation invariant operator. Q.E.D.

Theorem 7.13. *(Hall [Hal], Driver [D]) Let $u \in L^2(K, \rho_1)$. Then $e^{\Delta/2} u$ has a unique analytic continuation v to G. The map $u \to v$ is a unitary operator from $L^2(K, \rho)$ onto $\mathcal{H}L^2(G, \mu)$.*

Proof: Define $v = Vu$. By the previous corollary $v = e^{\Delta/2} u$ on K, so v is the unique [Hal] analytic continuation of $e^{\Delta/2} u$ to G. Since V is unitary the theorem follows. Q.E.D.

References

[AHK] Albeverio, S. and Hoegh-Krohn, R., *The energy representation of Sobolev Lie groups*, Compositio Math. **36** (1978), 37–52.

[AM] Airault, H. and Malliavin, P., *Integration geometrique sur l'espace de Wiener*, Bull. Sc. Mathematique **112** (1988), 3–52.

[BSZ] Baez, John C., Segal, Irving E., Zhou, Zhengfang, *"Introduction to Algebraic and Constructive Quantum Field Theory"*, Princeton University Press, Princeton, New Jersey, 1992.

[B1] Bargmann, V., *On a Hilbert space of analytic functions and an associated integral transform, Part I*, Communications of Pure and Applied Mathematics **24** (1961), 187–214.

[B2] Bargmann, V., *Remarks on a Hilbert space of analytic functions*, Proc. of the National Academy of Sciences **48** (1962), 199–204.

[B3] Bargmann, V., *Acknowledgement*, Proc. of the National Academy of Sciences **48** (1962), 2204.

[Ca] Cameron, R. H., *Some examples of Fourier–Wiener transforms of analytic functionals*, Duke Math.J. **12** (1945), 485–488.

[CM1] Cameron, R. H. and Martin, W. T., *Fourier–Wiener transforms of analytic functionals*, Duke Math. J. **12** (1945), 489–507.

[CM2] Cameron, R. H. and Martin, W. T., *Fourier–Wiener transforms of functionals belonging to L_2 over the space C*, Duke Math. J. **14** (1947), 99–107.

[Car] Carlen, Eric A., *Some integral identities and inequalities for entire functions and their applications to the coherent state transform*, J. of Funct. Anal. **97** (1991), 231–249.

[Co] Cook, J., *The mathematics of second quantization*, Trans. Amer. Math. Soc. **74** (1953), 222–245.

[D] Driver, B. K., *On the Kakutani–Itô–Segal–Gross and the Segal–Bargmann–Hall isomorphisms*, J. of Funct. Anal. **133** (1995), 69–128.

[Da] Davies, E.B., *"Heat kernels and spectral theory"*, Cambridge Univ. Press, New York, PortChester, Melbourne, Sydney, 1990.

[DG] Driver, B. K. and Gross, L., *Hilbert spaces of holomorphic functions on complex Lie groups,* (to appear in Proceedings of the 1994 Taniguchi Symposium).

[Dyn] Dynkin, E. B., *Markov processes and random fields,*, Bull. Amer. Math. Soc. **3** (1979), 975–999, [See Appendix].

[FR] Fang, Shizan et Ren, Jiagang, *Sur le squelette et les dèrivèes de Malliavin des fonctions holomorphes sur un espace de Wiener complexe*, J. of Math. Kyoto Univ. **33–3** (1993), 749–764.

[Fock1] Fock, V., *Verallgemeinerung und Lösung der Diracschen statistischen Gleichung*, Zeits. f. Phys. **49**, 339–357.

[Fock2] Fock, V., *Zur Quantenelektrodynamik*, Physikalische Zeitschrift der Sowjetunion **6**, 425–469.

[Fol] Folland, Gerald B., *Harmonic Analysis in Phase Space*, Ann. of Math. Studies, Princeton University Press, Princeton, NJ (1989).

[G1] Gross, L., *Potential theory on Hilbert space*, J. Funct. Anal. **1** (1967), 123–181.

[G2] Gross, L., *Logarithmic Sobolev inequalities on Lie groups*, Illinois J. of Math. **36** (1992), 447–490.

[G3] Gross, L., *Logarithmic Sobolev inequalities on loop groups*, J. of Funct. Anal. **102** (1992), 268–313.

[G4] Gross, L., *Uniqueness of ground states for Schrödinger operators over loop groups*, J. of Funct. Anal. **112** (1993), 373–441.

[G5] Gross, L., *The homogeneous chaos over compact Lie groups*, in "Stochastic Processes, A Festschrift in Honor of Gopinath Kallianpur" (S. Cambanis et al., eds.), Springer–Verlag, New York, 1993, pp. 117–123.

[G6] Gross, L., *Harmonic analysis for the heat kernel measure on compact homogeneous spaces*, in "Stochastic Analysis on Infinite Dimensional Spaces" (Kunita and Kuo, eds.), Longman House, Essex, England, 1994, pp. 99–110.

[GJ] Glimm, James and Jaffe, Arthur, *"Quantum physics: a functional integral point of view"*, Springer-Verlag, New York, 1987.

[Ha1] Hall, B., *The Segal–Bargmann "coherent state" transform for compact Lie groups*, J. of Funct. Anal. **122** (1994), 103–151.

[Ha2] Hall, B., *The inverse Segal–Bargmann transform for compact Lie groups* (to appear J. of Funct. Anal.).

[He] Hermite, C., *Sur un nouveau développement en série des fonctions*, See Ouvres de Charles Hermite, Tome II, p.293–308, Gauthier–Villars, Paris, 1908, originally in C. R. Acad. des Sci., t.LVIII (1864), 93 and 266.

[HKPS] Hida, T., Kuo, H.–H., Potthoff, J., Streit, L., *White Noise, an infinite dimensional calculus*, Kluwer Acad. Pub., Dordrecht/Boston, 1993.

[Hij1] Hijab, Omar, *Hermite functions on compact Lie groups I*, J. of Funct. Anal. **125** (1994), 480–492.

[Hij2] Hijab, Omar, *Hermite functions on compact Lie groups II*, J. of Funct. Anal. **133** (1995), 41–49.

[HP] Hille, E. and Phillips, R. S., *Functional analysis and semi–groups*, AMS Colloquium Publications **31** (1957), 109–116.

[Hit1] Hitsuda, Masayuki, *Formula for Brownian partial derivatives,*, Second Japan–USSR Symposium on Probability Theory, Kyoto (1972), 111–114.

[Hit2] Hitsuda, Masayuki, *Formula for Brownian partial derivatives*, Publ. Fac. of Integrated Arts and Sciences, Hiroshima University, Ser.3, v.4 (1978), 1–15.

[Ito1] Ito, K., *Multiple Wiener integral*, J. Math. Soc. Japan **3** (1951), 157-169.

[Ito2] Ito, K., *Complex multiple Wiener integral*, Japan J. Math. **22** (1953), 63–86.

[Ko] Kondratiev, Yu. G., *Spaces of entire functions of an infinite number of variables, connected with the rigging of Fock space*, Selecta Mathematica Sovietica **10** (1991, originally published in 1980), 165–180.

[KLPS] Kondratiev, Yu. G., Leukert, P., Potthoff, J., Streit, L., Westerkamp, W., *Generalized functionals in Gaussian spaces – the characterization theorem revisited* (to appear J. of Funct. Anal.).

[Kr1] Krée, Paul, *Solutions faibles d'equations aux dérivées fonctionelles*, Seminar Pierre Lelong I, Lecture Notes in Mathematics, see especially Sec. 3, vol. 410, Springer, New York/Berlin, 1972/1973, pp. 142–180.

[Kr2] Krée, Paul, *Solutions faibles d'equations aux dérivées fonctionelles*, Seminar Pierre Lelong II, Lecture Notes in Mathematics, see especially Sec. 5, vol. 474, Springer, New York/Berlin, 1973/1974, pp. 16–47.

[Kr3] Krée, Paul, *Calcul d'intégrales et de dérivées en dimension infinie*, J. of Funct. Anal. **31** (1979), 150–186.

[Ku] Kubo, I., *A direct setting of white noise calculus,*, Stochastic Analysis on Infinite Dimensional Spaces (Kunita and Kuo, eds.), Longman House, Essex, England, 1994, pp. 152–166.

[Kuo] Kuo, H. H., *Gaussian measures in Banach spaces*, Lecture Notes in Mathematics, vol. 463, Springer–Verlag, Berlin, New York, 1975.

[KT] Kusuoka, S. and Taniguchi, S., *Pseudoconvex domains in almost complex abstract Wiener spaces*, RIMS–preprint 891 (1992).

[Lee1] Lee, Yuh–Jia, *Analytic version of test functionals, Fourier transform, and a characterization of measures in white noise calculus*, J. of Funct. Anal. **100** (1991), 359–380.

[Lee2] Lee, Yuh-Jia, *Transformation and Wiener-Itô decomposition of white noise functionals,*, Bulletin of the Institute of Mathematics Academia Sinica **21** (1993), 279–291.

[MM] Malliavin, M–P. and Malliavin, P., *Integration on loop groups. I. Quasi invariant measures*, J. of Funct. Anal. **93** (1990), 207–237.

[Ni] Nielsen, T. T., *Bose algebras: The complex and real wave representations*, Lecture Notes in Mathematics, vol. 1472, Springer–Verlag, Berlin/New York, 1991.

[Ob] Obata, N., *White Noise Calculus and Fock Space*, Lecture Notes in Mathematics, vol. 1577, Springer–Verlag, Berlin/New York, 1994.

[Pan] Paneitz, S. M., Pedersen, J., Segal, I. E., and Zhou, Z., *Singular operators on Boson fields as forms on spaces of entire functions on Hilbert space*, J. of Funct. Anal. **100** (1990), 36–58.

[Ped] Pedersen, J., Segal, I. E., and Zhou, Z., *Nonlinear quantum fields in ≥ 4 dimensions and cohomology of the infinite Heisenberg group*, Trans. Amer. Math. Soc. **345** (1994), 73–95.

[Pot] Potthoff, J. and Streit, L., *A characterization of Hida distributions*, J. of Funct. Anal. **101** (1991), 212–229.

[R] Robinson, Derek W., *"Ellitpic operators and Lie groups"*, Clarendon Press, Oxford, New York, Tokyo, 1991.

[Sad] Sadasue, Gaku, *Equivalence-singularity dichotomy for the Wiener measures on path groups and loop groups,*, J. of Math. Kyoto Univ. **35** (to appear).

[Se1] Segal, I. E., *Tensor algebras over Hilbert spaces*, Trans. Amer. Math. Soc. **81** (1956), 106–134.

[Se2] Segal, I. E., *Mathematical characterization of the physical vacuum for a linear Bose–Einstein field*, Illinois J. Math. **6** (1962), 500–523.

[Se3] Segal, I. E., *Mathematical problems in relativistic quantum mechanics*, Proceedings of the AMS Summer Seminar on Applied Mathematics, Boulder, Colorado, Amer. Math. Soc. (1960), Providence, RI.

[Se4] Segal, I. E., *Construction of non–linear local quantum processes I*, Ann. of Math. **92** (1970), 462–481.

[Se5] Segal, I. E., *The complex wave representation of the free Boson field*, in "Topics in functional analysis: essays dedicated to M. G. Krein on the occasion of his 70th birthday", Advances in Mathematics: Supplementary Studies (I. Gohberg and M. Kac, eds.), vol. 3, Academic Press, New York, 1978, pp. 321–344.

[Shi1] Shigekawa, Ichiro, *Transformations of the Brownian motion on the Lie group*, in "Stochastic Analysis: Proceedings of the Taniguchi International Symposium on Stochastic Analysis" (K. Itô, ed.), Katata and Kyoto, 1982, Japan, North–Holland, Amsterdam/New York, 1984, pp. 409–422.

[Shi2] Shigekawa, Ichiro, *Itô-Wiener expansions of holomorphic functions on the complex Wiener space*, in "Stochastic Analysis" (E. Mayer, et al, eds.), Academic Press, San Diego, 1991, pp. 459–473.

[Sug1] Sugita, Hiroshi, *Various topologies in the Wiener space and Lévy's stochastic area*, Probab. Theory Relat. Fields **91** (1992), 283–296.

[Sug2] Sugita, Hiroshi, *Properties of holomorphic Wiener functions–skeleton, contraction, and local Taylor expansion*, Probab. Theory Relat. Fields **100** (1994), 117–130.

[Sug3] Sugita, Hiroshi, *Regular version of holomorphic Wiener function,* (to appear J. Math. Kyoto Univ.).

[Tan] Taniguchi, Setsuo, *On almost complex structures on abstract Wiener spaces*, preprint (1994).

[Zh] Zhou, Z., *The contractivity of the free Hamiltonian semigroup in L^p spaces of entire functions*, J. of Funct. Anal. **96** (1991), 407–425.

Lagrangian for pinned diffusion process

Keisuke Hara[1] * and Yoichiro Takahashi[2]

[1] Graduate School of Mathematical Sciences, University of Tokyo, Komaba, Meguro-ku, Tokyo 153, Japan
[2] Research Institute for Mathematical Science, Kyoto University, Kyoto 606-01, Japan

1. Introduction

In the 1960s Professor K. Itô tried to understand the Feynman path integral probabilistically and constructed its integral representation over time dependent Hilbert spaces ([11], also [10]). Near the end of the 1970s D.Fujiwara succeeded in proving the existence of the limit of finite dimensional path integrals for Schrödinger equations in a very strong sense [3], and later in showing "Itô's version" [4]. Inspired by their works and looking at the discussions on the effect of curvature among physicists, one of the authors started studying the problem of most probable path or the Lagrangian or the Onsager-Machlup function or the probability functional for diffusion process on manifolds (cf., e.g., [5],[16]), and which is completely determined by a collaboration with S. Watanabe [18] (see Theorem 1.2 below).

The purpose of the present paper is to give a similar result for pinned diffusion by purely probabilistic methods.

Let $x(t)$ be a nondegenerate conservative diffusion process on a manifold M. Assume that $g = (g_{ij})$ be a Riemannian metric tensor on M such that the diffusion $x(t)$ is associated with the drifted heat equation:

$$\frac{\partial u}{\partial t} = \frac{1}{2}\Delta u + fu \tag{1.1}$$

where Δ is the Laplace-Beltrami operator of (M, g) and f is a (smooth) vector field on M. Take a (smooth space-time) curve $\gamma : x = \gamma(t), 0 \leq t \leq T$, and consider the asymptotics of the probability that the diffusion $x(t)$ pinned at time T sojourns in a small tubular neighborhood of γ up to time T. Then it turns out to be the same as in the case of standard Euclidean Brownian motion $B(t)$ pinned at 0 at time T up to a constant, which is expressed by a certain Lagrangian (or the probability functional or the Onsager-Machlup function). Precisely, the following statement is true.

Theorem 1.1. *For the diffusion $x(t)$ starting from $\gamma(0)$ at time 0 and pinned at $\gamma(T)$ at time T the following asymptotics holds:*

---

* Supported by J.S.P.S. Research Fellowships for Young Scientists.

$$\lim_{\delta\to 0}\frac{P_{\gamma(0)}\left(\max_{0\le t\le T} d(x(t),\gamma(t))\le\delta \mid x(T)=\gamma(T)\right)}{P_0\left(\max_{0\le t\le T}|B(t)|\le\delta \mid B(T)=0\right)} = \exp\{-S(\gamma)\} \tag{1.2}$$

where d is the Riemannian distance,

$$S(\gamma)=\int_0^T L(\gamma(t),\dot\gamma(t))dt \tag{1.3}$$

and

$$L(x,v)=\frac{1}{2}|f-v|_x^2+\frac{1}{2}\mathrm{div}f(x)+\frac{1}{12}R(x). \tag{1.4}$$

Here, $|\ \ |_x$ is the Riemannian norm on the tangent space T_xM at x, div *stands for the divergence and $R(x)$ is the scalar curvature at x.*

Theorem 1.1 is proved by improving the methods used in Takahashi and Watanabe [18]. The improvement is drastic in the sense that the latter half, that is, the analytical part of the methods employed there is no more needed. In other words, simple calculations and naive estimates of certain Wiener functionals are now sufficient to obtain such asymptotics. The point is the discovery by the first author of a Besselizing drift which is smooth as well as natural [7]. The new method also makes the proof of the following older result much simpler and purely probabilistic:

Theorem 1.2. *(S. Watanabe-Y. Takahashi) For the diffusion $x(t)$ starting at $\gamma(0)$ at time 0 the following asymptotics holds:*

$$\lim_{\delta\to 0}\frac{P_{\gamma(0)}\left(\max_{0\le t\le T} d(x(t),\gamma(t))\le\delta\right)}{P_0\left(\max_{0\le t\le T}|B(t)|\le\delta\right)} = \exp\{-S(\gamma)\} \tag{1.5}$$

Comparing the above results with Weyl's volume formula for tubular neighborhood in differential geometry (cf. [6]), it is quite natural to ask what is the multidimensional generalization. But the authors have no answer.

The outline of the proof of Theorem 1 is as follows. In Section 2 we introduce the Fermi coordinate or the normal coordinate along the curve γ. Then our probability depends only on the radial part of $x(t)$. So it is quite natural to use the Cameron-Martin formula and to make the radial motion a Bessel process in order to compute the ratio stated in Theorem 1.1 and 1.2. On this stage the term $(1/2)|f-v|_x^2$ comes out from the Girsanov density. Here we use a new Besselization drift. The details will be discussed in Section 3. The next observation is the following:

Theorem 1.3. *Let α be a smooth differential 1-form on $[0,T] \times M$ which does not depend on dt and $Y(t)$ be a diffusion process whose radial part is a Bessel process. Then the following estimate holds for the stochastic line integral $\int_{Y_T} \alpha$ along the space time curve $Y_T = \{(t, Y(t)); 0 \le t \le T\}$:*

$$E\left[\exp \int_{Y_T} \alpha \;\middle|\; |Y(t)| \le \delta \right] = \exp O(\delta). \tag{1.6}$$

The proof of Theorem 1.3 will be given in Section 4, by using the stochastic Stokes Theorem and the Kunita-Watanabe theorem on the orthogonal martingale. Thus we rewrite the stochastic integral in the Girsanov density into a stochastic line integral. On this stage the divergence term $(1/2)\mathrm{div} f(x)$ and the curvature term $(1/12)R(x)$ appear as the difference of the Itô and the Stratonovich integrals.

Finally, let us give a brief historical remark. In the case where the space M is Euclidean, the curvature vanishes and the result was obtained directly from the application of the Girsanov theorem as was done by Dürr and Bach [1] and also by Hidemi Itô [9] independently. The situation is rather similar in the case of an Einstein space where curvature effect cancels out because the Ricci curvature R_{ij} is proportional to the metric g_{ij} ([17]). The general result is also obtained by Fujita and Kotani [2]. Their proof is analytic and they used a singular perturbation method for partial differential equation to obtain the asymptotics. The analytic part of the proof in Takahashi and Watanabe [18] to treat asymptotic behavior of a certain Wiener functional may be thought of as a substitution for that . But it may be worthy here to notice that the essence of the appearance of the scalar curvature term $(1/12)R(x)$ is a kind of ergodic theorem or averaging theorem due to the "rapid rotation" of the spherical motion as the tube shrinks. (A direct formulation of such an "asymptotic ergodic theorem" is possible and makes the older proof a little bit simpler but we omit it because it is still too tedious comparing with the present proof.) Our new method reduces such an ergodic theorem to a result of stochastic calculus. Concerning Theorem 1.3 we should refer to the result of Shepp and Zeitouni [14] which shows (1.6) holds for more general 1-form α if one replace $O(\delta)$ by $o(1)$.

2. Preliminary from Riemannian geometry

In this section, we introduce the Fermi coordinate and prepare some asymptotics of geometric quantities.

Let the diffusion (X_t, P_x) and the smooth curve $\gamma(t)$ on M be given as above. On the product manifold $[0,T] \times M$, let $\tilde{\gamma} = \tilde{\gamma}(t)$ be the space-time curve $t \in [0,T] \to (t, \gamma(t))$ and introduce a coordinate system in a neighborhood U of the curve $\tilde{\gamma}$ as follows. First, choose an orthonormal basis $e = \{e_1, e_2, \ldots, e_n\}$ in the tangent space $T_{\gamma(0)}M$ and let $e^t = \{e_1^t, e_2^t, \ldots, e_n^t\}$

be the orthonormal basis in $T_{\gamma(t)}$ obtained as the parallel translate of e along the curve $\gamma(t)$. Then there exists a neighborhood U of the curve $\tilde{\gamma}(t)$ in $[0,T] \times M$ such that the following mapping $(t,x) \in U \to (t,x^1,x^2,\ldots,x^n) \in [0,T] \times R^n$ is well defined :

$$x = \exp_{\gamma(t)} \sum_{i=1}^{n} (x^i e_i^t) \tag{2.1}$$

where $\exp_m$ is the exponential map.

In this coordinate system $(t,x^1,x^2,\ldots,x^n)$, we denote the components of the Riemannian metric tensor of M, its inverse, the Christoffel symbol and the vector field f by $g_{ij}(t,x), g^{ij}(t,x), \Gamma_{ij}^k(t,x)$ and $f^i(t,x)$ respectively.

The following properties are well-known but we shall give a sketch of the proof.

Lemma 2.1.

$$\sum_{j=1}^{n} g^{ij}(t,x)x^j = x^i, \tag{2.2}$$

$$g^{ij}(t,x) = \delta^{ij} + \frac{1}{3}\sum_{l,m=1}^{n} R_{iljm}(t,0)x^l x^m + O(|x|^3) \tag{2.3}$$

and

$$\Gamma_{jk}^i(t,x) = \frac{1}{3}\sum_{m=1}^{n} (R_{jmki}(t,0)x^m + R_{ijmk}(t,0)x^m) + O(|x|^2) \tag{2.4}$$

where $R_{ijkl}(t,x)$ is the component of the Riemann curvature tensor in the Fermi coordinate.

Proof. The first relation is the basic property of the normal coordinate system which follows directly from the definition.

By differentiating the both side of (2.2), one finds below

$$g^{ik}(t,x) = \delta^{ik} - \sum_j \frac{\partial}{\partial x^k} g^{ij}(t,x)x^j. \tag{2.5}$$

By differentiating them once more,

$$\frac{\partial}{\partial x^l} g^{ik}(t,x) + \frac{\partial}{\partial x^k} g^{il}(t,x) = -\sum_j \frac{\partial^2}{\partial x^k \partial x^l} g^{ij}(t,x)x^j. \tag{2.6}$$

Then one obtains the following identity:

$$g^{ij}(t,x) = \delta^{ij} + \frac{1}{2}\sum_{l,m} \frac{\partial^2}{\partial x^i \partial x^j} g^{lm}(t,x)x^l x^m. \tag{2.7}$$

From (2.5), (2.6) and (2.7), one can obtain the lemma by recalling the following definitions (see Spivak [15]) :

$$\begin{aligned} R_{ijkl} &= \frac{1}{2}\left(\frac{\partial^2}{\partial x^j \partial x^l} g^{ik} + \frac{\partial^2}{\partial x^i \partial x^k} g^{jl} - \frac{\partial^2}{\partial x^i \partial x^l} g^{jk} - \frac{\partial^2}{\partial x^j \partial x^k} g^{il}\right) \\ &\quad + \sum_{ab} g_{ab}(\Gamma^b_{ik}\Gamma^a_{jl} - \Gamma^b_{il}\Gamma^a_{jk}) \end{aligned} \tag{2.8}$$

and

$$\Gamma^l_{jk} = \frac{1}{2}\sum_m g^{lm}\left(\frac{\partial}{\partial x^j} g^{km} + \frac{\partial}{\partial x^k} g^{mj} - \frac{\partial}{\partial x^m} g^{jk}\right). \tag{2.9}$$

Now let the differential operator

$$\frac{\partial}{\partial t} + \frac{1}{2}\Delta + f \quad \text{on} \quad U \subset [0,T] \times M \tag{2.10}$$

be transformed to the differential operator

$$\frac{\partial}{\partial t} + \frac{1}{2}\sum_{i,j=1}^{n} g^{ij}(t,x)\frac{\partial^2}{\partial x^i \partial x^j} + \sum_{i=1}^{n} b^i(t,x)\frac{\partial}{\partial x^i} \quad \text{on} \quad V \subset [0,T] \times R^n. \tag{2.11}$$

Then it is easy to prove that

$$b^i(t,x) = f^i(t,x) - \dot{\gamma}^i(t) - \frac{1}{2}\sum_{k=1}^{n} \Gamma^i_{kk} + O(|x|^2), \tag{2.12}$$

where $\dot{\gamma}^i(t)$ are the components of the tangent vector $\dot{\gamma}(t) \in T_{\gamma(t)}M$. By the lemma above,

$$\sum_{k=1}^{n} \Gamma^i_{kk} = \frac{2}{3}\sum_{j=1}^{n} R_{ij}(t,0)x^j + O(|x|^2) \tag{2.13}$$

where $R_{ij}(t,x) = \sum_{m=1}^{n} R_{mimj}(t,x)$ is the Ricci tensor. Thus we have the following asymptotics.

Lemma 2.2.

$$b^i(t,x) = f^i(t,x) - \dot{\gamma}^i(t) - \frac{1}{3}\sum_{j=1}^{n} R_{ij}(t,0)x^j + O(|x|^2). \tag{2.14}$$

3. Proof of Theorems 1 and 2

Let $\sigma^i_j(t,x)$ be the square root of the matrix $g^{ij}(t,x)$. Then, the diffusion process $x(t)$ is governed by the following stochastic differential equation provided that it sojourns in the tubular neighborhood where the Fermi coordinate is defined:

$$dX^i(t) = \sum_{j=1}^{n} \sigma^i_j(t, X(t))dB^j(t) + b^i(t, X(t))dt. \tag{3.1}$$

Now let

$$c^i(t,x) = \frac{1}{2}\sum_{j=1}^{n} \frac{\partial}{\partial x^j} g^{ij}(t,x) \tag{3.2}$$

and consider the stochastic differential equation:

$$dY^i(t) = \sum_{j=1}^{n} \sigma^i_j(t, Y(t))dB^j(t) + c^i(t, Y(t))dt. \tag{3.3}$$

Lemma 3.1. *The radial part $R(t) = |Y(t)| = \sum(Y^i)^2$ of the process $Y(t)$ is a Bessel process with index $n = \dim M$,i.e.,*

$$dR(t) = 2R(t)dW_t + ndt \tag{3.4}$$

where W_t is a 1-dimensional standard Brownian motion.

Proof. From the identity

$$\sum_{j=1}^{n} g^{ij}(t,x)x^j = x^i, \tag{3.5}$$

it follows

$$\sum_{j=1}^{n} \sigma^i_j(t,x)x^j = x^i. \tag{3.6}$$

Hence by the Itô formula,

$$\begin{aligned} d(|Y(t)|^2) &= 2\sum_{i,j} Y^i\sigma^i_j dB^j + 2\sum_i Y^i c^i dt + \sum_{i,j} \sigma^i_j\sigma^i_j dt \\ &= 2\sum_i Y^i dB^i + (2\sum_i Y^i c^i + \sum_i g^{ii})dt. \end{aligned} \tag{3.7}$$

On the other hand, differentiating the both side of (3.5),

$$1 = g^{ii} + \sum_j x^j \frac{\partial}{\partial x^i} g^{ij}. \tag{3.8}$$

Then

$$2\sum_i x^i c^i(t,x) = \sum_i (1 - g^{ii}). \tag{3.9}$$

By (3.7) and (3.9),

$$d(|Y(t)|^2) = 2\sum_i Y(t)^i dB_t^i + ndt. \tag{3.10}$$

Therefore

$$d(|Y|^2) = 2|Y|dW_t + ndt, \tag{3.11}$$

where

$$W_t = \sum_i \int_0^t \frac{Y^i(s)}{|Y(s)|} dB_s^i \tag{3.12}$$

is a 1-dimensional standard Brownian motion. Consequently, $|Y(t)|$ is a Bessel process with index n.

Now we apply the Cameron-Martin formula to $X(t)$ with respect to $Y(t)$. Let us write, for vector fields u and v ,

$$g(u,v)(t,x) = \sum_{i,j=1}^{n} g_{ij}(t,x)u^i(t,x)v^j(t,x) \tag{3.13}$$

and set

$$\Phi(T) = \Phi_1(T) - \Phi_2(T) \tag{3.14}$$

where

$$\Phi_1(T) = \int_0^T g(b-c, dY)(t, Y(t)) \tag{3.15}$$

and

$$\begin{aligned} \Phi_2(T) &= \int_0^T g(b-c, cdt)(t, Y(t)) \\ &\quad + \frac{1}{2}\int_0^T g(b-c, b-c)(t, Y(t))dt. \end{aligned} \tag{3.16}$$

Then, for any set K measurable with respect to the radial motion R ,

$$\begin{aligned} &P_a\{\max d(x(t), \gamma(t)) \le \delta \mid K\} \\ &= P_0\{\max |X(t)| \le \delta \mid K\} \\ &= E[\exp \Phi(T);\ \max |Y(t)| \le \delta \mid K] \\ &= E[\exp \Phi(T) \mid \max |Y(t)| \le \delta, K]\ P\{\max |Y(t)| \le \delta \mid K\}. \end{aligned} \tag{3.17}$$

Our calculation of $\Phi_1(T)$ and $\Phi_2(T)$ is similar to that of [18]. The estimation of $\Phi_2(T)$ is easy.

Lemma 3.2. *On the set* $\{\max|Y(t)| \leq \delta\}$,

$$\begin{aligned} \Phi_2(T) &= \frac{1}{2}\int\{g(b,b)(t,0) - g(c,c)(t,0)\}dt + O(\delta) \\ &= \frac{1}{2}\int|f - \dot{\gamma}|^2 dt + O(\delta). \end{aligned} \tag{3.18}$$

Next we rewrite the Itô integral $\Phi_1(T)$ defined by (3.15) in terms of the stochastic line integral. Let $\Phi_0(T)$ be the stochastic line integral along the space-time curve $Y[0,T] : (t, Y(t)), 0 \leq t \leq T$, of the (space-time) 1-form α defined in the Fermi neighborhood by

$$\alpha = g(b - c, dx)(t, x). \tag{3.19}$$

Then, by definition,

$$\Phi_0(T) = \int g(b - c, \circ dY)(t, Y(t)). \tag{3.20}$$

Lemma 3.3. *On the set* $\{\max|Y(t)| \leq \delta\}$,

$$\Phi_0(T) - \Phi_1(T) = \frac{1}{2}\int \mathrm{div}\ \mathrm{f}(\gamma(\mathrm{t}))\mathrm{dt} + \frac{1}{12}\int \mathrm{R}(\gamma(\mathrm{t}))\mathrm{dt} + \mathrm{O}(\delta) \tag{3.21}$$

Proof. By definition,

$$\begin{aligned} \Phi_0 - \Phi_1 &= \int g(b - c, \circ dY) - \int g(b - c, dY) \\ &= \frac{1}{2}\int \mathrm{div}(b - c)dt. \end{aligned} \tag{3.22}$$

By (2.3), (2,15) and (3,2), we have the following expansion:

$$b^i(t,x) - c^i(t,x) = f^i(t,0) - \dot{\gamma}^i(t,0) + \sum_j\left(\frac{\partial f^i}{\partial x^j} + \frac{1}{6}R_{ij}\right)x^j + O(|x|^2). \tag{3.23}$$

Consequently the lemma follows from (3.22) and (3.23).

From Lemmas 3.2 and 3.3, we obtain that

$$\Phi(T) = \Phi_0(T) - S(\gamma) + O(\delta) \tag{3.24}$$

on the set $\{\max|Y(t)| \leq \delta\}$. Consequently,

$$\begin{aligned} &\frac{P_a\{\max d(x(t),\gamma(t)) \leq \delta \mid K\}}{P_a\{\max|B(t)| \leq \delta \mid K\}} \\ &= E[\exp\Phi(T) \mid \max|Y(t)| \leq \delta, K] \\ &= \exp\{-S(\gamma) + O(\delta)\}\ E[\exp\Phi_0(T) \mid \max|Y(t)| \leq \delta, K] \end{aligned} \tag{3.25}$$

and so Theorems 1.1 and 1.2 follow from Theorem 1.3.

4. Smallness of stochastic line integral confined in small ball

Let $Y(t)$ be the diffusion process considered in the previous section. Thus it is governed by the stochastic differential equation:

$$dY(t) = \sigma(t, Y(t))dB(t) + c(t, Y(t))dt \tag{4.1}$$

where the drift c was chosen so that the radial part $R(t) = |Y(t)|$ is a Bessel process with index $n = dimM$. Also recall the identity:

$$\sum_{j=1}^{n} \sigma^i_j(t,x)x^j = x^i. \tag{4.2}$$

Now let us consider the stochastic areas

$$\begin{aligned} A^{ij}(t) &= \int_0^t (Y^i \circ dY^j - Y^j \circ dY^i) \\ &= \int_0^t (Y^i dY^j - Y^j dY^i) \quad (1 \le i < j \le m), \end{aligned} \tag{4.3}$$

which can be written by using the spherical part $U(t) = Y(t)/R(t)$ as

$$\begin{aligned} A^{ij}(t) &= \int_0^t R(t)^2 (U^i \circ dU^j - U^j \circ dU^i) \\ &= \int_0^t R(t)^2 (U^i dU^j - U^j dU^i) \quad (1 \le i < j \le m). \end{aligned} \tag{4.4}$$

Lemma 4.1. *The quadratic variation processes satisfies the following relations on the set* $\{\max |Y(t)| \le \delta\}$:

$$\langle A^{ij}, R\rangle(t) = 0, \tag{4.5}$$

$$\langle A^{ij}, A^{kl}\rangle(t) = O(\delta^2 T). \tag{4.6}$$

Proof.

$$\begin{aligned} d\langle A^{ij}, R\rangle &= (Y^i dY^j - Y^j dY^i) \sum_l \frac{Y^l}{|Y|} dB^l \\ &= \sum_l \frac{Y^l}{|Y|} (Y^i \sigma^j_m \delta^{ml} dt - Y^j \sigma^i_m \delta^{ml} dt \\ &= \sum_l \frac{1}{|Y|} (Y^i Y^l \sigma^j_l - Y^j Y^l \sigma^i_l) dt \\ &= \frac{1}{|Y|} (Y^i Y^j - Y^j Y^i) dt = 0. \end{aligned} \tag{4.7}$$

$$
\begin{aligned}
d\langle A^{ij}, A^{kl}\rangle &= (Y^i dY^j - Y^j dY^i)(Y^k dY^l - Y^l dY^k) \\
&= (Y^i Y^k \sigma^j_m \sigma^l_n \delta^{mn} + Y^j Y^l \sigma^i_m \sigma^k_n \delta^{mn} - \ldots)dt \\
&= (Y^i Y^k g^{jl} + Y^j Y^l g^{ik} - Y^j Y^k g^{il} - Y^i Y^l g^{jk})dt \\
&= O(\delta^2)dt. \qquad (4.8)
\end{aligned}
$$

The most important consequence of Lemma 4.1 is that the martingale parts, say $M_{ij}(t)$, of the stochastic areas $A_{ij}(t)$ are orthogonal to the martingale part, say $W(t)$, of the radial motion $R(t)$. Consequently, one obtains the following fact from the Kunita-Watanabe theorem on orthogonal martingales ([12], also [8]).

Lemma 4.2. *The martingales $M_{ij}(t)$ are martingales even if they are conditioned by the σ-algebra $\mathcal{F}_R$ which is generated by $W(t), 0 \le t \le T$.*

On the other hand, the following is a direct consequence of the fact that the Stratonovich integral is the limit of the line integral along polygonal lines approximating the sample path (c.f. [8]).

Lemma 4.3. *(The stochastic Stokes Theorem) Let Y_T be the space-time random curve $(t, Y(t))$ $(0 \le t \le T)$ and Σ_T be the random surface $(t, sY(t))$ $(0 \le s \le 1, 0 \le t \le T)$. Then for each smooth space-time differential 1-form α the following stochastic Stokes formula holds:*

$$\int_{\partial \Sigma_T} \alpha = \int_{\Sigma_T} d\alpha. \qquad (4.9)$$

In other words, the stochastic line integral of α along $Y_{[0,T]}$ is given by

$$
\begin{aligned}
\int_{Y_T} \alpha &= \int_0^1 \sum_{i=1}^n \alpha_i(T, sY(T)) Y^i(T) ds + \int_0^T \alpha_0(t, 0) dt \qquad (4.10) \\
&+ \int_0^T \sum_{i=1}^n \alpha_{0i}(t, Y(t)) Y^i(t) dt + \int_0^T \sum_{i,j=1}^n \alpha_{ij}(t, Y(t)) \circ A^{ij}(t)
\end{aligned}
$$

where

$$\alpha_{ij}(t, x) = \frac{1}{2} \int_0^1 s \left(\frac{\partial \alpha_j}{\partial x^i} - \frac{\partial \alpha_i}{\partial x^j} \right)(t, sx) ds \qquad (4.11)$$

and

$$\alpha_{0i}(t, x) = \int_0^1 \left(\frac{\partial \alpha_0}{\partial x^i} - \frac{\partial \alpha_i}{\partial t} \right)(t, sx) ds. \qquad (4.12)$$

Since $\alpha_0 = 0$ in our case (see the definition (3.19)), it follows from the above lemma,

$$\int_{Y_T} \alpha = \int_0^T \sum_{i,j=1}^n \alpha_{ij} \circ A^{ij}(t) + O(\delta T). \tag{4.13}$$

Note that the right hand side of (4.13) above is not a martingale while the drift c in [18] was taken so that it is. But the treatment of the extra term is easy as follows.

Set

$$N(t) = \int_0^t \sum_{i,j,m}^n \alpha_{ij}(Y^i \sigma_m^j - Y^j \sigma_m^i) dB^m. \tag{4.14}$$

Then $N(t)$ is a martingale and, by Lemma 4.1,

$$\langle N, R \rangle = 0, \tag{4.15}$$

and

$$\langle N \rangle (T) = O(\delta^2 T) \quad \text{on} \quad \{\max |Y| \leq \delta\}. \tag{4.16}$$

Finally, on the set $\{\max |Y| \leq \delta\}$,

$$\begin{aligned} &\int_0^T \sum_{i,j=1}^n \alpha_{ij} \circ (Y^i \circ dY^j - Y^j \circ dY^i) \\ = \ & \int_0^T \sum_{i,j=1}^n \alpha_{ij} \circ (Y^i dY^j - Y^j dY^i) \\ = \ & N(T) + \frac{1}{2} \int_0^T \sum_{i,j=1}^n d\alpha_{ij}(Y^i dY^j - Y^j dY^i) \\ & + \int_0^T \sum_{i,j=1}^n \alpha_{ij}(Y^i c^j - Y^j c^i) dt \\ = \ & N(T) - \frac{1}{2} \langle N \rangle (T) + O(\delta T). \end{aligned} \tag{4.17}$$

Consequently,

$$\begin{aligned} & E[\exp(\int_{Y_T} \alpha \mid |Y(t)| \leq \delta] \\ = \ & E[\exp(N(T) - \frac{1}{2} \langle N \rangle (T) + O(\delta T)) \mid F_R] \\ = \ & \exp O(\delta T). \end{aligned} \tag{4.18}$$

We have completed the proof.

References

1. Dürr, D. and Bach, A. : *The Onsager-Machlup functions as Lagrangian for the most probable path of diffusion process*, Comm. Math. Phys. **60** (1978), 153–170.
2. Fujita, T. and Kotani, S. : *The Onsager-Machlup functions for diffusion processes*, J. Math. Kyoto. Univ., **22-1** (1982), 115-130.
3. Fujiwara, D. : *Remarks on convergence of Feynman path integrals*, Duke Math. J., **47** (1980), 559–600.
4. Fujiwara, D. : *Some Feynman path integral as oscillatory integrals over a Sobolev manifold*, Functional Analysis and Related Topics, Lecture Note in Math., **1540** (1992), 39–53.
5. Graham, R. : *Path integral formulation of general diffusion processes*, Z. Physik. B, **26** (1979), 281–290.
6. Gray, R. : *Tube*, Addison-Wesley Publishing Company, 1990.
7. Hara, K. : *Wiener functionals associated with joint distributions of exit time and position from small geodesic balls*, Ann. Probab., to appear.
8. Ikeda, N. and Watanabe,S. : *Stochastic differential equations and diffusion processes*, North Holland/Kodansha, 1981.
9. Itô, H. : *Probabilistic construction of Lagrangian of diffusion processes and its application*, Prog. Theor. Phys., **59** (1978), 725–741.
10. Itô, K. : *Wiener Integral and Feynman integral*, Proc. Fourth Berkeley Symp. Math. Stat. Probab. II (1960), 228–238 ; Selected Papers, 275–286.
11. Itô, K. : *Generalized uniform complex measures in the Hilbertian metric spaces with their application to the Feynman integral*, Proc. Fifth Berkeley Symp. Math. Stat. Probab. II (1955), 145–151 ; Selected Papers, 495–511.
12. Kunita, H. and Watanabe, S. : *On square integrable martingales*, Nagoya Math. J., **30** (1967), 209–245.
13. Onsager, L. and Machlup, S. : *Fluctuations and irreversible process I, II*, Phys. Rev. **91** (1953), 1505–1512, 1512–1515.
14. Shepp, L.A. and Zeitouni, O. : *A note on conditional moments and Onsager-Machlup functionals*, Ann. Probab., **20** (1992), 652–654.
15. Spivak, M. : *A Comprehensive Introduction to Differential Geometry*, Publish or Perish.
16. Stratonovich, R.L. : *On the probability functional of diffusion processes*, Select. Transl. in Math. Stat. Prob., **10** (1971), 273–286
17. Takahashi, Y. : *Most probable paths for diffusion processes*, (in Japanese), Diffusion Processes and Statistical Mechanics of Open Systems, R.I.M.S. Koukyuroku, **367** (1979), 215–228.
18. Takahashi, Y. and Watanabe, S. : *The probability functionals (Onsager-Machlup functions) of diffusion processes*, Proc. L.M.S. Symp. on Stochastic Integrals at Durham, Lecture note in Math., **851** (1980), 433–463.

Short Time Asymptotics and an Approximation for the Heat Kernel of a Singular Diffusion *

Yasuji Hashimoto[1], Shojiro Manabe[1] and Yukio Ogura[2]

[1] Department of Mathematics, Graduate School of Science, Osaka University, Toyonaka, Osaka 560, Japan
[2] Department of Mathematics, Saga University, Saga 840, Japan

1. Introduction

The class of diffusion processes is so wide that it includes not only the processes associated with elliptic operators with measurable coefficients but also those associated with the generators with distribution coefficients like measures or even derivatives of measures. That of one-dimensional ones is completely determined in 1950's and 1960's by many authors such as W. Feller, K. Itô, H. P. McKean and E.B. Dynkin, among others. The situation for multidimensional ones is however quite different and the problem is still open.

In this article, we take a diffusion process $X_0 = (X_0(t))$ on $\mathbb{R}^{N+1}$ associated with the symmetric form

$$(1.1)\quad \mathcal{E}_0(u,v) = \frac{1}{2}\int_{\mathbb{R}^{N+1}}\Big\{(1+\delta(\xi))\sum_{i,j=1}^{N} a^{ij}(x)\frac{\partial u}{\partial x^i}(x,\xi)\frac{\partial v}{\partial x^j}(x,\xi) + \frac{\partial u}{\partial \xi}(x,\xi)\frac{\partial v}{\partial \xi}(x,\xi)\Big\}dx d\xi, \quad u,v\in C_0^\infty(\mathbb{R}^{N+1})$$

on $L^2(\mathbb{R}^{N+1}, dxd\xi)$, where $(x,\xi)\in\mathbb{R}^{N+1}$ with $x\in\mathbb{R}^N$ and $\xi\in\mathbb{R}$, $dxd\xi$ is the Lebesgue measure, $\delta(\xi)$ is the Dirac delta function on $\mathbb{R}$ and $(a^{ij}(x))_{i,j=1}^N$ is a uniformly elliptic, symmetric, bounded and smooth matrix-valued function on $\mathbb{R}^N$ (see M. Fukushima *et al.* [3] for the correspondence between symmetric forms and diffusions). This process is proposed by N. Ikeda and S. Watanabe [4] in the study of 'very singular' diffusions in the case where $N = 1$ and $a^{11}(x) = 1$. Generally, the diffusion X_0 is obtained by a skew product of two mutually independent processes $Y = (Y(t))$ and $B = (B(t))$ as

$$X_0(t) = (Y(t+\phi(t)), B(t)),$$

where Y is the diffusion process on $\mathbb{R}^N$ generated by the second order differential operator $Lu = \sum_{i,j=1}^N \frac{\partial}{\partial x^i}(a^{ij}(x)\frac{\partial u}{\partial x^j})$ and B is the one-dimensional

* Research partially supported by Grant-in-Aid for Scientific Research No. 07640131.

Brownian motion with the local time $\phi(t)$ at 0: $\phi(t) = \lim_{\varepsilon\downarrow 0}(2\varepsilon)^{-1}\int_0^t I_{(-\varepsilon,\varepsilon)}(B(s))ds$. On the other hand, H. Pham Huy and E. Sanchez-Palencia [9] [11] studied the corresponding equation as a model of conductive thin layer, where $N = 2$ and (a^{ij}) is the identity matrix but the domain is a general bounded smooth domain.

We first obtain a Varadhan type short time asymptotic formula for the heat kernel $p_0(t,(x,\xi),(y,\eta))$ associated with (1.1) (see Theorem 2.1 below). The distance function d_0 is the minimum of the distance function d induced by the matrix $(a^{ij}(x))_{i,j=1}^N$ and the sum of the one-dimensional ones (see (2.1) for precise definition of d_0). The distance d_0 degenerates on the plane $\xi = 0$. Furthermore, when the two points $(x,0)$, $(y,0)$ belong to the plane $\xi = 0$, the normalizing ratio is $t^{1/3}$ rather than t and the distance is given by $d(x,y)^{2/3}$.

The symmetric form (1.1) is approximated by

$$(1.2)\quad \mathcal{E}_n(u,v) = \frac{1}{2}\int_{\mathbb{R}^{N+1}}\Big\{(1+\rho_n(\xi))\sum_{i,j=1}^N a^{ij}(x)\frac{\partial u}{\partial x^i}(x,\xi)\frac{\partial v}{\partial x^j}(x,\xi) + \frac{\partial u}{\partial \xi}(x,\xi)\frac{\partial v}{\partial \xi}(x,\xi)\Big\}dxd\xi, \quad u,v\in C_0^\infty(\mathbb{R}^{N+1}),$$

where $\{\rho_n\}$ is a δ-function approximation. It is shown in [9] [11] that the solutions of elliptic and of parabolic equations corresponding to (1.2) converge to those to (1.1) in their case mentioned above(see also Corollary 2.4 below). But, in connection with the theory of converging manifold, we are more interested in the convergence of distance functions and heat kernels(see [6] and [7]). Indeed, with the symmetric form (1.2), a Riemannian metric $g_n = (1+\rho_n(\xi))^{-1}g + d\xi^2$ is associated, where g is the Riemannian metric on $\mathbb{R}^N$ corresponding to $(a^{ij}(x))_{i,j=1}^N$. Thus it might be natural to associate the 'metric' $g_0 = (1+\delta(\xi))^{-1}g + d\xi^2$ with the symmetric form (1.1). In addition, the 'metric' g_0 can be considered to induce the distance function d_0. Indeed, we can show that the distance functions d_n induced by g_n converge to d_0(Proposition 2.2). Furthermore the corresponding heat kernels p_n converge to p_0 in some sense as $n\to\infty$ (Theorem 2.3).

The organization of this article is as follows. In the next Sect. 2, we state our results. In Sect. 3, we will give a proof of Theorem 2.1, which is based on the stochastic expression (3.2) of the heat kernel and due to the Laplace method. The proof of the rest assertions will be given in Sect. 4.

Acknowledgement. The authors would like to express their hearty thanks to Professor N. Ikeda who lead their interests to this problem and gave them valuable comments.

2. Statement of Results

Let $(a^{ij}(x))_{i,j=1}^N$ be as in Sect.1 and denote the associated Riemannian metric by g, that is, $g = \sum_{i,j=1}^N a_{ij}(x)dx^i dx^j$, where $(a_{ij}(x))_{i,j=1}^N$ is the inverse matrix of $(a^{ij}(x))_{i,j=1}^N$. This defines the distance function d, and put

$$d_0((x,\xi),(y,\eta)) := \left(d(x,y)^2 + (\xi-\eta)^2\right)^{1/2} \wedge (|\xi|+|\eta|) \tag{2.1}$$

for $(x,\xi),(y,\eta) \in \mathbb{R}^{N+1}$, where $a \wedge b$ stands for the minimum of a and b. Our first theorem is the following

Theorem 2.1. i) *For each $M > 0$, it holds that*

$$\lim_{t\downarrow 0} [-2t \log p_0(t,(x,\xi),(y,\eta))] = d_0((x,\xi),(y,\eta))^2, \tag{2.2}$$

uniformly in $d(x,y) \vee |\xi| \vee |\eta| \leq M$.
ii) *For each $x, y \in \mathbb{R}^N$, the asymptotic formula*

$$\lim_{t\downarrow 0} \left[-2t^{1/3} \log p_0(t,(x,0),(y,0))\right] = 3d(x,y)^{4/3} \tag{2.3}$$

holds.

Turning to the formula (1.2), we confirm our assumption ; ρ_n are nonnegative and smooth functions converging weakly to the δ-function, and their supports are included in the intervals $[-1/n, 1/n]$. It is then known that the corresponding heat kernel p_n satisfies the asymptotic formula (2.2) with the distance function d_0 replaced by d_n induced by $g_n = (1+\rho_n(\xi))^{-1}g + d\xi^2$(see Varadhan [12] and also Molchanov [8]). We further have the following

Proposition 2.2. *For each $M > 0$, it holds that*

$$\lim_{n\to\infty} d_n((x,\xi),(y,\eta)) = d_0((x,\xi),(y,\eta)), \tag{2.4}$$

uniformly in $|x| \vee |\xi| \vee |y| \vee |\eta| \leq M$.

Theorem 2.3. *For each $t > 0$, $(x,\xi) \in \mathbb{R}^{N+1}$ and $y \in \mathbb{R}^N$, it holds that*

$$\lim_{n\to\infty} \int_{\mathbb{R}} |p_n(t,(x,\xi),(y,\eta)) - p_0(t,(x,\xi),(y,\eta))|^2 d\eta = 0. \tag{2.5}$$

As an application of this theorem, we get

Corollary 2.4. *Suppose that f satisfies*

$$\int_{\mathbb{R}^{N+1}} |f(x,\xi)| e^{-\varepsilon\xi^2} dx d\xi < \infty, \quad \text{for any} \quad \varepsilon > 0. \tag{2.6}$$

Then, for each $t > 0$, and $(x,\xi) \in \mathbb{R}^{N+1}$, the functions

$$u_n(t,(x,\xi)) = \int_{\mathbf{R}^{N+1}} p_n(t,(x,\xi),(y,\eta))f(y,\eta)dyd\eta$$

converge to

$$u(t,(x,\xi)) = \int_{\mathbf{R}^{N+1}} p_0(t,(x,\xi),(y,\eta))f(y,\eta)dyd\eta.$$

Note that, if f is bounded and continuous, u_n is a solution to

$$\frac{\partial u}{\partial t} = \frac{1}{2}(1+\rho_n(\xi))\sum_{i,j=1}^{N}\frac{\partial}{\partial x^i}\left(a^{ij}(x)\frac{\partial u}{\partial x^j}\right) + \frac{1}{2}\frac{\partial^2 u}{\partial \xi^2} \quad \text{in } \mathbb{R}^{N+1},$$
$$u(0+,(x,\xi)) = f(x,\xi),$$

whereas, by Green's formula, u solves the equation

$$\frac{\partial u}{\partial t} = \frac{1}{2}\sum_{i,j=1}^{N}\frac{\partial}{\partial x^i}\left(a^{ij}(x)\frac{\partial u}{\partial x^j}\right) + \frac{1}{2}\frac{\partial^2 u}{\partial \xi^2} \quad \text{in } \mathbb{R}^{N+1}\setminus\{\xi=0\},$$
$$\sum_{i,j=1}^{N}\frac{\partial}{\partial x^i}\left(a^{ij}(x)\frac{\partial u}{\partial x^j}\right)(x,0) = \frac{\partial u}{\partial \xi}(x,0+) - \frac{\partial u}{\partial \xi}(x,0-) \quad \text{on } \{\xi=0\},$$
$$u(0+,(x,\xi)) = f(x,\xi).$$

3. Proof of Theorem 2.1

Let $(B(t), P_a)$ be a one-dimensional Brownian motion with the local time $\phi(t)$ at 0. Let also $q(t,x,y)$ be the transition probability density (with respect to the Lebesgue measure dy) of the diffusion process Y, i.e., the heat kernel corresponding to the generator L. We then have

$$\text{(3.1)} \quad p_0(t,(x,\xi),(y,\eta))d\eta = I(\xi,\eta)q(t,x,y)P_\xi(T_0>t,\ B(t)\in d\eta) + \int_0^\infty q(t+\sigma,x,y)P_\xi(T_0\le t,\ \phi(t)\in d\sigma,\ B(t)\in d\eta)$$

where T_0 is the first hitting time at the state 0, and

$$I(\xi,\eta) = \begin{cases} 1, & \xi\eta>0, \\ 0, & \text{otherwise.}\end{cases}$$

This with the formula [5, p.45 Problem 2.3.3] implies

$$\text{(3.2)} \quad p_0(t,(x,\xi),(y,\eta)) = I(\xi,\eta)q(t,x,y)(g(t,\xi-\eta)-g(t,\xi+\eta)) + \int_0^\infty q(t+\sigma,x,y)\frac{2(|\xi|+|\eta|+2\sigma)}{\sqrt{2\pi t^3}}\exp\left[-\frac{(|\xi|+|\eta|+2\sigma)^2}{2t}\right]d\sigma,$$

where $g(t,\xi) = \dfrac{1}{\sqrt{2\pi t}} e^{-\xi^2/2t}$. On the other hand, it is known that

$$q(t,x,y) \le \frac{C_1}{t^{N/2}}, \qquad t > 0, \tag{3.3}$$

for some $C_1 > 0$ (see [1]). Furthermore, for any $\varepsilon > 0$ and $C_2 > 0$, there exists a $t_0 > 0$ such that

$$\exp\left[-\frac{d(x,y)^2+\varepsilon}{t}\right] \le q(t,x,y) \le \exp\left[-\frac{d(x,y)^2-\varepsilon}{t}\right], \tag{3.4}$$

for all $0 < t \le t_0$ and $d(x,y) \le C_2$ (see [12]). Now we can proceed to

Proof of (2.2). We first assume that $\xi\eta \le 0$. It then follows from (3.2) that

$$\begin{aligned} &p_0(t,(x,\xi),(y,\eta)) \\ &= \int_0^\infty q(t+\sigma,x,y)\frac{2(|\xi|+|\eta|+2\sigma)}{\sqrt{2\pi t^3}} \exp\left[-\frac{(|\xi|+|\eta|+2\sigma)^2}{2t}\right] d\sigma \\ &= \int_0^\infty q(t+\sigma,x,y)\frac{d}{d\sigma}\left(-\frac{1}{\sqrt{2\pi t}}\exp\left[-\frac{(|\xi|+|\eta|+2\sigma)^2}{2t}\right]\right) d\sigma. \end{aligned}$$

This with (3.3) implies

$$p_0(t,(x,\xi),(y,\eta)) \le \frac{C_1}{t^{N/2}}\frac{1}{\sqrt{2\pi t}}\exp\left[-\frac{(|\xi|+|\eta|)^2}{2t}\right], \tag{3.5}$$

so that

$$\liminf_{t\downarrow 0}\Big[-2t\log p_0(t,(x,\xi),(y,\eta))\Big] \ge (|\xi|+|\eta|)^2. \tag{3.6}$$

For the upper estimate of $\limsup_{t\downarrow 0}[-2t\log p_0(t,(x,\xi),(y,\eta))]$, we note that

$$\begin{aligned} &p_0(t,(x,\xi),(y,\eta)) \\ &\ge \int_\delta^{2\delta} q(t+\sigma,x,y)\frac{2(|\xi|+|\eta|+2\sigma)}{\sqrt{2\pi t^3}} \exp\left[-\frac{(|\xi|+|\eta|+2\sigma)^2}{2t}\right] d\sigma \\ &\ge \frac{2(|\xi|+|\eta|+2\delta)}{\sqrt{2\pi t^3}}\exp\left[-\frac{(|\xi|+|\eta|+4\delta)^2}{2t}\right]\int_\delta^{2\delta} q(t+\sigma,x,y)d\sigma. \end{aligned}$$

Choose now $t > 0$ and $\delta > 0$ so that $t + 2\delta \le t_0$. Then, (3.4) ensures

$$\begin{aligned} &p_0(t,(x,\xi),(y,\eta)) \\ &\ge \frac{2(|\xi|+|\eta|+2\delta)}{\sqrt{2\pi t^3}}\delta\exp\left[-\frac{d(x,y)^2+\varepsilon}{\delta}\right]\exp\left[-\frac{(|\xi|+|\eta|+4\delta)^2}{2t}\right], \end{aligned}$$

which implies

$$\limsup_{t\downarrow 0}\left[-2t\log p_0(t,(x,\xi),(y,\eta))\right] \le (|\xi|+|\eta|+4\delta)^2.$$

Since δ can be taken to be arbitrarily small, we then get

$$\limsup_{t\downarrow 0}\left[-2t\log p_0(t,(x,\xi),(y,\eta))\right] \le (|\xi|+|\eta|)^2,$$

and, together with (3.6),

$$\lim_{t\downarrow 0}\left[-2t\log p_0(t,(x,\xi),(y,\eta))\right] = (|\xi|+|\eta|)^2.$$

But $d_0((x,\xi),(y,\eta)) = |\xi|+|\eta|$ provided $\xi\eta\le 0$. Hence we have (2.2) in this case.

We next assume $\xi\eta > 0$, and let first $\sqrt{d(x,y)^2+(\xi-\eta)^2} \le |\xi|+|\eta|$. Taking account of the estimate deriving (3.5), we then have

$$\begin{aligned} &p_0(t,(x,\xi),(y,\eta)) \\ &\le q(t,x,y)\frac{1}{\sqrt{2\pi t}}\exp\left[-\frac{(\xi-\eta)^2}{2t}\right] + \frac{C_1}{t^{N/2}}\frac{1}{\sqrt{2\pi t}}\exp\left[-\frac{(|\xi|+|\eta|)^2}{2t}\right] \\ &\le \frac{1}{\sqrt{2\pi t}}\left(\exp\left[-\frac{d(x,y)^2-\varepsilon+(\xi-\eta)^2}{2t}\right] + \frac{C_1}{t^{N/2}}\exp\left[-\frac{(|\xi|+|\eta|)^2}{2t}\right]\right), \end{aligned}$$

which implies that

$$\liminf_{t\downarrow 0}\left[-2t\log p_0(t,(x,\xi),(y,\eta))\right] \ge d(x,y)^2-\varepsilon+(\xi-\eta)^2.$$

Since ε is arbitrary and $\sqrt{d(x,y)^2+(\xi-\eta)^2} \le |\xi|+|\eta|$, we have

$$\liminf_{t\downarrow 0}\left[-2t\log p_0(t,(x,\xi),(y,\eta))\right] \ge d_0(x,y)^2$$

in this case. By the similar argument to that in the case of $\xi\eta\le 0$, we can also show that

$$\limsup_{t\downarrow 0}\left[-2t\log p_0(t,(x,\xi),(y,\eta))\right] \le d_0(x,y)^2$$

in this case, to obtain (2.2).

The proof for the case $|\xi|+|\eta| < \sqrt{d(x,y)^2+(\xi-\eta)^2}$ is similar and will be omitted.

The uniform convergence in $|x|\vee|\xi|\vee|y|\vee|\eta|\le M$ is also clear from the argument above. □

Proof of (2.3). *First step (lower bound).* Note that (3.2) is reduced to

$$(3.7) \qquad p_0(t,(x,0),(y,0)) = \int_0^\infty q(t+\sigma,x,y)\frac{4\sigma}{\sqrt{2\pi t^3}}\exp\left[-\frac{2\sigma^2}{t}\right]d\sigma$$
$$= \int_0^{t_0} + \int_{t_0}^\infty =: I(t) + II(t),$$

where t_0 is the constant for (3.4) determined according to arbitrary $\varepsilon > 0$ and C_2. It then follows from (3.4) that

$$I(t) \le \int_0^{t_0} \frac{4\sigma}{\sqrt{2\pi t^3}} e^{-f_1(t,\sigma)} d\sigma, \quad \text{where} \quad f_1(t,\sigma) = \frac{d(x,y)^2 - \varepsilon}{2(t+\sigma)} + \frac{2\sigma^2}{t}.$$

Simple calculation shows that

$$h(t,\sigma) := \frac{\partial}{\partial\sigma} f_1(t,\sigma) = -\frac{d(x,y)^2 - \varepsilon}{2(t+\sigma)^2} + \frac{4\sigma}{t}$$

and

$$h(t, ct^{1/3}) = \frac{-d(x,y)^2 + \varepsilon + 8c^3 + 16c^2t^{2/3} + 8ct^{4/3}}{2t^{2/3}(t^{2/3}+c)^2}.$$

We now assume that $d(x,y) > 0$ and $0 < \varepsilon < d(x,y)^2$. Then, for any constants $0 \le C_3 < (d(x,y)^2 - \varepsilon)^{1/3}/2 < C_4$, there exists a positive $t_1 (\le t_0)$ such that, for any $t \in (0, t_1]$, $f_1(t,\sigma)$ is decreasing in $\sigma \in (0, C_3 t^{1/3})$, and is increasing in $\sigma \in (C_4 t^{1/3}, \infty)$. It then follows that

$$(3.8) \qquad \int_0^{C_3 t^{1/3}} \frac{4\sigma}{\sqrt{2\pi t^3}} e^{-f_1(t,\sigma)} d\sigma \le e^{-f_1(t, C_3 t^{1/3})} \int_0^{C_3 t^{1/3}} \frac{4\sigma}{\sqrt{2\pi t^3}} d\sigma$$
$$= e^{-f_1(t, C_3 t^{1/3})} \frac{2}{\sqrt{2\pi}} {C_3}^2 t^{-5/6},$$

$$(3.9) \qquad \int_{C_4 t^{1/3}}^{t_0} \frac{4\sigma}{\sqrt{2\pi t^3}} e^{-f_1(t,\sigma)} d\sigma \le e^{-f_1(t, C_4 t^{1/3})} \frac{2}{\sqrt{2\pi}} (t_0^2 - {C_4}^2 t^{2/3}),$$

provided $0 < t < t_1^3/C_4$. Furthermore, we observe that

$$(3.10) \qquad \int_{C_3 t^{1/3}}^{C_4 t^{1/3}} \frac{4\sigma}{\sqrt{2\pi t^3}} e^{-f_1(t,\sigma)} d\sigma$$
$$\le \exp\left[-\min_{C_3 \le c \le C_4} f_1(t, ct^{\frac{1}{3}})\right] \frac{2}{\sqrt{2\pi}} ({C_4}^2 - {C_3}^2) t^{-5/6}.$$

Because

$$f_1(t, ct^{1/3}) = \frac{d(x,y)^2 - \varepsilon + 4c^3 + 4c^2 t^{2/3}}{2t^{1/3}(c + t^{2/3})},$$
$$\frac{\partial}{\partial c} f_1(t, ct^{1/3}) = \frac{-d(x,y)^2 + \varepsilon + 8c^3 + 16c^2 t^{2/3} + 8ct^{4/3}}{2t^{1/3}(c + t^{2/3})^2},$$

the point $c(t)$ which attains $\min_{C_3 \le c \le C_4} f_1(t, ct^{1/3})$ converges to $C_5 = (d(x,y)^2 - \varepsilon)^{1/3}/2$ as $t \downarrow 0$. Since $t^{1/3} f_1(t, ct^{1/3})$ is uniformly continuous in c near C_5, we have

$$\text{(3.11)} \qquad \lim_{t \downarrow 0} 2t^{1/3} f_1(t, c(t)t^{1/3}) = \left. \frac{d(x,y)^2 - \varepsilon + 4c^3}{c} \right|_{c=C_5} = 3(d(x,y)^2 - \varepsilon)^{2/3}/2.$$

We finally note that, with the aid of (3.3),

$$\text{(3.12)} \qquad \begin{aligned} II(t) &\le \int_{t_0}^{\infty} \frac{C_1}{(t+\sigma)^{N/2}} \frac{1}{\sqrt{2\pi t}} \frac{d}{d\sigma} \left(-\exp\left[-\frac{2\sigma^2}{t} \right] \right) d\sigma \\ &\le \frac{C_1}{(t+t_0)^{N/2}} \frac{1}{\sqrt{2\pi t}} \exp\left[-\frac{2t_0^2}{t} \right]. \end{aligned}$$

Now the formulas (3.7)–(3.12) ensure

$$\begin{aligned} &\liminf_{t \downarrow 0} \left[-2t^{1/3} \log p_0(t, (x,0), (y,0)) \right] \\ &\ge \left[\lim_{t \downarrow 0} 2t^{1/3} f_1(t, C_3 t^{1/3}) \right] \wedge \left[\lim_{t \downarrow 0} 2t^{1/3} f_1(t, C_4 t^{1/3}) \right] \wedge 3(d(x,y)^2 - \varepsilon)^{2/3} \\ &= 3(d(x,y)^2 - \varepsilon)^{2/3}. \end{aligned}$$

Since ε is arbitrary, we obtain the lower bound

$$\liminf_{t \downarrow 0} \left[-2t^{1/3} \log p_0(t, (x,0), (y,0)) \right] \ge 3d(x,y)^{4/3}.$$

In the case where $d(x,y) = 0$, we only have to employ (3.12) with $t_0 = 0$.

Second step (upper bound). Fix an arbitrary $\varepsilon > 0$ and choose two constants $0 \le C_6 < (d(x,y)^2 + \varepsilon)^{1/3}/2 < C_7$. We then note that

$$\begin{aligned} &p_0(t, (x,0), (y,0)) \\ &\ge \int_{C_6 t^{1/3}}^{C_7 t^{1/3}} q(t+\sigma, x, y) \frac{4\sigma}{\sqrt{2\pi t^3}} \exp\left[-\frac{2\sigma^2}{t} \right] d\sigma \\ &\ge \exp\left[-\left(f_2(t, C_6 t^{1/3}) \vee f_2(t, C_7 t^{1/3}) \right) \right] \frac{2}{\sqrt{2\pi}} (C_7{}^2 - C_6{}^2) t^{-5/6} \end{aligned}$$

for $t + C_7 t^{1/3} \le t_0$, where

$$f_2(t, \sigma) = \frac{d(x,y)^2 - \varepsilon}{2(t+\sigma)} + \frac{2\sigma^2}{t}.$$

This implies

$$\limsup_{t\downarrow 0}\left[-2t^{1/3}\log p_0(t,(x,0),(y,0))\right] \le \left[\frac{d(x,y)^2+\varepsilon+4{C_6}^3}{C_6}\right]\vee\left[\frac{d(x,y)^2+\varepsilon+4{C_7}^3}{C_7}\right].$$

Letting $C_6, C_7 \to (d(x,y)^2+\varepsilon)^{1/3}/2$ first, and then $\varepsilon \to 0$ in the above, we obtain the upper bound

$$\limsup_{t\downarrow 0}\left[-2t^{1/3}\log p_0(t,(x,0),(y,0))\right] \le 3d(x,y)^{4/3}. \qquad \square$$

4. Proof of the other assertions

Proof of Proposition 2.2. Suppose that ρ_n takes its maximum at $\xi_n \in [-1/n, 1/n]$. It is then clear that $\rho_n(\xi_n) \to \infty$ as $n \to \infty$. Take now the polygon combining $(x,\xi), (x,\rho_n(\xi_n)), (y,\rho_n(\xi_n))$ and (y,η), and measure its length with the metric g_n. It then follows that

$$d_n((x,\xi),(y,\eta)) \le |\xi|+|\eta|+\frac{2}{n}+\frac{d(x,y)}{\sqrt{1+\rho_n(\xi_n)}}.$$

Furthermore, the inequality $d_n((x,\xi),(y,\eta)) \le \sqrt{d(x,y)^2+(\xi-\eta)^2}$ clearly holds. We thus have

$$d_n((x,\xi),(y,\eta)) \le \left(|\xi|+|\eta|+\frac{2}{n}+\frac{d(x,y)}{\sqrt{1+\rho_n(\xi_n)}}\right)\wedge\sqrt{d(x,y)^2+(\xi-\eta)^2},$$

so that

$$\limsup_{n\to\infty} d_n((x,\xi),(y,\eta)) \le d_0((x,\xi),(y,\eta))$$

uniformly in $|x|\vee|\xi|\vee|y|\vee|\eta| \le M$.

For the converse inequality, it is enough to show that, for any $\varepsilon > 0$,

$$d_n(x,y) \ge (|\xi|+|\eta|-2\varepsilon)\wedge\sqrt{d(x,y)^2+(\xi-\eta)^2}, \qquad n \ge 1/\varepsilon. \tag{4.1}$$

Suppose first that $\xi\wedge\eta \ge \varepsilon$ and $n \ge 1/\varepsilon$. It is clear that the length of the segment connecting (x,ξ) and (y,η) is not less than $\sqrt{d(x,y)^2+(\xi-\eta)^2}$. Further, there is another geodesic curve from (x,ξ) and (y,η) which goes through two points (x',ε) and (y',ε). But $d_n((x,\xi),(x',\varepsilon)) \ge |\xi|-\varepsilon$ and $d_n((y,\eta),(y',\varepsilon)) \ge |\eta|-\varepsilon$. Hence we obtain (4.1).

Suppose next that $|\xi| < \varepsilon \le \eta$ and $n \ge 1/\varepsilon$. Then the geodesic curve from (x,ξ) to (y,η) passes through a point (z,ε) on the line $\xi=\varepsilon$, and $d_n((z,\varepsilon),(y,\eta)) \ge \eta-\varepsilon$. Thus formula (4.1) holds.

The proof of (4.1) for the remaining cases is similar, and will be omitted. $\square$

Proof of Theorem 2.3. Since the symmetric form $\mathcal{E}_n$ satisfies the Nash inequality

$$\|f\|_2^{2+2/(N+1)} \le C_8 \mathcal{E}_n(f,f)\|f\|_1^{2/(N+1)}, \qquad f \in D[\mathcal{E}_n] \bigcap L^1(\mathbb{R}^{N+1}),$$

it follows that

$$(4.2) \qquad p_n(t,(x,\xi),(y,\eta)) \le \frac{C_9}{t^{(N+1)/2}} e^{-C_{10} d_n((x,\xi),(y,\eta))^2/t},$$

where C_9 and C_{10} are constants independent of $t > 0$, $(x,\xi),(y,\eta) \in \mathbb{R}^{N+1}$ and $n \in \mathrm{N}$ (see [1]). On the other hand, for any $\varphi \in L^1(\mathbb{R}, d\xi)$ and $n \in \mathrm{N} \cup \{0\}$,

$$(4.3) \qquad \int_{\mathbb{R}} p_n(t,(x,\xi),(y,\eta))\varphi(\eta)d\eta = E_\xi[q(t+\phi_n(t),x,y)\varphi(B(t))],$$

where $\phi_0(t) = \phi(t)$ and

$$\phi_n(t) = \frac{1}{2}\int_{\mathbb{R}} \rho_n(\eta)\phi(t,\eta)d\eta = \frac{1}{2}\int_0^t \rho_n(B(s))ds, \quad n \in \mathrm{N},$$

$\phi(t,\eta)$ being the local time of $B(t)$ at η ; $\phi(t,\eta) = \lim_{\varepsilon\to 0} 1/(2\varepsilon)\int_0^t I_{(\eta-\varepsilon,\eta+\varepsilon)}$ $(B(s))ds$. Now,

$$\begin{aligned}
&\int_{\mathbb{R}} |p_n(t,(x,\xi),(y,\eta)) - p_0(t,(x,\xi),(y,\eta))|^2 d\eta \\
&= \int_{\mathbb{R}} (p_n - p_0)p_n d\eta - \int_{\mathbb{R}} (p_n - p_0)p_0 d\eta =: I_n + II_n
\end{aligned}$$

and

$$\begin{aligned}
I_n &= E_\xi[\{q(t+\phi_n(t),x,y) - q(t+\phi(t),x,y)\}p_n(t,(x,\xi),(y,B(t)))]^2 \\
&\le E_\xi[|q(t+\phi_n(t),x,y) - q(t+\phi(t),x,y)|^2]E_\xi[p_n(t,(x,\xi),(y,B(t)))^2].
\end{aligned}$$

We further note that $\phi_n(t) \to \phi(t)$ almost surely. Thereby, in view of (3.3), we see that

$$\lim_{n\to\infty} E_\xi[|q(t+\phi_n(t),x,y) - q(t+\phi(t),x,y)|^2] = 0$$

by the dominated convergence theorem. But, due to (4.2), $E_\xi[p_n(t,(x,\xi),$ $(y,B(t)))^2]$ is bounded in $n \in \mathrm{N}$. We thus obtain $\lim_{n\to\infty} I_n = 0$. The proof of $\lim_{n\to\infty} II_n = 0$ is similar and will be omitted. □

Proof of Corollary 2.4. It is clear that

$$d_n((x,\xi),(y,\eta)) \ge |\xi - \eta| - \frac{2}{n}.$$

Hence, we have from (4.2) that

$$(4.4) \qquad p_n(t,(x,\xi),(y,\eta)) \le \frac{C_9}{t^{(N+1)/2}} e^{-C_{11}\eta^2/t}, \qquad \eta \in \mathbb{R},\ n \in \mathbb{N},$$

for some positive constant C_{11} independent of $\eta \in \mathbb{R}$ and $n \in \mathbb{N}$. Now the assertion follows from Theorem 2.3 and the dominated convergence theorem. □

References

1. E. A. Carlen, S. Kusuoka and D. W. Stroock, Upper bounds for symmetric Markov transition functions, Ann. Inst. Henri Poincaré, sup. au n°2 (1987), 245-287.
2. K. Fukaya, Collapsing Riemannian manifolds and eigenvalues of Laplace operator, Invent. Math., **87**(1987), 517-547.
3. M. Fukushima, Y. Oshima and M. Takeda : Dirichlet Forms and Symmetric Markov Processes, Walter de Gruyter, 1994.
4. N. Ikeda and S. Watanabe : The local structure of a class of diffusions and related problems, in Proceedings of the Second Japan-USSR Symposium on Probability Theory, Edited by G. Maruyama and Yu. V. Prohorov, Lect. Notes in Math. **330**, Springer, 1973, 124-169.
5. K. Itô and H. P. McKean, Jr., Diffusion Processes and their Sample Paths, Second Edition, Springer, 1974.
6. A. Kasue and H. Kumura, Spectral convergence of Riemannian manifolds I,II, I; Tôhoku Math. J., **46**(1994), 147–179, II; Tôhoku Math. J., **48**(1996), 71–120.
7. A. Kasue, H. Kumura and Y. Ogura, Convergence of heat kernels on a compact manifolds, preprint.
8. S. A. Molchanov, Diffusion processes and Riemannian geometry, Russian Math. Survey, **30**(1975), 1–63.
9. H. Pham Huy and E. Sanchez-Palencia, Phénoménes de transmission â travers des couches minces de conductivité élevée, J. Math. Anal. Appl. Conference, **47**(1974), 284–309.
10. Qian Zhongmin and Wei Guoqiang, Large deviations for symmetric diffusion processes, Chin. Ann. Math., **13B**:4(1992), 430–439.
11. E. Sanchez-Palencia, Un type de perturbations singuliéres dans les problémes de transmission, C. R. Acad. Sci. Paris, Série A, **268**(1969), 1200–1203.
12. S. R. S. Varadhan, On the behavior of the fundamental solution of the heat equation with variable coefficients, Comm. Pure Appl. Math., **20**(1967), 431-455.
13. S. R. S. Varadhan, Diffusion processes in a small interval, Comm. Pure Appl. Math., **20**(1967), 659–685.

References

Van Vleck-Pauli formula for Wiener integrals and Jacobi fields

Nobuyuki Ikeda[1] and Shojiro Manabe[2]

[1] Department of Computer Science, Ritsumeikan University, Kusatsu, Shiga 525-77, Japan
[2] Department of Mathematics, Graduate School of Science, Osaka University, Toyonaka, Osaka 560, Japan

1. Introduction

In this paper, we are concerned with the explicit evaluation of Wiener integrals from the viewpoint of geometry of the space of paths. Our main purpose is to emphasize the close ties between explicit expressions of Wiener integrals associated with quadratic functionals and aspects of the theory of Jacobi fields.

In [13], we have studied Wiener integrals characterized by generalized Lagrangian L quadratic in coordinates and velocities and have obtained an analogous formula to the famous one of Van Vleck [39] and Pauli [31] for propagators in quantum mechanics which plays a basic role in the theory of semi-classical approximation, (for examples, see [6] – [8], [22],[24], and [32]). This implies that above Wiener integrals are expressed in terms of classical mechanics associated with L. Hence this fact is an analogue of the relation between classical and quantum mechanics, (for examples, see [4], [8] and [36]). On the other hand, in [3], C.DeWitt-Morette has investigated properties of the Van Vleck matrix by using techniques based on Jacobi fields (also, see [2]). Kac [20] and [21] also studied the closely related problems and added a number of interesting observations from a mathematical point of view, (also, see [1], [33]– [35]). We give a new proof of the result in [13] mentioned above and its extension, by combining the idea used in [3] with a decomposition of the symmetric Hilbert-Schmidt operators associated with quadratic Wiener functionals into a Volterra operator and an operator with finite dimensional range (see [12] and [13]).

The organization of the paper is as follows. In Section 2, we give generalities on the integrals associated with quadratic Wiener functionals F. In Section 3, we introduce the notion of generalized Lagrangians L corresponding to F given by (3.1) below and as in [3], we consider Jacobi fields associated with L. By using these notions, we give a sketch of the proof of Van Vleck-Pauli formula for the Wiener integrals associated with F. In Section 4, we show how one can obtain an extension of results of Section 3 in case of Gaussian processes. In Section 5, we give several remarks on Lévy's stochastic area formula from a geometric point of view. In Appendix, we give a sketch of the proof of Theorem 2.1 below and related remarks.

Before closing the introduction, we note that in [17], by using techniques based on generalized uniform complex measures in the Hilbertian metric space, K.Itô gave the explicit expression of the propagators with typical potentials in quantum mechanics.

2. General setting

For a fixed positive t, set $W_0^d = \{w \in C([0,t] \longrightarrow \mathbf{R}^d) : w(0) = 0\}$. Let $\{W_0^d, P\}$ be the d-dimensional Wiener space and H be the Cameron-Martin subspace of W_0^d. Let $\mathrm{HS}^{(s)}(H)$ be the space of symmetric Hilbert-Schmidt operators on H. The space $L^2(W_0^d, P)$ can be decomposed into the direct sum of mutually orthogonal subspaces known as homogeneous chaos in the sense of Wiener and Itô :

$$L^2(W_0^d, P) = \sum_{n=0}^{\infty} \oplus C_n,$$

([15]). For any quadratic Wiener functional $F \in C_2$, there corresponds a unique element B of $\mathrm{HS}^{(s)}(H)$ (for details, see [14]). From now on, we use the notation $F(w) = \langle Bw, w\rangle$. We first establish a general formula for the Wiener integral associated with F under some conditions. In order to state a result, we prepare several notations. For given linearly independent elements $h_1, \cdots, h_m$ in H, we set $\eta = \{h_1, \cdots, h_m\}$ and define $\eta(w) \in \mathbf{R}^m$ by

$$\eta(w) = \{\langle h_1, w\rangle, \cdots, \langle h_m, w\rangle\},$$

where

$$\langle h, w\rangle = \sum_{i=1}^{d} \int_0^t \dot{h}^i(t)dw^i(t) \quad \text{for } h = \begin{pmatrix} h^1 \\ h^2 \\ \vdots \\ h^d \end{pmatrix}.$$

We denote by $\eta\eta^*$ an $m \times m$ symmetric matrix given by

$$(\eta\eta^*)_{ij} = \langle h_i, h_j\rangle, \quad i, j = 1, \cdots, m.$$

Let δ_0 be the Dirac's delta function at the origin 0 in $\mathbf{R}^m$ and $\delta_0(\eta(w))$ be the pullback of δ_0 by $\eta(w)$ in the sense of Watanabe [40]. Then there exists a smooth nonnegative function $p(a\,; B)$ such that for any smooth function f with compact support on $\mathbf{R}$,

$$\int_{W_0^d} f(\langle Bw, w\rangle)\delta_0(\eta(w))P(dw) = \int_{\mathbf{R}} f(a)p(a\,; B)da,$$

(see [15],[26],[40]). If the integral of the right hand side of (2.1) below is well-defined, we use the following notation:

$$\int_{W_0^d} \exp\left[z\langle Bw, w\rangle\right]\delta_0(\eta(w))P(dw) = \int_{\mathbf{R}} \exp[za]p(a\,;\,B)da, \qquad z \in \mathbf{C}. \tag{2.1}$$

Then if the Wiener functional $\exp\left[z\langle Bw, w\rangle\right]$ satisfies some conditions mentioned in [40], the left hand side of (2.1) denotes the generalized expectation of generalized Wiener functional $\exp\left[z\langle Bw, w\rangle\right]\delta_0(\eta(w))$, (see [40]) and coincides with the integral of $\exp\left[z\langle Bw, w\rangle\right]$ by the measure:

$$\left((2\pi)^m \det(\eta\eta^*)\right)^{-1/2} P(\cdot|\eta(w) = 0).$$

For the above system $\eta = \{h_1, \cdots, h_m\}$, we define two subspaces of H by

$$H_\eta = \text{ closed linear subspace spanned by } h_1, \cdots, h_m$$

and

$$H_\eta^\perp = \text{ the orthogonal complement of } H_\eta \text{ in } H.$$

We define two operators according to above subspaces:

$$B^\# = P_{H_\eta^\perp} B P_{H_\eta^\perp} \quad \text{ and } \quad B_\eta = P_{H_\eta} B P_{H_\eta}.$$

In the rest of this paper, we use the comlexification of the d-dimensional Cameron-Martin space H and the natural extension of B,$B^\#$,B_η etc. on H. To avoid nonessential complexities of notation, we denote these by the same notations. For any $B \in \mathrm{HS}^{(s)}(H)$, we denote by $\det_2(I - 2zB)$ the regularized determinant of $I - 2zB$ and set $\sigma(B) = \max\{|\lambda_n| : \{\lambda_n\}$ the eigenvalues of $B\}$.

We can now state our first result.

Theorem 2.1. *If $z \in \mathbf{C}$ and $2|Re(z)|\sigma(B^\#) < 1$,*

$$\int_{W_0^d} e^{z\langle Bw, w\rangle}\delta_0(\eta(w))P(dw) \tag{2.2}$$

$$= \left(\frac{1}{2\pi}\right)^{m/2}\left[\det(\eta\eta^*)\det_2(I - 2zB^\#)\right]^{-1/2} \exp\left[-z\,\mathrm{tr}(\mathrm{B}_\eta)\right].$$

Remark 1. Let S be the space of all symmetric $n \times n$ matrices A with $\mathrm{Re}A$ positive definite. Then there exists a unique analytic branch of $S \ni A \longrightarrow (\det A)^{1/2} \in \mathbf{C}$ such that when A is real, $(\det A)^{1/2} > 0$, (for example, see

Hörmander [10], Chapter 3). In this paper, by using the same idea, we define $\left(\det_2(I-2zB)\right)^{1/2}$ for $B \in \mathrm{HS}^{(s)}(H)$ in standard manner.

For the proof of Theorem 2.1, see Appendix or Takanobu and Watanabe [38].

Remark 2. The formula (2.3) in [13] is not correct. It should be read as above formula (2.2). For details and related remarks, see Appendix.

In the rest of the paper, we always consider a subclass $\mathrm{HS}_V^{(s)}(H)$ of $\mathrm{HS}^{(s)}(H)$ where $\mathrm{HS}_V^{(s)}(H)$ is the totality of elements of satisfying the following assumption:

Assumption A B has the following decomposition:

$$B = B_0 + B_1,$$

where B_0 is a Volterra operator and B_1 is an operator with a finite dimensional range space.

According to this decomposition, let us take a system $\eta = \{h_1, h_2, \cdots, h_m\}$ of independent elements of H such that they span $\mathrm{Range}(B_1)$. We define H_η, $H_\eta^\perp$ etc. as before. Our next task is to reduce the calculation of $\det_2(I-2zB^\#)$ to that of finite dimensional determinant like Ikeda-Kusuoka-Manabe [12]. To this end, we define an operator $B_1^\#$ by

$$B^\# = B_0 + B_1^\#, \quad \text{i.e. }, B_1^\# = -P_{H_\eta} B_0.$$

Letting

$$Q(z) = I - 2zB_1^\#(I-2zB_0)^{-1}, \tag{2.3}$$

we see that

$$I - 2zB^\# = Q(z)(I-2zB_0).$$

Lemma 2.2. *Under the same condition as in Theorem 2.1, it holds that*

$$\begin{aligned} \det_2(I-2zB^\#) &= \det Q(z)\exp\left[2z\,\mathrm{tr}(B_1^\#)\right] \qquad (2.4)\\ &= \det Q(z)\exp\left[-2z\,\mathrm{tr}(P_{H_\eta}B_0)\right]. \end{aligned}$$

Proof. Because

$$I - 2zB^\# = \left\{I - 2zB_1^\#(I-2zB_0)^{-1}\right\}(I-2zB_0),$$

we have

$$\det_2(I-2zB^\#) = \det_2\left[\left\{I - 2zB_1^\#(I-2zB_0)^{-1}\right\}(I-2zB_0)\right]. \tag{2.5}$$

In order to calculate the right hand side, we approximate B_0 by a sequence of operators C_n with finite dimensional range spaces in Hilbert-Schmidt norm: $C_n \longrightarrow B_0$. Then, by noting the fact that $2zB_1^{\#}(I-2zC_n)^{-1}+2zC_n-2zB_1^{\#}(I-2zC_n)^{-1}2zC_n$ is nuclear operator,

$$\begin{aligned}
&\det{}_2\left[\left\{I-2zB_1^{\#}(I-2zC_n)^{-1}\right\}(I-2zC_n)\right] \\
&\quad=\det\left[I-\left\{2zB_1^{\#}(I-2zC_n)^{-1}+2zC_n-2zB_1^{\#}(I-2zC_n)^{-1}2zC_n\right\}\right] \\
&\qquad\times\exp\left[\operatorname{tr}\left\{2zB_1^{\#}(I-2zC_n)^{-1}+2zC_n-2zB_1^{\#}(I-2zC_n)^{-1}2zC_n\right\}\right] \\
&\quad=\det\left[I-2zB_1^{\#}(I-2zC_n)^{-1}\right]\det{}_2(I-2zC_n) \\
&\qquad\times\exp\left[\operatorname{tr}\left\{2zB_1^{\#}(I-2zC_n)^{-1}-2zB_1^{\#}(I-2zC_n)^{-1}2zC_n\right\}\right] \\
&\quad=\det\left[I-2zB_1^{\#}(I-2zC_n)^{-1}\right]\det{}_2(I-2zC_n)\exp\left[\operatorname{tr}(2zB_1^{\#})\right].
\end{aligned}$$

Combining this with (2.5) and letting $n\to\infty$, we have (2.4), where we have used the fact that $\det_2(I+T)$ is continuous in T with respect to Hilbert-Schmidt norm, see eg., Simon [37] .

By virtue of this lemma, it follows from Theorem 1.2 that

Corollary 2.3. *Under the same condition as in Theorem 2.1, it holds that*

$$\begin{aligned}
&\int_{W_0} e^{z\langle Bw,w\rangle}\delta_0(\eta(w))P(dw) \qquad (2.6)\\
&\quad=\left(\frac{1}{2\pi}\right)^{m/2}\left[\det(\eta\eta^*)\det Q(z)\right]^{-1/2}\exp\left[-z\operatorname{tr}(P_{H_\eta}B_1)\right]
\end{aligned}$$

Proof. Since $B_\eta=P_{H_\eta}B_0P_{H_\eta}+P_{H_\eta}B_1P_{H_\eta}$, we have $\operatorname{tr}(B_\eta)=\operatorname{tr}(P_{H_\eta}B_0)+\operatorname{tr}(P_{H_\eta}B_1)$. Hence by the above lemma, we have $\left[\det_2(I-2zB^{\#})\right]^{-1/2}e^{-z\operatorname{tr}(B_\eta)}$ $=\left[\det Q(z)\right]^{-1/2}e^{-z\operatorname{tr}(P_{H_\eta}B_1)}$, which completes the proof of (2.6).

3. Van Vleck-Pauli formula–classical case

As in [13], for simplicity we restrict ourselves in case of $d=2$ and we set $W_0=W_0^2$. Let us introduce the following quadratic Wiener functional:

$$F_t(w_x;\alpha,\beta) = \frac{\alpha}{2}\int_0^t \{w_x^1(s)dw_x^2(s) - w_x^2(s)dw_x^1(s)\} - \sum_{i=1}^{2}\beta_i\int_0^t (w_x^i(s))^2 ds,$$

where $w_x : W_0 \longrightarrow W$, $w_x(s) = x + w(s)$, $0 \le s < \infty$, $x \in \mathbf{R}^2$, $\alpha \in \mathbf{R}$ and $\beta = (\beta_1, \beta_2)$, $\beta_i \in \mathbf{R}$, $i = 1, 2$. It is useful to consider the generalized quadratic Lagrangian given by

$$L(x,\dot{x};z) = \frac{1}{2}\sum_{i=1}^{2}(\dot{x}^i)^2 - \frac{z\alpha}{2}(x^1\dot{x}^2 - x^2\dot{x}^1) + z(\beta_1(x^1)^2 + \beta_2(x^2)^2), \quad z \in \mathbf{C}. \tag{3.1}$$

Let Ω be the space of continuous functions $\phi : [0,t] \longrightarrow \mathbf{C}^2$ such that $\phi = \begin{pmatrix} \phi^1 \\ \phi^2 \end{pmatrix}$, ϕ^i, $i = 1, 2$: absolutely continuous with square integrable derivatives $\dot{\phi}^i$, $i = 1, 2$. Then $H = \{h \in \Omega;\, h(0) = 0\}$. The action integral $\mathcal{E}$ on Ω is defined by

$$\mathcal{E}[\phi] = \int_0^t L(\phi(s), \dot{\phi}(s); z)ds, \quad \phi \in \Omega. \tag{3.2}$$

We set $\Omega_{x,y} = \{\phi \in \Omega;\, \phi(0) = x, \phi(t) = y\}$ for $x, y \in \mathbf{R}^2$. An element $\phi_{cr} = \phi_{cr}(\cdot; x, y, z)$ of $\Omega_{x,y}$ is called a *critical path* (a *classical path* or a *geodesic*) if it is a critical point of $\mathcal{E}$ on $\Omega_{x,y}$. For a critical path ϕ_{cr}, we set

$$S_{cr}(t,x,y;z) = \mathcal{E}[\phi_{cr}].$$

Following [3], we now introduce some notions about Jacobi fields. A vector field Y along ϕ_{cr} is called a *Jacobi field* along ϕ_{cr} if Y is obtained by variation of ϕ_{cr}. We use the notation $Y(s,p)$, $0 \le s \le t$ for a Jacobi field with

$$Y(0,p) = 0, \quad \dot{Y}(0,p) = p \in \mathbf{R}^2.$$

We denote by $\mathcal{J}_t$ the mapping

$$\mathbf{R}^2 \ni p \longrightarrow Y(t,p) \in \mathbf{C}^2.$$

Two points $(0,x)$ and $(s, \phi_{cr}(s))$,$s > 0$ are called *conjugate* if there exists a Jacobi field Y along ϕ_{cr} such that $Y(0) = 0$, $Y(s) = 0$.

Remark. The studies of complete systems of vector fields obtained by the variation of critical paths go back to the papers of Jacobi, (for example, see [3],[19] and also [29], [30]).

Now let

$$\tilde{F}_t(w_x;\alpha,\beta) = F_t(w_x;\alpha,\beta) - \frac{t^2(\beta_1+\beta_2)}{2}.$$

Then as mentioned in Section 2, there exists a unique symmetric Hilbert-Schmidt operator B on H associated with $\tilde{F}_t$ given by

$$B = B_0 + B_1, \quad B_0 = \frac{\alpha}{2}\mathcal{I}[J] + \begin{pmatrix} \beta_1 & 0 \\ 0 & \beta_2 \end{pmatrix} \mathcal{I}^2,$$
$$B_1[h] = -\frac{\alpha}{4} J h(t)\phi_0 - \begin{pmatrix} \beta_1 & 0 \\ 0 & \beta_2 \end{pmatrix} \mathcal{I}[h](t)\phi_0, \ \phi_0(s) = s, \ J = \begin{pmatrix} 0 & -1 \\ 1 & 0 \end{pmatrix}$$

where

$$\mathcal{I}[h](s) = \begin{pmatrix} \mathcal{I}[h^1](s) \\ \mathcal{I}[h^2](s) \end{pmatrix}, \quad \mathcal{I}[h^i](s) = \int_0^s h^i(u)du, i = 1, 2, \quad \text{for } h = \begin{pmatrix} h^1 \\ h^2 \end{pmatrix}.$$

It is clear that the operator B satisfies the Assumption A. In this case, $\eta = \{h_1, h_2\}$, where $h_1 = \phi_0 \begin{pmatrix} 1 \\ 0 \end{pmatrix}$ and $h_2 = \phi_0 \begin{pmatrix} 0 \\ 1 \end{pmatrix}$. We set $H_1 = \text{Range}(B_1)$, so $H_\eta = H_1$. Then we can define the operator $Q(z)$ by (2.3).

Theorem 3.1 (Van Vleck-Pauli formula). *Let $\sigma^* = \inf\{s\,;\,(s, \phi_{cr}(s))$ is conjugate with $(0, x)\}$. Then, for $0 < t < \sigma^*$, it holds that*

$$\int_{W_0} e^{zF_t(w_x;\alpha,\beta)} \delta_y(w_x(t)) P(dw) \tag{3.3}$$
$$= \frac{1}{2\pi}\left[\det\left(\frac{\partial^2}{\partial x \partial y} S_{cr}(t, x, y; z)\right)\right]^{1/2} \exp\left[-S_{cr}(t, x, y; z)\right],$$
$$\mathcal{J}_t = t\,{}^tQ(z), \tag{3.4}$$
$$\mathcal{J}_t = \left(\frac{\partial^2}{\partial x \partial y} S_{cr}(t, x, y; z)\right)^{-1}, \tag{3.5}$$

where

$$\frac{\partial^2 S_{cr}(t, x, y)}{\partial x \partial y} = \left(\frac{\partial^2 S_{cr}(t, x, y)}{\partial x^i \partial y^j}\right)_{i,j=1,2}.$$

Proof. By the Cameron-Martin formula, we have

$$\int_{W_0} e^{z\tilde{F}_t(w_x;\alpha,\beta)} \delta_y(w_x(t)) P(dw)$$
$$= \int_{W_0} e^{z\tilde{F}_t(w;\alpha,\beta)} \delta_0(w(t)) P(dw) \exp\left[-S_{cr}(t, x, y; z)\right].$$

In view of Corollary 2.3, it is sufficient to show that (3.4),(3.5) and

$$\text{tr}(P_{H_1} B_1 P_{H_1}) = -\frac{t^2(\beta_1 + \beta_2)}{2}. \tag{3.6}$$

Let ϕ_{cr} be a critical path such that $\phi_{cr} \in \Omega_{x,y}$ and X_1, X_2 be elements of the tangent space $T_{\phi_{cr}}(\Omega_{x,y})$ of $\Omega_{x,y}$ at ϕ_{cr}, (see [29]). We choose a smooth 2-parameter variation of $\phi_{cr} = \phi(s, x, y)$, $0 \le s \le t$:

$$\gamma : U \times [0,t] \longrightarrow \mathbf{R}^2, \qquad U \text{ is a neighborhood of } (0,0) \text{ in } \mathbf{R}^2$$

such that

$$\gamma(0,0;s) = \phi(s,x,y), \quad \frac{\partial\gamma}{\partial u_1}(0,0;s) = X_1(s), \quad \frac{\partial\gamma}{\partial u_2}(0,0;s) = X_2(s),$$
$$\frac{\partial^2}{\partial u_2 \partial u_1}\gamma(0,0;0) = \frac{\partial^2}{\partial u_2 \partial u_1}\gamma(0,0;t) = 0.$$

Define $\bar{\gamma}(u_1,u_2) \in \Omega$ by

$$\bar{\gamma}(u_1,u_2)(s) = \gamma(u_1,u_2;s), \quad 0 \le s \le t.$$

Then, as in [29], by repeating integration by parts, we obtain the following second variation formula for the action integral $\mathcal{E}$ given by (3.2):

$$\begin{aligned}\frac{\partial^2}{\partial u_2 \partial u_1}\mathcal{E}[\bar{\gamma}(u_1,u_2)]|_{(u_1,u_2)=(0,0)} &= \sum_{i=1}^{2}[X_1^i(t)\dot{X}_2^i(t) - X_1^i(0)\dot{X}_2^i(0)] \\ &-\frac{z\alpha}{2}[X_1^2(t)X_2^1(t) - X_1^2(0)X_2^1(0) - X_1^1(t)X_2^2(t) + X_1^1(0)X_2^2(0)] \quad (3.7)\\ &-\sum_{i=1}^{2}\int_0^t X_1^i(s)\left[\frac{d^2}{ds^2}X_2^i(s) - z\alpha\frac{d}{ds}(J[X_2(s)])^i - 2z\beta_i X_2^i(s)\right]ds.\end{aligned}$$

Setting, for $p,q = 1,2$,

$$\gamma_{pq}(u_1,u_2;s) = \phi_{cr}\Big(s,(x^1+\delta_p^1 u_1, x^2+\delta_p^2 u_1),(y^1+\delta_p^1 u_2, y^2+\delta_p^2 u_2)\Big),$$

$$X_{1p}(s) = \frac{\partial}{\partial y^p}\phi_{cr}(s,x,y) \quad \text{and} \quad X_{2q}(s) = \frac{\partial}{\partial x^q}\phi_{cr}(s,x,y), \quad 0 \le s \le t,$$

where δ_i^p is the Kronecker's delta, we have

$$\frac{\partial^2}{\partial u_1 \partial u_2}\gamma_{pq}(0,0;0) = \frac{\partial^2}{\partial u_1 \partial u_2}\gamma_{pq}(0,0;t) = 0,$$

$$X_{1p}(0) = 0, \quad X_{2q}(t) = 0, \quad X_{2p}^i(0) = \delta_p^i, \quad X_{1q}^i(t) = \delta_q^i, \quad p,q = 1,2.$$

Combining this with (3.7), we have

$$\begin{pmatrix} \frac{\partial^2}{\partial x^i \partial y^1}S_{cr}(t,x,y) \\ \frac{\partial^2}{\partial x^i \partial y^2}S_{cr}(t,x,y)\end{pmatrix} = \mathcal{J}_t^{-1}\begin{pmatrix}\delta_i^1 \\ \delta_i^2\end{pmatrix}, \quad i = 1,2,$$

This completes the proof of (3.5). The proof of (3.4) proceeds as in Section 5, so the details will be omitted. Finally (3.6) can be verified by a direct calculation, which completes the proof.

Before closing this section, we note that as in the case of the theory of Feynman path integrals, the formulae (3.3), (3.4) and (3.5) play an important role in the study of asymptotics and related topics based on techniques of integrals on function space, (for examples, see [6], [11], [14], [22], [27], [28], [36] and [38]).

4. Van Vleck-Pauli formula for a class of Gaussian process

We give a generalization of Van Vleck-Pauli formula for a class of Gaussian processes discussed in [12]. For simplicity, in this section we restrict ourselves in case of the Wiener space $\{W_0^d, P\}$ with $t = 1$ and $d = 2$. First of all, we summarize the framework in [12]. We set, for $w \in W_0$,

$$X_n(t,w) = \mathcal{I}^n[w](t), \quad X_0(t,w) = w(t), \quad 0 \le t \le 1, \qquad \text{for } w \in W_0,$$

where $\mathcal{I}^k$ ($\mathcal{I}$ was defined in Section 3) is the operator given by

$$\mathcal{I}^k = \mathcal{I}\mathcal{I}^{k-1}, \quad k \ge 1, \quad \mathcal{I}^0 \ : \ \text{identity}.$$

We consider a quadratic Wiener functional $F_n(w)$ given by, for a fixed positive integer n,

$$F_n(w) = (-1)^n \int_{X_n(w)[0,1]} \omega,$$

where ω is the differential 1-form on $\mathbf{R}^2$ given by

$$\omega = \frac{1}{2}(x^1 dx^2 - x^2 dx^1) \tag{4.1}$$

and $X_n(w)[0,1]$ is the curve in $\mathbf{R}^2$ given by $\{X_n(t,w)\,;\, 0 \le t \le 1\}$. Set, for $t, s \in [0,1]$,

$$a(t,s) = \frac{1}{2(2n)!}\left[\sum_{k=0}^{n-1} \binom{2n}{k} s^{2n-k}(-t)^k + \frac{1}{2}\binom{2n}{n} s^n(-t)^n\right],$$

and we define the operators B_0 and B_1 on H by

$$\frac{d}{dt}B_0[h](t) = \frac{1}{2}\mathcal{I}^{2n}[J[h]](t), \quad \frac{d}{dt}B_1[h](t) = -\int_0^1 a(1-t, 1-s)J[\dot{h}](s)ds.$$

Set $B = B_0 + B_1$. Then it is easy to see that $B \in \mathrm{HS}_V^{(s)}(H)$. We consider a system $\eta = \{h_0, h_1, \cdots, h_{2n+1}\}$ in H given by

$$h_{2i} = \mathcal{I}[\phi_i]\begin{pmatrix} 1 \\ 0 \end{pmatrix}, \quad h_{2i+1} = \mathcal{I}[\phi_i]\begin{pmatrix} 0 \\ 1 \end{pmatrix},$$

where $\phi_i(t) = t^i$, $i = 0, 1, \cdots, n$. The range space $H_1 = R(B_1)$ of B_1 on H is the subspace H_η of H generated by the system η in H. We define $H_1^\perp$, $B^\#$ etc, as in Section 2. We consider a quadratic form $F_B[\cdot; z]$ on H given

$$F_B[h;z] = \langle (I - 2zB)h, h\rangle, \quad h \in H, \ z \in \mathbf{C}.$$

For given $x \in \mathbf{R}^{2(n+1)}$, set

$$H(\eta, x) = \{h \in H\,;\, \eta(h) = x\}$$

where $\eta(h) = (\langle h_0, h\rangle, \langle h_1, h\rangle, \cdots, \langle h_{2(n+1)}, h\rangle)$. We call $\phi \in H(\eta, x)$ a *critical path* (a *classical path* or a *geodesic*) of $F_B[\cdot; z]$ on $H(\eta, x)$ if for every $h \in H_1^{\perp}$

$$DF_B[\phi; z](h) = 0,$$

i.e.,

$$\langle (I - 2zB)\phi, h\rangle = 0, \quad \text{for every } h \in H_1^{\perp},$$

where D denotes the Fréchet derivative.

This implies that if ϕ is a critical path of $F_B[\cdot; z]$ on $H(\eta, x)$, then

$$(I - 2zB)\phi \in H_1.$$

Hence we have

$$\left(\frac{d}{dt}\right)^{2(n+1)} \phi(t) - zJ[\dot{\phi}](t) = 0, \quad 0 \le t \le 1. \tag{4.2}$$

This can be regarded as the Euler-Lagrange equation in our case. Since (4.2) is a linear equation with constant coefficients, the corresponding Jacobi equation in our case, is the same type as (4.2). Hence if a vector field $\{\phi(t)\,;\, 0 \le t \le 1\}$ along ϕ_{cr} satisfies (4.2), then it is called the Jacobi field along ϕ_{cr}.

We consider the operator $Q(z)$ on H_1 defined by (2.3). Then we have the following

Lemma 4.1. *It holds that*

$$\mathrm{tr}(P_{H_1} B_0) = 0.$$

It follows that

$$\det{}_2(I - 2zB^{\#}) = \det Q(z).$$

Proof. This lemma follows at once from the matrix representation of $P_{H_1} B_0$ and Lemma 2.2 .

Next we define the mapping $\mathcal{J}_1$ by

$$\mathcal{J}_1 : \mathbf{R}^{2(n+1)} \ni p = \begin{pmatrix} p_0 \\ p_1 \\ \vdots \\ p_{2n+1} \end{pmatrix} \longrightarrow \begin{pmatrix} \langle \mathcal{J}(\cdot; p), h_0\rangle \\ \langle \mathcal{J}(\cdot; p), h_1\rangle \\ \vdots \\ \langle \mathcal{J}(\cdot; p), h_{2n+1}\rangle \end{pmatrix} \in \mathbf{C}^{2(n+1)} \tag{4.3}$$

where $\mathcal{J}(\cdot; p)$ is the Jacobi field, (i.e., the solution of (4.2)) such that

$$\begin{aligned} \mathcal{J}(0; p) &= 0 \\ \left(\frac{d}{dt}\right)^k \mathcal{J}(t; p)|_{t=0} &= k! \begin{pmatrix} p_{2(k-1)} \\ p_{2(k-1)+1} \end{pmatrix}, \quad k = 1, 2, \cdots, n+1 \\ \left(\frac{d}{dt}\right)^k \mathcal{J}(t; p)|_{t=0} &= 0, \quad k = n+2, \cdots, 2n+1. \end{aligned} \tag{4.4}$$

We set

$$S_{cr}(x; z, B) = F_B[\phi_{cr}; z].$$

Theorem 4.2 (Van Vleck-Pauli formula). *Assume that* $z \in \mathbf{C}$, $2|\mathrm{Re}(z)|\sigma(B^{\#}) < 1$. *Then*

(i)

$$^tQ(z) = K_1, \quad where \quad K_1 = A^{-1}\mathcal{J}_1, \quad A = (\langle h_i, h_j\rangle). \tag{4.5}$$

(ii) *For* $x \in \mathbf{R}^{2(n+1)}$,

$$\int_{W_0} e^{zF_n(w)}\delta_x(\eta(w))P(dw) \tag{4.6}$$
$$= \Big(\frac{1}{2\pi}\Big)^{n+1}\Big(\det \mathcal{J}_1\Big)^{-1/2} \exp[-\frac{1}{2}S_{cr}(x; z, B)].$$

Proof. The conclusion (ii) follows from (i),(2.6), Lemma 4.1 and the formula (2.6) in [13]. Hence it is sufficient to show (i). We set

$$\tilde{\mathcal{J}}(t;p) = (I - 2zB^{\#})^{-1}[h^*](t), \quad h^* = \sum_{i=0}^{2n+1} p_i h_i.$$

Then, for every $h \in H_1^{\perp}$,

$$\begin{aligned}
&\langle (I - 2zB)\tilde{\mathcal{J}}(\cdot;p), h\rangle \\
&= \langle (I - 2zB^{\#})\tilde{\mathcal{J}}(\cdot;p), h\rangle + 2z\langle (B_1^{\#} - B_1)\tilde{\mathcal{J}}(\cdot;p), h\rangle \\
&= \langle (I - 2zB^{\#})\tilde{\mathcal{J}}(\cdot;p), h\rangle \\
&= \langle h^*, h\rangle \\
&= 0,
\end{aligned}$$

which means that $\{\tilde{\mathcal{J}}(\cdot;p), 0 \le t \le 1\}$ is a Jacobi field. For $k = 1, \cdots, n+1$

$$\begin{aligned}
&\Big(\frac{d}{dt}\Big)^k \tilde{\mathcal{J}}(t;p)|_{t=0} \\
&= \Big(\frac{d}{dt}\Big)^k (I - 2zB_0)\tilde{\mathcal{J}}(t;p)|_{t=0} \\
&= \Big(\frac{d}{dt}\Big)^k \Big[(I - 2zB_0)(I - 2zB_0)^{-1}Q(z)^{-1}[h^*]\Big]|_{t=0} \\
&= \sum_{i=0}^{2n+1} p_i \Big(\frac{d}{dt}\Big)^k \Big(\sum_{j=0}^{2n+1} (Q(z)^{-1})_{ij} h_j(t)\Big)|_{t=0}
\end{aligned}$$

where

$$Q(z)^{-1}[h_i] = \sum_{j=0}^{2n+1} (Q(z)^{-1})_{ij} h_j(t).$$

Hence

$$\left(\frac{d}{dt}\right)^k \tilde{\mathcal{J}}(t;p)|_{t=0}$$

$$= k! \sum_{i=0}^{2n+1} p_i \left[\left(Q(z)^{-1}\right)_{i,2(k-1)} \begin{pmatrix} 1 \\ 0 \end{pmatrix} + \left(Q(z)^{-1}\right)_{i,2(k-1)+1} \begin{pmatrix} 0 \\ 1 \end{pmatrix} \right]$$

$$= k! \sum_{i=0}^{2n+1} p_i \begin{pmatrix} \left(Q(z)^{-1}\right)_{i,2(k-1)} \\ \left(Q(z)^{-1}\right)_{i,2(k-1)+1} \end{pmatrix}$$

$$= k! \begin{pmatrix} \left({}^t\left(Q(z)^{-1}\right)\right)(p)_{2(k-1)} \\ \left({}^t\left(Q(z)^{-1}\right)\right)(p)_{2(k-1)+1} \end{pmatrix}.$$

We thus have

$$\begin{pmatrix} \frac{d}{dt} \tilde{\mathcal{J}}(t;p)|_{t=0} \\ \frac{1}{2!}\left(\frac{d}{dt}\right)^2 \tilde{\mathcal{J}}(t;p)|_{t=0} \\ \vdots \\ \frac{1}{(n+1)!}\left(\frac{d}{dt}\right)^{n+1} \tilde{\mathcal{J}}(t;p)|_{t=0} \end{pmatrix} = \begin{pmatrix} \left({}^t\left(Q(z)^{-1}\right)\right)(p)_0 \\ \left({}^t\left(Q(z)^{-1}\right)\right)(p)_1 \\ \vdots \\ \left({}^t\left(Q(z)^{-1}\right)\right)(p)_{2n+1} \end{pmatrix} = {}^t\left(Q(z)^{-1}\right)(p).$$

It is clear that

$$\left(\frac{d}{dt}\right)^k \tilde{\mathcal{J}}(t;p)|_{t=0} = 0, \quad k = n+2, \cdots, 2n+1.$$

Hence setting

$$\mathcal{J}(t;p) = \tilde{\mathcal{J}}(t, {}^tQ(z)(p)),$$

we obtain a solution of (4.2) with boundary condition (4.4). For $i = 0, 1, \cdots, 2n+1$,

$$\langle \tilde{\mathcal{J}}(\cdot;p), h_i \rangle = \langle (I - 2zB^{\#})\tilde{\mathcal{J}}(\cdot;p), h_i \rangle = \langle \sum_{j=0}^{2n+1} p_j h_j, h_i \rangle = A(p)_i.$$

This implies that

$$\begin{pmatrix} \langle \mathcal{J}(\cdot;p), h_0 \rangle \\ \langle \mathcal{J}(\cdot;p), h_1 \rangle \\ \vdots \\ \langle \mathcal{J}(\cdot;p), h_{2n+1} \rangle \end{pmatrix} = A\left({}^tQ(z)\right)(p)$$

and hence

$$\mathcal{J}_1 = A\,{}^tQ(z),$$

which completes the proof.

5. Lévy's stochastic area formula

We consider the case of (3.1) with $\alpha = 1$ and $\beta_1 = \beta_2 = 0$, i.e., the case of the uniform magnetic field. Then, by repeating the straightforward calculation for $\mathcal{J}_t$ and $S_{cr}(t, 0, y; z)$ of the Van Vleck-Pauli formula, we obtain the famous Lévy's stochastic area formula ([13] and [25]): if $|\mathrm{Re}(z)|t < 2\pi$,

$$\int_{W_0} \exp\left[z \int_{w[0,t]} \omega\right] \delta_x(w(t)) P(dw)$$
$$= \frac{1}{2\pi t} \frac{\frac{zt}{2}}{\sin \frac{zt}{2}} \exp\left[-\left(\frac{\frac{zt}{2}}{\sin \frac{zt}{2}} \cos \frac{zt}{2}\right) \frac{|x|^2}{2t}\right] \tag{5.1}$$

where ω is the differential 1-form given by (4.1) and $w[0,t]$ is the curve in $\mathbf{R}^2$ given by $\{w(s)\,;\, 0 \le s \le t\}$. As pointed out in [13] and [16], if $z \in \mathbf{R}$, and $|z|t \ge 2\pi$

$$\int_{W_0} \exp\left[z \int_{w[0,t]} \omega\right] \delta_x(w(t)) P(dw) = \infty.$$

Let $\{\lambda_n\}$ be the eigenvalues of $B^{\#}$ on $H_1^{\perp}$. Then

$$\lambda_n = \frac{1}{2} \frac{1}{2n\pi}, \; n \in \mathbf{Z}^*, \quad \mathbf{Z}^* = \mathbf{Z} \setminus \{0\},$$

multiplicity of λ_n , $n \in \mathbf{Z}^*$, are 2. Then by the eigenfunction expansion of the Brownian motion with respect to the system of eigenfunctions of $B^{\#}$ ([18]), we obtain that the left hand side of (5.1) with $t = 1$ equals to

$$\frac{1}{2\pi} \left(\prod_{n \in \mathbf{Z}^*} (1 - \frac{2z}{2} \frac{1}{2n\pi}) \exp\left[\frac{2z}{2} \frac{1}{2n\pi}\right]\right)^{-1}$$
$$\times \exp\left[-\left(1 + 2(\frac{z}{2\pi})^2 \sum_{n=1}^{\infty} \frac{1}{(\frac{z}{2\pi})^2 - n^2}\right) \frac{|x|^2}{2}\right]. \tag{5.2}$$

Combining this with the Van Vleck-Pauli formula and (5.1), we have the Euler formula for the sine function [5]:

$$\prod_{n \in \mathbf{Z}^*} \left(1 - \frac{2z}{2} \frac{1}{2n\pi}\right) \exp\left[\frac{2z}{2} \frac{1}{2n\pi}\right] = \frac{\sin \frac{z}{2}}{\frac{z}{2}} \tag{5.3}$$

and the characterization of the double sine function in terms of differential equation:

$$\left(\frac{\mathcal{S}_2'}{\mathcal{S}_2}\right)(\frac{z}{2\pi}) = \frac{z}{2} \cot \frac{z}{2}, \quad \mathcal{S}_2(0) = 1, \tag{5.4}$$

where $\mathcal{S}_2(x)$ is the double sine function such that

$$\mathcal{S}_2(x) = e^x \prod_{n \in \mathbf{Z}^*} P_2\left(\frac{x}{n}\right)^n, \quad P_2(x) = (1 - x) \exp\left[x + \frac{x^2}{2}\right].$$

(see [9] and [23]). Hence we start from the Van Vleck-Pauli formula and obtain the proof of basic formulae (5.3) and (5.4) in Analysis in terms of classical mechanics.

6. Appendix

As we mentioned in Section 2, the formula (2.3) in the page 283 in [13] should be corrected. According with this change, we must also modify the lines form the bottom 3 of the page 285 to the top 3 of the page 286 in [13].

A sketch of the proof of Theorem 2.1 above. Letting $\tilde{\eta}(w)(s) = \sum_{j=1}^{m} h_j \langle h_j, w \rangle$, we have

$$\int_{W_0} e^{z\langle Bw,w\rangle} \delta_0(\eta(w)) P(dw)$$

$$= \left(\frac{1}{2\pi}\right)^{m/2} \left[\det(\eta\eta^*)\right]^{-1/2} \int_{W_0} e^{z\langle B(w-\tilde{\eta}(w)),w-\tilde{\eta}(w)\rangle} P(dw).$$

Combining this with the equality

$$\langle B(w - \tilde{\eta}(w)), w - \tilde{\eta}(w)\rangle = \langle B^{\#} w, w\rangle - \mathrm{tr}(B_\eta),$$

we obtain (2.2).

References

1. Dashen,R.F., Hasslacher,B. and Neveu,N.: Nonperturbation methods and extended-hadron models in field theory I, semiclassical functional methods, Phys.Rev. D **10** (1974) 4114–4129.
2. DeWitt,B : Supermanifolds, Second ed., Cambridge Univ.Press, 1992.
3. DeWitt-Morette,C. : The semiclassical expansion, Ann.Phys. **97** (1976) 367–399.
4. Dowker,J.S.: When is the 'sum over classical paths' exact?, J. Phys. A : Gen .Phys. **3**(1970) 431–461.
5. Euler,L.: De summis serierum reciprocarum, Comm. acad. scient. Petropolitane. **7** (1734/35) 123–134.
6. Feynman,R.P. and Hibbs,A.R. : Quantum Mechanics and Path Integrals, McGraw-Hill,1965.
7. Gutzwiller,M.C.: Periodic orbits and classical quantization conditions, J. Math. Phys.**12** (1971) 343–354.
8. Gutzwiller,M.C. : Chaotic classical and quantum mechanics, Springer, 1990.
9. Hölder,O.: Über eine transcendente Function, Göttingen Nachrichten (1886) 514–522.
10. Hörmander,L.: The analysis of linear partial differential operators I, Springer, 1983 .
11. Ikeda,N. : Probabilistic methods in the study of asymptotics, in École d'Été de Probabilités de Saint-Flour XVIII-1988, Lecture Notes in Math.**1427** (1990) 195–325, Springer.
12. Ikeda,N., Kusuoka,S. and Manabe,S. : Lévy's stochastic area formula for Gaussian processes, Comm.Pure Appl.Math.**47**(1994) 329–360.
13. Ikeda,N., Kusuoka,S. and Manabe,S.: Lévy's stochastic area formula and related problems, in Stochastic Analysis, eds. Cranston and Pinsky,M., Proc.Symp. Pure Math.**57** (1995) 281–305, AMS.

14. Ikeda,N. and Manabe,S. : Asymptotic formulae for stochastic oscillatory integrals, in Asymptotic Problems in Probability Theory: Wiener functionals and asymptotics ,136–155, eds. Elworthy,D. and Ikeda,N., Pitman Research Notes in Math.Series, **284** Longman, 1993.
15. Ikeda,N. and Watanabe,S. : Stochastic differential equations and diffusion processes, Second ed., North-Holland/Kodansha, 1989.
16. Itô,K. and McKean,H.P.: Diffusion processes and their sample paths, Springer, 1965.
17. Itô,K. : Generalized uniform complex measures in the Hilbertian metric space with their application to the Feynman integrals, in Fifth Birkeley Symp.Math. Statist.Probab. II (1965) 145–161.
18. Itô,K. and Nisio.M : On the convergence of sums of independent Banach space valued random variables, Osaka J.Math. **5**(1968) 35–48.
19. Jacobi,C.J.: Zur Theorie Variations-Rechnung Und Der Differential-Gleichungen, J. Reine Angew. Math.**17** (1837) 68–82.
20. Kac,M.: Semiclassical quantum mechanics and Morse's theory, in Trends in application of pure Math.to Mech. vol.2, 163-170 ed. H.Zorski,1979.
21. Kac,M.: Integration in function spaces and some of its applications, Scuola Normale Superiore Pub.,1980.
22. Khandekar,D.C., Lawande,S.V. and Bhagwat,K.V.: Path-integral methods and their applications, World Scientific, 1993.
23. Kurokawa,N. : Multiple sine functions and Selberg Zeta functions, Proc.Japan Acad.**67**(1991) 61–64.
24. Levit,S. and Smilansky,U.: A new approach to Gaussian path integrals and the evaluation of the semiclassical propagator, Ann.Phys **103** (1977) 198–207.
25. Lévy,P. : Wiener's random function and other Laplacian random functions, in Proc.Second Berkeley Symp.Math.Stat.Prob. II,171–186, 1950.
26. Malliavin,P. : Stochastic calculus of variations and hypoelliptic operators, in Proc.Intern.Symp. SDE.195–263, ed. K.Itô, Kinokuniya/Wiley,1976.
27. Matsumoto,H. : Semiclassical asymptotics of eigenvalues for Schrödinger operators with magnetic fields, J.Funct.Anal. **129**(1995) 719–759.
28. Matsumoto,H. : Quadratic Hamiltonians and associated orthogonal polynomials, to appear in J.Funct.Anal.
29. Milnor,J.: Morse theory, Princeton Univ.Press, 1963.
30. Morse,M. : Variational analysis, John Wiley, 1973.
31. Pauli, W: Lectures on Physics, MIT Press, 1973.
32. Papadopoulos,G.J.: Gaussian path integrals, Phys.Rev. **D11** (1975) 2870–2875.
33. Rezende,J.: Quantum systems with time-dependent harmonic part and the Morse index, J.Math.Phys.**25** (1984) 3264–3269.
34. Schulman,L.S. : Caustics and multi-valuedness, in Functional Integration and its Applications, ed. A.M.Arthurs, 1974.
35. Schulman,L.S. : Caustics and the semi-classical propagator for chaotic systems, in Path integrals from meV to MeV, 81–96, Proc.4 th. Intern.Conf. eds. Grabert,H.,Inomata,A., Schulman,L.S. and Weiss,U. World Sci.,Singapore.
36. Schulman,L.S.: Techniques and applications of path integration, John Wiley,1981.
37. Simon,B.: Notes on infinite determinants of Hilbert space operators, Adv.in Math. **24** (1977) 244–273.
38. Takanobu,S. and Watanabe,S. : Asymptotic expansion formulas of the Schilder type for a class of conditional Wiener functional integrations, in Asymptotic Problems in Probability Theory: Wiener functionals and asymptotics ,194–241,

eds. Elworthy,D. and Ikeda,N., Pitman Research Notes in Math.Series, **284** Longman, 1993.
39. Van Vleck,J.H.: The correspondence principle in the statistical interpretation of quantum mechanics, Proc.Nat.Acad.Sci.U.S.A.**14** (1928) 178–188.
40. Watanabe,S.: Analysis of Wiener functionals (Malliavin calculus) and its applications to heat kernels, Ann.Prob.**15** (1987) 1–39.

Some recent developments in nonlinear filtering theory

G. Kallianpur

Center for Stochastic Processes, University of North Carolina at Chapel Hill, Phillips Hall, Chapel Hill, NC 27599-3260, USA

1. Introduction

One of Professor Norbert Wiener's beloved projects in his later years was to develop a theory of nonlinear prediction. It was Professor Kiyosi Itô's discovery of stochastic calculus (now subsumed under the more inclusive title of stochastic analysis) that made such a development possible. This is not the place to describe in detail Professor Itô's continuing contributions to the subject he helped to create. However, one incident which illustrates his eagerness to explore and assimilate new ideas stands out in my mind.

In the early 70's, having heard of Don Fisk's 1963 thesis (which had not been published in its entirety and was then unavailable), Professor Itô came to Minneapolis to talk about Fisk's work and to borrow my copy of his thesis. What followed was typical of Professor Itô's scientific personality. He not only studied Fisk's and Stratonovich's work but in two papers published in 1974 pointed out the advantages of Fisk's integral over the integral that bears his name by giving two examples, one in connection with Stroock and Varadhan's support theorem and the second in defining stochastic parallel displacement, [6, 7]. In the present article, devoted to a discussion of some recent results in nonlinear filtering theory, I shall present another example: a chaos expansion in terms of multiple Fisk-Stratonovich integrals for the solution of the Zakai equation and point out its usefulness in obtaining approximations to the solution.

Another part of this article discusses the uniqueness of the solution of the Zakai equation (and the Kushner-Fujisaki-Kallianpur-Kunita equation) when the signal process takes values in an arbitrary complete, separable metric space. This work was directly motivated by the need to solve filtering problems when the unobserved process is infinite dimensional, e.g. when it is the solution of a stochastic differential equation (SDE) in an infinite dimensional space or of a stochastic partial differential equation (SPDE). SDE's in infinite dimensional spaces were studied by Professor Itô in pioneering papers culminating in a CBMS monograph in 1984 [8].

In a short section of the paper I give a brief review of an entirely different approach to nonlinear filtering (when the signal and observation noise in the conventional model are independent). This theory, developed in collaboration with R.L. Karandikar, has been presented in detail in our book [10]. Since the basis of this theory is finitely additive Gaussian white noise in Hilbert space, reference to this work has necessarily to be cursory. Nonetheless it has at least some pleasing features that the measure-valued or density valued SDE's of the conventional filtering theory are replaced by "ordinary" PDE's in which the randomness contributed by the observations occurs as a 'parameter' in the coefficients of the equation. Perhaps because this work has not been extended to correlated signal and noise models or because of the unfamiliarity with finitely additive probabily theory, our approach has not become popular and has found favor with some engineers.

The results presented in Sections 2 and 4 were obtained recently in collaboration with my colleagues, A. Bhatt, R.L. Karandikar and A. Budhiraja during their stay at the Center for Stochastic Processes. Section 3 describes recent results obtained by the first two of these authors and are an improvement of earlier theorems given in the book by Karandikar and myself.

2. Uniqueness of solution of the measure valued filter equations

We begin with the standard filtering model

$$Y_t = \int_0^t h_s(X_s)ds + W_t, \quad 0 \leq t \leq T \tag{2.1}$$

or, in Itô differential form

$$dY_t = h_t(X_t)dt + dW_t. \tag{2.2}$$

The basic assumptions on (2.1) are the following: $X = (X_t)$, called the signal process, is assumed to be a Markov process taking values in a complete, separable, metric space S. The values of X are not observed. Instead, we have the observation process Y whose values provide information about X. $W = (W_t)$ is a k-dimensional Wiener process which plays the part of "noise" in the observation apparatus. It is assumed that all these processes are defined on a probability space $(\Omega, \mathcal{F}, P)$. Although the results of this section have been proved for correlated signal and noise, in order to conserve space, here I shall restrict myself to the important special case when X and W are independent.

The nonlinear filtering problem is to estimate X_t the "present" state of X) given the nonlinear past of the observation process Y, that is, given the σ-field $\mathcal{F}_t^Y = \sigma\{Y_s, 0 \leq s \leq t\}$ augmented with all P-null sets. The solution is given by $\pi_t(\cdot) := P[X_t \in (\cdot)|\mathcal{F}_t^Y]$, the conditional distribution, or equivalently by $\sigma_t(\cdot)$, the unnormalized conditional distribution. The two are related by

$$< \pi_t, f > = \frac{< \sigma_t, f >}{< \sigma_t, 1 >} \tag{2.3}$$

where $f \in C_b(S)$, the space of bounded continuous functions and $<\mu, f>$ denotes $\int_S f(x)\mu(dx)$. Furthermore, in order to facilitate recursive computation, one wishes to obtain them as the unique solutions of the respective measure valued SDE's.

The following basic assumptions will be made concerning the model (2.1):

(A.1) X_t is a Markov process taking values in a complete, separable, metric space S. Further, X is characterized by a martingale problem posed by an operator A_0 where A_0 = the restriction of the infinitesimal generator of X to a suitable domain, $\mathcal{D}(A_0)$.

(A.2) $E\int_0^T |h(X_s)|^2 ds < \infty$; $h : S \to \mathbf{R}^d$ is continuous.

(A.3) X and W are stochastically independent.

Introduce a new probability measure P_0 (called the reference probability) by

$$\frac{dP_0}{dP} = \exp\left\{-\sum_{i=1}^{k}\int_0^T h^i(X_s)dY_s^i + \frac{1}{2}\sum_{i=1}^{k}\int_0^T [h^i(X_s)]^2 ds\right\}.$$

It is well known that under P_0, the observation process Y is a Wiener process and the law of X under P_0 = law of X under P.

Then, using the Bayes formula,

$$<\sigma_t, f> = \mathbf{E}_{P_0}\left\{f(X_t)\frac{dP}{dP_0}|\mathcal{F}_t^Y\right\}$$

it can be shown that σ_t satisfies the measure-valued stochastic differential equation known as the Zakai equation

$$\begin{aligned} <\sigma_t, f> &= <\sigma_0, f> + \int_0^t <\sigma_s, A_0 f> ds \\ &+ \int_0^t \sum_{i=1}^{k} <\sigma_s, h^i f> dY_s^i, \forall f \in \mathcal{D}(A_0). \end{aligned} \tag{2.4}$$

The conditional distribution $\pi_t(\cdot)$ satisfies a nonlinear measure-valued equation called the Kushner equation, also studied in detail by Fujisaki, Kallianpur and Kunita (See [2] for references):

$$\begin{aligned} <\pi_t, f> &= <\pi_0, f> + \int_0^t <\pi_s, A_0 f> ds \qquad (2.5) \\ &+ \int_0^t \sum_{i=1}^{k} \{<\pi_s, h^i f> - <\pi_s, h^i><\pi_s, f>\} d\nu_s^i, \\ &\forall f \in \mathcal{D}(A_0) \end{aligned}$$

where

$$\nu_t^i = Y_t^i - \int_0^t <\pi_s, h^i> ds \tag{2.6}$$

is the innovation Wiener process.

The uniqueness of solution of the basic SDE's of nonlinear filtering theory, especially of the Zakai equation, has been studied by many authors. Most of the effort has been naturally concentrated on signal processes which are Markov processes with a finite dimensional Euclidean state space. The growth of the theory of infinite dimensional dynamical systems has brought filtering problems to the forefront in which the unobserved, signal process is the solution of an SPDE or an infinite dimensional SDE. An example of such a problem is discussed in recent work by [2] where also the uniqueness problem is settled in great generality. For the sake of brevity, we shall present here, the results obtained in [2] for the important special case when signal and observation noise are independent.

To establish the uniqueness of solution of the SDE's (2.4) and (2.5) we need the signal process X to satisfy the following conditions:

(A.4) X is the unique solution to the martingale problem for (A_0, π_0) where $\pi_0 \in \mathcal{P}(S)$ (the class of Borel probability measures on S). That is,

(i) $\pi \circ X_0^{-1} = \pi_0$;

(ii) $\mathcal{D}(A_0)$ is an algebra that separates points in S and contains the constant functions;

(iii) There exists $\Phi \in C(S)$ such that for $f \in \mathcal{D}(A_0)$ and $x \in S$, $|(A_0 f)(x)| \leq C_f \Phi(x)$, ($C_f$, a constant);

(iv) $\{(f, \Phi^{-1} A_0 f) : f \in \mathcal{D}(A_0)\}$ is contained in the closure (under bounded pointwise convergence) of a countable set;

(v) $\int_0^T E\Phi(X_s)ds < \infty$;

(vi) $f(X_t) - \int_0^t A_0 f(X_s)ds$ is a martingale $\forall f \in \mathcal{D}(A_0)$;

(vii) If X' is another process satisfying (i) - (vi), then the law of X' = the law of X.

Let $D([0,T], S)$ be the Skorohod space of paths from $[0,T]$ to S.

(A.5) The $D([0,T], S)$ martingale problem for (A_0, S_z) is well posed for every Dirac measure δ_z, $z \in S$, in the sense that there exists a cadlag solution (Z_t) $0 \leq t \leq T$ to the martingale problem and any two solutions with cadlag paths are identical in law;

(A.6) For all $\mu \in \mathcal{P}(S)$, any progressively measurable solution to the martingale problem for (A_0, μ) has a cadlag modification.

The unique solution for the Zakai equation is given by the following result.

Theorem 1 *Assume conditions (A.1) - (A.6). Suppose that ρ_t is an $\mathcal{F}_t^Y$-adapted, $\mathcal{M}_+(S)$-valued cadlag process satisfying*

$$
\begin{aligned}
< \rho_t, f > &= < \pi_0, f > + \int_0^t < \rho_s, A_0 f > ds \\
&\quad + \sum_{i=1}^k \int_0^t < \rho_s, h^i f > dY_s^i, (f \in \mathcal{D}(A_0))
\end{aligned}
$$

and (2.7)

$$E_{P^0}\int_0^T < \rho_t, \Phi > \, dt < \infty.$$

Then $\rho_t = \sigma_t$ for all $t, 0 \leq t \leq T$ a.s. where σ_t is the unnormalized conditional measure defined by (2.7).

Uniqueness in law also holds for (2.3). Using (2.3), uniqueness can similarly be proved for Equation (2.5).

Details of the proof of the above result and of a more general result for the correlated signal and noise model are given in [2].

Applications of the above results are discussed in [2]. The following result derives a robustness property for the solution of (2.4).

Robustness. Consider a sequence $X^n \Longrightarrow X$, (convergence in law) and let Y^n be the observation process corresponding to X^n. Further suppose that

$$|A_0^n f(x)| \leq C_f \Phi^n(x) \quad \forall n$$

and

$$\sup_n \mathbf{E}_{P_0^n} \int_0^T \{\Phi^n(X^n)\}^a \, dt = C_1 < \infty \quad \text{forsome} \quad a > 1.$$

X^n, X are not assumed to be defined on the same probability space and are solutions of the martingale problems for A_0^n and A_0.

Let $h^n \to h$ uniformly on compacts of S and let

$$\sup_n E_{P^n} \exp\left\{ b \int_0^T |h^n(X_s^n)|^2 ds \right\} = C_2 < \infty \quad \text{forsome} \quad b > 1;$$

$$\sup_n \sup_{0 \leq t \leq T} E_{P^n} |h^n(X_t^n)|^2 = C_3 < \infty.$$

Since Y^n and Y are Brownian motions under P_0^n and P_0, σ_t and σ_t^n can be written as measure-valued Wiener functionals:

$$\sigma_t = F_t(Y), \quad \sigma_t^n = F^n(Y^n).$$

Using Skorohod's representation theorem it is possible to redefine the functionals F^n and F on the Wiener space $(\Omega^0, \mathcal{F}^0, \mu)$ where

$$\Omega^0 = C([0,T], \mathbf{R}^k), \mathcal{F}^0 = \text{Borel} \quad \sigma\text{-field}$$

on Ω^0 and $\mu =$ Wiener measure. The following result expresses the robustness property.

Theorem 2 *$F^n \to F$ in μ-probability as $D([0,T], \mathcal{M}_+(S))$-valued random variables, where $\mathcal{M}_+(S)$ is the space of positive finite measures and $D([0,T], \mathcal{M}_+(S))$ is the Skorohod space of $\mathcal{M}_+(S)-$ valued processes on $[0,T]$.*

3. A white noise approach to nonlinear filtering

As mentioned in the Introduction, we present only a very brief summary of this theory, emphasizing the motivation rather than the techniques.

As in the conventional theory presented in Section 2, the signal process X is an S-valued Markov process, S being a complete separable metric space. The filtering model is no longer given by (2.1) or (2.2) but what some engineers would consider a more realistic model:

$$y_t = h_t(X_t) + e_t \tag{3.1}$$

In (3.1), $\mathcal{H}$ is a separable Hilbert space and $h : [0,T] \times S \to \mathcal{H}$ is a continuous function such that

$$\mathbf{E} \int_0^T \| h_s(X_s) \|_{\mathcal{H}}^2 < \infty.$$

$(e_t)_{0 \leq t \leq T}$ is an $\mathcal{H}$-valued white noise independent of X, and y_t is the observation at time t. (e_t) is defined on $\mathcal{H}$ endowed with the finitely additive canonical Gauss measure. It will be seen that (3.1) corresponds to the intuitive model

$$\text{observation} = \text{signal} + \text{noise}.$$

For the relevant definitions of the conditional distribution, conditional expectation, etc. in this theory we refer the reader to [10]. The aim, as before, is to obtain a recursive method for computing $F_t(y)$, the conditional distribution of X_t given $\{y_s, 0 \leq s \leq t\}$. If f is a real valued bounded, measurable function on S, then

$$E\left[f(X_t) | y_s, 0 \leq s \leq t\right] = < f, F_t(y) >$$

where $< f, \mu >$ denotes $\int f d\mu$. For G, a Borel set in S and $y \in H := L^2([0,T], \mathcal{H})$, we have

$$F_t(y)(G) = \frac{\Gamma_t(y)(G)}{\Gamma_t(y)(S)}$$

where the unnormalized conditional distribution of X_t given $\{y_s, 0 \leq s \leq t\}$,

$$\Gamma_t(y)(G) = \mathbf{E}\left\{1_G(X_t) \exp \int_0^t c_s^y(X_s) ds\right\}$$

and

$$c_s^y(x) = (h_s(x), y_s)_{\mathcal{H}} - \frac{1}{2} \| h_s(x) \|_{\mathcal{H}}^2 .$$

In [10], Kallianpur and Karandikar had derived a measure-valued equation for $F_t(y)$ (resp. $\Gamma_t(y)$) and proved the uniqueness of solution under the assumption that the domain of the full generator of X is known (together with additional restrictions on h). This condition is difficult to fulfill in practice and has been considerably relaxed in a recent paper by Bhatt and Karandikar

[1]. For reasons of space, it is convenient to state the result when X is a diffusion in $\mathbf{R}^d$. Let $\mathcal{D} = C_c^2(\mathbf{R}^d)$, the space of twice continuously differentiable functions with compact support in $\mathbf{R}^d$, and let

$$A_t f(x) = \frac{1}{2}\sum_{i,j=1}^{d} a_{ij}(t,x)\frac{\partial^2 f(x)}{\partial x_i \partial x_j} + \sum_{i=1}^{d} b_i(t,x)\frac{\partial f(x)}{\partial x_i}.$$

Here a and b are the diffusion and drift coefficients, and $f \in \mathcal{D}$. Further it is assumed that $a = (a_{ij}) : [0,T] \times \mathbf{R}^d \to \mathcal{S}_+^d$ and $b = (b_i) : [0,T] \times \mathbf{R}^d \to \mathbf{R}^d$ are continuous. ($\mathcal{S}_+^d$ = the class of positive definite $d \times d$ matrices).

Theorem 3 *Suppose that (X_t) is the unique solution to the martingale problem for (A_t, μ_0), μ_0 being the initial probability measure. Then for $y \in C([0,T],\mathcal{H})$ and $0 \leq t_0 \leq T$,*
$\{\Gamma_t(y), 0 \leq t \leq t_0\}$ is the unique solution of the equation

$$< f, \rho_t > = < f, \rho_0 > + \int_0^t < A_s f + c_s^y(\cdot) f, \rho_s > ds,$$

$0 \leq t \leq t_0$, $f \in \mathcal{D}$. The uniqueness is in the class of measures $\{\rho_t\} \subset \mathcal{M}_+(S)$ satisfying the following conditions:

For every Borel set $E \subset S, t \to \rho_t(E)$ is measurable for all

$$\rho \in \mathcal{M}_+(S), \tag{3.2}$$

$$\int_0^T \int_S \| h_s(x) \|_{\mathcal{H}}^2 \, d\rho_s(x) ds < \infty. \tag{3.3}$$

Similarly, the conditional distribution $\{F_t(y);\ \ 0 \leq t \leq t_0\}$ is the unique solution of the equation

$$\begin{aligned} < f, \pi_t > &= < f, \mu_0 > + \int_0^t < A_s f + c_s^y(\cdot) f, \pi_s > ds \\ &\quad - \int_0^t < c_s^y, \pi_s > < f, \pi_s > ds, \end{aligned}$$

$0 \leq t \leq t_0, f \in \mathcal{D}$. The uniqueness is in the class of measures $\{\pi_t\} \subset \mathcal{P}(S)$ satisfying the conditions (3.2) and (3.3) with ρ_t replaced by π_t and $\mathcal{M}_+(S)$ replaced by $\mathcal{P}(S)$.

For $y \in L^2([0,T],\mathcal{H})$, $\Gamma_t(y)$ and $F_t(y)$ can be determined from the above results since $C([0,T],\mathcal{H})$ is dense in $L^2([0,T],\mathcal{H})$ and $\Gamma_t(y)$ and $F_t(y)$ are known to be continuous in y.

4. Multiple Stratonovich integral expansion for the conditional density and approximations to the solution of the Zakai equation

From a purely theoretical point of view, the nonlinear filtering problem can be considered to be solved once it is shown that the Zakai equation has a unique solution. This problem has been discussed at a very general level in the preceding sections. The question addressed in this section is concerned with applications: how to obtain suitable approximations to the solution of the Zakai equation for the unnormalized conditional density. In recent years, there has been a considerable literature on this subject.

As far as I know, Krylov and Veretennikov were the first to use multiple Wiener integrals (MWI) to prove the existence and uniqueness of solution of a SDE for which they also obtained a series representation. A functional series approach to the filtering problem was initiated by Ocone who derived MWI expansions separately for the numerator and the denominator of the Kallianpur-Striebel formula. Lo and Ng used Cameron-Martin expansions to approximate the conditional density. (See [3] for references to these papers.)

An important result in which a MWI representation for $u(t,x)$, the solution of the Zakai equation that exploits the connection between MWI and iterated Itô integrals has been obtained by Kunita [11]. An independent derivation is given in ([3], Theorem 2.5) which uses the fact that the kernels in the MWI representation of $u(t,x)$ satisfy an infinite system of integro-differential equations or equivalently, the Fourier coefficients of the kernels satisfy an infinite system of PDE's. (Also see [12] for a spectral approach).

The Stratonovich or Fisk-Stratonovich stochastic integral was originally introduced for its computational facility since it obeyed the usual rules of integral and differential calculus. In recent years, multiple Stratonovich integrals (MSI) have been defined and studied by several authors from many different points of view. We shall obtain in this section a chaos expansion for $u(t,x)$ in terms of MSI. A difficulty that arises in developing a square integrable Wiener functional in an L^2-convergent series of MSI is that MSI of different orders are not orthogonal. However, as we shall see, in the case of the solution of the Zakai equation we can still obtain an L^2-convergent series expansion in terms of (mutually non orthogonal) MSI.

We begin by reviewing some results on MSI which we will be using later in this section. Let $g_p \in L_s^2[0,T]^p$ where $L_s^2[0,T]^p$ is the class of real valued square integrable, symmetric functions on $[0,T]^p$. Then g_p has a MSI w.r.t. Y_t if and only if the following sum

$$\sum_{i_1,\ldots,i_p=1}^{m} \frac{1}{\Delta_{i_1}\ldots\Delta_{i_p}} \left(\int_{\Delta_{i_1}\times\ldots\times\Delta_{i_p}} g_p(s_1,\ldots,s_p)ds_1\ldots ds_p \right)$$

$$Y(\Delta_{i_1})\ldots Y(\Delta_{i_p})$$

converges in $L^2(P_0)$ as $|\pi| \to 0$ where $\pi = \{0 = t_1 < \ldots < t_{m+1} = T\}$, $\Delta_{i_j} = (t_{i_j} - t_{i_j})$, $Y(\Delta_{i_j}) := Y(t_{i_j+1}) - Y(t_{i_j})$ and $|\pi| = \max(t_{i+1} - t_i)$. The limit of the above sum is called the MSI of g_p and denoted by $\delta_p(g_p)$. The above definition is due to Solé and Utzet (see [4] for references).

In order to derive a relation between MSI and MWI we need first to introduce the notion of Hilbert space valued k-traces. The idea was introduced by Hu and Meyer in a paper in which they gave a heuristic derivation of such a relation which is well known as the Hu-Meyer formula. We give below a rigorous definition of k-traces due to Johnson and Kallianpur [9]. Other definitions have also been given by Budhiraja and Kallianpur and by Solé and Utzet [4].

Definition Let $f_p \in L_s^2[0,T]^p$. Fix $k, 1 \leq k \leq [p/2]$. Suppose that for every complete orthonormal sequence (CONS) $\{\phi_i\}$ of $L^2[0,T]$, the series

$$\sum_{i_1,\ldots,i_k=1}^{N} \sum_{i_{2k+1},\ldots,i_p=1}^{N} < f_p, \varphi_{i_1} \otimes \varphi_{i_1} \otimes \ldots \otimes \varphi_{i_k} \otimes \varphi_{i_k} \otimes \varphi_{i_{2k+1}} \otimes \ldots \otimes \varphi_{i_p} > \cdot$$

$$\cdot \varphi_{i_{2k+1}} \otimes \ldots \otimes \varphi_{i_p}$$

converges in $L^2[0,T]^{p-2k}$ to a limit which is independent of the choice of the CONS $\{\varphi_i\}$. The limit of the above series is defined to be the *limiting k-trace* of f_p and is denoted by $\overset{\rightarrow k}{Tr} f_p$. $\overset{\rightarrow 0}{Tr} f_p$ is defined to be f_p.

Hu-Meyer Formula. Let $f_p \in L_s^2[0,T]^p$. Assume that $\overset{\rightarrow k}{Tr} f_p$ exists for all k, $1 \leq k \leq [p/2]$. Then the MSI $\delta_p(f_p)$ exists and

$$\begin{aligned} \delta_p(f_p) &= \sum_{k=0}^{[p/2]} c_{p,k} I_{p-2k}(\overset{\rightarrow k}{Tr} f_p) \\ &= \sum_{i_1,\ldots,i_p=1}^{\infty} a_{i_1\ldots i_p} I_1(\varphi_{i_1}) \ldots I_1(\varphi_{i_p}) \end{aligned}$$

where $c_{p,k} = \frac{p!}{(p-2k)!k!2^k}$ and $a_{i_1\ldots i_p}$ are the coefficients in the expansion of f_p with respect to the tensorial CONS $\{\varphi_{i_1} \otimes \ldots \otimes \varphi_{i_p}\}$. The symbol I in the above formula denotes the multiple Wiener integral.

In this section we assume that the signal process X is the unique solution of the SDE

$$dX_t = b(t, X_t)dt + a(t, X_t)dW_t$$

where W is a Wiener process independent of B, the Wiener process in the observation model. The initial value X_0 and W are independent, and $(X_0, W) \perp\!\!\!\perp B$. The following further conditions are assumed.

The coefficients $a(t,x)$, $b(t,x)$, $h(t,x)$ are in $C_b^{1,\infty}$ $:= C_b^{1,\infty}([0,T] \times \mathbf{R})$ (i.e., once continuously dfferentiable in t and infinitely differentiable in x with all derivatives bounded).

X_0 has a density $q(x)$ belonging to $C_b^\infty(\mathbf{R})$. Let $\pi_t(dy)$ denote the conditional distribution of X_t given $\mathcal{F}_t^Y$. A well known result of nonlinear filtering theory tells us that

$$\pi_t(f) = \frac{\rho_t(f)}{\rho_t(1)}$$

where $\rho_t(f)$ is a continuous stochastic process with values in the set of finite positive measures on $\mathbf{R}$.

Under the stated assumptions it can be shown that $\rho_t(\cdot)$ is absolutely continuous w.r.t. Lebesgue measure. The density $u(t,x)$ is referred to as the unnormalized conditional density. Moreover, under the above conditions $u(t,x)$ is the unique solution to the stochastic partial differential equation (SPDE) (the Zakai equation)

$$du(t,x) = \mathcal{L}(t)^* u(t,x)dt + h(t,x)u(t,x)dY_t \tag{4.1}$$

where for $f \in C_b^2$,

$$\begin{aligned}\mathcal{L}(t)^* f(x) \quad := \quad & \frac{1}{2}a^2(t,x)\frac{\partial^2 f}{\partial x^2}(x) + \left[\frac{\partial a(t,x)}{\partial x} - b(t,x)\right]\frac{\partial f}{\partial x}(x) \\ & + \left[\frac{1}{2}\frac{\partial^2 a^2(t,x)}{\partial x^2} - \frac{\partial b(t,x)}{\partial x}\right] f(x).\end{aligned}$$

If we transform the measure P by defining $dP_0 = Z_T dP$ where

$$Z_t = \exp\left[-\int_0^t h(s,X_s)dB_s - \frac{1}{2}\int_0^t h^2(s,X_s)ds\right]$$

then we use the well known fact that under P_0, Y and X are independent, the law of X under P and under P_0 is the same and most importantly, under P_0, Y in the SPDE (4.1) is a Wiener process.

Having formulated the set up for the Zakai equation, our aim is to obtain an $L^2(P_0)$-convergent chaos expansion for $u(t,x)$ in terms of MSI. The motivation for this is that the facility of computation of MSI's might make such an approximation more amenable to obtaining a computable approximation to $u(t,x)$.

(a) **MSI chaos expansion for $u(t,x)$:** We use Kunita's technique and his method of constructing the solution of the Zakai equation. Let $(\Omega_1, \mathcal{F}_1, Q)$ be a probability space on which a Brownian motion B^* is defined. Define the following process on $(\Omega \otimes \Omega_1, \mathcal{F} \otimes \mathcal{F}_1, P_0 \otimes Q)$:

$$\begin{aligned}& \hat{\phi}_{s,t}(y,\omega,\omega_1) \\ &= \exp\left\{\int_s^t h(r,\hat{\zeta}_{r,t}(y,\omega_1)) \circ \hat{d}Y_r(\omega) + \int_s^t k(r,\hat{\zeta}_{r,t}(y,\omega_1))dr\right\}\end{aligned}$$

where $0 \le s \le t$, $o\hat{d}$ denotes the backward Stratonovich integral, $k(t,x) = \frac{1}{2}\left(\frac{\partial^2 a^2}{\partial x^2} - \frac{\partial b}{\partial x}\right)(t,x)$ and $\hat{\zeta}_{s,t}(x,\omega_1)$, ($\omega_1 \in \Omega_1$ and is sometimes suppressed for convenience) is the unique solution of the backward SDE

$$\hat{d}\hat{\zeta}_s = -\left(\frac{\partial a(s,\hat{\zeta}_s)}{\partial x} - b(s,\hat{\zeta}_s)\right) ds - a(s,\hat{\zeta}_s) \circ \hat{d}B^*_s, \quad 0 \le s < t$$

with $\hat{\zeta}_{tt}(x,\omega_1) = x$. Then

$$u(t,x,\omega) = E_Q\left[q(\hat{\zeta}_{0,t}(x,\omega_1))\hat{\phi}_{0,t}(x,\omega,\omega_1)\right], \quad \omega \in \Omega.$$

Now $\{Y_t, \mathcal{F}_t^Y \otimes \mathcal{F}_1, P_0 \otimes Q\}$ is a Wiener martingale independent of $\hat{\zeta}$ (Here, we set $Y_t(\omega,\omega_1) = Y_t(\omega)$). Hence the above expression for the solution of the Zakai equation can be rewritten as

$$u(t,x,\omega)$$
$$= E_Q\left[q(\hat{\zeta}_{0,t}(x,\omega_1)) \exp\left\{\int_0^t h(r,\hat{\zeta}_{r,t}(x))dY_r + \int_0^t k(r,\hat{\zeta}_{r,t}(x))dr\right\}\right].$$

To obtain the desired expansion, the key step is the introduction of the Feynman-Stratonovich semigroup defined as follows: For $0 \le s \le t \le T$,

$$T_{s,t}f(x) := E_Q\left[f(\hat{\zeta}_{s,t}(x)) \exp\{\int_s^t k(r,\hat{\zeta}_{r,t}(x))dr\}\right]$$

where $x \in \mathbf{R}$ and $f \in C_b^2$. Writing $h^{t\otimes p}(\omega_1)$ for

$$h(s_1,\hat{\zeta}_{s_1,t}(\omega_1)) \ldots h(s_p,\hat{\zeta}_{s_p,t}(\omega_1)) 1_{[0,t]^p}(s_1,\ldots,s_p)$$

we show that, for $p \ge 1$ and $k = 0,1,2,\ldots,[p/2]$, the limiting trace

$$\overrightarrow{Tr}^k \; E_Q\left[q(\hat{\zeta}_{0,t}(x,\omega_1)) \exp\{\int_0^t k(r,\hat{\zeta}_{r,t}(x,\omega_1))dr\} h^{t\otimes p}(\omega_1)\right]$$

exists and equals

$$E_Q[q(\hat{\zeta}_{0,t}(x,\omega_1)) \exp\left\{\int_0^t k(r,\hat{\zeta}_{r,t}(x,\omega_1))dr\right\}$$
$$(\int_0^t h^2(r,\hat{\zeta}_{r,t}(x,\omega_1))dr)^k h^{t\otimes p-2k}].$$

Moreover, the above k-traces are consistent. For fixed (t,x), if we let

$$f_p(t,x|s_1,\ldots,s_p) = 1_{[0,t]^p}(s_1,\ldots,s_p) E_Q\left[q(\hat{\zeta}_{0,t}(x,\omega_1)) \exp\right.$$
$$\left.\left\{\int_0^t k(r,\hat{\zeta}_{r,t}(x,\omega_1))dr\right\} \cdot h(s_1,\hat{\zeta}_{s_1,t}(x,\omega_1)) \ldots h(s_p,\hat{\zeta}_{s_p,t}(x,\omega_1))\right]$$

it follows from the above that the traces $\overrightarrow{Tr}^{k} f_p(t, x|s_1, \ldots, s_p)$ exist, so that the MSI $\delta_p(f_p(t, x|\cdot))$ exists.

It is easy to see that $u(t, x, \cdot)$ is a square integrable Wiener functional belonging to $\mathbf{L}^2 := \mathbf{L}^2(\Omega, \mathcal{F}_T^Y, P_0)$ and we are able to prove the following result.

Theorem 4

$$u(t, x, \cdot) = T_{0,t}q(x) + \sum_{p=1}^{\infty} \frac{1}{p!} \delta_p(f_p(t, x|\cdot))$$

where the series on the right side converges in $L^2(P_0)$.

Note that, in contrast to the Wiener chaos expansion in terms of MWI, the MSI of different orders are not orthogonal. Using the expressions for f_p, a Stratonovich analog of the Kunita-Krylov-Veretennikov representation for $u(t, x)$ is obtained.

Theorem 5 *For every $(t, x) \in [0, T] \times \mathbf{R}$, we have P_0-a.s.,*

$$u(t, x) = \tag{4.2}$$
$$T_{0,t}q(x) + \sum_{p=1}^{\infty} \int_0^t \Big(\int_0^{t_1} \cdots \Big(\int_0^{t_{p-1}} T_{t_1,t}(h(t_1, \cdot) T_{t_2,t}(h(t_2, \cdot) \ldots$$
$$T_{t_p,t_{p-1}} h(t_p, \cdot) T_{0,t_p} f) \ldots)))(x) \circ dY_{t_p}) \circ \ldots \circ dY_{t_2}) \circ dY_{t_1}.$$

The series on the right side converges in $L^2(P_0)$. Further the integrals in each term are iterated Stratonovich integrals. Two steps are involved in obtaining (4.2): the first, in showing that the kernels can be expressed in terms of the semigroup $T_{s,t}$ and the second, in using a result that gives a MSI as an iterated Stratonovich integral (Theorem 3.4 of [4]) as follows.

$$\delta_p(g_p) = p! \int_0^T \Big(\int_0^{t_p} \Big(\cdots \int_0^{t_2} g_p(t_1, \ldots, t_p) \circ dW_{t_1}) \ldots \circ dW_{t_{p-1}}) \circ dW_{t_p}).$$

where $g_p \in L_s^2[0, T]^p$ and $0 < t_1 \leq t_2 \leq \ldots \leq t_p \leq T$.

The practical application which we wish to make of these expansions is to obtain approximations to $u(t, x)$ with an error bound. It is convenient, first, to give a Fourier series type expansion for $u(t, x)$ together with an infinite system of PDE's which determine the coefficients in a unique manner. To do this, the semigroup $T_{t,s}$ and its generator $\mathcal{A}(t)$ are used in an essential manner. For $f \in C_b^2$ we have

$$\mathcal{A}(t)f(x) :=$$
$$a^2(t, x) \frac{\partial^2 f}{\partial x^2}(x) + \left\{ \frac{\partial a(t, x)}{\partial x} - b(t, x) \right\} \frac{\partial f}{\partial x}(x) + (k - \frac{1}{2} h^2)(t, x) f(x).$$

The Cauchy problem

$$\frac{\partial w(t,x)}{\partial t} = \mathcal{A}(t)w(t,x), \quad w(0,x) = g(x),$$

$g \in C_b^2$ has the uniqué solution $w(t,x) = (T_{0,t}g)(x)$ which belongs to the class $C_b^{1,2}$. From the expansion given in Theorem 5 one deduces the following result which is the first step towards the approximation of $u(t,x)$.

(b) Approximation to the solution of the Zakai equation.

Theorem 6

$$u(t,x) =$$

$$T_{0,t}q(x) + \sum_{p=1}^{\infty} \sum_{i_1,\ldots,i_p=1}^{\infty} a_{i_1,\ldots,i_p}(t,x) \left\{ \sum_{k=0}^{[p/2]} c_{p,k} I_{p-2k}^t(\varphi_{i_1} \otimes \ldots \otimes \varphi_{i_{p-2k}}) \right\}$$

where $I_{p-2k}^t(\cdot)$ is the MWI over $[0,t]^{p-2k}$ and the coefficients $a_{i_1,\ldots,i_p}(t,x)$ uniquely solve a system of PDE's in the class $C_b^{1,2}$, determined by $\mathcal{A}(t)$.

The expression in $\{\ldots\}$ in Theorem 6 can be replaced by $(\int_0^t \varphi_{i_1}(s)dY_s)\ldots(\int_0^t \varphi_{i_p}(s)dY_s)$ which is more amenable to computation. As an approximation to $u(t,x)$ we take

$$u_{n,N}(t,x) := \sum_{p=1}^{n} \frac{1}{p!} \sum_{i_1,\ldots,i_p=1}^{N} a_{i_1\ldots i_p}(t,x) \int_0^t \varphi_{i_1}(s)dY_s \ldots \int_0^t \varphi_{i_p}(s)dY_s.$$

To facilitate the computation of the single integrals, we choose a special CONS, namely, the generalized Haar functions in $L^2[0,T]$ introduced by Nualart and Zakai in a different context. (See [5] for reference).

Let $\{\Pi_N\}$ be a sequence of partitions with $\Pi_1 = \{0,T\}$ and for $N \geq 2$, Π_N, a refinement of Π_{N-1} by only one point. Define $e_1 := \frac{1}{T}$. For $m \geq 1$, e_{m+1} is defined as follows: Assume that α the point of refinement in Π_{m+1} lies in the interval $(d_1,d_2]$, $d_1,d_2 \in \Pi_M$. Then for $t \in (d_1,d_2]$,

$$e_{m+1}(t) :=$$

$$\left\{\frac{d_2-\alpha}{(\alpha-d_1)(d_2-d_1)}\right\}^{1/2} 1_{(d_1,\alpha]}(t) - \left\{\frac{\alpha-d_1}{(d_2-\alpha)(d_2-d_1)}\right\}^{1/2} 1_{(\alpha,d_2]}(t)$$

and unchanged from $e_m(t)$ in the other intervals. Then $\{e_m\}_1^\infty$ is a CONS and for each $N \geq 1$, the linear span of $\{e_1,\ldots,e_N\}$ is identical to that of $\{|\Delta_1|^{-1/2}1_{\Delta_1},\ldots,|\Delta_N|^{-1/2}1_{\Delta_N}\}$ where $\Delta_i = (\tau_i,\tau_{i+1}]$, $|\Delta_i| = \tau_{i+1}-\tau_i$ and $\{0 = \tau_1 < \ldots < \tau_{N+1} = T\}$ is the partiion Π_N. Writing $\hat{u}_{n,N}(t,x)$ for $u_{n,N}(t,x)$ with the φ_i replaced by e_i it can be shown that

$$E_{P_0}\left[u(t,x) - \hat{u}_{n,N}(t,x)\right]^2 \leq \frac{K_1(2K_2t)^{n+1}e^{2K_2t}}{(n+1)!} + Ce^{3K_2t}(|\Pi_N| \wedge t) \quad (4.3)$$

Here K_1, K_2 and C are constants not depending on n, N, t or x. Finally, estimation of the single Wiener integrals in $\hat{u}_{n,N}(t,x)$ by computing increments of the observation process leads to the following result.

Theorem 7 *Let $\Pi_N = \{t_j^N\}$ be as above. Then*

$$E_{P_0}[u(t,x) - \sum_{p=1}^{n} \sum_{i_1,\dots,i_p=1}^{N} b_{i_1,\dots,i_p}(t,x)(Y_{t_{i_1+1}^N \wedge t} - Y_{t_{i_1}^N \wedge t}) \cdots$$
$$\cdots \left(Y_{t_{i_p+1}^N \wedge t} - Y_{t_{i_p}^N \wedge t}\right)]^2$$

has the same bound as in (4.3) of Theorem 6. The coefficients $b_{i_1,\dots,i_p}(t,x)$ are symmetric in $(i_1,\dots,i_p)$ and can be determined recursively. (See [5]).

References

1. Bhatt, A.G., Karandikar, R.L., (1995): Evolution equations for Markov processes: Applications to the white noise theory of filtering, *Applied Mathematics and Optimization* **31**, 326-348
2. Bhatt, A.G., Kallianpur, G., Karandikar, R.L. (1995): Uniqueness and robustness of solution of measure valued equations of nonlinear filtering, to appear in *Annals of Probability*
3. Budhiraja, A., Kallianpur, G. (1995): Approximations to the solution of the Zakai equations using multiple Wiener and Stratonovich integral expansions, University of North Carolina Center for Stochastic Processes Technical Report No. 447, January 1995. To appear in *Stochastics and Stochastic Reports*
4. Budhiraja, A., Kallianpur, G. (1995): Two results on multiple Stratonovich integrals, University of North Carolina Center for Stochastic Processes Technical Report No. 457, June 95. Submitted for publication
5. Budhiraja, A. and Kallianpur, G. (1995): The Feynman-Stratonovich semigroup and Stratonovich integral expansions in nonlinear filtering, University of North Carolina Center for Stochastic Processes Technical Report No. 466, July 1995. Submitted for publication
6. Itô, K. (1974): Stochastic differentials, *Applied Mathematics and Optimization*, **1**, 374-380
7. Itô, K. (1974): Stochastic parallel transport, Probabilistic Methods in Stochastic Differential Equations, *Lecture Notes in Mathematics* **451**, Springer-Verlag, 1-7
8. Itô, K. (1984): Foundations of Stochastic Differential Equations in Infinite Dimensional Spaces, *CBMS Notes*, Baton Rouge (1983), SIAM
9. Johnson, G.W., Kallianpur, G. (1993): Homogeneous chaos, *p*-forms, scaling and the Feynman integral, *Transactions of the American Mathematical Society* **340**, 503-548
10. Kallianpur, G., Karandikar, R.L. (1988): White noise theory of prediction, filtering and smoothing, Gordon and Breach, New York
11. Kunita, H. (1981): Stochastic partial differential equations connected with nonlinear filtering, *Nonlinear Filtering and Stochastic Control, Proceedings*, Cortona 1981, *Lecture Notes in Mathematics* **972**
12. Lototsky, S., Mikulevicius, R., Rozovskii, B.L. (1995): Nonlinear Filtering Revisited: A Spectral Approach, CAMS Report 95-3

Detecting a single defect in a scenery by observing the scenery along a random walk path

Harry Kesten

Department of Mathematics, Cornell University, White Hall, Ithaca NY 14853-7901, USA

Summary. A *scenery* on $\mathbb{Z}$ is a map $\xi : \mathbb{Z} \to \{0, \dots, k-1\}$; we think of ξ as a coloring of $\mathbb{Z}$, which assigns to each point of $\mathbb{Z}$ one of k colors. For a given scenery ξ, denote by $\widehat{\xi}$ a scenery obtained from ξ by changing $\xi(0)$ only. Let $\{S_n\}_{n\geq 0}$ be a simple symmetric random walk on $\mathbb{Z}$, starting at the origin. Assume that we observe one of the two sequences $\{\xi(S_n)\}_{n\geq 0}$ or $\widehat{\xi}(S_n)\}_{n\geq 0}$, without being told which of the two sequences is observed. If ξ is known, can we decide (with zero probability of error) on the basis of these observations which of the two sequences was observed ? We prove that this can be done for 'almost all' ξ, when $k \geq 5$.

1. Introduction

This is a continuation of Benjamini and Kesten (1995), which also gives some background and references for the problem discussed here. Benjamini and Kesten (1995) investigated a general analogue of the problem described in the summary, namely to distinguish between two sceneries ξ and η on a general graph $\mathcal{G}$, by observing either $\{\xi(S_n)\}_{n\geq 0}$ or $\{\eta(S_n)\}_{n\geq 0}$ for a simple random walk $\{S_n\}_{n\geq 0}$ on $\mathcal{G}$. In this paper we restrict ourselves to the case $\mathcal{G} = \mathbb{Z}$, and for $\{S_n\}$ we always take a simple symmetric random walk with $S_0 = 0$. A scenery ξ is an element of

$$\Xi = \Xi^{(k)} := \{0, 1, \dots, k-1\}^{\mathbb{Z}}.$$

We denote our sequence of observations by $\{X_n\}_{n\geq 0}$; thus $\{X_n\}_{n\geq 0} = \{\xi(S_n)\}_{n\geq 0}$ or $\{\eta(S_n)\}_{n\geq 0}$. The sequence $\{X_n\}_{n\geq 0}$ takes its values in the space

$$\mathcal{X} := \{0, 1, \dots, k-1\}^{\mathbb{N}}.$$

Denote a generic point of $\mathcal{X}$ by $x = (x_0, x_1, \dots)$ and let

$$\mathcal{F}_\ell^m = \sigma - \text{field generated by the coordinate functions } x_n \text{ with } \ell \leq n \leq m,$$

$$\mathcal{F}_\ell = \mathcal{F}_l^\infty = \bigvee_{m\geq \ell} \mathcal{F}_\ell^m, \quad \mathcal{F} = \mathcal{F}_0 = \mathcal{F}_0^\infty.$$

1991 Mathematics Subject Classification 28A99, 60J15.
Key words and phrases Random walk, scenery, distinguishability,mutually singular measures.

Finally, let Q^ξ and Q^ξ_ℓ be the two probability measures induced on $\mathcal{F}$ by $\{\xi(S_n)\}_{n\geq 0}$ and on $\mathcal{F}_\ell$ by $\{\xi(S_n)\}_{n\geq \ell}$, respectively. We say that two sceneries ξ and η are *distinguishable* if Q^ξ_ℓ and Q^η_ℓ are mutually singular for all ℓ.

It was originally conjectured by Benjamini, and independently by den Hollander and Keane (private communications) that ξ and η are distinguishable if and only if η cannot be obtained from ξ by a reflection and/or an even translation, that is, if and only if there does not exist a $j \in \mathbb{Z}$ for which

$$\begin{aligned} \eta(n) &= \xi(n+2j), \quad n \in \mathbb{Z}, \text{ or} \\ \eta(n) &= \xi(-n+2j), \quad n \in \mathbb{Z}. \end{aligned} \tag{1.1}$$

It is not difficult to see (compare Howard (1995a)) that if ξ and η do satisfy (1.1) for some $j \in \mathbb{Z}$, then they are not distinguishable. However, it is not known whether the conjecture holds in the other direction. If it does, then for 'most' ξ, ξ must be distinguishable from a $\widehat{\xi}$ which differs from ξ only at the origin. Benjamini and Kesten (1995) explicitly raised the problem when such ξ and $\widehat{\xi}$ are distinguishable. Our result here is that for 'almost all' ξ, ξ and $\widehat{\xi}$ are distinguishable when $k \geq 5$. We would not be surprised, though, if the result (and hence the conjecture of Benjamini and of den Hollander and Keane) fails when $k = 2$.

To state our result precisely we introduce the product measure

$$\rho = \rho^{(k)} = \prod_{n\in\mathbb{Z}} \rho_n^{(k)}$$

on $\Xi^{(k)}$, where each $\rho_n^{(k)}$ is the uniform measure on $\{0, 1, \ldots, k-1\}$, that is

$$\rho_n^{(k)}(j) = \frac{1}{k}, \quad 0 \leq j \leq k-1.$$

We shall prove the following result.

Theorem. *If $k \geq 5$, then for almost all ξ $[\rho^{(k)}]$, ξ is distinguishable from any $\widehat{\xi}$ with*

$$\widehat{\xi}(n) = \xi(n), \quad n \in \mathbb{Z}\setminus\{0\}, \text{ and } \widehat{\xi}(0) \neq \xi(0). \tag{1.2}$$

Remark. The title of this paper is based on the terminology of Howard (1995b) which calls a finite number of changes in a *periodic* scenery 'defects'. Even though this terminology is less appropriate for a general, non-periodic scenery, we find it convenient to use the term defect for the change from ξ to $\widehat{\xi}$.

Outline of proof. The reason why it is difficult to distinguish two sceneries ξ and η is that we do not observe the value of S_n. In other words, we do not know where the random walk is at a given time. We must distil this kind of information from the observations $\{X_n\}$. For the case $\eta = \widehat{\xi}$ we shall do this for the times $0 < \tau_1 < \tau_2 < \cdots$, which are the successive times τ for which

$$(1.3) \qquad X_{\tau - L_n + i} = \xi(i), \quad 1 \le i \le L_n,$$

for a suitable L_n. If $L_n \sim 2C_1 \log n$ for a suitable constant C_1, then with high probability we will have

$$(1.4) \qquad |S_{\tau_j}| \le 2L_n \text{ for all } \tau_j \le n.$$

That is, the block of colors $\xi(1), \dots, \xi(L_n)$ is seen only at times when $S_.$ is within distance $2L_n$ from the origin (see Lemma 2.1 for a precise statement). This step works as long as we choose L_n such that

$$(1.5) \qquad \frac{n}{k^{L_n}} \to 0, \quad n \to \infty.$$

Under this condition then, we know the locations of S_{τ_j} reasonably well (with high probability) for those τ_j which are $\le n$. We now try to distinguish ξ and $\widehat{\xi}$ on the basis of

$$(1.6) \qquad \widehat{N}_n := \#\{j \le n^{\kappa} : X_{\tau_j + M_n} = \xi(0)\},$$

where $\#A$ denotes the cardinality of A,

$$(1.7) \qquad M_n = 2L_n^7,$$

and κ will be determined later. Since $\widehat{\xi}(0) \ne \xi(0), X_{\tau_j + M_n}(\xi)$ should have a greater probability of being equal to $\xi(0)$ than $X_{\tau_j+M_n}(\widehat{\xi})$, and since S_{τ_j} is expected to be 'close to' the origin (see (1.4)) this can be expected to make $\widehat{N}_n$ noticeably larger when ξ is the true scenery, than when $\widehat{\xi}$ is the true scenery. This will indeed be shown to be the case, basically when there are at least n^{κ} values of $\tau_j \le n$, that is, if $\widehat{N}_n$ only counts τ_j-values which are $\le n$. As we shall see, this will be the case with high probability when

$$(1.8) \qquad n^{1/2} 2^{-L_n} \ge n^{2\kappa} \text{ eventually}.$$

Conditions (1.5) and (1.8) can both be satisfied for some $\kappa > 0$ if and only if $k > 4$. This is the reason why we have to take $k \ge 5$ in our proof. (Actually, $n^{1/2} 2^{-L_n} \ge$ (some high power of $\log n$) suffices instead of condition (1.8); this does not lead to a better value for k, though.)

Acknowledgement. This research was supported by the NSF through a grant to Cornell University. The author also wishes to thank the Mittag-Leffler Institute in Djursholm, Sweden, for its hospitality while this paper was being completed. The author is indebted to Itai Benjamini for many stimulating discussions about the problem of this paper.

2. Details

C_i will denote a strictly positive finite constant throughout this section. We take

(2.1) $$L_n = 2\lfloor C_1 \log n \rfloor,$$

where C_1 is chosen such that (1.5) and (1.8) hold for some $\kappa > 0$; we fix such a κ for the remainder of this paper. Note that L_n is even. For technical reasons we replace the τ_j by the times σ_j defined as follows for any fixed $\ell \geq 1$:

$$\begin{aligned} \sigma_1 = \quad \sigma_1(\xi, n, \ell) \quad &= \quad \min\{t \geq \ell + L_n : t \text{ is even and} \\ &\qquad X_{t-L_n+i} = \xi(i),\ 1 \leq i \leq L_n\}, \\ \sigma_{j+1} = \quad \sigma_{j+1}(\xi, n, \ell) &= \quad \min\{t > \sigma_j + M_n : t \text{ is even and} \\ &\qquad X_{t-L_n+i} = \xi(i),\ 1 \leq i \leq L_n\}. \end{aligned}$$

The difference with the previous τ_js is that the separation between σ_j and σ_{j+1} is at least M_n, (see (1.7) for M_n) and that the σ_j are even and $\geq \ell + L_n$. Note that for given ξ, the σ_j are functions of the observations $\{X_n\}$; in fact if ξ and $\widehat{\xi}$ satisfy (1.2), then the functions $\sigma_j(\xi, n, \ell)$ and $\sigma_j(\widehat{\xi}, n, \ell)$ on $\mathcal{X}$ are the same, because $\xi(i) = \widehat{\xi}(i)$, $1 \leq i \leq L_n$, under (1.2).

We further work with

(2.2) $$N_n = N_n(\xi, \ell) := \#\{1 \leq j \leq n^\kappa : X_{\sigma_j(\xi,n,\ell)+M_n} = \xi(0)\},$$

instead of $\widehat{N}_n$ defined in (1.6). This N_n is a function of the observations $\{X_n\}_{n \geq \ell}$ and $\xi(0), \xi(1), \ldots, \xi(L_n)$ only. In the remainder we shall often suppress the ℓ from the notation.

The probability space on which the random walk $\{S_n\}$ is defined is not of great importance. For convenience we take it to be

$$\Omega := \mathbb{Z}^{\mathbb{N}}.$$

The probability measure on Ω which governs the random walk will be denoted by P. We remind the reader that $P\{S_0 = 0\} = 1$. When both ξ and $\{S_n\}$ are taken as random, then we take them independent, so that the pair $(\xi, \{S_n\})$ is governed by the measure $\rho \times P$ on $\Xi^{(k)} \times \Omega$. For expection with respect to a probability measure μ we shall also use the symbol μ; that is if Y is a random variable, rather than an event, then $\mu\{Y\} = \int Y d\mu$. This convention cannot lead to harmful confusion.

Lemma 2.1. *When ξ and $\{S_n\}$ are random, then*

(2.3) $$\begin{aligned} &\rho \times P\{\exists j \in [L_n, n] \text{ with } |S_j| > 2L_n \text{ and such that} \\ &\qquad \xi(S_{j-L_n+i}) = \xi(i),\ 1 \leq i \leq L_n\} \\ &\leq nk^{-L_n} \end{aligned}$$

and

$$(2.4) \quad \rho \times P\{\exists \widehat{\xi} \text{ satisfying (1.2) and a } L_n \le j \le n \text{ with } |S_j| > 2L_n \text{ and such that } \widehat{\xi}(S_{j-L_n+i}) = \widehat{\xi}(i) (= \xi(i)), 1 \le i \le L_n\} \le nk^{-L_n+1}.$$

Proof. First pick the random walk path $S_0, S_1, \dots, S_n$, according to the measure P. Next choose $\xi(j)$ for $j \notin \{1, \dots, L_n\}$, independent of each other and of $\{S_j\}_{j\ge 0}$, and uniform on $\{0, \dots, k-1\}$. Then for any $\widehat{\xi}$ which satisfies (1.2), we also know $\widehat{\xi}(j)$ for $j \notin \{1, \dots, L_n\} \cup \{0\}$; for $\widehat{\xi}(0)$ we can still take any of the $k-1$ values in $\{0, \dots, k-1\}$ other than $\xi(0)$. Now for any $j \le n$ with $|S_j| > 2L_n$ it must be that $|S_{j-L_N+i}| \ge |S_j| - |L_n - i| > 2L_n - L_n = L_n$ for $1 \le i \le L_n$. Thus, also the possible sequences $\{\xi(S_{j-L_n+i}), 1 \le i \le L_n\}$, of length L_n which can occur for some $j \in [L_n, n]$ with $|S_j| > 2L_n$ have been determined. These are also the possible sequences for $\{\widehat{\xi}(S_{j-L_n+i}), 1 \le i \le L_n\}$, for such j. Since $j \le n$, there are at most n such possibilities for $\{\xi(S_{j-L_n+i}), 1 \le i \le L_n\}$, and at most kn for $\{\widehat{\xi}(S_{j-L_n+i}), 1 \le i \le L_n\}$. Finally we choose $\xi(1) = \widehat{\xi}(1), \xi(2) = \widehat{\xi}(2), \dots, \xi(L_n) = \widehat{\xi}(L_n)$. These are independent of each other and of the $\xi(j)$ with $j \notin \{1, \dots, L_n\}$. Again they are uniform in $\{0, \dots, k-1\}$. Therefore the ρ-probability that $\{\xi(1), \dots, \xi(L_n)\}$ will equal one of the possible sequences $\{\xi(S_{j-L_n+i}), 1 \le i \le L_n\}$ with $L_n \le j \le n, |S_j| > 2L_n$, is at most nk^{-L_n}. This proves (2.3);(2.4) follows in the same way. □

The next lemma will be used to approximate the distribution of N_n; it tells us that we only make a small error if we act as if $S_{\sigma_i} = 0$. (If $S_{\sigma_i} = 0$ actually always held, then N_n would have a binomial distribution, but we shall not use this fact.) We define for $q \in \{0, \dots, k-1\}$

$$(2.5) \qquad \Lambda_q = \Lambda_q(\xi) = \{j \in \mathbb{Z} : \xi(j) = q\}.$$

Lemma 2.2. *For $M = M_n = 2L_n^7$ it holds that*

$$(2.6) \quad \rho\Big\{ \sup_{\substack{|p| \le L_n + 2 \\ q \in \{0,\dots,k-1\}}} |P\{2p + S_M \in \Lambda_q(\xi)\} - P\{S_M \in \Lambda_q(\xi)\}| \ge \frac{L_n^{3/2}}{M^{3/4}}\Big\} \to 0 \quad (n \to \infty).$$

Also

$$(2.7) \quad \rho\Big\{\text{for some } \widehat{\xi} \text{ satisfying (1.2) one has } \sup_{\substack{|p| \le L_n + 2 \\ q \in \{0,\dots,k-1\}}} \Big|P\{2p + S_M \in \Lambda_q(\widehat{\xi})\} - P\{S_M \in \Lambda_q(\widehat{\xi})\}\Big| \ge \frac{L_n^{3/2}}{M^{3/4}}\Big\} \to 0 \quad (n \to \infty).$$

Proof. We only prove (2.6); the proof of (2.7) is essentially the same.

$$\begin{aligned}(2.8)\qquad & P\{2p+S_M \in \Lambda_q\} - P\{S_M \in \Lambda_q\} \\ &= \sum_{r\in\mathbb{Z}} I_{\Lambda_q}(r)\,[P\{S_M = r-2p\} - P\{S_M = r\}] \\ &= \sum_{r\in\mathbb{Z}} I_{\Lambda_q}(r)\gamma(r,p) = \sum_{r\in\mathbb{Z}} \left[I_{\Lambda_q}(r) - \frac{1}{k}\right]\gamma(r,p),\end{aligned}$$

where I_Λ is the indicator function of Λ,

$$\gamma(r,p) = P\{S_M = r-2p\} - P\{S_M = r\};$$

the last equality in (2.8) uses that

$$\sum_{r\in\mathbb{Z}} \gamma(r,p) = \sum_{r\in\mathbb{Z}} P\{S_M = r-2p\} - \sum_{r\in\mathbb{Z}} P\{S_M = r\} = 0.$$

Note that under ρ, the random variables $I_{\Lambda_q} - 1/k$ are i.i.d. and take the values $1-1/k$ and $-1/k$ with probability $1/k$ and $1-1/k$, respectively. In particular

$$\rho\{I_{\Lambda_q} - \frac{1}{k}\} = 0.$$

We shall apply Bernstein's inequality (see for instance ex. 4.3.14 in Chow and Teicher (1986)) to the right hand side of (2.8). In order to do this we need a bound on $\gamma(r,p)$. Since we took M even, $\gamma(r,p) = 0$ if r is odd, while for even r and $p > 0$,

$$\begin{aligned}\gamma(r,p) &= \binom{M}{(M+r-2p)/2} 2^{-M} - \binom{M}{(M+r)/2} 2^{-M} \\ &= \binom{M}{(M+r)/2} 2^{-M} \\ &\quad \times \left[\frac{(M+r-2p+2)(M+r-2p+4)\cdots(M+r)}{(M-r+2)(M-r+4)\cdots(M-r+2p)} - 1\right] \\ &= P\{S_M = r\}\left[\frac{(M+r-2p+2)(M+r-2p+4)\cdots(M+r)}{(M-r+2)(M-r+4)\cdots(M-r+2p)} - 1\right].\end{aligned}$$

A similar formula holds for $p < 0$. From this and the local central limit theorem (Feller (1968), Theorem VII.3.1) one obtains that there exist some constants C_i such that for $|r| \le M^{5/8}$, $|p| \le L_n + 2 \le M^{2/8}$

$$\begin{aligned}|\gamma(r,p)| &\le C_2|p|\frac{|r|+|p|}{M} P\{S_M = r\} \\ &\le C_2 L_n \frac{|r|+L_n}{M} P\{S_M = r\} \\ &\le C_3 L_n \frac{|r|+L_n}{M^{3/2}} \exp(-C_4 r^2/M).\end{aligned}$$

For any r we also have

$$\sum_{|s|\geq|r|} P\{S_M = s\} \leq C_5 \exp(-C_6 r^2/M)$$

by standard tail estimates for the binomial distribution (e.g., Bernstein's inequality, ex. 4.3.14 in Chow and Teicher (1988), can be used for this). Consequently for any r, and $|p| \leq L_n$,

$$\begin{aligned} |\gamma(r,p)| &\leq 2 \sum_{|s|\geq|r|-2L_n} P\{S_M = s\} \\ &\leq 2C_5 \exp[-C_6(|r| - 2L_n)^2/M]. \end{aligned}$$

It follows that for $|p| \leq L_n$

$$\sup_r |\gamma(r,p)| \leq C_7 \frac{L_n}{M}$$

and

$$\begin{aligned} &\textstyle\sum_r |\gamma(r,p)|^2 \leq \sup_r |\gamma(r,p)| \sum_r |\gamma(r,p)| \\ &\leq C_8 \tfrac{L_n}{M} \Big[\textstyle\sum_{|r|\leq M^{1/2}} L_n \tfrac{|r|+L_n}{M} P\{S_M = r\} \\ &\qquad + \textstyle\sum_{k=1}^{\infty} L_n \tfrac{2^k M^{1/2}}{M} P\{2^k M^{1/2} \leq |S_M| < 2^{k+1} M^{1/2}\Big] \\ &\leq C_9 \tfrac{L_n^2}{M^{3/2}}. \end{aligned} \tag{2.9}$$

Bernstein's inequality (ex. 4.3.14 in Chow and Teicher (1988)) now gives

$$\begin{aligned} &\rho\Big\{ \big|\textstyle\sum_{r\in\mathbb{Z}} [I_{A_q}(r) - \tfrac{1}{k}]\gamma(r,p)\big| \geq L_n^{3/2} M^{-3/4} \Big\} \\ &\leq 2\exp\left[-\frac{(L_n^{3/2}/M^{3/4})^2}{2\big(C_9 L_n^2/M^{3/2} + C_7 L_n^{5/2}/M^{7/4}\big)} \right] \\ &\leq 2\exp[-C_{10} L_n]. \end{aligned}$$

By virtue of (2.8) this proves (2.6). □

We also need to make sure that (with high probability) there are at least n^{κ} values of j with $\sigma_j \leq n/2$, so that N_n counts only js with $\sigma_j + M_n \leq n$. The next lemma provides the necessary estimate. We define

$$\nu_n = \nu_n(\xi) = \nu_n(\xi, \ell) = \max\{j : \sigma_j(\xi, n, \ell) \leq \frac{n}{2}\}.$$

As with the σ_j, we have for all $\xi, \widehat{\xi}$ satisfying (1.2),

$$\nu_n(\xi) = \nu_n(\widehat{\xi}).$$

Lemma 2.3. *For κ satisfying (1.8) and for any fixed ξ and $\widehat{\xi}$ satisfying (1.2),*

$$Q^{\xi}\{\nu_n(\xi) \le n^{\kappa}\} \to 0 \quad (n \to \infty) \tag{2.10}$$

and

$$Q^{\widehat{\xi}}\{\nu_n(\widehat{\xi}) \le n^{\kappa}\} \to 0 \quad (n \to \infty). \tag{2.11}$$

Proof. Once again we only prove (2.10); the same proof works for (2.11). Now let ξ be the scenery from which the observations come, so that $X_j = \xi(S_j)$. Define

$$\begin{aligned} \lambda_1 &= \min\{t \ge \ell + L_n : S_t = 0\}, \\ \lambda_{j+1} &= \min\{t \ge \lambda_j + 2M_n : S_t = 0\}. \end{aligned}$$

Then $\lambda_j < \infty$ a.e. $[P]$ for all j, and the random vectors $S_{\lambda_j}, S_{\lambda_j+1}, \dots, S_{\lambda_j+L_n}$, $j \ge 1$, are i.i.d. Moreover, by the periodicity of $S_.$, each λ_j is even. Furthermore,

$$\{S_{\lambda_j+i} = i \text{ for } 1 \le i \le L_n\} \subset \{\xi(S_{\lambda_j+i}) = \xi(i) \text{ for } 1 \le i \le L_n\}. \tag{2.12}$$

From this it follows that

$$\begin{aligned} &P\{\xi(S_{\lambda_j+i}) = \xi(i),\ 1 \le i \le L_n\} \\ &\quad \ge P\{S_{\lambda_j+i} = i \text{ for } 1 \le i \le L_n\} = 2^{-L_n}. \end{aligned} \tag{2.13}$$

Also, because the difference between two successive λs exceeds M_n, one proves by induction on j that σ_j is at most equal to

$$\ell + (j+1)L_n + 1 + (\text{the} j\text{-th value of } \lambda_r \text{ with } S_{\lambda_r+i} = i \text{ for } 1 \le i \le L_n).$$

Consequently, for large n,

$$\nu_n \ge \#\{j : \lambda_j + (j+1)L_n \le \frac{n}{4}, S_{\lambda_j+i} = i,\ 1 \le i \le L_n\},$$

so that

$$\begin{aligned} Q^{\xi}\{\nu_n \le n^{\kappa}\} &\le P\left\{\#\{j : \lambda_j \le n/8\} \le n^{1/2}(\log n)^{-8}\right\} \\ &+ P\left\{\sum_{j \le n^{1/2}(\log n)^{-8}} I[S_{\lambda_j+i} = i,\ 1 \le i \le L_n] \le n^{\kappa}\right\}. \end{aligned} \tag{2.14}$$

From the independence property noted before (2.12),

$$\sum_{j \le n^{1/2}(\log n)^{-8}} I[S_{\lambda_j+i} = i,\ 1 \le i \le L_n]$$

has a binomial distribution corresponding to $\lfloor n^{1/2}(\log n)^{-8} \rfloor$ trials and successprobability 2^{-L_n}. Since, by (1.8),

$$n^{1/2}(\log n)^{-8}2^{-L_n} \geq n^{2\kappa}(\log n)^{-8},$$

the second probability in the right hand side of (2.14) tends to 0 (e.g., by Chebychev's inequality).

Next note that (by induction on j)

$$\lambda_j \leq \text{ time of } (2jM_n)\text{-th visit after time } \ell + L_n \text{ by } S_. \text{ to } 0,$$

so that the first probability in the right hand side of (2.14) is at most

$$P\{S_. \text{ visits the origin fewer than } 2n^{1/2}(\log n)^{-8}M_n \sim 2^8 C_1^7 n^{1/2}(\log n)^{-1} \text{ times during } [\ell + L_n, \frac{n}{8}]\}.$$

It is well known (Révész (1990), Sect. 9.2) that this last probability tends to 0 as $n \to \infty$. □

Proof of Theorem. We say that ξ has *property* $A(n)$ if

$$\text{(2.15)} \quad Q_\ell^\xi\{\exists j \text{ with } \sigma_j(\xi, n, \ell) \leq \frac{n}{2} \text{ and } |S_{\sigma_j(\xi,n,\ell)}| > 2L_n\} \leq (nk^{-L_n})^{1/2}$$

and if for each $\widehat{\xi}$ satisfying (1.2)

$$\text{(2.16)} \quad Q_\ell^{\widehat{\xi}}\{\exists j \text{ with } \sigma_j(\widehat{\xi}, n, \ell) \leq \frac{n}{2} \text{ and } |S_{\sigma_j(\widehat{\xi},n,\ell)}| > 2L_n\} \leq (nk^{-L_n})^{1/2}.$$

We shall say that ξ has *property* $B(n)$ if

$$\text{(2.17)} \quad \sup_{\substack{|p| \leq L_n + 2 \\ q \in \{0,\ldots,k-1\}}} \left|P\{2p + S_{M_n} \in \Lambda_q(\xi)\} - P\{S_{M_n} \in \Lambda_q(\xi)\}\right| \leq L_n^{3/2} M_n^{-3/4}$$

and if for each $\widehat{\xi}$ satisfying (1.2)

$$\text{(2.18)} \quad \sup_{\substack{|p| \leq L_n + 2 \\ q \in \{0,\ldots,k-1\}}} \left|P\{2p + S_{M_n} \in \Lambda_q(\widehat{\xi})\} - P\{S_{M_n} \in \Lambda_q(\widehat{\xi})\}\right| \leq L_n^{3/2} M_n^{-3/4}.$$

If ξ is the observed scenery, then by definition of σ_j,

$$\xi(S_{\sigma_j(\xi,n)-L_n+i}) = \xi(i),\ 1 \leq i \leq L_n.$$

The event in (2.15) is therefore contained in

$$\{\exists j \in [L_n, n] \text{ with } |S_j| > 2L_n \text{ and such that } \xi(S_{j-L_n+i}) = \xi(i),\ 1 \leq i \leq L_n\},$$

so that the left hand side of (2.15) is at most

$$P\{\exists j \in [L_n, n] \text{ with } |S_j| > 2L_n \text{ and such that } \xi(S_{j-L_n+i}) = \xi(i),\ 1 \le i \le L_n\},$$

By virtue of Lemma 2.1 and Markov's inequality we therefore have

$$\begin{aligned}
&\rho\{\xi \text{ does not have property (2.15)}\} \\
&\le (nk^{-L_n})^{-1/2}\rho\{P\{\exists j \in [L_n, n] \text{ with } |S_j| > 2L_n \text{ and} \\
&\qquad \text{such that } \xi(S_{j-L_n+i}) = \xi(i),\ 1 \le i \le L_n\}\} \\
&\le (nk^{-L_n})^{1/2} \to 0 \quad (\text{see}(1.5)).
\end{aligned}$$

Similarly

$$\rho\{\xi \text{ does not have property (2.16) for some } \widehat{\xi} \text{ which satisfies (1.2)}\} \le k(nk^{-L_n})^{1/2} \to 0.$$

Therefore, there exists a sequence $n_1 < n_2 < \cdots$ such that almost all $\xi\,[\rho]$ have property $A(n_q)$ for all large q. By Lemma 2.2 we may further take the n_q such that almost all $\xi\,[\rho]$ have property $B(n_q)$ for large q. Finally define

$$\mu_n = \mu_n(\xi) = \lfloor n^{\kappa} \rfloor P\{S_{M_n} \in \Lambda_{\xi(0)}(\xi)\}. \tag{2.19}$$

We shall prove that for any ξ which has properties $A(n_q)$ and $B(n_q)$ for all large q, it holds that

$$Q_\ell^{\xi}\{N_{n_q} - \mu_{n_q}(\xi) \le -2n_q^{\kappa} L_{n_q}^{3/2} M_{n_q}^{-3/4}\} \to 0\ (q \to \infty), \tag{2.20}$$

and for each $\widehat{\xi}$ satisfying (1.2) for this ξ

$$Q_\ell^{\widehat{\xi}}\{N_{n_q} - \mu_{n_q}(\xi) \le -2n_q^{\kappa} L_{n_q}^{3/2} M_{n_q}^{-3/4}\} \to 1\ (q \to \infty). \tag{2.21}$$

(Note that the events in (2.20) and (2.21) are the same; only the probability measures differ.) We shall only prove (2.20) in detail, but before doing this we show that (2.20) plus (2.21) imply our theorem. If (2.20) and (2.21) hold for all ξ which have properties $A(n_q)$ and $B(n_q)$ eventually, then we can thin the sequence $\{n_q\}$ (if necessary) so that for almost all $\xi\,[\rho]$, the left hand side of (2.20) is at most $1/q^2$ and the left hand side of (2.21) is at least $1 - 1/q^2$, for all large q. Then for the event

$$C := \{N_{n_q} - \mu_{N_{n_q}}(\xi) \le -2n_q^{\kappa} L_{n_q}^{3/2} M_{n_q}^{-3/4} \text{ for infinitely many } q\},$$

we have for almost all $\xi\,[\rho]$ that

$$Q_\ell^{\xi}\{C\} = 0 \text{ and } Q_\ell^{\widehat{\xi}}\{C\} = 1 \tag{2.22}$$

for all $\widehat{\xi}$ satisfying (1.2). Note that N_{n_q} is $\mathcal{F}_\ell^{\infty}$-measurable by our definitions of $\sigma_j = \sigma_j(\xi, n, \ell)$ and $N_n = N_n(\xi, \ell)$. Because ℓ can be taken as large as desired throughout this proof, (2.22) implies the Theorem.

Now for the proof of (2.20). Fix ξ. Let $\{U_j^{(n)}\}_{j\geq 1} = \{U_j^{(n)}(\xi)\}_{j\geq 1}$ be i.i.d. random variables with

$$P\{U_j^{(n)} = 1\} = 1 - P\{U_j^{(n)} = 0\} = P\{S_{M_n} \in \Lambda_{\xi(0)}(\xi)\}.$$

We take the $\{U_j^{(n)}\}$ also independent of $\{S_r\}_{r\geq 0}$. For any ℓ we write $R_{\ell,n}^{\xi}$ for the joint distribution of $\{X_r\}_{r\geq \ell}$ and $\{U_j^{(n)}\}_{j\geq 1}$. Thus, if we take $U_j^{(n)}$ as the j-th coordinate function on $\{0,1\}^{[1,2,\dots)}$, then $R_{\ell,n}^{\xi}$ is a measure on $\mathcal{F}_\ell \times \mathcal{G}$, where $\mathcal{G}$ is the σ-field generated by the coordinate functions on $\{0,1\}^{[1,2,\dots)}$.

Also define

$$V_j^{(n)} = I[S_{\sigma_j(\xi,n,\ell)+M_n} \in \Lambda_{\xi(0)}(\xi)]$$

on the event

$$D_j^{(n)} := \{\sigma_j(\xi,n,\ell) \leq \frac{n}{2} \text{ and } |S_{\sigma_j(\xi,n,\ell)}| \leq 2L_n\},$$

while $V_j^{(n)} = U_j^{(n)}$ on the complement of $D_j^{(n)}$. We first note that if the scenery is ξ, then on the event $D_j^{(n)}$,

$$V_j^{(n)} = I[X_{\sigma_j(\xi,n,\ell)+M_n} = \xi(0)].$$

Moreover, $D_j^{(n)}$ and $V_j^{(n)}$ are $\mathcal{F}_\ell \times \mathcal{G}$-measurable. Therefore

$$\begin{aligned}
&R_{\ell,n}^{\xi}\{N_n \neq \sum_{1\leq j\leq n^\kappa} V_j^{(n)}\} \\
&\leq R_{\ell,n}^{\xi}\{\nu_n(\xi) \leq n^\kappa\} \\
&\qquad + R_{\ell,n}^{\xi}\{\exists j \text{ with } \sigma_j(\xi,n,\ell) \leq \frac{n}{2} \text{ and } |S_{\sigma_j(\xi,n,\ell)}| > 2L_n\} \\
&= Q_\ell^{\xi}\{\nu_n(\xi) \leq n^\kappa\} \\
&\qquad + Q_\ell^{\xi}\{\exists j \text{ with } \sigma_j(\xi,n,\ell) \leq \frac{n}{2} \text{ and } |S_{\sigma_j(\xi,n,\ell)}| > 2L_n\}.
\end{aligned}$$

Therefore, if ξ has property $A(n_q)$ eventually, then (by (2.10) and (2.15))

$$R_{\ell,n}^{\xi}\{N_{n_q} \neq \sum_{j=1}^{n_q^\kappa} V_j^{(n_q)}\} \to 0 \quad (q \to \infty). \tag{2.23}$$

Next we want to show that if ξ has the property (2.17), then the $V_j^{(n_q)}, j \geq 1$, are 'almost i.i.d.' To this end consider

$$R_{\ell,n}^{\xi}\{V_j^{(n)} = 1 | \sigma_r, r \leq j, S_s, s \leq \sigma_j, U_t, t \leq j-1\}. \tag{2.24}$$

On the complement of $D_j^{(n)}$, this conditional probability is just

$$(2.25) \qquad P\{U_j^{(n)} = 1\} = P\{S_{M_n} \in \Lambda_{\xi(0)}(\xi)\}.$$

On the event $D_j^{(n)} \cap \{S_{\sigma_j} = 2p\}$, the strong Markov property for the random walk $\{S_n\}$ shows that (2.24) equals

$$(2.26) \qquad P\{2p + S_{M_n} \in \Lambda_{\xi(0)}(\xi)\}$$

(note that S_{σ_j} must be even because we took σ_j even). For $n = n_q$ we have on $D_j^{(n)} \cap \{S_{\sigma_j} = 2p\}$ that (2.26), and hence (2.24), differs from (2.25) by at most $L_n^{3/2} M_n^{-3/4}$ (by (2.17) for $n = n_q$). Note that $V_{j-1}^{(n)}$ is measurable with respect to the σ-field,$\mathcal{H}_{j-1}$ say, generated by $\sigma_1, \ldots, \sigma_j, S_s, s \le \sigma_j, U_t, t \le j-1$, and we have just shown that for $n = n_q$ and q large,

$$\begin{aligned}
&\left| R_{\ell,n}^{\xi}\{V_j^{(n)} = 1 | \mathcal{H}_{j-1}\} - P\{S_{M_n} \in \Lambda_{\xi(0)}(\xi)\}\right| \\
&= \left| R^{\xi}{}_{\ell}, n\{V_j^{(n)} = 1 | \mathcal{H}_{j-1}\} - P\{U_j^{(n)} = 1\}\right| \\
&\le L_n^{3/2} M_n^{-3/4} = 2^{-3/4} L_n^{-15/4} \sim 2^{-9/2}[C_1 \log n]^{-15/4}.
\end{aligned}$$

It now follows easily from Freedman (1973) or Bernstein's inequality (Chow and Teicher (1986), ex. 4.3.14) that

$$\begin{aligned}
(2.27) \quad & R_{\ell,n}^{\xi}\{\sum_{j \le n^\kappa} V_j^{(n)} \le \mu_n(\xi) - 2n^\kappa L_n^{3/2} M_n^{-3/4}\} \\
&\le R_{\ell,n}^{\xi}\Big\{\sum_{j \le n^\kappa} [V_j^{(n)} - E\{V_j^{(n)} | \mathcal{G}_{j-1}\}] \le -n^\kappa L_n^{3/2} M_n^{-3/4}\Big\} \\
&\to 0 \text{ as } n \to \infty \text{ through } \{n_1, n_2, \ldots\}.
\end{aligned}$$

This, together with (2.23) shows that

$$\begin{aligned}
&Q_\ell^{\xi}\{N_n \le \mu_n(\xi) - 2n^\kappa L_n^{3/2} M_n^{-3/4}\} \\
&= R_{\ell,n}^{\xi}\{N_n \le \mu_n(\xi) - 2n^\kappa L_n^{3/2} M_n^{-3/4}\} \\
&\to 0 \text{ as } n \to \infty \text{ through } \{n_1, n_2, \ldots\}.
\end{aligned}$$

Thus (2.20) holds.

Essentially the same arguments work for (2.21). One now obtains

$$\begin{aligned}
& Q_\ell^{\widehat{\xi}}\Big\{ \textstyle\sum_{j \le n^\kappa} I[S_{\sigma_j(\widehat{\xi}, n, \ell) + M_n} \in \Lambda_{\xi(0)}(\widehat{\xi})] \\
(2.28) \qquad & \qquad \le \lfloor n^\kappa \rfloor P\{S_{M_n} \in \Lambda_{\xi(0)}(\widehat{\xi})\} + 2n^\kappa L_n^{3/2} M_n^{-3/4}\}\Big\} \\
& \to 1 \text{ as } n \to \infty \text{ through } \{n_1, n_2, \ldots\}.
\end{aligned}$$

The crucial point in obtaining (2.21) now is that

$$(2.29) \qquad \begin{aligned} &P\{S_{M_n} \in \Lambda_{\xi(0)}(\widehat{\xi})\} \\ &= P\{S_{M_n} \in \Lambda_{\xi(0)}(\xi)\} - P\{S_{M_n} = 0\} \\ &= P\{S_{M_n} \in \Lambda_{\xi(0)}(\xi)\} - \frac{2+o(1)}{\sqrt{2\pi M_n}}, \end{aligned}$$

because by definition of Λ_q and (1.2),

$$\Lambda_{\xi(0)}(\xi) \triangle \Lambda_{\xi(0)}(\widehat{\xi}) = \Lambda_{\xi(0)}(\xi) \setminus \Lambda_{\xi(0)}(\widehat{\xi}) = \{0\}.$$

By (2.29) and (1.7), for large n,

$$\begin{aligned} &\lfloor n^\kappa \rfloor P\{S_{M_n} \in \Lambda_{\xi(0)}(\widehat{\xi})\} + 2n^\kappa L_n^{3/2} M_n^{-3/4} \\ &= \lfloor n^\kappa \rfloor P\{S_{M_n} \in \Lambda_{\xi(0)}(\xi)\} - (2+o(1)) n^\kappa (2\pi M_n)^{-1/2} + 2n^\kappa L_n^{3/2} M_n^{-3/4} \\ &\leq \mu_n - 2n^\kappa L_n^{3/2} M_n^{-3/4}. \end{aligned}$$

Thus (2.21) follows from (2.28) and an analogue of (2.23).

References

Benjamini,I. and Kesten, H. (1995): Distinguishing sceneries by observing the scenery along a random walk path, submitted to J. d'Anal. Math.

Chow, Y.S. and Teicher, H. (1988): Probability Theory. Independence, Interchangeability, Martingales, 2nd ed., Springer-Verlag.

Feller, W. (1968): An Introduction to Probability Theory and Its Applications, vol. I, 3rd ed., John Wiley & Sons.

Freedman, D. (1973): Another note on the Borel-Cantelli lemma and the strong law, with the Poisson approximation as a by-product, Ann. Probab. 1, 910-925.

Howard, C.D. (1995a): The orthogonality of measures induced by random walks with scenery, Ph.D. Thesis, Courant Inst. of Math. Sciences, New York NY.

Howard, C.D. (1995b): Detecting defects in periodic scenery by random walks on $\mathbb{Z}$, to appear in Random Structures and Algorithms.

Révész, P. (1990): Random Walk in Random and Non-Random Environments, World Scientific.

Analytic approach to Yor's formula of exponential additive functionals of Brownian motion

Shin-ichi Kotani

Department of Mathematics, Graduate School of Science, Osaka University, Toyonaka, Osaka 560, Japan

Yor [4] obtained an exact formula for a one-dimensional Brownian motion $\{B_t\}$:

$$E_0\left(f\left(\int_0^t e^{2B_s}ds\right)g(e^{B_t})\right)$$
$$= c(t)\int_0^\infty dy\int_0^\infty dz g(y)f\left(\frac{1}{z}\right)\exp\left\{-\frac{z(1+y^2)}{2}\psi_{yz}(t)\right\},$$

$c(t) = (2\pi^3 t)^{-\frac{1}{2}}\exp(\frac{\pi^2}{2t}), \psi_r(t) = \int_0^\infty \exp(-\frac{u^2}{2t} - r\cosh u)\sinh u \sin(\frac{\pi u}{t})du.$

This formula was used by Kawazu-Tanaka [2] to show an asymptotic behavior of the tail probability of the maximum of a diffusion moving in random media. In this paper, we try to get this asymptotic form analytically.

Simultaneously, we can generalize the additive functionals. Replace e^{2x} by $V(x) \geq 0$ satisfying

$$\text{(A)} \quad \int_{-\infty} |x|V(x)dx < \infty,$$
$$\text{(B)} \quad V(0) > 0 \text{ and } V(x) \text{ is non-decreasing on } \mathbf{R}_+ = [0,\infty).$$

Assume $f(t)$ has an expression

$$f(t) = \int_0^\infty e^{-\alpha t}\sigma(d\alpha),$$

with a non-negative measure on $(0,\infty)$ satisfying

$$\int_1^\infty \sigma(d\alpha) < \infty.$$

Let $g_\alpha(x,y)$ be the Green function of $\frac{1}{2V}\frac{d^2}{dx^2}$ on $\mathbf{R}$ and $g_\alpha^+(x,y)$ be the Green function of $\frac{1}{2V}\frac{d^2}{dx^2}$ on $\mathbf{R}_+$ with Neumann boundary condition at $x = 0$. Set

$$\theta(x) = \int_0^1 g_\alpha(0,0)^2 g_\alpha^+(0,0) g_\alpha^+(x,0)\sigma(d\alpha).$$

Assume $\theta(x) < \infty$ for each $x \geq 0$. The conditions on $g(x) \geq 0$ are

$$\int_{-\infty} g(x)|x|dx < \infty, \int^{+\infty} g(x)x^{\frac{3}{2}}\theta(x)dx < \infty.$$

Then we have

Theorem . *As* $t \to \infty$,

$$t^{\frac{3}{2}} E_0 f\left(\int_0^t V(B_s)ds\right) g(B_t) \longrightarrow \int_{\mathbf{R}} A(0,x)g(x)dx,$$

where $A(x,y) = \frac{1}{\sqrt{2\pi}} \int_0^\infty g_\alpha(-\infty,x) g_\alpha(-\infty,y)\sigma(d\alpha)$.

If we introduce the heat kernel $p(t,x,y)$ of the diffusion $\{X_t\}$ with generator $\frac{1}{2V}\frac{d^2}{dx^2}$, then

$$A(x,y) = \frac{1}{\sqrt{2\pi}} \int_0^\infty f(t)Q(t,x,y)dt,$$

where $Q(t,x,y) = \int_0^t p(t-s,-\infty,x)p(s,-\infty,y)ds$. For $\{X_t\}$, $-\infty$ is the entrance boundary.

1. Upper estimates of heat kernels.

For a suitable real valued Borel measurable function $V(x)$ on $\mathbf{R}$, let $p_V(t,x,y)$ be the heat kernel of a heat equation

$$\frac{\partial u}{\partial t} = Lu, \text{ with } Lu = \frac{1}{2}u'' - Vu.$$

In this section, we give upper estimates of $p_V(t,x,y)$. Suppose V is a non-negative and locally integrable function on $\mathbf{R}$. For $\lambda \geq 0$, define $f_\pm(x,\lambda,V)$ as a unique solution of

$$\begin{aligned} &Lf = \lambda f \text{ on } \mathbf{R}, \quad f(0) = 1 \\ &f_+(f_-) \text{ is decreasing (increasing).} \end{aligned} \tag{1.1}$$

Probabilistically, $f_\pm$ have representations

$$f_\pm(x,\lambda,V) = E_x\left(e^{-\int_0^{\tau_0} V(B_s)ds} e^{-\lambda\tau_0}\right) \text{ for } x \in \mathbf{R}_\pm, \tag{1.2}$$

where $\tau_0 = \inf\{t > 0; B_t = 0\}$. The Green function $g_\lambda(x,y,V)$ of $\lambda - L$ is given by

$$\begin{aligned} g_\lambda(x,y,V) &= g_\lambda(y,x,V) \\ &= (h_+(\lambda,V) + h_-(\lambda,V))^{-1} f_+(x,\lambda,V) f_-(y,\lambda,V) \end{aligned} \tag{1.3}$$

if $x \geq y$. Here we define

$$h_\pm(\lambda,V) = \mp f'_\pm(0,\lambda,V).$$

Lemma 1. *If* $V_1(x) \geq V_2(x)$ *on* $\mathbf{R}_+$, *then*

$$h_+(\lambda, V_1) - h_+(0, V_1) \leq h_+(\lambda, V_2) - h_+(0, V_2) \tag{1.4}$$

$$h_+(\lambda, V_1) \geq h_+(\lambda, V_2). \tag{1.5}$$

Proof. For $x \geq 0$, (1.2) implies an inequality

$$f_+(x, 0, V_1) - f_+(x, \lambda, V_1) \leq f_+(x, 0, V_2) - f_+(x, \lambda, V_2).$$

Differentiating this at $x = 0$ shows (1.4). (1.5) follows from an inequality $f_+(x, \lambda, V_1) \leq f_+(x, \lambda, V_2)$. ∎

Lemma 2.

$$0 \leq g_0(0, 0, V) - g_\lambda(0, 0, V) \leq 2\sqrt{2\lambda} g_0(0, 0, V)^2.$$

Proof. Since we have

$$g_\lambda(x, y, V) = \int_0^\infty e^{-\lambda t} p_V(t, x, y) dt,$$

$g_\lambda(0, 0, V)$ is decreasing in $\lambda > 0$. On the other hand (1.4) shows

$$\begin{aligned} h_\pm(\lambda, V) - h_\pm(0, V) &\leq h_\pm(\lambda, 0) - h_\pm(0, 0) \\ &= \sqrt{2\lambda}. \end{aligned}$$

Therefore, we see

$$\begin{aligned} g_0(0, 0, V) - g_\lambda(0, 0, V) &= g_0(0, 0, V) g_\lambda(0, 0, V) \times \\ &\quad (h_+(\lambda, V) - h_+(0, V) + h_-(\lambda, V) - h_-(0, V)) \\ &\leq 2\sqrt{2\lambda} g_0(0, 0, V)^2. \end{aligned}$$ ∎

Now we give a general lemma on Laplace transformation.

Lemma 3. *Let* σ *be a non-negative measure on* $\mathbf{R}_+$. *Set*

$$p(t) = \int_0^\infty e^{-\lambda t} \sigma(d\lambda), \; g(\lambda) = \int_0^\infty e^{-\lambda t} p(t) dt.$$

Then

$$p(2t) \geq \frac{1}{t}\Big(g(0) - g(\frac{1}{t})\Big).$$

This can be proved easily by Jensen's inequality. Lemmas 2 and 3 show

Lemma 4.

$$p_V(t, 0, 0) \leq 8t^{-\frac{3}{2}} g_0(0, 0, V)^2.$$

Let

$$p(t,x)=(2\pi t)^{-\frac{1}{2}}e^{-\frac{x^2}{2t}},\ F_x(t)=E_x(e^{-\int_0^{\tau_0}V(B_s)ds};\tau_0\le t).$$

Then Feynman-Kac formula implies

Lemma 5.

$$p_V(t,x,y)\le p(t,x-y), \tag{1.6}$$

$$F_x'(t)\ge p(t,x)\frac{|x|}{t}. \tag{1.7}$$

Combining these lemmas we obtain the first estimate of $p_V(t,x,0)$ from the above.

Proposition 6. *These exists a constant C independent of V such that*

$$p_V(t,x,0)\le C(1+|x|)t^{-\frac{3}{2}}(g_0(0,0,V)^2+1) \tag{1.8}$$

holds for all $x\in\mathbf{R}$ and $t>0$.

Proof. The first passage time decomposition of $p_V(t,x,0)$ gives

$$\begin{aligned}p_V(t,x,0) &= \int_0^t p_V(t-s,0,0)dF_x(s)\\ &= \int_0^1 p_V(t-s,0,0)dF_x(s)+\int_1^{t-1} p_V(t-s,0,0)dF_x(s)\\ &\quad+\int_{t-1}^t p_V(t-s,0,0)dF_x(s)\\ &= \text{I+II+III}.\end{aligned}$$

The first term can be estimated simply by Lemma 4 in the following way:

$$\begin{aligned}\text{I} &\le Cg_0(0,0,V)^2\int_0^1(t-s)^{-\frac{3}{2}}dF_x(s)\\ &\le Cg_0(0,0,V)^2t^{-\frac{3}{2}}.\end{aligned}$$

Lemmas 4 and 5 imply

$$\begin{aligned}\text{II} &\le Cg_0(0,0,V)^2|x|\int_1^{t-1}(t-s)^{-\frac{3}{2}}s^{-\frac{3}{2}}ds\\ &\le Cg_0(0,0,V)^2|x|t^{-\frac{3}{2}},\end{aligned}$$

and Lemma 5 implies

$$\begin{aligned}\text{III} &\le C|x|\int_{t-1}^t(t-s)^{-\frac{1}{2}}s^{-\frac{3}{2}}ds\\ &\le C|x|t^{-\frac{3}{2}},\end{aligned}$$

which completes the proof. ∎

Now we assume the condition (A) on V. Under this condition we try to obtain a sharper upper estimate of $p_V(t,x,0)$ for $x \in \mathbf{R}_+$.

Lemma 7. *For $x \geq y \geq 0$, we have*

$$f_+(x,\lambda,V)f_+(y,\lambda,V)^{-1} \leq e^{-(x-y)h_+(\lambda,V)}.$$

Proof. To show this estimate, first note an identity

$$f_+(x+y,\lambda,V) = f_+(x,\lambda,V_y)f_+(y,\lambda,V), \tag{1.9}$$

where $V_y(\cdot) = V(\cdot + y)$. This comes from the Markov property of B_t. Differentiating (1.9) at $x = 0$, we have

$$f'_+(y,\lambda,V) = -h_+(\lambda,V_y)f_+(y,\lambda,V),$$

which shows

$$f_+(x,\lambda,V) = e^{-\int_0^x h_+(\lambda,V_y)dy}. \tag{1.10}$$

On the other hand, the monotonicity of V and Lemma 1 imply $h_+(\lambda,V_y) \geq h_+(\lambda,V)$ for any $y \geq 0$. Hence the lemma follows. ∎

Lemma 8.

$$E_x(e^{-\int_0^{\tau_0} V(B_s)ds}\tau_0) = \int_0^\infty g(x,y)f_+(y,0,V)dy \equiv f(x). \tag{1.11}$$

$$E_x(e^{-\int_0^{\tau_0} V(B_s)ds}\tau_0^2) = 2\int_0^\infty f(y)dy \tag{1.12}$$

where $g(x,y)$ is the 0-th order Green function of $-L$ imposed Dirichlet boundary condition at $x=0$.

Proof. These identities can be immediately obtained by differentiating the equation (1.1) of $f_+(x,\lambda,V)$ with respect to λ. ∎

Lemma 9. *Suppose $h_+(0,V)^{-1} \geq C_1 > 0$. Then there exists a constant C depending only on C_1 such that*

$$E_x(e^{-\int_0^{\tau_0} V(B_s)ds}; \tau_0 \geq t) \leq Ch_+(0,V)^{-2}f_+(x,0,V)\left(\frac{x}{t}\right)^{\frac{3}{2}}$$

holds.

Proof. Let $\psi(x)$ be the solution of

$$L\psi = 0,\ \psi(0) = 0,\ \psi'(0) = 1.$$

For simplicity, set $f_+(x) = f_+(x,0,V)$. Then

$$g(x,y) = g(y,x) = \psi(x)f_+(y)$$

if $x \le y$. However an identity

$$\psi(x) = f_+(x)\int_0^x \frac{dy}{f_+(y)^2}$$

gives

$$\int_0^\infty g(x,y)f(y)dy = f_+(y)\int_0^x \frac{dy}{f_+(y)^2}\int_y^\infty f_+(z)f(z)dz. \tag{1.13}$$

Setting $f = f_+$ in (1.13) and applying Lemma 7, we see

$$\int_0^\infty g(x,y)f_+(y)dy \le \frac{1}{2}h_+(0,V)^{-1}f_+(x)x. \tag{1.14}$$

Substituting the right hand side of (1.14) into (1.13) and applying Lemma 7 again, we have

$$2\int_0^\infty g(x,y)f(y)dy \le \frac{1}{4}h_+(0,V)^{-2}x(h_+(0,V)^{-1}+x)f_+(x).$$

Schwartz inequality shows

$$\begin{aligned} E_x(e^{-\int_0^{\tau_0} V(B_s)ds}\tau_0^{\frac{3}{2}}) &\le Cf_+(x)h_+(0,V)^{-\frac{3}{2}}x(h_+(0,V)^{-1}+x)^{\frac{1}{2}} \\ &\le Cf_+(x)h_+(0,V)^{-2}x^{\frac{3}{2}}, \end{aligned}$$

which concludes the lemma. ∎

Now a sharper estimate of $p_V(t,x,0)$ is possible.

Proposition 10. *Fix $a > 0$. Suppose*

$$g_0(a,a,V)g_0(0,0,V) \ge C_1,\ h_+(0,V_a)^{-1} \ge C_1.$$

Then there exists a constant C depending only on C_1 and a such that

$$\begin{aligned} p_V(t,x,0) &\le Cg_0(a,a,V)g_0(0,0,V)h_+(0,V_a)^{-2}\times \\ &\quad f_+(x-a,0,V_a)(x-a)^{\frac{3}{2}}t^{-\frac{3}{2}} \end{aligned} \tag{1.15}$$

holds for $x \le 2a$ and $t > 0$.

Proof. The first passage time decomposition implies for $x > a$

$$p_V(t,x,0) = \int_0^t p_V(t-s,a,0)dF_{x,a}(s),$$

where

$$F_{x,a}(t) = E_x(e^{-\int_0^{2a} V(B_s)ds}; \tau_a \le t)$$

and τ_a is the first hitting time of B_t at a. Separating the above integration into three parts as in the proof of Prop. 6 and set

$$\begin{aligned} \mathrm{I} &= \int_0^1 p_V(t-s,a,0)dF_{x,a}(s), \\ \mathrm{II} &= \int_1^{t-1} p_V(t-s,a,0)dF_{x,a}(s), \\ \mathrm{III} &= \int_{t-1}^{t} p_V(t-s,a,0)dF_{x,a}(s). \end{aligned}$$

Since $a \neq 0$, Lemma 5 implies that $p_V(t-s,a,0)$ $(t-1 \leq s \leq t)$ can be dominated from the above by a constant C. Hence

$$\mathrm{III} \leq C(F_{x,a}(+\infty) - F_{x,a}(t-1)).$$

Lemma 9 shows that (1.15) is valid for the term III. To estimate I and II, note

$$\begin{aligned} p_V(t,a,0) &\leq p_V(t,a,a)^{\frac{1}{2}} p_V(t,0,0)^{\frac{1}{2}} \\ &\leq g_0(a,a,V)g_0(0,0,V)t^{-\frac{3}{2}}. \end{aligned} \tag{1.16}$$

We have applied Lemma 4 in the above. Then integration by parts gives

$$\begin{aligned} \mathrm{II} \leq\ & Cg_0(a,a,V)g_0(0,0,V)\Big\{t^{-\frac{3}{2}}(F_{x,a}(+\infty) - F_{x,a}(t-1)) \\ &+ \int_1^{t-1} (t-s)^{-\frac{5}{2}}(F_{x,a}(+\infty) - F_{x,a}(s))ds\Big\}. \end{aligned}$$

Lemma 9 shows (1.15) is valid for the term II. Using (1.16) again, we see easily

$$\mathrm{I} \leq Cg_0(a,a,V)g_0(0,0,V)F_{x,a}(+\infty)t^{-\frac{3}{2}},$$

which completes the proof. ∎

2. Asymptotics of heat kernels as $t \to +\infty$ for fixed space variables.

In this section, we investigate $t^{-\frac{3}{2}}$ asymptotics of $p_V(t,x,y)$ near $t = \infty$ under the condition (B) on V. The lemma below enables us to obtain asymptotics $p_V(t,x,y)$ in terms of the Green function $g_\lambda(x,y,V)$.

Lemma 11. *For any fixed $x, y \in \mathbf{R}$, suppose*

$$\lambda^{-\frac{1}{2}}(g_0(x,y,V) - g_\lambda(x,y,V)) \longrightarrow A(x,y)$$

as $\lambda \to 0$. Then

$$t^{\frac{3}{2}} p_V(t,x,y) \longrightarrow \frac{1}{2\sqrt{\pi}} A(x,y).$$

as $t \to \infty$.

Proof. To show this lemma, it is sufficient to see the following. Suppose $\mathbf{P}(t) \leq \mathbf{P}(s)$ is a 2×2 positive definite matrix satisfying for $t > s \geq 0$. Let

$$\mathbf{C}(\lambda) = \int_0^\infty e^{-\lambda t}\mathbf{P}(t)dt,$$

and assume that $\mathbf{C}(0)$ is finite. If

$$\lambda^{-\frac{1}{2}}(\mathbf{C}(0) - \mathbf{C}(\lambda)) \longrightarrow \mathbf{A}$$

as $\lambda \to 0$, then

$$t^{\frac{3}{2}}\mathbf{P}(t) \longrightarrow \frac{1}{2\sqrt{\pi}}\mathbf{A} \quad \text{as } t \to \infty.$$

Since, in the scalar case, the above is a conclusion of a Tauberian theorem, a matrix version follows simply by taking an inner product $(\mathbf{P}(t)z, z)$ with arbitrary $z \in \mathbf{C}^2$. ∎

From now on we omit the dependence of V in the notations of heat kernels and Green functions.

Lemma 12. *Under the condition (B), we have as* $\lambda \to 0$

$$\lambda^{-\frac{1}{2}}(h_-(\lambda) - h_-(0)) \longrightarrow \sqrt{2}f_-(-\infty, 0)^2, \tag{2.1}$$

$$\lambda^{-\frac{1}{2}}(f_+(x,\lambda) - f_-(x,0)) \to \sqrt{2}f_-(-\infty,0)^2\psi_0(x). \tag{2.2}$$

Proof. Let $\varphi(x)$ be a solution of

$$\varphi(x) = 1 + \int_{-\infty}^x (x-y)\varphi(y)V(y)dy.$$

The condition (B) guarantees the existence and uniqueness of the solution. φ satisfies

$$L\varphi = 0,\ \varphi(-\infty) = 1,\ \varphi'(-\infty) = 0.$$

It is easy to see that

$$\varphi(x) = f_-(x,0)f_-(-\infty,0)^{-1}. \tag{2.3}$$

We make a harmonic transformation of L, that is,

$$\hat{L}f = \varphi^{-1}L(f\varphi) = \frac{1}{2}f'' + \frac{\varphi'}{\varphi}f'.$$

Introducing

$$S(x) = \int_0^x \varphi(y)^{-2}dy,\ M(x) = 2\int_0^x \varphi(y)^2 dy.$$

We define $\hat{M}(s) = M(x(s))$ with the inverse function $x(s)$ of $S(x)$. Then

$$\hat{L} = \frac{d}{dM(x)}\frac{d}{dS(x)} = \frac{d}{d\hat{M}(s)}\frac{d}{ds}.$$

Note

$$\lim_{s\to-\infty}\frac{\hat{M}(s)}{s} = \lim_{x\to-\infty}\frac{M(x)}{S(x)} = 2. \tag{2.4}$$

We introduce an increasing solution $\hat{f}_-(x,\lambda)$ for $\hat{L}$ similarly as in the case of L. Define $\hat{h}_-(\lambda) = \hat{f}'_-(0,\lambda)$. Since

$$\frac{\varphi(x)}{\varphi(0)}\hat{f}_-(S(x),\lambda) = f_-(x,\lambda),$$

we see

$$h_-(\lambda) = \hat{h}_-(\lambda)\varphi(0)^{-2} + \frac{\varphi'(0)}{\varphi(0)}.$$

Noting $\hat{h}_-(0) = 0$, we have an identity

$$h_-(x) - h_-(0) = \hat{h}_-(\lambda)\varphi(0)^{-2}. \tag{2.5}$$

Since the asymptotic behavior of $\hat{M}$ near $s = -\infty$ is known in (2.4), Kac [1] implies as $\lambda \to 0$

$$\lambda^{-\frac{1}{2}}\hat{h}_-(\lambda) \to \sqrt{2}.$$

This combined with (2.3) and (2.5) shows (2.1). To prove (2.2) we introduce another solution $\varphi_\lambda(x)$ of $Lf = \lambda f$ satisfying $f(0) = 1, f'(0) = 0$. Then as functions of λ, $\varphi_\lambda(x)$, $\psi_\lambda(x)$ become holomorphic on $\mathbf{C}$. f_- is nothing but

$$f_-(x,\lambda) = \varphi_\lambda(x) + h_-(\lambda)\psi_\lambda(x).$$

Therefore making use of (2.1) we have (2.2). ∎

Lemma 13. *Under the condition (A), unless V vanishes on $\mathbf{R}_+$, $h_+(\lambda)$ and $f_+(x,\lambda)$ are holomorphic in a neighborhood of $\lambda = 0$.*

Proof. Under the condition of the lemma, we have $\liminf_{x\to\infty} V(x) > 0$, therefore L restricted on $\mathbf{R}_+$ with Dirichlet or Neumann boundary condition at $x = 0$ has no spectrum in a neighbourhood of 0, which implies the analyticity of their Green functions in λ and hence that of h_+ and $f_-(x,\cdot)$. ∎

Lemmas 11, 12 and 13 together with (1.3) immediately show

Proposition 14. *Under the conditions (A) and (B), we have as $t \to \infty$*

$$t^{\frac{3}{2}}p_V(t,x,0) \longrightarrow \frac{1}{\sqrt{2\pi}}g_0(0,0)^2 f_-(-\infty,0)^2 f_+(x,0).$$

3. Proof of Theorem

First we obtain an upper bound of $p_{\alpha V}(t,x,0)$ for $\alpha \in (0,1]$, $x \geq 2a$, $t \geq 1$, by applying (1.15). a is a positive fixed number. (1.3) gives

$$g_0(a,a,\alpha V) = g_0(0,0,\alpha V) f_+(a,0,\alpha V) f_-(a,0,\alpha V),$$

and (1.9) shows

$$f_+(a,0,\alpha V) f_+(x-a,0,\alpha V_a) = f_+(x,0,\alpha V).$$

On the other hand, $f_+(a,0,\alpha V)$ is bounded on $\alpha \in (0,1]$ and $h_+(0,\alpha V_a) \geq h_+(0,\alpha V)$, which, in conclusion, shows

$$p_{\alpha V}(t,x,0) \leq C g_0(0,0,\alpha V)^2 h_+(0,\alpha V)^{-2} f_+(x,0,\alpha V) \left(\frac{x}{t}\right)^{\frac{3}{2}}, \qquad (3.1)$$

for $t > 0$, $x \geq 2a$, $\alpha \in (0,1]$ with a constant C. The Green function $g_\alpha^+(x,y)$ of $\frac{1}{2V}\frac{d^2}{dx^2}$ on $\mathbf{R}_+$ with Neumann boundary condition at $x=0$ has an expression

$$g_\alpha^+(x,0) = g_\alpha^+(0,x) = h_+(0,\alpha V)^{-1} f_+(x,0,\alpha V)$$

and the Green function $g_\alpha(x,y)$ of the same operator defined on the whole line $\mathbf{R}$ is described by $f_\pm$ as (1.3): for $x \geq y$

$$g_\alpha(x,y) = (h_+(0,\alpha V) + h_-(0,\alpha V))^{-1} f_+(x,0,\alpha V) f_-(y,0,\alpha V).$$

Therefore (3.1) can be replaced by a more intrinsic form:

$$p_{\alpha V}(t,x,0) \leq C g_\alpha(0,0)^2 g_\alpha^+(0,0) g_\alpha^+(x,0) \left(\frac{x}{t}\right)^{\frac{3}{2}}. \qquad (3.2)$$

Set

$$\rho(\alpha) = C g_\alpha(0,0)^2 g_\alpha^+(0,0) \int_0^\infty g_\alpha^+(0,x) x^{\frac{3}{2}} g(x)\,dx.$$

Then for $\alpha \in (0,1]$

$$t^{\frac{3}{2}} \int_{2a}^\infty p_{\alpha V}(t,0,x) g(x)\,dx \leq \rho(\alpha).$$

The assumption on θ implies $\int_0^1 \rho(\alpha)\sigma(d\alpha) < \infty$. Pick up $\alpha_0 \in (0,1]$ such that $\rho(\alpha_0) < \infty$. Then for $\alpha \geq 1$,

$$t^{\frac{3}{2}} \int_{2a}^\infty p_{\alpha V}(t,0,x) g(x)\,dx \leq \rho(\alpha_0) < \infty.$$

Then the dominated convergence theorem shows

$$\int_0^\infty \left(t^{\frac{3}{2}} \int_{2a}^\infty p_{\alpha V}(t,0,x)g(x)dx \right) \sigma(d\alpha)$$
$$\longrightarrow \int_{2a}^\infty A(0,x)g(x)dx,$$

as $t \to \infty$. For the region $x \in (-\infty, 2a]$, the computation is much easier. We have only to note (1.8), (B) and the condition $\theta(0) < \infty$. $\theta(0) < \infty$ implies $\int_0^1 g_\alpha(0,0)^2 \sigma(d\alpha) < \infty$. This completes the proof of the theorem. ∎

Now we give a simpler upper bound of $\theta(x)$. We have a trivial inequality:

$$g_\alpha(0,0)^2 g_\alpha^+(0,0) g_\alpha^+(x,0) \le h_+(0,\alpha V)^{-4} f_+(x,0,\alpha V).$$

However we have

$$h_+(0,\alpha V)^{-1} \le 2U^{-1}(\frac{1}{\alpha}),$$

where $U(x) = x \int_0^x V(y)dy$. For the proof see Kotani-Watanabe [3]. On the other hand we see

$$f_+(x,0,\alpha V) \le \psi_0'(x,0,\alpha V)^{-1} \le \left(1 + \alpha \int_0^x yV(y)dy\right)^{-1}.$$

Therefore setting

$$\mu(t) = \int_0^1 U^{-1}(\frac{1}{\alpha})^4 \frac{\sigma(d\alpha)}{1+\alpha t},$$

we see

Lemma 15.

$$\theta(x) \le 16\mu\Big(\int_0^x yV(y)dy\Big).$$

References

1. Kac, I. S., Generalization of an asymptotic formula of V.A.Marčenko for spectral functions of a second order boundary value problem, Math. USSR. Izv. 7 (1973), 422-436.
2. Kawazu, K., and Tanaka, H., On the maximum of a diffusion process in a drifted Brownian environment, Séminaire de Probabilités, LMN 1557, 78-85.
3. Kotani, S., and Watanabe, S., Krein's spectral theory of strings and generalized diffusion processes, Proceedings of Functional Analysis in Markov processes, ed. Fukushima, M., LMN 923, 235-259.
4. Yor, M., On some exponential functionals of brownian motion, Adv. Appl. Probab. 24 (1992), 509-531.

Stochastic differential equations with jumps and stochastic flows of diffeomorphisms

Hiroshi Kunita

Graduate School of Mahtematics, Kyushu University, Hakozaki, Fukuoka 812, Japan

1. Introduction

After fundamental works of K. Itô in 1940s, theory of stochastic differential equations (SDE) has been studied extensively. The flow property of the solution of SDE was studied around 1980 by Elworthy, Bismut, Ikeda-Watanabe, Kunita, Meyer etc. It was proved that under the Lipschitz condition of the coefficients of the equation, the solution of any SDE driven by a Brownian motion or a continuous semimartingale admits a version of a stochasic flows of homeomorphisms. Further if the coefficients are smooth, it admits a version of a stochastic flow of diffeomorphisms. Details are found in Kunita's book [11]. In this paper, we will be mainly concerned with SDE driven by a Lévy process or a semimartingale with jumps and discuss the flow property of the solutions. Before we introduce our SDE, let us briefly recall the relation between SDE driven by a Brownian motion or a continuous semimartingle and a stochastic flow of homeomorphisms.

Consider Itô's SDE on $\mathbf{R}^d$ based on an m-dimensional Bronwian motion or continuous semimartinale $Z(t) = (Z^1(t), ..., Z^m(t))$:

$$d\xi(t) = \sum_{j=1}^{m} v_j(\xi(t))dZ^j(t), \tag{1.1}$$

where $v_1, ..., v_m$ are Lipschitz continuous maps from $\mathbf{R}^d$ into itseslf. For any $s \geq 0$ and $x \in \mathbf{R}^d$, the equation has a unique global solution starting from x at time s. We denote it by $\xi_{s,t}(x), t \geq s$. It has a version which is continuous in (t, x) and the maps $\xi_{s,t} : \mathbf{R}^d \to \mathbf{R}^d$ are homeomorphisms for any $s < t$ a.s. Further if $v_1, ..., v_m$ are smooth, the maps are diffeomorphisms a.s.

SDE can also be defined on a manifold. As is pointed out by Itô, the change of local coordinates require a special rule for representing the SDE using the Itô integral. Later, it was recognized, perhaps by several persons independently, that the use of the Stratonovich integral provides a coordinate free representation of SDE, though we have to often rewrite it with the Itô integral in order to prove some useful results. The Stratonovich version of SDE on the manifold M can be written as

$$d\xi(t) = \sum_{j=1}^{m} v_j(\xi(t)) \circ dZ^j(t), \tag{1.2}$$

where $v_1, ..., v_m$ are smooth vector fields on the manifold. The solution can be regarded as a stochastic integral curve of the vector field valued process $\sum_j Z^j(t)v_j$. It may explode in a finite time, but its maximal solution $\xi_{s,t}(p), p \in M, s \leq t < \sigma(s,p)$ ($\sigma(s,p)$ is the explosion time of $\xi_{s,t}(p)$) defines a stochastic flow of local diffeomorphisms. Another advantage of using the Stratonovich integral is that the solution flow has some nice symmetric properties with respect to the time inverse, which are known for the deterministic flows generated by ordinary differential equations. For example, the inverse flow $\xi_{s,t}^{-1}$ satisfies the backward Stratonovich SDE of the same type. Details are found in [11].

Now Itô's SDE on $\mathbf{R}^d$ driven by a Lévy process or a semimartingale with jumps $Z(t) = (Z^1(t), ..., Z^m(t))$ can be written as (1.1). Assuming that $v_1, ..., v_m$ are Lipschitz continuous, the solution of (1.1) admits a version of a stochastic flow of continuous maps, i.e., a version $\xi_{s,t}(x)$ is continuous in $x \in \mathbf{R}^d$ and cadlag in $t \in [s, \infty)$. However, the maps $\xi_{s,t}$ are not homeomorphisms in general, since the maps $\varphi : x \to x + \sum_j \Delta Z^j(t)v_j(x)$ caused by the jumps $\Delta Z^j(t) = Z^j(t) - Z^j(t-)$ of the driving process may not be homeomorphisms. By the same reason, the solutions of Stratonovich equation (1.2) do not define a stochastic flow of local diffeomorphisms, either. Furthermore, (1.2) is no longer a coordinate free representation of SDE, if the driving process $Z(t)$ has jumps.

We shall consider another representation of SDE, which is due to Marcus [14]. The equation is written as

$$d\xi(t) = \sum_{j=1}^{m} v_j(\xi(t)) \circ dZ_c^j(t) + \sum_{j=1}^{m} v_j(\xi(t-))dZ_d^j(t)$$
$$+\{\mathrm{Exp}(\sum_j \Delta Z^j(t)v_j)(\xi(t-)) - \xi(t-) - \sum_j \Delta Z^j(t)v_j(\xi(t-))\}, \quad (1.3)$$

where $Z_c(t)$ and $Z_d(t)$ are the continuous part and the discontinuous part of the semimartingale $Z(t)$, respectively. $\varphi(t,x) = \mathrm{Exp}(tv)(x)$ is the solution flow of the differential equation

$$\frac{d\varphi(t)}{dt} = v(\varphi(t)), \varphi(0) = x. \quad (1.4)$$

Equation (1.3) is a coordinate free formulation of SDE with jumps. We shall call it a *canonical SDE driven by a vector field valued semimartingale* $X(t) = \sum_j Z^j(t)v_j$.

The above equation looks complicated. But the probabilistic meaning is simple. At the jumping time t of the driving process $Z(t)$, the solution flow flies from the state $\xi_{s,t-}(x)$ along with the integral curve $\mathrm{Exp}(rv), 0 \leq r \leq 1$ with the infinite speed, where $v = \sum_j \Delta Z^j(t)v_j$ and lands at the position of $r = 1$, i.e., it jumps to the state $\xi_{s,t}(x) = \mathrm{Exp}(\sum_j \Delta Z^j(t)v_j)(\xi_{s,t-}(x))$. Therefore if the map $\xi_{s,t-}$ is a homeomorphisms of the state, the map $\xi_{s,t}$

should also be a homeomorphism, since Exp$v$ ia a homeomorphism for any Lipschitz continuous vector field v.

The above canonical form of SDE has been studied by several authors. See Fujiwara [5], Estrade [4], Applebaum-Kunita [1], Kurtz-Pardoux-Protter [12]. In this paper we shall discuss canonical SDE's in more general forms. Instead of the finite dimensional vector field valued driving process $X(x,t) = \sum_j Z^j(t)v_j(x)$, we consider a (infinite dimensional) vector field valued process. Let $\mathcal{V}$ be the space of vector fields with certain smoothness conditions. Let $X(t) = X(x,t)$ be a Lévy process or a semimartingale with values in $\mathcal{V}$. Denote by $X_c(t)$ the continuous part of $X(t)$ and by $X_d(t)$ the discontinuous part of $X(t)$ such that $X(t) = X_c(t) + X_d(t)$. Then the canonical SDE on $\mathbf{R}^d$ is defined by

$$
\begin{aligned}
d\xi(t) &= X_c(\xi(t), \circ dt) + X_d(\xi(t-), dt) \\
&\quad + \mathrm{Exp}(\Delta X(t))(\xi(t-)) - \xi(t-) - \Delta X(\xi(t-), t),
\end{aligned}
\tag{1.5}
$$

where $\int_0^t X_d(\xi(s-), ds)$ and $\int_0^t X_c(\xi(s), \circ ds)$ are the nonlinear Itô integral and nonlinear Stratonovich integral, respectively. These will be defined by (2.7) and (2.8).

SDE represented by nonlinear integrals was introduced by Le Jan, who discussed the SDE driven by a vector field valued Brownian motion. Later it was developed with full details by Kunita, restricting to *continuous* semimartingales or Brownian motions. The study of SDE with jumps represented by nonlinear integral was initiated by Fujiwara-Kunita [6], where the Itô type SDE driven by a Lévy process is considered. It was extended to the Itô SDE driven by a special semimartingale by Carmona-Nualart [2]. We shall reformulate these works in the framework of the canonical SDE mentioned above.

2. Semimartingales with spatial parameters and the associated three nonlinear stochastic integrals

Let $(\Omega, \mathcal{F}, P)$ be a probability space carrying a filtration $(\mathcal{F}_t)_{t>0}$ of a right continuous increasing family of sub σ-fields of $\mathcal{F}$. Let $\{X(x,t), t \geq 0\}_{x\in\mathbf{R}^d}$ be a family of $\mathbf{R}^d$-valued stochastic processes with spatial parameter $x \in \mathbf{R}^d$ defined on $(\Omega, \mathcal{F}, P)$. If $X(x,t)$ is continuous in x for each t a.s., we can regard it as a C-valued process, where $C := C(\mathbf{R}^d; \mathbf{R}^d)$ is the space of continuous maps form $\mathbf{R}^d$ into itself equipped with the compact uniform topology. We denote it by $X(t) = X(x,t), t \geq 0$.

Suppose that $X(t), t \geq 0$ is a C-valued cadlag process (right continuous with left hand limits). It is called a *semimartingale* if for each $x \in \mathbf{R}^d$, $X(x,t), t \geq 0$ is an $\mathbf{R}^d$-valued semimartingale adapted to the filtration $(\mathcal{F}_t)_{t>0}$. For such $X(t)$, we set

$$N((s,t] \times E) = \sharp\{r \in (s,t]; \Delta X(r) \in E\}, \tag{2.1}$$

where $\sharp\{\cdots\}$ denotes the number of points of the set $\{\cdots\}$, $\Delta X(r) = X(r) - X(r-)$ and E is a Borel subset of $C - \{0\}$. There exists a predictable measure-valued process $\nu_t(dv)$ on C and a predictable strictly increasing process A_t such that the real valued process

$$\tilde{N}((s,t] \times E) := N((s,t] \times E) - \int_s^t \nu_r(E)dA_r, \quad t \geq s \tag{2.2}$$

is a local martingale for any s and E. We assume that $X(t)$ has the following properties. $X(t)$ has the decomposition $X(t) = X_c(t) + X_d(t)$ where $X_c(t) = X_c(x,t), t \geq 0$ is a continuous semimartingale for each x and $X_d(t) = X_d(x,t), t \geq 0$ is a discontinuous semimartingale for each x, represented by

$$X_d(x,t) = \int_{\mathcal{U}} v(x)\tilde{N}((0,t], dv) + \int_{\mathcal{U}^c} v(x)N((0,t], dv), \tag{2.3}$$

where $\mathcal{U}$ is a bounded Borel subset of C. Further, $X_c(t) = M_c(t) + B_c(t)$, where $M_c(t) = M_c(x,t), t \geq 0$ is a continuous local martingale and $B_c(t) = B_c(x,t), t \geq 0$ is a continuous process of bounded variation. Then there exists a predictable process $(a(x,y,t,\omega), b(x,t,\omega)), t \geq 0$ and a strictly increasing predictable process $A_t, t \geq 0$ such that

$$\langle M_c^i(x,t), M_c^j(y,t)\rangle = \int_0^t a^{ij}(x,y,r)dA_r, \quad \forall t > 0, \tag{2.4}$$

$$B_c^i(x,t) = \int_0^t b^i(x,r)dA_r, \quad \forall t > 0, \tag{2.5}$$

where $\langle \cdot, \cdot \rangle$ denotes the quadratic variational process. The triple $a(t) = (a^{ij}(t)), b(t) = (b^i(t))$ and ν_t are called the *characteristics* of $X(t)$ (with respect to A_t).

Before we proceed to defining nonlinear stochastic integrals based on the C-valued semimartingale, let us introduce some function spaces. Let m be a positive integer and let $0 < \delta \leq 1$. For a multi-index $\alpha = (\alpha_1, ..., \alpha_d)$ of nonnegative integers, we set $D^\alpha = \partial^{|\alpha|}/(\partial x_1)^{\alpha_1} \cdots (\partial x_d)^{\alpha_d}$, where $|\alpha| = \alpha_1 + \cdots + \alpha_d$. For an m-times continuously differentiable function v, we define a norm

$$\|v\|_{m+\delta} := \sum_{0 \leq |\alpha| \leq m} \|D^\alpha v\| + \sum_{|\alpha| = m} \sup_{x \neq y} \frac{|D^\alpha v(x) - D^\alpha v(y)|}{|x-y|^\delta},$$

where $\| \ \|$ is the supremum norm. Let C^m be the set of all m-times continuously differentiable functions on $\mathbf{R}^d$. Set $C_b^{m+\delta} = \{v \in C^m : \|v\|_{m+\delta} < \infty\}$. Next let $\tilde{C} = C(\mathbf{R}^d \times \mathbf{R}^d : \mathbf{R}^d \otimes \mathbf{R}^d)$, where $\mathbf{R}^d \otimes \mathbf{R}^d$ is the space of $d \times d$-matrices. For an m-times continuously differentiable function w, we define a norm by

$$\|w\|^{\sim}_{m+\delta} = \sum_{0\le|\alpha|\le m} \|\tilde{D}^\alpha w\|^{\sim}$$
$$+ \sum_{|\alpha|=m} \sup_{x\neq y} \frac{|\tilde{D}^\alpha w(x,x) - \tilde{D}^\alpha w(x,y) - \tilde{D}^\alpha w(y,x) + \tilde{D}^\alpha w(y,y)|}{|x-y|^{2\delta}},$$

where $\| \ \|^{\sim}$ is the supremum norm on $\tilde{C}$ and $\tilde{D}^\alpha = D_x^\alpha D_y^\alpha$. We set $\tilde{C}_b^{m+\delta} = \{w \in \tilde{C} : \|w\|^{\sim}_{m+\delta} < \infty\}$.

Now let $X(t), t \ge 0$ be a C-valued semimartingale with characteristics $(a(t), b(t), \nu_t)$. We introduce:

Condition A. (1) $a(x,y,t)$ is a continuous $\tilde{C}_b^{1+1}$-valued process satisfying $\|a(t)\|^{\sim}_{1+1} \le K_t$.
(2) $b(t)$ is a continuous C_b^{0+1}-valued process satisfying $\|b(t)\|_{0+1} \le K_t$.
(3) The measures $\nu_t(\cdot)$ are supported by C_b^{1+1}. There exists a Borel set $\mathcal{U}$ of C_b^{1+1} such that $\|v\|_{1+1} \le c$ for all $v \in \mathcal{U}$ for some $c > 0$ and

$$\nu_t(\mathcal{U}^c) \le K_t, \quad \int_{\mathcal{U}} \|v\|^2_{1+1} \nu_t(dv) \le K_t. \tag{2.6}$$

Here, $K_t, t \ge 0$ is a positive predictable process satisfying $\int_0^T K_t dA_t < \infty, \quad \forall T > 0$.

Let $\eta_t, t \ge 0$ be an $\mathbf{R}^d$-valued cadlag process adapted to $(\mathcal{F}_t)$. We will define three nonlinear integrals of η_t based on a C-valued semimartingale. The first is the *Itô integral*:

$$\int_s^t X(\eta_{r-}, dr) := \lim_{|\Delta|\to 0} \sum_{k=1}^n \left(X(\eta_{t_{k-1}}, t_k) - X(\eta_{t_{k-1}}, t_{k-1})\right), \tag{2.7}$$

where $\Delta = \{s = t_0 < t_1 < \cdots < t_n = t\}$ are the partitions of $[s,t]$ and $|\Delta| = \max_k(t_k - t_{k-1})$. It is well defined and is an $\mathbf{R}^d$-valued semimartingale.

Next suppose that $\eta_t, t \ge 0$ is an $\mathbf{R}^d$-valued semimartingale. Then we can define the *Stratonovich integral* by

$$\int_s^t X(\eta_r, \circ dr) := \lim_{|\Delta|\to 0} \sum_{k=1}^{n-1} \frac{1}{2}\{X(\eta_{t_{k+1}}, t_{k+1}) + X(\eta_{t_k}, t_{k+1})$$
$$-X(\eta_{t_{k+1}}, t_k) - X(\eta_{t_k}, t_k)\}. \tag{2.8}$$

Thirdly we consider the ordinary differential equation (1.4), where $v(x)$ is a Lipschitz continuous vector field (elements of C_b^{0+1}). It has a unique global solution $\varphi_t(x), t \in (-\infty, \infty)$, which we denote by $\mathrm{Exp}tv(x)$. Then $\{\mathrm{Exp}tv\}_{t\in(-\infty,\infty)}$ is a one papameter group of homeomorphisms of $\mathbf{R}^d$, i.e., the map $\mathrm{Exp}tv : \mathbf{R}^d \to \mathbf{R}^d$ is an onto homeomorphism for any t and satisfies $\mathrm{Exp}tv \circ \mathrm{Exp}sv = \mathrm{Exp}(t+s)v$ for any $s, t \in (-\infty, \infty)$. In view of Condition A(3), we can show that

$$\int_{\mathcal{U}} \|\mathrm{Exp}v - I - v\| \nu_t(dv) \le cK_t, \quad \forall t > 0,$$

where c is a positive constant. Then the possibly infinite sum

$$\sum_{s \le t} \Big(\mathrm{Exp}\Delta X(s)(x) - x - \Delta X(x,s) \Big)$$

is absolutely convergent a.s. The *canonical integral* of a cadlag semimartingale η_t (or *Marcus's canonical extension of the Itô integral*) based on the vector field valued semimartingale $X(t)$ is defined by

$$\begin{aligned} \int_s^t X(\eta_r, \circ dr) \quad &:= \quad \int_s^t X_c(\eta_r, \circ dr) + \int_s^t X_d(\eta_{r-}, dr) \\ &+ \sum_{s \le r \le t} \Big(\mathrm{Exp}\Delta X(r)(\eta_{r-}) - \eta_{r-} - \Delta X(\eta_{r-}, r) \Big) \end{aligned} \tag{2.9}$$

3. SDE on Euclidean space based on vector field valued semimartingales and stochastic flows of homeomorphisms

Let $X(t) = X(x,t), t \ge 0$ be a C-valued semimartingale whose characteristics satisfy Condition A. In the followings, we will assume that $X(t)$ is quasi-left continuous, i.e., we assume that the increasing process $A(t)$ is continuous in t a.s. The SDE's associated with general (non-quasi-left continuous) semimartingales will be discussed in [7].

In this section we regard $X(t)$ as a vector field valued semimartingale. We shall consider the SDE represented by the canonical integral:

$$\xi_t = x + \int_s^t X(\xi_r, \circ dr), \tag{3.1}$$

where $0 \le s < t$. It is equivalent to the following SDE represented by the Itô integral:

$$\xi_t = x + \int_s^t \tilde{X}(\xi_{r-}, dr), \tag{3.2}$$

where

$$\tilde{X}(x,t) := X(x,t) + \int_0^t c(x,s) dA_s + \sum_{0 < s \le t} \Big(\mathrm{Exp}\Delta X(s)(x) - x - \Delta X(x,s) \Big), \tag{3.3}$$

and $c(t) = (c^i(x,t))$ is defined by

$$c^i(x,t) = \frac{1}{2} \sum_j \frac{\partial a^{ij}}{\partial x^j}(x,y,t) \Bigg|_{y=x}. \tag{3.4}$$

Indeed, let ξ_t be a solution of Itô equation (3.2). Then

$$\int_s^t X_c(\xi_r, \circ dr) = \int_s^t X_c(\xi_{r-}, dr) + \int_s^t c(\xi_r, r)dA_r. \tag{3.5}$$

Therefore the canonical integral (2.9) is transformed to the Itô integral $\int_s^t \tilde{X}(\xi_{r-}, dr)$. The process ξ_t is called a *solution of the canonical SDE (3.1) driven by the vector field valued semimartingale* $X(t)$.

An advantage of representing SDE by the canonical integral is that the solution admits the property of the stochastic flow of homeomorphisms. As an example we first consider the canonical SDE driven by $X(t) = X(x,t) = Z(t)v(x)$, where $v(x)$ is a Lipschitz continuous vector field on $\mathbf{R}^d$ and $Z(t)$ is a one dimensional (scalor) semimartingale. Then the solution of the canonical SDE driven by $X(t) = Z(t)v$ is given by $\xi_{s,t}(x) = \text{Exp}((Z(t) - Z(s))v)(x)$. (Doss's representation of the solution.) The fact can be verified strightforward using Itô's formula for semimartingales with jumps. We will prove that the same fact is valid for a more general semimartingale $X(t)$.

It should be noted that not all SDE's represented by Itô integrals based on C-valued semimartingales are transformed to canonical SDE's. The solutions of Itô SDE's (3.2) will not define flows of homeomorphisms, if the jumps $\Delta\tilde{X}(t)$ are not homeomorphisms.

The following theorem was first proved by Fujiwara-Kunita [6] in the case where the driving process $X(t)$ is a Lévy process. An extension to semimartingale driving process is done by Carmona-Nurlart [2].

Theorem 3.1. *Assume that the characteristics of the C-valued semimartingale $X(t)$ satisfy Condition A. Then the canonical SDE (3.1) has a unique solution*

$\xi_{s,t}(x), t \geq s$ for any s, x. Further, a certain version $\xi_{s,t}(x)$ of the solution admits the following properties:
(i) $\xi_{s,u}(x) = \xi_{t,u}(\xi_{s,t}(x))$ holds for all $x \in \mathbf{R}^d$ and $s < t < u$, a.s.
(ii) The map $\xi_{s,t} : \mathbf{R}^d \to \mathbf{R}^d$ is an onto homeomorphism for all $s < t$ a.s.
(iii) $\xi_{s,t}$ is a C-valued cadlag processes in both s and t.

The above $\xi_{s,t}$ is called the *stochastic flow of homeomorphisms generated by* $X(t)$.

Proof. (Outline) We first consider the case where $\nu_t(\mathcal{U}^c) = 0, K_t = 1, A_t = t$ hold for all $t > 0$. Set $\tilde{v} = \text{Exp}v - I$. Then in view of Condition A(3), we can prove that there exists a positive constant c such that $\int_{\mathcal{U}} \|\tilde{v}\|_{0+1}^2 \nu_t(dv) \leq c$. Therefore, $\nu_t, t \geq 0$ satisfies

$$\int_{\mathcal{U}} |\tilde{v}(x) - \tilde{v}(y)|^p \nu_t(dv) \leq |x-y|^p \int_{\mathcal{U}} \|\tilde{v}\|_{0+1}^p \nu_t(dv) \leq c|x-y|^p, \quad \forall x, y \in \mathbf{R}^d$$

for any $p \geq 2$. Then the Itô equation (3.2) has a unique solution for any initial data s, x. We denote the solution by $\xi_{s,t}(x), t \in [s, \infty)$. Further, for any $p > 2$ and $T > 0$, there exists a positive constant M depending on p, T only such that $\xi_{s,t}(x)$ satisfies

$$E[\sup_{s\le r\le t} |\xi_{s,r}(x)-x-\xi_{s,r}(y)+y|^p] \le M(t-s)|x-y|^p, \quad \forall x,y \in \mathbf{R}^d, s,t \in [0,T],$$

$$E[\sup_{s\le r\le t} |\xi_{s,r}(x) - x|^p] \le M(t-s)(1+|x|)^p, \quad \forall x \in \mathbf{R}^d, s,t \in [0,T].$$

(c.f. Lemma 2.1 in [6]). Then Kolmogorov's theorem implies that $\xi_{s,t}, t \in [s,\infty)$ has a version of a right continuous C-valued process for any fixed s. The property (i) is immediate from the uniqueness of the solution of Itô SDE (3.2).

Now in view of Condition A(3), we can prove $\sup_{v\in\mathcal{U}} \|\mathrm{Exp} v - I\|_{0+1} < \infty$ and $\sup_{v\in\mathcal{U}} \|\mathrm{Exp}(-v) - I\|_{0+1} < \infty$. Hence for any $p > 2$ and $T > 0$, there exists a positive constant M depending on p, T only such that

$$E[\sup_{s\le r\le t} |\xi_{s,r}(x) - \xi_{s,r}(y)|^{-p}] \le M|x-y|^{-p}, \quad \forall x \ne y \in \mathbf{R}^d, s,t \in [0,T],$$

$$E[\sup_{s\le r\le t} (1+|\xi_{s,r}(x)|)^{-p}] \le M(1+|x|)^{-p} \quad \forall x \in \mathbf{R}^d, s,t \in [0,T]$$

by Theorem 2.6 in [6]. Then $\sup_{s\le r\le t} |\xi_{s,r}(x) - \xi_{s,r}(y)|^{-1}$ is continuous in $(x,y) \in \mathbf{R}^d \times \mathbf{R}^d - \{(x,x) : x \in \mathbf{R}^d\}$ by Kolmogorov's theorem. Hence $\xi_{s,r} : \mathbf{R}^d \to \mathbf{R}^d$ is one to one for any $r \in [s,t]$ a.s. for any s. Further, we have $\liminf_{|x|\to\infty} \inf_{r\in[s,t]} |\xi_{s,r}(x)| = \infty$ a.s. for any s and t. Hence $\xi_{s,r} : \mathbf{R}^d \to \mathbf{R}^d$ is onto for any $r \in [s,t]$ a.s. for any s. This proves the property (ii).

Now let $\xi_{s,t}^{-1}$ be the inverse map of $\xi_{s,t}$. Since $\xi_{s,t}$ is cadlag with respect to t, the inverse $\xi_{s,t}^{-1}$ is also cadlag with respect to t. Take a version of $\xi_{s,t}$ as $\xi_{0,s}^{-1}\xi_{0,t}$. Then it is cadlag both in s and t, proving the property (iii).

Suppose next that $0 < \int_0^T \nu_t(\mathcal{U}^c)dA_t < \infty, \forall T > 0$. Let $s = \sigma_0 < \sigma_1 < \cdots \sigma_n < \cdots$ be a sequence of jumping times of the counting process $Y(t) := N((s,t], \mathcal{U}^c)$. Then it holds $\sigma_n \to \infty$ as $n \to \infty$. Set

$$X'(t) := X(t) - \sum_{\Delta X(s)\in\mathcal{U}^c, s\le t} \Delta X(s).$$

Let $\xi'_{s,t}$ be the solution of the canonical equation driven by $X'(t)$. It has a version (denoted again by $\xi'_{s,t}$) satisfying (i)-(iii) of the theorem. We define $\xi_{s,t}(x), t \in [s,\infty)$ by

$$\xi_{s,t}(x) := \xi'_{\sigma_n,t} \circ \mathrm{Exp}\Delta X(\sigma_n) \circ \xi'_{\sigma_{n-1},\sigma_n} \circ \cdots \circ \mathrm{Exp}\Delta X(\sigma_1) \circ \xi'_{s,\sigma_1}(x),$$

if $t \in [\sigma_n, \sigma_{n+1})$. Since $\mathrm{Exp}\Delta X(\sigma_n)$ etc are homeomorphisms of $\mathbf{R}^d$, $\xi_{s,t}$ satisfies (i)-(iii) of the theorem. Further it is a solution of the canonical equation (3.1).

Finally in the case where $K_t \ne 1$ or $A_t \ne t$, let τ_t be the inverse function of the strictly increasing process $\int_0^t K_s dA_s$ and let $\hat{X}(t) := X(\tau_t)$. It is again a C-valued semimartingale with respect to $\hat{\mathcal{F}}_t := \mathcal{F}_{\tau_t}$, whose characteristics are given by $\hat{a}(t) = a(\tau_t), \hat{b}(t) = b(\tau_t)$ and $\hat{\nu}_t = \nu_{\tau_t}$ with respect to $\hat{A}_t = t$. Further the process $\hat{K}_t$ is given by $\hat{K}_t = 1$. Let $\hat{\xi}_{s,t}$ be the solution of the canonical SDE driven by $\hat{X}(t)$. It is a stochastic flow of homeomorphism. We define $\xi_{s,t} = \hat{\xi}_{A_s,A_t}$. Then it is a solution of the canonical equation (3.1) and in fact a flow of homeomorphisms. The proof is complete.

Suppose that we are given a two parameter family of sub σ-fields of $\mathcal{F}$, $(\mathcal{F}_{s,t})_{0\le s<t<\infty}$ such that $\mathcal{F}_{s,t} \subset \mathcal{F}_{s',t'}$ if $t' \ge t$ and $s' \le s$, and $\cap_{\epsilon>0}\mathcal{F}_{s,t+\epsilon} = \mathcal{F}_{s,t}$, $\cap_{\epsilon>0}\mathcal{F}_{s-\epsilon,t} = \mathcal{F}_{s,t}$ for any $s < t$. A C-valued cadlag process $X(t), t \ge 0$ is called a *forward-backward semimartingle* if $X(t) - X(s), t \in [s,\infty)$ is a forward semimartingale adapted to the filtration $(\mathcal{F}_{s,t})_{t\in[s,\infty)}$ for any s and also $X(t) - X(s), s \in [0,t]$ is a backward semimartingale adapted to the filtration $(\mathcal{F}_{s,t})_{s\in[0,t]}$ for any t. Any C-valued Lévy process is a forward-backward semimartingale.

Now let $\eta_s, 0 \le s \le t$ (t is fixed) be a cadlag process adapted to the filtration $(\mathcal{F}_{s,t})$. The *backward Itô integral of η_s based on a forward-backward semimartingale $X(x,t)$* is defined by

$$\int_s^t X(\eta_r, \hat{d}r) := \lim_{|\Delta|\to 0} \sum_{k=1}^{n} (X(\eta_{t_k}, t_k) - X(\eta_{t_k}, t_{k-1})). \tag{3.6}$$

It is a backward cadlag semimartingale with respect to s. The canonical backward integral $\int_s^t X(\eta_r, \circ\hat{d}r)$ can be defined similarly.

We obtain a backward SDE satisfied by the inverse flow $\xi_{s,t}^{-1}$.

Theorem 3.2 (cf [9]). *Let $X(t)$ be the C-valued semimartingale of Theorem 3.1. Assume that $X(t)$ is a forward-backward semimartingale. Then the inverse flow $\xi_{s,t}^{-1}$ is a cadlag C-valued process both in s and t. Further, it is a backward semimartingale and satisfies the following Itô's backward SDE.*

$$\xi_{s,t}^{-1}(y) - y = \int_s^t \hat{X}(\xi_{r,t}^{-1}(y), \hat{d}r), \tag{3.7}$$

where

$$\hat{X}(x,t) = -X(x,t) + \sum_{s\le t} \Big(\mathrm{Exp}(-\Delta X(s))(x) - x + \Delta X(x,s)\Big) + \int_s^t c(x,s)dA_s. \tag{3.8}$$

Thus $\xi_{s,t}^{-1}$ is represented as a solution of a canonical backward SDE driven by $-X$, i.e.,

$$\xi_{s,t}^{-1}(y) = y + \int_s^t (-X)(\xi_{r,t}^{-1}(y), \circ\hat{d}r). \tag{3.9}$$

Proof. Since $\xi_{s,t}$ is a cadlag C-valued process both in s and t, the inverse $\xi_{s,t}^{-1}$ should have the similar property. We will prove (3.7). In view of equation (3.2), $\xi_{s,t}(x)$ satisfies

$$\begin{aligned} \xi_{s,t}(x) - x = & \int_s^t X_c(\xi_{s,r-}(x), dr) + \int_s^t c(\xi_{s,r-}(x), r) dA_r \qquad (3.10) \\ & + \int_s^t \int_{\mathcal{U}} (\mathrm{Exp} v(\xi_{s,r-}(x)) - \xi_{s,r-}(x)) \tilde{N}(drdv) \\ & + \int_s^t \int_{\mathcal{U}^c} (\mathrm{Exp} v(\xi_{s,r-}(x)) - \xi_{s,r-}(x)) N(drdv) \\ & + \int_s^t \int_{\mathcal{U}} (\mathrm{Exp} v(\xi_{s,r-}(x)) - \xi_{s,r-}(x) - v(\xi_{s,r-}(x))) \nu_r(dv) dA_r. \end{aligned}$$

Substitute $x = \xi_{s,t}^{-1}(y)$ at each term of the above. It holds

$$\int_s^t X_c(\xi_{s,r-}(x), dr)\Big|_{x=\xi_{s,t}^{-1}(y)} = \int_s^t X_c(\xi_{r,t}^{-1}(y), \hat{d}r) - 2\int_s^t c(\xi_{r,t}^{-1}(y), r) \hat{d}A_r,$$

$$\begin{aligned} & \int_s^t \int_{\mathcal{U}^c} (\mathrm{Exp} v(\xi_{s,r-}(x)) - \xi_{s,r-}(x)) N(drdv)\Big|_{x=\xi_{s,t}^{-1}(y)} \\ & \qquad = \int_s^t \int_{\mathcal{U}^c} (\xi_{r,t}^{-1}(y) - \mathrm{Exp}(-v)(\xi_{r,t}^{-1}(y)) N(\hat{d}rdv), \end{aligned}$$

$$\begin{aligned} & \int_s^t \int_{\mathcal{U}} (\mathrm{Exp} v(\xi_{s,r-}(x)) - \xi_{s,r-}(x)) \tilde{N}(drdv)\Big|_{x=\xi_{s,t}^{-1}(y)} \\ & = \int_s^t \int_{\mathcal{U}} (\xi_{r,t}^{-1}(y) - \mathrm{Exp}(-v)(\xi_{r,t}^{-1}(y)) \tilde{N}(\hat{d}rdv) \\ & \quad - \int_s^t \int_{\mathcal{U}} (\mathrm{Exp} v(\xi_{r,t}^{-1}(y)) + \mathrm{Exp}(-v)(\xi_{r,t}^{-1}(y)) - 2\xi_{r,t}^{-1}(y)) \nu_r(dv) \hat{d}A_r. \end{aligned}$$

Therefore we obtain (3.7).

4. Smoothness and the diffeomorphic properties of the stochastic flow

We introduce conditions for the characteristics of C-valued semimartingale $X(t)$.

Condition $A^{m+\delta}$. (1) $a(t)$ is a predictable continuous $\tilde{C}_b^{m+1+\delta}$-valued-process such that $\|a(t)\|_{m+\delta}^{\sim} \le K_t$.
(2) $b(t)$ is a predictable continuous $C_b^{m+\delta}$-valued-process with $\|b(t)\|_{m+\delta} \le$

K_t.
(3) $\int_{\mathcal{U}} \|v\|^2_{m+1+\delta}\nu_t(dv) \le K_t$ holds for any t. Here K_t is a positive predictable process such that $\int_0^T K_t dA_t < \infty$ a.s. for any $T > 0$.

The next theorem is an improvement of the result of Kunita [10].

Theorem 4.1. *Assume that the characteristics of the C-valued semimartingale $X(t)$ satisfy Condition $A^{m+\delta}$ for some $m \ge 1$ and $\delta > 0$. Then the stochastic flow $\xi_{s,t}$ generated by $X(t)$ has a version of C^m-diffeomorphisms, i.e., it satisfies (i) of Theorem 3.1 and*
(ii') The map $\xi_{s,t} : \mathbf{R}^d \to \mathbf{R}^d$ is an onto C^m-diffeomorphism for any $s < t$ a.s.
(iii') $\xi_{s,t}$ is a C^m-valued cadlag process in both s and t.

The above $\xi_{s,t}$ is called the *stochastic flow of C^m-diffeomorphism generated by $X(t)$*.

Proof. (Outline) We show the differentiability of $\xi_{s,t}(x)$ in the case $m = 1$. We will only consider the case where $K_t = 1, \nu_t(\mathcal{U}^c) = 0, A_t = t$ for all $t > 0$. The general case can be verified similarly as in the proof of Theorem 3.1. Set

$$N_{s,t}(x,y) := \frac{1}{y}\{\xi_{s,t}(x + ye_i) - \xi_{s,t}(x)\}, \quad y \in \mathbf{R}^1 - \{0\},$$

where $e_i = (0, \cdots, 1, 0, \cdots, 0)$ (1 appears only at the i-th component). Then for any $p > 2$ and $T > 0$, there exists a positive constant M depending on p, T only such that

$$E[\sup_{s<r\le t} |N_{s,r}(x,y) - N_{s,r}(x',y')|^p] \le M\{|x - x'|^{p\delta} + |y - y'|^{p\delta}\}(t - s)$$

holds for any $x, x' \in \mathbf{R}^d$, $y, y' \in \mathbf{R}^1 - \{0\}$ and $s, t \in [0, T]$. Proof can be carried out similarly as in Lemma 2.2 in [6]. Take p such that $p\delta > d + 1$. Then by Kolmogorov's theorem, $N_{s,t}(x,y)$ has a continuous extension at $y = 0$ and $N_{s,t}(x,0), t \in [s,\infty)$ is a C-valued cadlag process. Consequently for each fixed s, $\xi_{s,t}, t \in [s,\infty)$ is a C^1-valued cadlag process.

We will prove the diffeomorphic property of the map $\xi_{s,t} : \mathbf{R}^d \to \mathbf{R}^d$. It is sufficient to prove that the Jacobian matrix of $\xi_{s,t}$ is invertible. We denote by $D\xi_{s,t}$ and $DX(t)$ the Jacobian matrix of $\xi_{s,t}$ and $X(t)$, respectively. Then Itô equation (3.2) implies

$$D\xi_{s,t}(x) = I + \int_s^t D\tilde{X}(\xi_{s,r-}(x), dr) D\xi_{s,r-}(x). \tag{4.1}$$

In what follows we fix a point $x \in \mathbf{R}^d$ and write $\xi_{s,t}(x)$ as $\xi_{s,t}$ etc. Consider the Itô SDE for the matrix valued process:

$$U_{s,t} = I - \int_s^t U_{s,r-} D\tilde{X}(\xi_{s,r-}, dr) + \int_s^t U_{s,r-} d(\xi_{s,r-}, r) dA_r + \int_s^t U_{s,r-} dJ_r, \tag{4.2}$$

where $d(x,r) = D_x D_y a(x,y,r)|_{y=x}$ and

$$J_t := \sum_{s\le t} (I + \Delta D\tilde{X}(\xi_{s,r-}, r))^{-1} \Delta D\tilde{X}(\xi_{s,r-}, r)^2.$$

The above J_t is well defined and is a semimartingale. Indeed, we have the equality $I + \Delta D\tilde{X}(\xi_{s,r-}, r) = D\mathrm{Exp}(\Delta X(r))(\xi_{s,r-})$ and the right hand side is is invertible, since the Jacobian matrix $D\mathrm{Exp}v$ is invertible for any v.

Now equation (4.2) has a unique solution $U_{s,t}$. Leandre [13] shows that $U_{s,t} D\xi_{s,t} = I$, by checking $d_t(U_{s,t} D\xi_{s,t}) = 0$. (c.f. Theorem V.6.3 in [16].) This proves that the Jacobian matrix $D\xi_{s,t}$ is invertible. Consequently $\xi_{s,t} : \mathbf{R}^d \to \mathbf{R}^d$ is an onto diffeomorphism for all $t \in [s,\infty)$ a.s. for any $s > 0$. Now the inverse map $\xi_{s,t}^{-1}$ is also a C^1-valued cadlag process with respect to t. Therefore, $\xi_{s,t} = \xi_{0,s}^{-1} \circ \xi_{0,t}$ is a cadlag process with respect to s.

The m-times differentiability of $\xi_{s,t}(x)$ can be verified by induction. It is actually a C^m-valued cadlag process in both s and t. The proof is complete.

5. Stochastic differential equations on manifolds and stochastic flows of local diffeomorphisms

Let M be a connected paracompact C^∞-manifold of dimension d. We denote by $C^\infty(M)$ the set of all C^∞-functions over M. Let $\mathcal{V}^\infty(M)$ be the set of C^∞-vector fields on M. In this section, we often regard $v \in \mathcal{V}^\infty(M)$ as a first order differential operator. Then at each coordinate neighborhood (U,η), v is represented by $vf(p) = \sum_{i=1}^d v^i(p)(\partial f/\partial x_i)(p)$. If v is a complete vector field, it generates a one parameter group of diffeomorphisms $\{\varphi_t\}_{t\in\mathbf{R}^d}$ of M such that $df(\varphi_t(p))/dt = vf(\varphi_t(p))$ holds for any $f \in C^\infty(M)$. We denote φ_t by $\mathrm{Exp}tv$ as before. We define $\|v\|_m^{(U)}$ by $\|v\|_m^{(U)} := \sum_{|\alpha|=0}^m \sum_{i=1}^d \sup_{p\in U} |D^\alpha v^i(p)|$.

A $\mathcal{V}^\infty(M)$-valued process $X(t), t \ge 0$ is called a *semimartingale*, if $X(t)f$ ia a C^∞-valued semimartingale for any $f \in C^\infty(M)$.

Given a $\mathcal{V}^\infty(M)$-valued semimartingale $X(t)$, we define $N((s,t],dv)$ similarly as in (2.1). Then there exists a predictable measure-valued process ν_t on $\mathcal{V}^\infty(M)$ and a predictable strictly increasing process A_t such that (2.2) is a local martingale. Then $X(t)$ has the decomposition $X(t) = X_c(t) + X_d(t)$, where $X_c(t)$ is a continuous semimatingale and $X_d(t)$ is a discontinuous one represented by

$$X_d(t) = \int_{\mathcal{U}} v\tilde{N}((0,t],dv) + \int_{\mathcal{U}^c} vN((0,t],dv),$$

where $\mathcal{U}$ is a Borel subset of $\mathcal{V}^\infty(M)$ such that $\sup_{v\in\mathcal{U}} \|v\|_m^{(U)} < \infty$ for any U. Further, $X_c(t) = M_c(t) + B_c(t)$, where $M_c(t)f$ is a C^∞-valued local martingale and $B_c(t)f$ is a continuous process of bounded variation. We can choose the process A_t such that there exists a $\mathcal{V}^\infty(M) \otimes \mathcal{V}^\infty(M)$-valued process $a(t), t \geq 0$ and $\mathcal{V}(M)$-valued rocess $b(t), t \geq 0$ satisfying

$$\begin{aligned} \langle M_c(t)f(p), M_c(t)g(q)\rangle &= \int_0^t a(s)(f,g)(p,q)dA_s, \\ B_c(t)f(p) &= \int_0^t b(s)f(p)dA_s, \end{aligned}$$

for all $f, g \in C^\infty(M)$. With local coordinate $\eta = (x_1, ..., x_d)$, $a(t)(f,g)(p,q)$ is represented by $\sum_{ij} a^{ij}(p,q,t)(\partial f/\partial x_i)(p)(\partial g/\partial x_j)(q)$. The triple $(a(t), b(t), \nu_t)$ are called the characteristics of $X(t)$. We assume

Condition B. (1) $a^{ij}(p,q,t)$ are infinitely differentiable with respect to p, q and the derivatives are continuous with respect to (p,q,t).
(2) $b(p,t)$ is infinitely continuously differentiable.
(3) ν_t are supported by complete vector fields. It holds $\int_{\mathcal{U}}(\|v\|_m^{(U)})^2\nu_t(dv) < \infty$ for any positive integer m and coordinate neighborhood U.

Let $\xi_t, t \geq 0$ be an M-valued semimartingale. Then the canonical integral of ξ_t based on $X(t)$ is defined by

$$\begin{aligned} \int_s^t X(\circ dr)f(\xi_r) &:= \int_s^t X_c(\circ dr)f(\xi_r) + \int_s^t X_d(dr)f(\xi_{r-}) \\ &+ \sum_{s \leq r \leq t} \{f(\mathrm{Exp}\Delta X(r)(\xi_{r-})) - f(\xi_{r-}) - \Delta X(r)f(\xi_{r-})\}. \end{aligned}$$

We shall consider a canonical SDE driven by $X(t)$: An M-valued cadlag process $\xi_t, t \geq 0$ adapted to $(\mathcal{F}_t)$ is called a *global solution* of a canonical SDE driven by the vector field valued semimartingale $X(t)$, if it satisfies

$$f(\xi_t) = f(p) + \int_s^t X(\circ dr)f(\xi_r), \quad \forall f \in C^\infty(M). \tag{5.1}$$

The SDE may not have a global solution, even if the driving process $X(t)$ takes values in complete vector fields. The solution might explode in a finite time. We shall define a local solution of the above equation. Let $\xi_t, t \in [0, \sigma)$ be an M-valued cadlag process, where σ is an accessible stopping time. It is called a *local process*. Since σ is accessible, there exists an increasing sequence of stopping times such that $\sigma_n < \sigma$ and $\sigma_n \to \sigma$. If each stopped process $\xi_t^n = \xi_{t\wedge\sigma_n}, t \geq 0$ is adapted to $(\mathcal{F}_t)$ and satisfies the above equation (5.1), the local process $\xi_t, t \in [0, \sigma)$ is called a *local solution* of the canonical SDE (5.1) driven by $X(t)$. The solution is called *maximal* if $\lim_{t\to\sigma} \xi_t = \infty$, where ∞ is the infinity of M (one point compactification), whenever $\sigma < \infty$.

We shall rewrite the canonical SDE (5.1) using Itô integral. Set

$$L_c(t)f(p) := \frac{1}{2}\sum_{i,j} a^{ij}(p,p,t)\frac{\partial^2}{\partial x^i \partial x^j}f(p) + \sum c^i(p,t)\frac{\partial}{\partial x^i}f(p), \tag{5.2}$$

where $c^i(p,t)$ is defined by (3.4). It is a coordinate free (not depending on the choice of local coordinates) second order differential operator. Set

$$\begin{aligned}\tilde{X}(t)f(p) &:= X(t)f(p) + \int_0^t L_c(s)f(p)dA_s \\ &\quad + \sum_{0<s\leq t} \{f(\text{Exp}\Delta X(s)(p)) - f(p) - \Delta X(s)f(p)\}. \quad (5.3)\end{aligned}$$

Let $\xi_t, t \in [0,\sigma)$ be an M-valued local process adapted to $(\mathcal{F}_t)$ satisfying the Itô SDE

$$f(\xi_t) = f(p) + \int_s^t \tilde{X}(dr)f(\xi_{r-}), \quad \forall f \in C^\infty(M). \quad (5.4)$$

Then it satisfies the canonical SDE (5.1). Conversely if ξ_t is a local solution of the canonical SDE. Then it satisfies the above Itô equation.

Theorem 5.1. *Assume that the characteristics of the vector field valued semimartingale $X(t)$ satisfy Condition B. Then the canonical SDE (5.1) driven by $X(t)$ has a unique maximal solution $\xi_{s,t}(p), t < \sigma(s,p)$. Further it has a version (denoted by the same notation) with the following property.*
(i) It has the flow property $\xi_{s,u}(p) = \xi_{t,u}(\xi_{s,t}(p))$ if $s < t < u < \sigma(s,p)$.
(ii) Set $\mathbf{D}_{s,t}(\omega) = \{p : \sigma(s,p,\omega) > t\}$. Then $\mathbf{D}_{s,t}(\omega)$ is an open subset of M for any $s < t$. The map $\xi_{s,t}(\omega) : \mathbf{D}_{s,t}(\omega) \to M$ is an into C^∞-diffeomorphism for every $s < t$ a.s.

The above $\xi_{s,t}$ is called the *stochastic flow of local C^∞-diffeomorphisms generated by $\mathcal{V}^\infty(M)$-valued semimartingale $X(t)$.*

If $\sigma(s,p) = \infty$ holds for all p for any $s > 0$ in the above theorem, the SDE or $X(t)$ is called *strongly complete.* If $X(t)$ is strongly complete, the solution $\xi_{s,t}$ defines a diffeomorphism from M into M. Then the assertion of Theorem 3.2 is valid for SDE on manifolds. Indeed, we have the following Corollary.

Corollary 5.1. *Suppose that $X(t)$ is a forward-backward $\mathcal{V}^\infty(M)$-valued semimartingale. If both $X(t)$ and $-X(t)$ are strongly complete, the solution $\xi_{s,t}$ has a version of a stochastic flow of C^∞-diffeomorphisms.*

The proof of Theorem 5.1 and its Corollary are omitted. It will appear in [7].

Remark. It is not simple to conclude whether the maximal solutions define a stochastic flow of global diffeomorphisms. Here we list several special cases where the maximal solutions define a global flow.
(1) The case $M = \mathbf{R}^d$. The global Lipschitz condition for the process $X(t)$ implies a global stochastic flow as we have seen in Sections 3 and 4. In case where the global Lipschitz condition is not satisfied, the problem is complicated. Recently another sufficient condition was given by Xue-Mei-Li [13] in the case of Brownian flow.
(2) The case where M is a bounded domain of $\mathbf{R}^d$. A sufficient condition for a global stochastic flow was obtaind by Taniguchi [17] in case of Brownian

flows.
(3) The case where M is a compact manifold. See Fujiwara [5].
(4) The case M is a Lie group and $X(t)$ is a semimartingale with values in invariant vector fields. (Estrade [4], Applebaum-Kunita [1].)

References

1. D. Applebaum, H. Kunita, Lévy flows on manifolods and Lévy processes on Lie groups, J. Math. Kyoto Univ. 33(1993), 1103-1123.
2. R.N. Carmona, D. Nualart, *Nonlinear stochastic integrators, equations and flows*, Stochastic Monographs 6(1990), Gordon and Breach.
3. S. Cohen, Géométrie différentielle stochastique avec sauts, C. R. Acad. Sci. Paris,314, Serie I (1992), 767-770.
4. A. Estrade, Exponentielle stochastiques et intégrale multiplicative discontinues, Ann. Inst. Henri Poincaré, Probabilites et Statistiques, 28(1992), 107-129.
5. T. Fujiwara, Stochastic differential equations of jump type on manifolds and Lévy flows, J. Math. Kyoto Univ. 31(1991), 99-119.
6. T. Fujiwara, H. Kunita, Stochastic differential equations of jump type and Lévy processes in diffeomorphsms group, J. Math. Kyoto Univ. 25(1985), 71-106.
7. T. Fujiwara, H. Kunita, Canonical stochastic differential equations driven by semimartingales with spatial parameters and stochastic flows of diffeomorphisms, in preparation.
8. N. Ikeda, S. Watanabe, *Stochastic differential equations and diffusion processes*, 2nd ed., North-Holland/Kodansha (1989).
9. H. Kunita, Tightness of probability measures in $D([0,T];C)$ and $D([0,T];D)$, J. Math. Soc. Japan 38(1986), 309-334.
10. H. Kunita, Convergence of stochastic flows with jumps and Lévy processes in diffeomorphisms group, Ann. Inst. Henri Poincaré, Probabilites et Statistiques 22(1986), 287-321.
11. H. Kunita, *Stochastic flows and stochastic differential equations*, Cambridge Univ. Press, Cambridge (1990).
12. T.G. Kurtz, E. Pardoux, P. Protter, Stratonovich stochastic differential equations driven by general semimartingales, Ann.Inst. Henri Poincaré, Probabilites et Statistiques, 31(1995), 351-377.
13. R. Leandre, Flot d'une équation différentielle stochastique avec semimartingale directrice discontinue, Séminaire Probab. XIX, Lect. Notes in Math. 1123(1985), 271-274.
14. Xue-Mei Li, Strong p-completeness of stochastic differential equations and the existence of smooth flows on noncompact manifolds, PTRF 100(1994), 485-511.
15. S.I. Marcus, Modelling and approximation of stochastic differential equations driven by semimartingales, Stochastics 4(1981), 223-245.
16. P. Protter, *Stochastic integration and differential equations, a new aproach*, Springer-Verlag (1990).
17. S. Taniguchi, Stochastic flows of diffeomorphisms on an open set of $\mathbf{R}^n$, Stochastics and Stochastics Reports, 28(1989), 301-315.

A Remark on American Securities

Shigeo Kusuoka

Graduate School of Mathematical Sciences, University of Tokyo, Komaba, Meguro-ku, Tokyo 153, Japan

1. Fundamental Theorems

In this paper, we discuss about American securities. To simplify the notions, we only discuss the case that the free risk spot rate is zero and the maturity (or the horizon) is 1 (So the price of free risk bond is constant). Also we assume that there is no dividend or no transaction cost and that there is no restriction on short sale.

Let $(\Omega, \mathcal{F}, P)$ be a complete probability space and $\{\mathcal{F}_t\}_{t\in[0,1]}$ is a right continuous filtration such that $\{A \in \mathcal{F}; P(A) = 0\} \subset \mathcal{F}_0$. One can regard $\mathcal{F}_t$ as available information at time t, $t \in [0,1]$. Let us assume that there are N securities other than the risk free bond and the market price of the i-th security at time t is given by $S_t^i, t \in [0,1]$, $i = 1,\dots,N$. Then $S_t = (S_t^1,\dots,S_t^N)$ must be $\mathcal{F}_t$-measurable $\mathbf{R}^N$-valued random variable. We assume that $\{S_t\}_{t\in[0,1]}$ is a locally bounded cádlág process.

The most primitive trading strategy is the following. Let $0 = \tau_0 \leq \tau_1 \leq \tau_2 \leq \dots \leq \tau_n \leq 1$ be stopping times. The trader takes the portfolio position η_0 at time zero. Also at time τ_k, $k = 1,\dots,n$, the trader changes the portfolio position from η_{k-1} to η_k. Here η_k is an $\mathbf{R}^{\mathbf{N}}$-valued random variable, and to take the portfolio position $\eta_k = (\eta_k^1,\dots,\eta_k^N)$ means to hold the i-th security by the amount of η_k^i. Since there is no restriction on short sale in our model, η_k^i can be negative. Since the available information at time τ_k is $\mathcal{F}_{\tau_k}$, η_k must be $\mathcal{F}_{\tau_k}$-measurable. The capital gain V_t at time $t, t \in [0,1]$, by this strategy is given by

$$V_t = \sum_{k=1}^{n+1} \eta_{k-1} \cdot (S_{t\wedge\tau_k} - S_{t\wedge\tau_{k-1}}), \quad t \in [0,1]$$

Here $\tau_{n+1} = 1$.

Let $\xi_t, t \in [0,1]$, be the prior portfolio position at time t. Then we see that $\xi_t = \eta_k, t \in (\tau_k, \tau_{k+1}]$, $k = 0,1,\dots,n$, and $\xi_0 = 0$. We call predictable processes of this form simple integrands. If $\{S_t\}_{t\in[0,1]}$ is a semimartingale, we have

$$V_t = \int_0^t \xi_s dS_s (= \int_{[0,t]} \xi_s dS_s).$$

So even if $\{S_t\}_{t\in[0,1]}$ is not a semimartingale, we write $\int_0^t \xi_s dS_s$ for V_t.

The following fundamental result is due to Delbaen and Schachermayer [2].

Theorem 1.1. *Suppose that there does not exist any sequence* $\{\xi^{(n)}\}_{n=1}^{\infty}$ *of simple integrands satisfying the following three conditions.*
(1) $\int_0^t \xi_s^{(n)} dS_s \geq -1, \quad t \in [0,1],\ n \geq 1,\ P-a.s.$
(2) $\liminf_{n\to\infty} \text{ess.inf} \int_0^1 \xi_s^{(n)} dS_s \geq 0.$
(3) $V = \lim_{n\to\infty} \int_0^1 \xi_s^{(n)} dS_s$ *exists P-a.s. and* $P(V > 0) > 0$.
Then $\{S_t\}_{t\in[0,1]}$ *is a semimartingale.*

If $\{S_t\}_{t\in[0,1]}$ is a semi-martingale, we can think of more general trading strategies for any $\mathbf{R}^N$-valued S-integrable predictable process $\{\xi_t\}_{t\in[0,1]}$ such that the prior portfolio position at time t is ξ_t, $t \in [0,1]$. Delbaen and Schachermayer [2] also proved the following fundamental theorem.

Theorem 1.2. *Suppose that* $\{S_t\}_{t\in[0,1]}$ *is a semimartingale. Suppose moreover that that there does not exist any sequence* $\{\xi^{(n)}\}_{n=1}^{\infty}$ *of S-integrable* $\mathbf{R}^N$*-valued predictable processes satisfying the following three conditions.*
(1) $\int_0^t \xi_s^{(n)} dS_s \geq -1,\ P-a.s. \quad t \in [0,1],\ n \geq 1.$
(2) $\liminf_{n\to\infty} \text{ess.inf} \int_0^1 \xi_s^{(n)} dS_s \geq 0.$
(3) $V = \lim_{n\to\infty} \int_0^1 \xi_s^{(n)} dS_s$ *exists* $P-a.s.$ *and* $P(V > 0) > 0$.
Then there is an equivalent martingale measure, i.e., there is a probability measure Q *on* $(\Omega, \mathcal{F})$ *equivalent to the measure* P *such that* $\{S_t\}_{t\in[0,1]}$ *is a martingale under the measure* Q.

The above theorems show that there is an equivalent martingale measure if there is no trading strategy by which one can obtain benefit from the security market without any risk.

We say that the market is *complete* if there is a unique equivalent martingale measure. By Jacod [6] and Harrison-Pliska [5], we have the following.

Theorem 1.3. *Suppose that there is an equivalent martingale measure. Then the market is complete if and only if the following condition* (C) *holds.*
(C) *For any bounded* $\mathcal{F}_1$*-measurable random variable* Y*, there is a bounded* $\mathcal{F}_0$*-measurable random variable* Y_0 *and an S-integrable* $\mathbf{R}^N$*-valued predictable process* $\{\xi_t\}_{t\in[0,1]}$ *such that*

$$Y = Y_0 + \int_0^1 \xi_t dS_t,\ P-a.s.$$

We discuss only complete markets in this paper. See Kramkov [8] for incomplete markets. Also, see Duffie [3] for the argument from a viewpoint of mathematical economics. In this paper, we assume that $\{S_t\}_{t\in[0,1]}$ is a P-martingale for simplicity, and assume that there is a unique equivalent martingale measure.

2. Preliminary Results

In this section, we assume that $(\Omega, \mathcal{F}, P)$ is a complete probability space and $\{\mathcal{F}_k\}_{k=0}^K$ is a filtration. Then we have the following.

Theorem 2.1. *Let $X : \{0,1,\ldots,K\} \times \Omega \to \mathbf{R}$ be an $\mathcal{F}_k$-adapted process such that $E[|X(k)|] < \infty$, $k = 0,1,\ldots,K$. Let Δ_k, $k = 0,1,\ldots,K$, be $\mathcal{F}_k$-measurable random variables inductively given by*

$$\begin{aligned}\Delta_K &= 0,\\ \Delta_{k-1} &= \max\{X(k-1) - E[X(K) + \sum_{j=k}^{K} \Delta_j | \mathcal{F}_{k-1}], 0\},\end{aligned}$$

$k = K, K-1, \ldots, 1$. Let $Z = X(K) + \sum_{j=0}^K \Delta_j$. Then we have the following.
(1) *$E[Z|\mathcal{F}_\tau] \geq X(\tau)$ $\quad P - a.s.$ for any $\mathcal{F}_k$-stopping time τ with $\tau \leq K$.*
(2) *There is an $\mathcal{F}_k$-stopping time σ satisfying $\sigma \leq K$ such that*

$$E[Z|\mathcal{F}_\sigma] = X(\sigma).$$

In particular,

$$E[Z|\mathcal{F}_0] = \sup\{E[X(\tau)|\mathcal{F}_0]; \tau \text{ is an } \mathcal{F}_k\text{-stopping time with } \tau \leq K\}$$

for $P - a.s.\omega$. Here we take a regular conditional expectation for $E[\cdot|\mathcal{F}_0]$.
(3) *For any $p \in (1,\infty)$, there is a constant $C(p)$ depending only on p such that*

$$E[|Z|^p] \leq C(p) E[(\max\{X(k); k = 0,1,\ldots,K\} + |X(K)|)^p].$$

(4) *$\Delta_{k-1} = \tilde{Z}(k-1) - E[\tilde{Z}(k)|\mathcal{F}_{k-1}]$, $\quad k = 1,\ldots,K$, where $\tilde{Z}(k) = \sup\{E[X(\tau)|\mathcal{F}_k];\ \tau$ is a stopping time with $k \leq \tau \leq K\}$, $\quad k = 0,1,\ldots,K$.*

Theorem 2.1 is an immediate consequence of the following Lemma.

Lemma 2.1. *Let $X : \{0,1,\ldots,K\} \times \Omega \to \mathbf{R}$ and $\Delta_k : \Omega \to \mathbf{R}$, $k = 0,1,\ldots,K$, be as in Theorem 2.1. Then we have the following.*
(1) *If an integrable random variable $Y : \Omega \to \mathbf{R}$ satisfies*

$$E[Y|\mathcal{F}_k] \geq X(k) \quad P - a.s., \quad k = 0,1,\ldots,K,$$

then

$$E[Y|\mathcal{F}_k] \geq E[X(K) + \sum_{j=k}^{K} \Delta_j | \mathcal{F}_k] \quad P - a.s., \quad k = 0,1,\ldots,K.$$

(2) *$E[X(K) + \sum_{j=k}^K \Delta_j | \mathcal{F}_k] \geq X(k) \quad P - a.s., \quad k = 0,1,\ldots,K.$*

(3) *$E[X(K) + \sum_{k=0}^K \Delta_k | \mathcal{F}_\tau] \geq X(\tau) \quad P - a.s.$ for any $\mathcal{F}_k$-stopping time τ*

with $\tau \leq K$.
(4) $\min\{E[X(K) + \sum_{j=0}^{K} \Delta_j|\mathcal{F}_k] - X(k); k = 0, 1, \ldots, K\} = 0 \quad P-a.s.$
(5) *There is an* $\mathcal{F}_k$*-stopping time* σ *satisfying* $\sigma \leq K$ *such that*

$$E[X(K) + \sum_{k=0}^{K} \Delta_k|\mathcal{F}_\sigma] = X(\sigma).$$

In particular,

$$\begin{aligned} &E[X(K) + \sum_{k=0}^{K} \Delta_k|\mathcal{F}_0] \\ = \ &\sup\{E[X(\tau)|\mathcal{F}_0];\ \tau \text{ is an } \mathcal{F}_k\text{-stopping time with } \tau \leq K\}. \end{aligned}$$

(6) $E[(\sum_{k=0}^{K} \Delta_k)^2] \leq 4E[(\max\{X(k) - X(K); k = 0, 1, \ldots, K\})^2]$.
(7) $E[(\sum_{k=0}^{K} \Delta_k)^p]$
$\leq p^{p/(p-1)}(\frac{p}{p-1})^p E[(\max\{X(k) - X(K); k = 0, 1, \ldots, K\})^p]$ *for any* $p \in (1, 2)$.
(8) $E[(\sum_{k=0}^{K} \Delta_k)^p] \leq 2^{2p(p-2)}(\frac{p}{p-1})^p p^{p(4p-7)} E[(\max\{X(k) - X(K);\ k = 0, 1, \ldots, K\})^p]$ *for any* $p \in (2, \infty)$.
(9) $\Delta_{k-1} = \tilde{Z}(k-1) - E[\tilde{Z}(k)|\mathcal{F}_{k-1}]$, $k = 1, \ldots, K$.

Proof. (1) We prove this assertion by induction. It is obvious when $k = K$. Assume that the assertion is valid for k. Let $U = Y - X(K) - \sum_{j=k}^{K} \Delta_j$. Then we see that $E[U|\mathcal{F}_k] \geq 0$. Also we see that

$$E[U|\mathcal{F}_{k-1}] \geq X(k-1) - E[X(K) + \sum_{j=k}^{K} \Delta_j|\mathcal{F}_{k-1}].$$

So we see that $E[U|\mathcal{F}_{k-1}] \geq \Delta_{k-1}$. This implies our assertion for $k-1$.
(2) The assertion is valid for $k = K$. It is easy to see that

$$\begin{aligned} E[X(K) + \sum_{j=k-1}^{K} \Delta_j|\mathcal{F}_{k-1}] &= \Delta_{k-1} + E[E[X(K) + \sum_{j=k}^{K} \Delta_j|\mathcal{F}_k]|\mathcal{F}_{k-1}] \\ &\geq X(k-1) \end{aligned}$$

for $k = 1, \ldots, K$. So we have our assertion.
(3) Let τ be an $\mathcal{F}_k$-stopping time with $\tau \leq K$. Then

$$E[W|\mathcal{F}_\tau] = \sum_{k=0}^{K} 1_{\{\tau = k\}} E[W|\mathcal{F}_k]$$

for any integrable random variable W. Thus we have our assertion from Assertion (2).
(4) Let $Y = X(K) + \sum_{j=0}^{K} \Delta_j$. Then by Assertion (2) we have $E[Y|\mathcal{F}_k] \geq$

$X(k)$ $P-a.s.$, $k=0,1,\dots,K$. Let $\xi=\min\{E[Y|\mathcal{F}_k]-X(k); k=0,1,\dots,K\}$ ≥ 0. Then we see that $E[Y|\mathcal{F}_k]-X(k)\geq\xi$, and so $E[Y|\mathcal{F}_k]-X(k)\geq E[\xi|\mathcal{F}_k]$. This implies that $E[Y-\xi|\mathcal{F}_k]\geq X(k)$ $P-a.s.$, $k=0,1,\dots,K$. Therefore by Assertion (1), we have $E[Y-\xi|\mathcal{F}_0]\geq E[X(K)+\sum_{k=0}^K \Delta_k|\mathcal{F}_0]$. So we have $E[Y-\xi]\geq E[Y]$, which implies that $E[\xi]\leq 0$. Since $\xi\geq 0$, we see that $\xi=0$, $P-a.s.$ This proves our assertion.

(5) Let $\sigma=\min\{k\geq 0; E[X(K)+\sum_{j=0}^K \Delta_j|\mathcal{F}_k]-X(k)=0\}$. Then by Assertion (4), we see that σ is an $\mathcal{F}_k$-stopping time satisfying $\sigma\leq K$ $P-a.s.$ Moreover, we have $E[X(K)+\sum_{j=0}^K \Delta_j|\mathcal{F}_\sigma]=X(\sigma)$ $P-a.s.$, and so $E[X(K)+\sum_{j=0}^K \Delta_j|\mathcal{F}_0]=E[X(\sigma)|\mathcal{F}_0]$. By Assertion (3), we have $E[X(K)+\sum_{k=0}^K \Delta_k|\mathcal{F}_0]\geq E[X(\tau)|\mathcal{F}_0]$ for any $\mathcal{F}_k$-stopping time τ with $\tau\leq K$. This proves our assertion.

Now let us prove Assertions (6), (7) and (8). Let $Y=\max\{X(k)-X(K);\ k=0,1,\dots,K\}$. Then it is obvious that $E[Y+X(K)|\mathcal{F}_k]\geq X(k)$, $k=0,1,\dots,K$. So by Assertion (1) we see that

$$E[\sum_{j=k}^K \Delta_j|\mathcal{F}_k]\leq E[Y|\mathcal{F}_k],\quad k=0,1,\dots,K.$$

Let $Z_k=\sum_{j=k}^K \Delta_j$, $k=0,1,\dots,K$. Then we see that $Z_K=0$ and that

$$Z_{k-1}^p-Z_k^p=\int_0^1\frac{d}{dt}(Z_k+t\Delta_{k-1})^p dt\leq pZ_{k-1}^{p-1}\Delta_{k-1}.$$

Note that

$$0\geq\delta_k\geq\max\{E[X(k)-X(K)|\mathcal{F}_k],0\}\geq E[Y|\mathcal{F}_k].$$

So by Doob's inequality we see that $E[Z_0^p]<\infty$ if $E[Y^p]<\infty$, $p\in(1,\infty)$.

Note that

$$E[Z_0^2]\leq 2E[\sum_{k=1}^K \Delta_{k-1}E[Z_{k-1}|\mathcal{F}_{k-1}]]$$

$$\leq\sum_{k=0}^K 2E[\Delta_k E[Y|\mathcal{F}_k]]=2E[Z_0Y]\leq 2E[Z_0^2]^{1/2}E[Y^2]^{1/2}.$$

This implies Assertion (6).

Let $p\in(1,2)$. Then we have

$$E[Z_0^p]\leq pE[\sum_{k=1}^K \Delta_{k-1}E[Z_{k-1}^{p-1}|\mathcal{F}_{k-1}]]$$

$$\leq \sum_{k=0}^K pE[\Delta_k E[Z_k|\mathcal{F}_k]^{p-1}]\leq\sum_{k=0}^K pE[\Delta_k E[Y|\mathcal{F}_k]^{p-1}]$$

$$\leq pE[Z_0\{\max\{E[Y|\mathcal{F}_k]; k=0,1,\dots,K\}^{p-1}]$$

$$\leq pE[Z_0^p]^{1/p}E[\{\max\{E[Y|\mathcal{F}_k]; k=0,1,\dots,K\}^p]^{1-1/p}.$$

This and Doob's inequality imply that

$$\begin{aligned} E[Z_0^p] &\le p^{p/(p-1)} E[\{\max\{E[Y|\mathcal{F}_k]; k = 0, 1, \ldots, K\}^p] \\ &\le p^{p/(p-1)} (\frac{p}{p-1})^p E[Y^p]. \end{aligned}$$

This implies Assertion (7).

Let $p \in (2, \infty)$. Then we have

$$\begin{aligned} & E[Z_0^p] \le pE[\sum_{k=1}^{K} \Delta_{k-1} E[Z_{k-1}^{p-1} | \mathcal{F}_{k-1}]] \\ \le & \sum_{k=0}^{K} pE[\Delta_k E[Z_k|\mathcal{F}_k]^{1/(2p-3)} E[Z_k^{(2p-1)/2} | \mathcal{F}_k]^{(2p-4)/(2p-3)}] \\ \le & pE[Z_0 \{\max\{E[Y|\mathcal{F}_k]; k = 0, 1, \ldots, K\}^{1/(2p-3)} \\ & \times \{\max\{E[Z_0^{(2p-1)/2} | \mathcal{F}_k]; k = 0, 1, \ldots, K\}^{(2p-4)/(2p-3)}]. \\ \le & pE[Z_0^p]^{1/p} E[\{\max\{E[Y|\mathcal{F}_k]; k = 0, 1, \ldots, K\}^p]^{1/p(2p-3)} \\ & \times E[\{\max\{E[Z_0^{(2p-1)/2} | \mathcal{F}_k]; k = 0, 1, \ldots, K\}^{2p/(2p-1)}]^{(2p-1)(p-2)/p(2p-3)} \\ \le & p(\frac{p}{p-1})^{1/(2p-3)} (2p)^{2(p-2)/(2p-3)} E[Z_0^p]^{(2p^2-3p-1)/p(2p-3)} E[Y^p]^{1/p(2p-3)}. \end{aligned}$$

This implies Assertion (8).

By Assertion (5), we see that

$$E[X(K) + \sum_{j=k}^{K} \Delta_j | \mathcal{F}_k] \ = \ \tilde{Z}(k), \quad k = 0, 1, \ldots, K.$$

This implies Assertion (9).

This completes the proof.

Remark By the proof of Lemma 2.1 (6), (7), (8), we see the following. Let $p \in (1, \infty)$ and assume that

$$E[(\max\{X(k); k = 0, 1, \ldots, K\} + |X(K)|)^p] \ < \ \infty.$$

If an integrable random variable $Y : \Omega \to \mathbf{R}$ satisfies

$$E[Y|\mathcal{F}_k] \ \ge \ X(k) \ P-a.s., \quad k = 0, 1, \ldots, K,$$

then we have the following.

(1) $E[(\sum_{k=0}^{K} \Delta_k)^2] \le 4E[(Y - X(K))^2]$.

(2) $E[(\sum_{k=0}^{K} \Delta_k)^p] \le p^{p/(p-1)} (\frac{p}{p-1})^p E[(Y - X(K))^p]$, $p \in (1, 2)$.

(3) $E[(\sum_{k=0}^{K} \Delta_k)^p] \le 2^{2p(p-2)} (\frac{p}{p-1})^p p^{p(4p-7)} E[(Y - X(K))^p]$, $p \in (2, \infty)$.

3. Price of American Securities

In this section, we assume that $(\Omega, \mathcal{F}, P)$ is a complete probability space and $\{\mathcal{F}_t\}_{t\in[0,1]}$ is a right-continuous filtration. Then we have the following.

Theorem 3.1. *Let $X : [0,1] \times \Omega \to \mathbf{R}$ be an $\mathcal{F}_t$-adapted cádlág process such that*

$$E[(\sup\{X(t); t \in [0,1]\} + |X(1)|)^p] \ < \ \infty$$

for some $p \in (1, \infty)$. Then there is an $\mathcal{F}_1$-measurable random variable $Z : \Omega \to \mathbf{R}$ satisfying the following.
(1) $E[Z^p] \le C(p)E[(\sup\{X(t); t \in [0,1]\} + |X(1)|)^p] < \infty$.
(2) $E[Z|\mathcal{F}_\tau] \ge X(\tau)$ $P-a.s.$ *for any $\mathcal{F}_t$-stopping time τ with $\tau \le 1$.*
(3) $E[Z] = \sup\{E[X(\tau)]; \tau$ *is an $\mathcal{F}_t$-stopping time with $\tau \le 1\}$.*

Before proving Theorem 3.1, we make some preparations. Let us think of the situations in Theorem 3.1. For each $n \ge 1$, let $\mathcal{G}_k^{(n)} = \mathcal{F}_{k/n}$ and $X^{(n)}(k) = X(k/n)$, $k = 0, 1, \ldots, n$, and apply Theorem 2.1. Then we see that there is an $\mathcal{F}_1$-measurable random variable Z_n such that

$$E[Z_n|\mathcal{F}_{k/n}] \ge X(k/n) \ P-a.s., \quad k = 0, 1, \ldots, n, \tag{3.1}$$

$$E[Z_n] \le \sup\{E[X(\tau)]; \tau \text{ is an } \mathcal{F}_t\text{-stopping time with } \tau \le 1\}. \tag{3.2}$$

and

$$E[Z_n^p] \le C(p)E[(\sup\{X(t); t \in [0,1]\} + |X(1)|)^p] < \infty. \tag{3.3}$$

Let $\{Z_n(t); t \in [0,1]\}$ be the cádlág version of $E[Z_n|\mathcal{F}_t], t \in [0,1]$. Then we have

$$E[Z_n + \sup_{t\in[0,1]} (X(t) - Z_n(t))|\mathcal{F}_t] \ge X(t) \ \ P-a.s., \quad t \in [0,1].$$

Let us take an $\eta > 0$ and fix it for a while. Let $\sigma_n = \inf\{t \in [0,1]; X(t) > Z_n(t) + \eta\}$. Also let $\sigma_n' = (([n\sigma_n] + 1)/n) \wedge 1$. Then σ_n and σ_n' are stopping times and $\sigma_n < \sigma_n' \le \sigma_n + 1/n$. Since $X(\sigma_n') \le Z_n(\sigma_n')$, we have

$$\begin{aligned} & 1_{\{\sigma_n<1\}}X(\sigma_n) \ge 1_{\{\sigma_n<1\}}(Z(\sigma_n) + \eta) \\ \ge \ & 1_{\{\sigma_n<1\}}(E[Z(\sigma_n')|\mathcal{F}_{\sigma_n}] + \eta) \ge 1_{\{\sigma_n<1\}}(E[X(\sigma_n')|\mathcal{F}_{\sigma_n}] + \eta). \end{aligned}$$

So we have

$$1_{\{\sigma_n<1\}}\eta \le 1_{\{\sigma_n<1\}}E[\sup\{X(\sigma_n) - X((\sigma_n + s) \wedge 1); s \in [0,1/n]\}|\mathcal{F}_{\sigma_n}],$$

which implies that

$$\begin{aligned} & P(\sigma_n < 1)\eta \\ \le \ & P(\sigma_n < 1)^{1/p'}E[E[\sup\{X(\sigma_n) - X((\sigma_n + s) \wedge 1); s \in [0,1/n]\}|\mathcal{F}_{\sigma_n}]^p]^{1/p}. \end{aligned}$$

Here 1/p + 1/p' = 1. Thus we have the following.

Proposition 3.1.

$$P(\sup\{X(t) - Z_n(t); t \in [0,1]\} > \eta) \le C\eta^{-p}, \quad \eta > 0,$$

where $C = \sup\{E[E[\sup\{X(\tau) - X((\tau+s)\wedge 1);\ s \in [0,1/n]\}|\mathcal{F}_\tau]^p];$ τ *is an* $\mathcal{F}_t$*-stopping time with* $\tau \le 1\}$.

Now let us prove Theorem 3.1. By the (3.3), there is a subsequence $\{n_k\}_{k=1}^{\infty} \subset \{2^n; n \ge 1\}$ and $Z \in L^p(\Omega; \mathcal{F}_1, dP)$ such that $Z_{n_k} \to Z$ weakly in $L^p(\Omega; \mathcal{F}_1, dP)$. Then we see from (3.1) and (3.2) that

$$E[Z|\mathcal{F}_{2^{-n}k}] \ge X(2^{-n}k) \quad P-a.s., \quad n \ge 0,\ k = 0,1,\dots,2^n, \tag{3.4}$$

and

$$E[Z] \le \sup\{E[X(\tau)];\ \tau \text{ is an } \mathcal{F}_t\text{-stopping time with } \tau \le 1\}. \tag{3.5}$$

Let τ be an $\mathcal{F}_t$-stopping time with $\tau \le 1$. Let $\tau_n = 2^{-n}([2^n\tau]+1)\wedge 1$. Then by (3.4) we have

$$E[Z|\mathcal{F}_{\tau_n}\ ] \ \ge\ X(\tau_{n_k}) \ \ P-a.s., \quad k \ge 0.$$

Letting $k \to \infty$, we have Assertion (2). Then Assertion (3) follows from Assertion (2) and (3.5).

This completes the proof of Theorem 3.1.

Now let us explain the meaning of Theorem 3.1 in mathematical finance. Let us think of the situation in Section 1. We assume that the process $\{S_t\}_{t\in[0,1]}$ is a P-martingale and that the market is complete. Now let us think of an American security for which the holder can obtain cash by the amount of X_t from the writer when the holder exercises it at time t. The first question is how much the price π of this American security should be. The answer is the following. For simplicity, we assume that $P(B) = 0$ or 1, $B \in \mathcal{F}_0$.

Suppose that the stochastic process $\{X_t\}_{t\in[0,1]}$ satisfies the assumption in Theorem 3.1. Let Z be that in Theorem 3.1. Since the market is complete, by Theorem 1.3 and Burkholder's inequality, there is a predictable process $\{\xi_t\}_{t\in[0,1]}$ such that $E[Z|\mathcal{F}_t] = E[Z] + \int_0^t \xi_s dS_s$, $t \in [0,1]$. Then we see that

$$E[Z] + \int_0^t \xi_s dS_s \ge X_t, \ t \in [0,1],\ P-a.s. \tag{3.6}$$

and

$$E[Z] = \sup\{E[X(\tau)]; \tau \text{ is an } \mathcal{F}_t\text{-stopping time with } \tau \le 1\}. \tag{3.7}$$

If $\pi > E[Z]$, then by taking the trading strategy $\{\xi_t\}_{t\in[0,1]}$, the writer hedge the claim from the holder without any risk by the initial cost $E[Z]$. So the writer take the arbitrage by the amount of $\pi - E[Z]$ without taking any risk.

On the other hand, suppose that $\pi < \mathrm{E}[\mathrm{Z}]$. Then there is a stopping time σ with $\sigma \leq 1$ such that $E[X_\sigma] > \pi$. Since the market is complete, there is a predictable process $\{\eta_t\}_{t\in[0,1]}$ such that $X_\sigma = E[X_\sigma]$ $+ \int_0^\sigma \eta_s dS_s$. Then by taking the trading strategy $\{-\eta_t\}_{t\in[0,\sigma]}$, the holder can replicate the contingent claim $-X_\sigma$ at time σ without taking any risk by the initial cost $-E[X_\sigma]$. So the holder take the arbitrage $E[X_\sigma] - \pi$ without taking any risk.

Both cases are not admissible in mathematical finance. So the price π should be equal to $E[Z]$.

Remark. (1) Theorem 3.1 is a corollary to the theory of optimal stopping time problems. See Zabczyk [9] and Karatzas [7] for this point of view.
(2) In the above discussion, we only use the facts (3.6) and (3.7) for a random variable Z. Such a random variable Z satisfying (3.6) and (3.7) is not unique in general. So the hedging strategy of the writer is not unique, even if the market is complete. This fact causes some problems in the discussion below.

Theoretically the price of the American security is given by $\sup\{E[X(\tau)];$ τ is an $\mathcal{F}_t$-stopping time with $\tau \leq 1\}$ and the hedging strategy is given by the martingale representation of the random variable Z. However, it is not so easy to compute them in general. So we will think of some approximation scheme in the remainder of the paper.

4. A Certain Metric

In this section, $\mathcal{P}(M)$ denotes the set of all probability measures on a Polish space M. Let N be a metric space. Let $\mathcal{X}_{M;N}$ denotes the set of all pairs (X,μ) for which X is a measurable map from M into N and $\mu \in \mathcal{P}(M)$. Let $Dis_{M;N} : \mathcal{X}_{M;N} \times \mathcal{X}_{M;N} \to [0,\infty)$ be given by

$$\begin{aligned} & Dis_{M;N}((X_1,\mu_1),(X_2,\mu_2)) \\ = \ & \inf\{\int_{M\times M}((dis_M(x_1,x_2) + dis_N(X_1(x_1),X_2(x_2)))\wedge 1)\nu(dx_1,dx_2); \\ & \nu \in \mathcal{P}(M\times M), \nu\circ\pi_1^{-1} = \mu_1, \nu\circ\pi_2^{-1} = \mu_2\} \end{aligned}$$

Here dis_M, dis_N are distance functions on M, N respectively. Also, π_i : $M \times M \to M$, $i = 1,2$, be canonical projections given by $\pi_1(x_1,x_2) = x_1$, $\pi_2(x_1,x_2) = x_2$, $x_1, x_2 \in M$. Then one can easily see the following.

Proposition 4.1. (1) $Dis_{M;N}((X_1,\mu_1),(X_2,\mu_2)) = 0$, *if and only if* $\mu_1 = \mu_2$ *and* $X_1(x) = X_2(x)$ $\mu_1 - a.s.x$ *for* $(X_1,\mu_1),(X_2,\mu_2) \in \mathcal{X}_{M;N}$.
(2) $Dis_{M;N}((X_1,\mu_1),(X_2,\mu_2)) = Dis_{M;N}((X_2,\mu_2),(X_1,\mu_1))$
and

$$\begin{aligned} & Dis_{M;N}((X_1,\mu_1),(X_3,\mu_3)) \\ \leq \ & Dis_{M;N}((X_1,\mu_1),(X_2,\mu_2)) + Dis_{M;N}((X_2,\mu_2),(X_3,\mu_3)) \end{aligned}$$

for any $(X_i, \mu_i) \in \mathcal{X}_{M;N}$, $i = 1, 2, 3$.

Definition Let $(X_n, \mu_n), (X, \mu) \in \mathcal{X}_{M;N}, n \geq 1$. We say that $(X_n, \mu_n) \to (X, \mu)$ in $\mathcal{X}_{M;N}$ if $Dis_{M;N}((X_n, \mu_n), (X, \mu)) \to 0$.

Proposition 4.2. *Let* $(X_n, \mu_n), (X, \mu) \in \mathcal{X}_{M;N}$, $n \geq 1$, *and suppose that* $(X_n, \mu_n) \to (X, \mu)$ *in* $\mathcal{X}_{M;N}$. *Then there are a probability space* $(\Omega, \mathcal{F}, P)$ *and* M*-valued random variables* Z_n, $n \geq 1$, *and* Z *such that*
(1) $P \circ Z_n^{-1} = \mu_n, n \geq 1$, *and* $P \circ Z^{-1} = \mu$,
(2) $Z_n \to Z$ *in probability,*
and
(3) $X_n \circ Z_n \to X \circ Z$ *in probability.*

Proof. By the assumption, there are $\nu_n \in \mathcal{P}(M \times M)$, $n \geq 1$, such that $\nu_n \circ \pi_1^{-1} = \mu$, $\nu_n \circ \pi_2^{-1} = \mu_n$, $n \geq 1$, and

$$\int_{M \times M} ((dis_M(x_1, x_2) + dis_N(X(x_1), X_n(x_2))) \wedge 1) \nu_n(dx_1, dx_2) \to 0.$$

Since M is Polish, there exist measurable maps $\rho_n : M \to \mathcal{P}(M)$, $n \geq 1$, such that $\nu_n(dx_1, dx_2) = \mu(dx_1)\rho_n(x_1)(dx_2)$. Let $\Omega = M^{\{0\} \cup \mathbf{N}}$, $\mathcal{F}$ be a Borel algebra of Ω and $P(dx) = \mu(dx_0) \otimes \otimes_{n=1}^{\infty} \rho_n(x_0)(dx_n)$. Let $Z : \Omega \to M$, $Z_n : \Omega \to M$, $n \geq 1$, be given by $Z(x) = x_0$, $Z_n(x) = x_n$, $n \geq 1, x = (x_0, x_1, \ldots)$. Then we have $P \circ Z^{-1} = \mu$, $P \circ Z_n^{-1} = \mu_n$, and

$$\begin{aligned} &E^P[(dis_N(X \circ Z, X_n \circ Z_n) + dis_M(Z, Z_n)) \wedge 1] \\ = &\int_{M \times M} ((dis_M(x_1, x_2) + dis_N(X(x_1), X_n(x_2))) \wedge 1) \nu_n(dx_1, dx_2) \to 0. \end{aligned}$$

So, we have our assertion.

Proposition 4.3. *Let* $(X_n, \mu_n), (X, \mu) \in \mathcal{X}_{M;N}$, $n \geq 1$, *and suppose that there are a probability space* $(\Omega, \mathcal{F}, P)$ *and* M*-valued random variables* Z_n, $n \geq 1$, *and* Z *such that*
(1) $P \circ Z_n^{-1} = \mu_n$, $n \geq 1$, *and* $P \circ Z^{-1} = \mu$,
(2) $Z_n \to Z$ *in probability,*
and
(3) $X_n \circ Z_n \to X \circ Z$ *in probability.*
Then $(X_n, \mu_n) \to (X, \mu)$ *in* $\mathcal{X}_{M;N}$.

Proof. Let $\nu_n \in \mathcal{P}(M \times M)$ given by $\nu_n = P \circ (Z, Z_n)^{-1}$. Then we have

$$\begin{aligned} &Dis_{M;N}((X, \mu), (X_n, \mu_n)) \\ \leq &\int_{M \times M} ((dis_M(x_1, x_2) + dis_N(X(x_1), X_n(x_2))) \wedge 1) \nu_n(dx_1, dx_2) \\ = &E^P[(dis_N(X \circ Z, X_n \circ Z_n) + dis_M(Z, Z_n)) \wedge 1] \to 0. \end{aligned}$$

Thus we see that $(X_n, \mu_n) \to (X, \mu)$ in $\mathcal{X}_{M;N}$.

Lemma 4.1. *Let $(X_n, \mu_n), (X, \mu) \in \mathcal{X}_{M;N}$, $n \geq 1$. Then $(X_n, \mu_n) \to (X, \mu)$ in $\mathcal{X}_{M;N}$, if and only if $\mu_n \to \mu$ weakly as $n \to \infty$, and*

$$\inf\{\limsup_{n\to\infty} E^{\mu_n}[dis_N(X_n, G) \wedge 1] + E^{\mu}[dis_N(X, G) \wedge 1];\ G \in C(M; N)\} = 0.$$

Here $C(M; N)$ denotes the set of all continuous maps from M into N.

Proof. (if part) Since $\mu_n \to \mu$ weakly, by Skorohod's theorem there are a probability space $(\Omega, \mathcal{F}, P)$ and M-valued random variables Z, Z_n, $n \geq 1$, such that $P \circ Z_n^{-1} = \mu_n$, $n \geq 1$, $P \circ Z^{-1} = \mu$, and $Z_n \to Z$ in probability. For any $\varepsilon > 0$, there is a $G \in C(M; N)$ such that

$$\limsup_{n\to\infty} E^{\mu_n}[dis_N(X_n, G) \wedge 1] + E^{\mu}[dis_N(X, G) \wedge 1] < \varepsilon.$$

Then

$$\begin{aligned} & \limsup_{n\to\infty} E^P[dis_N(X_n \circ Z_n, X \circ Z) \wedge 1] \\ \leq\ & E^P[dis_N(X \circ Z, G \circ Z) \wedge 1] + \limsup_{n\to\infty} E^P[dis_N(X_n \circ Z_n, G \circ Z_n) \wedge 1] \\ <\ & \varepsilon. \end{aligned}$$

So we see that $X_n \circ Z_n \to X \circ Z$ in probability. Thus we have our assertion from Proposition 4.3.
(only if part) Let $(\Omega, \mathcal{F}, P)$, $Z_n, n \geq 1$, and Z be as in Proposition 4.2. Then we see that $\mu_n \to \mu$ weakly. For any $\varepsilon > 0$, there is a $G \in C(M; N)$ such that $E^{\mu}[dis_N(X, G) \wedge 1] < \varepsilon$. Then

$$\begin{aligned} & \limsup_{n\to\infty} E^{\mu_n}[dis_N(X_n, G) \wedge 1] \\ =\ & \limsup_{n\to\infty} E^P[dis_N(X_n \circ Z_n, G \circ Z_n) \wedge 1] \\ =\ & E^P[dis_N(X \circ Z, G \circ Z) \wedge 1] < \varepsilon. \end{aligned}$$

This implies our assertion.

Theorem 4.1. *Let $(X_n, \mu_n), (X, \mu) \in \mathcal{X}_{M;N}, n \geq 1$, and suppose that there are a probability space $(\Omega, \mathcal{F}, P)$ and M-valued random variables Z_n, $n \geq 1$, and Z such that*
(1) $P \circ Z_n^{-1} = \mu_n, n \geq 1$, and $P \circ Z^{-1} = \mu$,
and
(2) $Z_n \to Z$ in probability.
Then $X_n \circ Z_n \to X \circ Z$ in probability, if and only if $(X_n, \mu_n) \to (X, \mu)$ in $\mathcal{X}_{M;N}$.

Proof. The only if part follows from Proposition 4.3. So it is sufficient to prove the if part. By Lemma 4.1 for any $\varepsilon > 0$, there is a $G \in C(M;N)$ such that

$$E^{\mu}[dis_N(X,G) \wedge 1] + \limsup_{n\to\infty} E^{\mu_n}[dis_N(X_n,G) \wedge 1] < \varepsilon.$$

Then we have

$$\begin{aligned} &\limsup_{n\to\infty} E^P[dis_N(X \circ Z, X_n \circ Z_n) \wedge 1] \\ \le\ & E^P[dis_N(X \circ Z, G \circ Z) \wedge 1] + \limsup_{n\to\infty} E^P[dis_N(X_n \circ Z_n, G \circ Z_n) \wedge 1] \\ & \qquad + \limsup_{n\to\infty} E^P[dis_N(G \circ Z, G \circ Z_n) \wedge 1] \\ <\ & \varepsilon. \end{aligned}$$

Thus we have our assertion.

Let us remind again that if $\mu_n \to \mu$ weakly, then by Skorohod's theorem there are a probability space (Ω, F, P) and measurable maps $Z_n : \Omega \to M$, $n \ge 1$, and $Z : \Omega \to M$ such that $P \circ Z_n^{-1} = \mu_n$, $n \ge 1$, and $P \circ Z^{-1} = \mu$ and that $Z_n \to Z$, $n \to \infty$, $P - a.s.$

The following is easy consequence of Theorem 4.1 and Skorohod's theorem.

Proposition 4.4. *Suppose that $\mu, \mu_n \in \mathcal{P}(M)$, $n \ge 1$.*
(1) *Let N be a Polish space and let $X_n : M \to N$, $X : M \to N$, $X_n^{(k)} : M \to N$ and $X^{(k)} : M \to N$, $n, k \ge 1$, be measurable maps. If $(X_n^{(k)}, \mu_n) \to (X^{(k)}, \mu)$ in $\mathcal{X}_{M;N}$, $k = 1, 2, \ldots$, and if*

$$\limsup_{k\to\infty} \sup_n E^{\mu_n}[dis_N(X_n, X_n^{(k)}) \wedge 1] = 0$$

and

$$\limsup_{k\to\infty} E^{\mu}[dis_N(X, X^{(k)}) \wedge 1] = 0,$$

then $(X_n, \mu_n) \to (X, \mu)$ in $\mathcal{X}_{M;N}$.
(2) *Let N_1 and N_2 be Polish spaces, $X_n : M \to N_1$, $n \ge 1$, and $X : M \to N_1$, be measurable maps and let $f : N_1 \to N_2$ be a continuous map. If $(X_n, \mu_n) \to (X, \mu)$ in $\mathcal{X}_{M;N_1}$, then $(f \circ X_n, \mu_n) \to (f \circ X, \mu)$ in $\mathcal{X}_{M;N_2}$.*
(3) *Let $X_n : M \to \mathbf{R}$ $n \ge 1$, and $X : M \to \mathbf{R}$, be measurable functions. Suppose that $(X_n, \mu_n) \to (X, \mu)$ in $\mathcal{X}_{M;\mathbf{R}}$ and that there is a non-decreasing continuous function $g : [0,\infty) \to [0,\infty)$ such that $\lim_{x\uparrow\infty} \frac{g(x)}{x} = \infty$ and $\sup_n E^{\mu_n}[g(|X_n|)] < \infty$. Then $E^{\mu}[g(|X|)] < \infty$ and $\lim_{n\to\infty} E^{\mu_n}[X_n] = E^{\mu}[X]$.*

Remark. Suppose that $\mu_n = \mu$, $n = 1, 2, \ldots$. Then $(X_n, \mu_n) \to (X, \mu)$ in $\mathcal{X}_{M;N}$, if and only if $X_n \to X$ in probability with respect to μ.

5. Basic Results

In this section, we denote $D([0,1];\mathbf{R}^d)$ by W. Let $\tilde{\mathcal{F}}_t$ and $\mathcal{F}_t$, $t \in [0,1]$, be σ-algebras over W given by $\tilde{\mathcal{F}}_t = \sigma\{w(s); s \le t\}$, $t \in [0,1]$, $\mathcal{F}_t = \cap_{s>t}\tilde{\mathcal{F}}_s$, $t \in [0,1)$ and $\mathcal{F}_1 = \tilde{\mathcal{F}}_1$. Let $\mathcal{M}^p$, $p \in (1,\infty)$, be the set of all $\mathcal{F}_t$-martingale measures μ on W such that $E^{\mu}[\sup_{t\in[0,1]} |w(t)|^p] < \infty$.

Let $p \in (1,\infty)$ and fix it. We assume that μ_n, $n = 1,2,\ldots,\infty$, are elements of $\mathcal{M}^p$ satisfying the following.
(A-1) $\mu_n \to \mu_\infty$ weakly as $n \to \infty$.
(A-2) $\sup_n E^{\mu_n}[\sup_{t\in[0,1]} |w(t)|^p] < \infty$.
(A-3) $\mu_\infty(w(t) = w(t-)) = 1$, $t \in (0,1]$.
(A-4) μ_∞ satisfies the martingale representation property, i.e., for any bounded random variable $Y : W \to \mathbf{R}$, there are predictable process $f : [0,1] \times W \to \mathbf{R}^d$ such that

$$E^{\mu_\infty}[\{\int_0^1 (\sum_{i,j=1}^d f^i(t,w) f^j(t,w) d[w^i,w^j]_t)\}^{p/2}] < \infty, \quad \mu_\infty - a.s.w$$

and

$$Y = E^{\mu_\infty}[Y] + \int_0^1 f(t,w)dw(t), \quad \mu_\infty - a.s.$$

Proposition 5.1. *Let $t \in [0,1]$, and let $X(w) = w(t)$, $w \in W$. Then $(X,\mu_n) \to (X,\mu_\infty)$ in $\mathcal{X}_{W;\mathbf{R}^d}$.*

Proof. Let $w_\infty \in W$ such that $w_\infty(t-) = w_\infty(t)$. Then if $\{w_n\}_{n=1}^\infty$ is a sequence in W converging to w_∞, then $w_n(t) \to w_\infty(t)$. So by the assumption (A-3), Theorem 4.1 and Skorohood's theorem, we have our assertion.

Proposition 5.2. *For any $G \in C_b(W;\mathbf{R})$, there are $g_{n,k} \in C_b(W;\mathbf{R}^d), n \ge 1$, $k = 1,\ldots,2^n$, satisfying the following.*
(1) $g_{n,k}$ is $\mathcal{F}_{(k-1)2^{-n}}$-measurable.
(2) $E^{\mu}[|G-(E^{\mu}[G]+\sum_{k=1}^{2^n} g_{n,k}\cdot(w(k2^{-n})-w((k-1)2^{-n})))|] \to 0$ as $n \to \infty$.
Here $C_b(W;\mathbf{R})$ denotes the set of all bounded continuous functions on W.

Proof. Let $\mathcal{G}$ be a set of bounded predictable processes $g : [0,1] \times W \to \mathbf{R}^d$ such that there are $n \ge 1$ and $\mathcal{F}_{(k-1)2^{-n}}$-measurable $g_{n,k} \in C_b(W;\mathbf{R}^d)$, $k = 1,2,\ldots,2^n$, for which $g(t,w) = \sum_{k=1}^{2^n} g_{n,k}(w)1_{((k-1)2^{-n},k2^{-n}]}(t)$, $t \in [0,1]$, $w \in W$. Then $\mathcal{G}$ is an algebra.

Let $f : [0,1] \times W \to \mathbf{R}^d$ is an arbitrary bounded progressively measurable process such that $f(t,w)$ is left continuous in t. Let $f_n(t,w) = \sum_{k=2}^{2^n} f((k-2)2^{-n},w)1_{((k-1)2^{-n},k2^{-n}]}(t)$, $t \in [0,1], w \in W$, $n \ge 1$. Then we see that $f_n(t,w) \to f(t,w)$, $n \to \infty$, $t \in (0,1], w \in W$.

So the above observation, the monotone theorem and Assumption (A-4) imply our assertion.

Proposition 5.3. *For any $F, G \in C_b(W;\mathbf{R})$ and $t \in [0,1]$,*

$$E^{\mu_n}[E^{\mu_n}[F|\mathcal{F}_t]G]] \to E^{\mu_\infty}[E^{\mu_\infty}[F|\mathcal{F}_t]G], \quad n \to \infty.$$

Proof. Let us take $g_{n,k} \in C(W;\mathbf{R}^d)$, $n \geq 1$, $k = 1,\dots,2^n$, as in Proposition 5.2. Let $G_n = E^\mu[G] + \sum_{k=1}^{2^n} g_{n,k} \cdot (w(k2^{-n} \wedge t) - w((k-1)2^{-n} \wedge t)$ and $G'_n = \sum_{k=1}^{2^n} g_{n,k} \cdot (w(k2^{-n} \vee t) - w((k-1)2^{-n} \vee t)$. Then we see that $E^\mu[|G - (G_n + G'_n)|] \to 0$ as $n \to \infty$. Note that $E^{\mu_n}[G'_m|\mathcal{F}_t] = 0$. So we have

$$\begin{aligned} & |E^{\mu_n}[E^{\mu_n}[F|\mathcal{F}_t]G] - E^{\mu_n}[FG_m]| \\ = \; & |E^{\mu_n}[E^{\mu_n}[F|\mathcal{F}_t]\{G - (G_m + G'_m)\}]| \\ \leq \; & \|F\|_\infty E^{\mu_n}[|G - (G_m + G'_m)|] \end{aligned}$$

Therefore we have

$$\limsup_{m\to\infty} \limsup_{n\to\infty} |E^{\mu_n}[E^{\mu_n}[F|\mathcal{F}_t]G] - E^{\mu_n}[FG_m]| = 0.$$

Similarly we have

$$\limsup_{m\to\infty} |E^{\mu_\infty}[E^{\mu_\infty}[F|\mathcal{F}_t]G] - E^{\mu_\infty}[FG_m]| = 0.$$

Since $E^{\mu_n}[FG_m] \to E^{\mu_\infty}[FG_m]$ for each $m \geq 1$, we obtain our assertion.

Proposition 5.4. *Let $X_n : W \to \mathbf{R}$, $n = 1, 2, \dots, \infty$ be measurable functions. Suppose that $(X_n, \mu_n) \to (X_\infty, \mu_\infty)$ in $\mathcal{X}_{W;\mathbf{R}}$ and that there is an $M > 0$ such that $|X_n| \leq M$, $n = 1, 2, \dots, \infty$. Then $(E^{\mu_n}[X_n|\mathcal{F}_t], \mu_n) \to (E^{\mu_\infty}[X_\infty|\mathcal{F}_t], \mu_\infty)$ in $\mathcal{X}_{W;\mathbf{R}}$ for any $t \in [0,1]$.*

Proof. From the assumption and Lemma 4.1, for any $\varepsilon > 0$, there are $G'_m \in C(W;\mathbf{R}), m \geq 1$, such that

$$\limsup_{m\to\infty} \{\limsup_{n\to\infty} E^{\mu_n}[|X_n - G'_m| \wedge 1] + E^{\mu_\infty}[|X_\infty - G'_m| \wedge 1]\} = 0.$$

Let $G_m = (G'_m \wedge M) \vee (-M) \in C(W;\mathbf{R})$, $m \geq 1$. Then it is easy to see that

$$\limsup_{m\to\infty} \{\limsup_{n\to\infty} E^{\mu_n}[|X_n - G_m|^2] + E^{\mu_\infty}[|X_\infty - G_m|^2]\} = 0.$$

Then we have

$$\limsup_{m\to\infty} \limsup_{n\to\infty} E^{\mu_n}[|E^{\mu_n}[X_n|\mathcal{F}_t] - E^{\mu_n}[G_m|\mathcal{F}_t]|^2] = 0 \tag{5.1}$$

Also, there are $F_\ell \in C(W;\mathbf{R})$, $\ell \geq 1$, such that $|F_\ell| \leq M$, $\ell \geq 1$, and

$$\limsup_{\ell\to\infty} E^{\mu_\infty}[|E[X_\infty|\mathcal{F}_t] - F_\ell|^2] = 0. \tag{5.2}$$

Note that

$$\begin{aligned}
& E^{\mu_n}[|E^{\mu_n}[G_m|\mathcal{F}_t] - F_\ell|^2] \\
= \quad & E^{\mu_n}[E^{\mu_n}[G_m|\mathcal{F}_t]G_m] - 2E^{\mu_n}[E^{\mu_n}[G_m|\mathcal{F}_t]F_\ell] + E^{\mu_n}[F_\ell^2] \\
\to \quad & E^{\mu_\infty}[E^{\mu_\infty}[G_m|\mathcal{F}_t]G_m] - 2E^{\mu_\infty}[E^{\mu_\infty}[G_m|\mathcal{F}_t]F_\ell] + E^{\mu_\infty}[F_\ell^2], \quad n \to \infty \\
= \quad & E^{\mu_\infty}[|E^{\mu_\infty}[G_m|\mathcal{F}_t] - F_\ell|^2].
\end{aligned}$$

So we see that

$$\limsup_{\ell\to\infty} \limsup_{m\to\infty} \limsup_{n\to\infty} E^{\mu_n}[|E^{\mu_n}[G_m|\mathcal{F}_t] - F_\ell|^2] = 0.$$

Combining this and (5.1), we have

$$\limsup_{\ell\to\infty} \limsup_{n\to\infty} E^{\mu_n}[|E^{\mu_n}[X_n|\mathcal{F}_t] - F_\ell|^2] = 0.$$

This, (5.2) and Lemma 4.1 imply our assertion.

Proposition 5.5. *Let $X_n : W \to \mathbf{R}$, $n \geq 1$, and $X_\infty : W \to \mathbf{R}$ be measurable functions. Suppose that $(X_n, \mu_n) \to (X_\infty, \mu_\infty)$ in $\mathcal{X}_{W;\mathbf{R}}$ and that there is an increasing function $g : [0,\infty) \to [0,\infty)$ such that $\liminf_{x\uparrow\infty} \frac{g(x)}{x} = \infty$ and that $\sup_n E^{\mu_n}[g(|X_n|)] < \infty$. Then $(E^{\mu_n}[X_n|\mathcal{F}_t], \mu_n) \to (E^{\mu_\infty}[X|\mathcal{F}_t], \mu_\infty)$ in $\mathcal{X}_{W;\mathbf{R}}$ for any $t \in [0,1]$.*

Proof. By Proposition 4.4, we see that $((X_n \wedge m) \vee (-m), \mu_n) \to ((X_\infty \wedge m) \vee (-m), \mu_\infty)$ in $\mathcal{X}_{W;\mathbf{R}}$, $m \geq 1$. Also we see from the assumption that

$$\lim_{m\to\infty} \sup_n E^{\mu_n}[|X_n - (X_n \wedge m) \vee (-m)|] = 0.$$

This implies that

$$\lim_{m\to\infty} \sup_n E^{\mu_n}[|E^{\mu_n}[X_n|\mathcal{F}_t] - E^{\mu_n}[(X_n \wedge m) \vee (-m)|\mathcal{F}_t]|] = 0.$$

So by this and Proposition 4.4, we have our assertion.

Now let C be a nonempty set of $\mathbf{R}^N$-valued progressively measurable functions satisfying the following.

(1) There are $0 = t_0 < t_1 < \ldots < t_M = 1$ such that $f(t,w) = f(t_k, w)$, $t \in (t_{k-1}, t_k]$, $w \in W$, $f \in C$, $k = 1, \ldots, M$.

(2) $\sup\{|f(t,w)|; f \in C, w \in W, t \in [0,1]\} < \infty$.

(3) If $w_n \to w$ in W, $n \to \infty$, then $\sup\{|f(t,w) - f(t,w_n)|; f \in C\} \to 0$, $n \to \infty$, $t \in [0,1]$.

(4) If g is a $\mathbf{R}^N$-valued progressively measurable function and $f_n \in C$, $n = 1, 2, \ldots$, and if $f_n(t,w) \to g(t,w)$, $n \to \infty$, $t \in [0,1]$, $w \in W$, then $g \in C$.

(5) C is convex, *i.e.*, $sf + (1-s)g \in C$ for any $s \in [0,1]$, $f, g \in C$.

Let $\mathcal{K} = \{c + \int_0^1 f(t,w)dw(t);\ c \in \mathbf{R}, f \in C\}$ and $q \in (1,p)$. Then $\mathcal{K}$ is a closed convex set in $L^q(W, \mathcal{F}_1, \mu_n)$, $n = 1, 2, \ldots, \infty$. Let

$$d_{n,q}(Y,\mathcal{K}) = \inf\{\|Y-F\|_{L^q(W,\mathcal{F}_1,d\mu_n)};\ F\in\mathcal{K}\},\ Y\in L^q(W,\mathcal{F}_1,d\mu_n).$$

Since L^q-norm is uniformly convex, there is a unique element $\Phi_{n,q}(Y)\in\mathcal{K}$ such that

$$\|Y-\Phi_{n,q}(Y)\|_{L^q(W,\mathcal{F}_1,d\mu_n)} = d_{n,q}(Y,\mathcal{K})$$

for each $Y\in L^q(W,\mathcal{F}_1,d\mu_n)$, $n=1,2,\ldots,\infty$.

Lemma 5.1. *Suppose that* $(Y_n,\mu_n)\to(Y_\infty,\mu_\infty)$ *in* $\mathcal{X}_{W;\mathbf{R}}$, *and that* $\sup_n E^{\mu_n}[|Y_n|^r]<\infty$ *for some* $r\in(q,\infty)$. *Then we have the following.*
(1) $(\Phi_{n,q}(Y_n),\mu_n)\to(\Phi_{\infty,q}(Y_\infty),\mu_\infty)$ *in* $\mathcal{X}_{W;\mathbf{R}}$.
(2) *Let* $c_n\in\mathbf{R}$, $f_n\in\mathcal{C}$, $n=1,2,\ldots,\infty$, *such that*

$$\Phi_{n,q}(Y_n)=c_n+\int_0^1 f_n(t,w)dw(t).$$

Then

$$E^{\mu_\infty}[|(c_\infty+\int_0^1 f_\infty(t,w)dw(t))-(c_n+\int_0^1 f_n(t,w)dw(t))|^q]\to 0,\quad n\to\infty.$$

Proof. Since

$$\sup_n \|Y_n\|_{L^q(W,\mathcal{F}_1,d\mu_n)}<\infty$$

and

$$\sup_n \|\int_0^1 f_n(t,w)dw(t)\|_{L^q(W,\mathcal{F}_1,d\mu_n)}<\infty,$$

we see that $\sup_n|c_n|<\infty$. Let $\{m_k\}_{k=1}^\infty$ be an arbitrary increasing sequence of integers. Then by Ascoli-Arzelá's lemma, there are $c\in\mathbf{R}$, $f\in\mathcal{C}$ and a subsequence $\{n_k\}_{k=1}^\infty$ of $\{m_k\}_{k=1}^\infty$ such that $c_{n_k}\to c$ and $f_{n_k}(t,w)\to f(t,w)$, $t\in[0,1]$, $w\in W$. Then by Proposition 5.1 we have $(\int_0^1 f_{n_k}(t,w)dw(t),\mu_n)\to(\int_0^1 f(t,w)dw(t),\mu_\infty)$ in $\mathcal{X}_{W;\mathbf{R}}$. Note that

$$\sup_n\sup\{E^{\mu_n}[|\int_0^1 g(t,w)dw(t)|^p];\ g\in\mathcal{C}\}<\infty.$$

So by Proposition 4.4 we see that

$$E^{\mu_{n_k}}[|Y_{n_k}-\Phi_{n_k,q}(Y_{n_k})|^q]\to E^{\mu_\infty}[|Y_\infty-(c+\int_0^1 f(t,w)dw(t))|^q].$$

Also, we see that $(\int_0^1 f_\infty(t,w)dw(t),\mu_n)\to(\int_0^1 f_\infty(t,w)dw(t),\mu_\infty)$ in $\mathcal{X}_{W;\mathbf{R}}$, and

$$E^{\mu_{n_k}}[|Y_{n_k}-(c_\infty+\int_0^1 f_\infty(t,w)dw(t))|^q]$$
$$\to E^{\mu_\infty}[|Y_\infty-(c_\infty+\int_0^1 f_\infty(t,w)dw(t))|^q].$$

These implies that

$$E^{\mu_\infty}[|Y_\infty - (c + \int_0^1 f(t,w)dw(t))|^q] \leq E^{\mu_\infty}[|Y_\infty - (c_\infty + \int_0^1 f_\infty(t,w)dw(t))|^q].$$

So we see that $\Phi_{\infty,q}(Y_\infty) = c + \int_0^1 f(t,w)dw(t)$, $\mu_\infty - a.s.w.$ Therefore we see that
$(\Phi_{n_k,q}(Y_{n_k}), \mu_{n_k}) \to (\Phi_\infty(Y_\infty), \mu_\infty)$ in $\mathcal{X}_{W;\mathbf{R}}$. This implies Assertion (1).

Also, we see that $E^{\mu_\infty}[|c + \int_0^1 f(t,w)dw(t)) - (c_{n_k} + \int_0^1 f_{n_k}(t,w))dw(t))|^q]$ $\to 0$, $k \to \infty$. This implies Assertion (2).

This completes the proof.

6. Approximation of Price

In this section, we think of the situation and the assumptions (A-1), (A-2), (A-3), (A-4) as given in the previous section. Let $q, r \in (1, \infty)$ with $q < r$ and $q < p$, and let $X_n : [0,1] \times W \to \mathbf{R}$, $n = 1, 2, \ldots, \infty$, be $\mathcal{F}_t$-adapted cádlág processes. We assume the following conditions, moreover.

(C-1) $\sup_n E^{\mu_n}[(\sup\{X_n(t); t \in [0,1]\} + |X_n(1)|)^r] < \infty$.
(C-2) $(X_n(t), \mu_n) \to (X_\infty(t), \mu_\infty)$ in $\mathcal{X}_{W;\mathbf{R}}$ for any $t \in [0,1]$.
(C-3) $\limsup_{m\to\infty} \limsup_{n\to\infty} \varepsilon_{n,m} = 0$. Here

$$\varepsilon_{n,m} = \sup\{E^{\mu_n}[(\sup_{s\in[0,1/m]} (X_n(\tau) - X_n((\tau + s) \wedge 1)))^r];$$

$$\tau \text{ is an } \mathcal{F}_t\text{-stopping time with } \tau \leq 1\}.$$

For each $n = 1, 2, \ldots, \infty$ and $m \geq 1$, let

$$\pi_n = \sup\{E^{\mu_n}[X_n(\tau)];\ \tau \text{ is an } \mathcal{F}_t\text{-stopping time with } \tau \leq 1\},$$

$$\Delta_{n,m,m} = 0,$$

$$\Delta_{n,m,k-1} = \max\{X_n((k-1)/m) - E^{\mu_n}[X_n(1) + \sum_{j=k}^m \Delta_{n,m,j} | \mathcal{F}_{(k-1)/m}], 0\},$$

$k = m, m-1, \ldots, 1$, and

$$Z_{n,m} = X_n(1) + \sum_{j=0}^m \Delta_{n,m,j}.$$

Then we have the following.

Theorem 6.1. (1) $\pi_n \to \pi_\infty$, $n \to \infty$.
(2) *Let* $c_{n,m} \in \mathbf{R}$ *and* $f_{n,m} \in C$, $m \geq 1$, $n = 1, 2, \ldots, \infty$, *satisfy*

$$\Phi_{n,q}(Z_{n,m}) = c_{n,m} + \int_0^1 f_{n,m}(t,w)dw(t), \quad \mu_n - a.s.w.$$

Then

$$\limsup_{m\to\infty} \limsup_{n\to\infty} E^{\mu_\infty}[| \sup_{t\in[0,1]} \{X_\infty(t) - (c_{n,m} + \int_0^t f_{n,m}(t,w)dw(t))\}|^q]^{1/q}$$

$$\leq \frac{q}{q-1} \limsup_{m\to\infty} \inf\{\|Z_{\infty,m} - u\|_{L^q(W,\mathcal{F}_1,d\mu_\infty)};\ u \in \mathcal{K}\}.$$

Proof. By Propositions 4.4 and 5.5, we see inductively $(\Delta_{n,m,k}, \mu_n) \to (\Delta_{\infty,m,k}, \mu_\infty)$ in $\mathcal{X}_{W;\mathbf{R}}$, $k = m, m-1, \ldots, 0$, which implies $(Z_{n,m}, \mu_n) \to (Z_{\infty,m}, \mu_\infty)$ in $\mathcal{X}_{W;\mathbf{R}}$. Therefore by Proposition 5.5, we have $(E^{\mu_n}[Z_{n,m}|\mathcal{F}_t], \mu_n) \to (E^{\mu_\infty}[Z_{\infty,m}|\mathcal{F}_t], \mu_\infty)$ in $\mathcal{X}_{W;\mathbf{R}}$, $t \in [0,1]$, $m \geq 1$. By Theorem 3.1, we see that $\sup_n E^{\mu_n}[|Z_{n,m}|^r] < \infty$, $m \geq 1$. It is easy to see that

$$E^{\mu_n}[Z_{n,m}] \leq \pi_n \leq E^{\mu_n}[Z_{n,m} + \sup_{t\in[0,1]} (X_n(t) - E^{\mu_n}[Z_{n,m}|\mathcal{F}_t])].$$

Also, by Proposition 3.1, we have

$$E^{\mu_n}[| \sup_{t\in[0,1]} (X_n(t) - E^{\mu_n}[Z_{n,m}|\mathcal{F}_t])|^q] \leq \frac{1}{r-q} \varepsilon_{n,m}^{q/r}.$$

Then one can easily see that

$$E^{\mu_\infty}[| \sup_{t\in[0,1]} (X_\infty(t) - E^{\mu_\infty}[Z_{\infty,m}|\mathcal{F}_t])|^q]$$

$$\leq \limsup_{n\to\infty} E^{\mu_n}[| \sup_{t\in[0,1]} (X_n(t) - E^{\mu_n}[Z_{n,m}|\mathcal{F}_t]|^q].$$

and

$$\limsup_{m\to\infty} \limsup_{n\to\infty} E^{\mu_n}[| \sup_{t\in[0,1]} (X_n(t) - E^{\mu_n}[Z_{n,m}|\mathcal{F}_t])|^q] = 0.$$

Since $E^{\mu_n}[Z_{n,m}] \to E^{\mu_\infty}[Z_{\infty,m}]$, $n \to \infty$, we have Assertion (1).

Also, by Lemma 5.1, we see that

$$\limsup_{n\to\infty} E^{\mu_\infty}[| \sup\{X_\infty(t) - (c_{n,m} + \int_0^t f_{n,m}(t,w)dw(t));\ t \in [0,1]\}|^q]^{1/q}$$

$$\leq E^{\mu_\infty}[| \sup\{X_\infty(t) - (c_{\infty,m} + \int_0^t f_{\infty,m}(t,w)dw(t));\ t \in [0,1]\}|^q]^{1/q}$$

So by Doob's inequality, we have Assertion (2).

7. Example

Let $N = 1$ and $\{B(t)\}_{t\in[0,1]}$ be a standard Brownian motion. Also, let $\sigma > 0$. Let μ_∞ be a probability measure on $C([0,1];\mathbf{R}) \subset W$ such that μ is the law of $\{\exp(\sigma B(t) - \sigma^2 t/2)\}_{t\in[0,1]}$. Let $\{U_n\}_{n=1}$ be identically distributed independent random variables such that $P(U_1 = \pm 1) = 1/2$. Let $\mu_n, n \geq 1$, be a probability measure on W such that μ_n is the law of $\{\exp((\sigma/n)\sum_{k=1}^{[nt]} U_k$ - $[nt]a_n)$; $t \in [0,1]\}$, where $a_n = \log\cosh(\frac{\sigma}{n})$. Then it is well known that the assumptions (A-1), (A-2), (A-3) and (A-4) are satisfied. Let $F : [0,1] \times W \to [0,\infty)$ be a measurable functions such that $F(t,\cdot) : W \to [0,\infty)$ is $\mathcal{F}_t$-measurable, $t \in [0,1]$, and that $F(t_n, w_n) \to F(t_\infty, w_\infty)$, if $(t_n, w_n) \to (t_\infty, w_\infty)$ in $[0,1] \times W$ and $w_\infty \in C([0,1];\mathbf{R})$. Also, we assume that there is an $r > 1$ such that $\sup_n E^{\mu_n}[\sup_{t\in[0,1]} F(t,w)^r] < \infty$. Let $X_n(t,w) = F(t,w)$, $(t,w) \in [0,1] \times W$, $n = 1, 2, \ldots, \infty$. Then it is easy to show that the conditions (C-1), (C-2) and (C-3) are satisfied. So we have

$$\begin{aligned} &\pi_n = \sup\{E^{\mu_n}[F(\tau,w)];\ \tau \text{ is an } \mathcal{F}_t\text{-stopping time with } \tau \leq 1\} \\ \to\ \ &\pi_\infty = \sup\{E^{\mu_\infty}[F(\tau,w)];\ \tau \text{ is an } \mathcal{F}_t\text{-stopping time with } \tau \leq 1\}. \end{aligned}$$

So the price π_∞ can be approximated by π_n. Since to compute π_n is the problem in finite state probability space, it is computable in principle. Thus we have a method to compute the price π_∞ approximately.

However, we still have a problem what hedging strategy the writer should take. We will discuss this problem in the forthcoming paper.

References

1. Ansel, J.P., and Stricker, C., *Couverture des actifs contingents*, Ann. Inst. Henri Poincaré **30**(1994), 303-315
2. Delbaen, F., and Schachermayer, W., *A general version of the fundamental theorem of asset pricing*, Math. Ann. **300**(1994), 463-520
3. Duffie, D., *Dynamic Asset Pricing Theory*, Princeton University Press, Princeton, 1992
4. Harrison, M., and Kreps, D.M., *Martingales and Arbitrage in Multiperiod Securities Markets*, Journal of Economic Theory **20**(1979), 381-408
5. Harrison, M., and Pliska, S.R., *Martingales and stochastic integrals in the theory of continuous trading*, Stoch. Processes Appl. **11**(1981), 215-260
6. Jacod, J., *Calcul stochastique et probleme de martingales*, Lec. Notes in Math. vol.714 Springer, Berlin Heidelberg, 1979
7. Karatzas, I., *On the pricing ofAmerican options*, Appl. Math. Optm. 17(1988), 37-60
8. Kramkov, D.O., *Optional decomposition of supermartingales and hedging contingent claims in incomplete security markets*, Preprint
9. Zabczyk, J., *Stopping problems in stochastic control*, Proc. Int. Congress of Math. Warszawa 1983

Calculus for multiplicative functionals, Itô's formula and differential equations

T. J. Lyons and Z. M. Qian

Department of Mathematics, Imperial College of Science, Technology & Medicine
180 Queen's Gate, London SW7 2BZ, UK

1. Introduction

The theory of stochasic integrals and stochastic differential equations was established by K Itô [3, 4] (also see [2]). In past four decade years, Itô's stochastic analysis has established for itself the central role in modern probability theory. Itô's theory of stochastic differential equations has been one of the most important tools. However, Itô's construction of stochastic integrals over Brownian motion possesses an essentially random characterization, and is meaningless for a single Brownian path. The Itô map obtained by solving Itô's stochastic differential equations is nowhere continuous on the Wiener space.

It is shown in Lyons [5] (also see [6, 7]) that if we revise the notion of a path slightly, it is possible to give a pathwise treatment to stochastic integrals, and Itô's map obtained via this kind of integral is continuous in an appropriate topology. The idea is based on correctly identifying the "differential" of a rough path.

To define an integral of the form $\int \alpha(X)dX$, where α is a 1-form and X is a path, it is sufficient to give the meanings to multi-integrals

$$X^i_{st} \widehat{=} \int_{s<t_1<\cdots<t_i<t} dX_{t_1}\cdots dX_{t_i}$$

of low degree $i \leq [p]$, where p is related to the roughness of the path X.

If X is a continuous path with finite variation in a Banach space V, it is well known that X^i_{st}, $i \geq 2$, are uniquely determined by $X^1_{st} = X_t - X_s$. In this case we can define X^k_{st} to be a tensor in $V^{\otimes k}$ by

$$X^k_{st}(\xi_1 \otimes \cdots \otimes \xi_k) = \int_{s<t_1<\cdots<t_k<t} d\langle X_{t_1}, \xi_1\rangle \cdots d\langle X_{t_k}, \xi_k\rangle \tag{1}$$

for any $\xi_1, \cdots, \xi_k \in V^*$, the right hand side is the conventional integral. In other words, we have the recursive definition,

$$X^k_{st} = \lim_{m(D)\to 0} \sum_l \sum_{\substack{i+j=k \\ i,j\geq 1}} X^i_{st_{l-1}} \otimes X^j_{t_{l-1}t_l},$$

where $D = \{s = t_0 \le \cdots \le t_r = t\}$ and $m(D) = \max_i(t_i - t_{i-1})$. Moreover, we have

$$\sup_D \sum_l \left|X^k_{t_{l-1}t_l}\right|^{\frac{1}{k}} < \infty, \quad k = 1, 2, \cdots,$$

where the sup takes over all dissection D of $[s,t]$.

It is shown in Lyons [7] that these observations have a generalization to rough paths. If X has finite p-variation (i.e. $\frac{1}{p}$-Hölder's continuous after a time change), i.e.

$$\sup_D \sum_l |X_{t_l} - X_{t_{l-1}}|^p < \infty,$$

and if X^k_{st} satisfies the following analytic condition that

$$\sup_D \sum_l \left|X^k_{t_{l-1}t_l}\right|^{\frac{p}{k}} < \infty, \quad k = 1, \cdots, \tag{2}$$

then X^k_{st} is uniquely determined by $X^1_{st}, \cdots, X^{[p]}_{st}$ for $k \ge [p]+1$ (see Th.1 below).

So it is reasonable to regard $(X^i_{st})_{i=1}^{[p]}$ as a "differential" of X. The basic property of these differentials are summarized in the K T Chen's formula (see [1]). Put $X_{st} \widehat{=} (1, X^1_{st}, \cdots, X^{[p]}_{st})$, and regard X_{st} as an element in the tensor algebra $T^{[p]}(V)$,

$$T^{[p]}(V) = \sum_{k=0}^{[p]} V^{\otimes k}, \quad V^{\otimes 0} = \mathrm{R}.$$

Then

$$X_{st} \otimes X_{tu} = X_{su}, \quad \forall s \le t \le u,$$

where the tensor product $\otimes$ is taken in the truncated tensor algebra $T^{[p]}(V)$, i.e. if $a = (a_0, \cdots, a_{[p]})$, $b = (b_0, \cdots, b_{[p]})$, then

$$(a \otimes b)_k = \sum_{i+j=k} a_i \otimes b_j, \quad k = 0, 1, \cdots, [p].$$

We note that the analytic requirement (2) is crucial for the uniqueness. Indeed, even for a smooth path $t \to X_t$, if $3 > p > 2$, the choice of X^2_{st} is never unique. For example, we can choose

$$X^2_{st} \widehat{=} \int_{s<t_1<t_2<t} dX_{t_1} dX_{t_2},$$

the coventional double integral. Then we can see that $(1, X^1_{st}, X^2_{st})$ is a "differential", i.e. $(1, X^1_{st}, X^2_{st})$ satisfies K T Chen's formula, and

$$\sup_D \sum_l \left|X^i_{t_{l-1}t_l}\right|^{\frac{p}{i}} < \infty, \quad i = 1, 2. \tag{3}$$

However for any constant K, and

$$X(K)^2_{st} \widehat{=} X^2_{st} + K(t-s)A,$$

where A is an element of $V \otimes V$, $(1, X^1_{st}, X(K)^2_{st})$ will be a "differential" and satisfy (3).

Note that almost all Brownian motion paths (W_t) are $\frac{1}{p}$-Hölder continuous for any $2 < p < 3$, so that for almost all σ,

$$\sup_D \sum_l |W_{t_l}(\sigma) - W_{t_{l-1}}(\sigma)|^p < \infty,$$

and not $\frac{1}{2}$-Hölder continuous. Thus above arguments yield that the definition of integrals along Brownian path is no longer unique. This explained why we have different meanings of stochastic integrals along a Brownian motion, e.g. Itô's integral and Stratonovich's integral.

In this paper we study the consequences of perturbation of a multiplicative functional by an additive functional. In Lyons [7], a calculus for geometric multiplicative functionals (i.e. "Stratonovich's integral") is established. In certain sense, present work is a study of the connection between Itô's integral and Stratonovich's integral. In particular we obtain an Itô's formula for multiplicative functionals. The paper is organised as follows. In section 2 we recall several basic notions and results obtained in [7]. In section 3, we shall give an Itô's formula for multiplicative functionals. In section 4 we disscuss a differential equation driven by a multiplicative functional.

2. Differentiation and integration

In this section, we recall several basic notions and results established in Lyons [7]. Throughout this paper, let I be a bounded interval $[0, T]$, and let Δ denote the set of all pairs (s, t) such that $s \leq t$, $s, t \in I$. A continuous function ω on Δ with values in $[0, \infty)$ is called a control function if ω is subadditive, i.e.

$$\omega(s,t) + \omega(t,u) \leq \omega(s,u), \quad \forall (s,t),(t,u) \in \Delta,$$

and $\lim_{t \to s} \omega(s,t) = 0$ for any $s \in I$. Given a real separable Banach space V and a natural number n, we will use $T^{(n)}(V)$ to denote the tensor algebra over V of degree n, i.e.

$$T^{(n)}(V) = \sum_{k=0}^{n} V^{\otimes k}, \quad V^{\otimes 0} \widehat{=} \mathrm{R},$$

where we endow $V^{\otimes k}$ with any compatible Banach tensor norm.

We say a map $X: \Delta \to T^{(n)}(V)$ is of finite p-variation if

$$X_{st} = (1, X^1_{st}, \cdots, X^n_{st}), \quad X^k_{st} \in V^{\otimes k},$$

and

$$|X^i_{st}| \le \frac{\omega(s,t)^{\frac{i}{p}}}{\beta\left(\frac{i}{p}\right)!}, \quad \forall (s,t) \in \Delta, \ i = 1, \cdots, n,$$

for some control function ω, where β is a fixed constant depending only on p. In this case we say X is of finite p-variation controlled by ω. We say such a function X is an almost multiplicative functional if in addition $n \ge [p]$ and

$$\left|(X_{st} \otimes X_{tu} - X_{su})^i\right| \le K_1 \omega(s,u)^\theta, \quad \forall (s,t),(t,u) \in \Delta, i = 1, \cdots, n,$$

for some constants $K_1 \ge 0$ and $\theta > 1$, where the product $\otimes$ is taken in the truncated tensor algebra $T^{(n)}(V)$, and the index i indicates the component in $V^{\otimes i}$ of an element in $T^{(n)}(V)$.

A map $X : \Delta \to T^{(n)}(V)$ is called a multiplicative functional if $X_{st} = (1, X^1_{st}, \cdots, X^n_{st})$ for each pair $(s,t) \in \Delta$, and X satisfies K T Chen's formula,

$$X_{st} \otimes X_{tu} = X_{su}, \quad \text{in } T^{(n)}(V), \forall (s,t),(t,u) \in \Delta.$$

If $X : \Delta \to T^{(n)}(V)$ is a multiplicative functional, and $X_{st} = (1, X^1_{st}, \cdots, X^n_{st})$, then we also use either X^1_t or X_t to denote X^1_{0t}, and X to denote the path $t \to X_t$, if no confusion arises.

Multiplicative functionals appear naturally as the iterated integral sequences given by

$$X^i_{st} = \int_{s<t_1<\cdots<t_i<t} dX_{t_1} \cdots dX_{t_i},$$

where the integral on the right hand side could be the standard one, or the Itô's one or any other reasonable choice.

The following is one of the results in Lyons [7].

Theorem 1. *Let $X : \Delta \to T^{(n)}(V)$ be a multiplicative functional with finite p-variation controlled by ω, and let $n \ge [p]$. Then there is a unique map $X^k : \Delta \to V^{\otimes k}$ for each $k \ge n+1$, such that*

$$X_{st} \widehat{=} (1, X^1_{st}, X^2_{st}, \cdots, X^k_{st}, \cdots)$$

is a multiplicative functional in $T^{(\infty)}(V)$ with finite p-variation. In addition, X is of finite p-variation controlled by ω, that is

$$|X^k_{st}| \le \frac{\omega(s,t)^{\frac{k}{p}}}{\beta\left(\frac{k}{p}\right)!}, \quad \forall (s,t) \in \Delta, \ k = 1, \cdots,$$

and if $k \ge [p] + 1$, then

$$X^k_{st} = \lim_{m(D)\to 0} \sum_{l=1}^{r} \sum_{\substack{i+j=k \\ i,j\ge 1}} X^i_{st_{l-1}} \otimes X^j_{t_{l-1}t_l},$$

where $D = \{s = t_0 \le t_1 \le \cdots \le t_r = t\}$, and $m(D) = \max_i(t_i - t_{i-1})$.

Moreover if $X, Y\colon \Delta \to T^{(n)}(V)$ are two multiplicative functionals with finite p-variation controlled by ω, and

$$\left|X^i_{st} - Y^i_{st}\right| \le \varepsilon\omega(s,t)^{\frac{i}{p}}, \quad \forall i = 1, \cdots, [p], \ (s,t) \in \Delta, \tag{4}$$

then

$$\left|X^i_{st} - Y^i_{st}\right| \le K_2\varepsilon\omega(s,t)^{\frac{i}{p}}, \quad \forall i = 1, \cdots, \ (s,t) \in \Delta,$$

where K_2 is a constant only depending on p, $\max\omega$ and I.

Another result obtained in [7] is the following

Theorem 2. *Let $X\colon \Delta \to T^{(n)}(V)$ be an almost multiplicative functional with finite p-variation controlled by ω and let $n \ge [p]$. Then there is a unique multiplicative functional $\widehat{X}$ with finite p-variation controlled by $K_3\omega$ for some constant K_3, such that*

$$\left|\widehat{X}^i_{st} - X^i_{st}\right| \le K_4\omega(s,t)^{\theta}, \quad \forall (s,t) \in \Delta, \ i = 1, \cdots, [p].$$

Indeed,

$$\widehat{X}^1_{st} = \lim_{m(D)\to 0} \sum_l X^1_{t_{l-1}t_l},$$

and more generally,

$$\widehat{X}^k_{st} = \lim_{m(D)\to 0} \sum_l \left\{X^k_{t_{l-1}t_l} + \sum_{\substack{i+j=k\\ i,j\ge 1}} \widehat{X}^i_{st_{l-1}} \otimes \widehat{X}^j_{t_{l-1}t_l}\right\}. \tag{5}$$

Moreover the map $X \to \widehat{X}$ is continuous in the following sense that if X, Y are two almost multiplicative functionals with finite p-variation controlled by ω, and X, Y satisfy (4), then

$$\left|\widehat{X}^i_{st} - \widehat{Y}^i_{st}\right| \le K(\varepsilon)\omega(s,t)^{\frac{i}{p}}, \quad \forall (s,t) \in \Delta, \ i = 1, \cdots, [p],$$

where $K(\varepsilon)$ is a constant depending on ε satisfying $\lim_{\varepsilon\to 0} K(\varepsilon) = 0$.

By (5), it is easily seen that if X, Y are two almost multiplicative functionals with finite p-variation controlled by ω, and

$$|X^i_{st} - Y^i_{st}| \le K_5\omega(s,t)^{\theta}, \quad \forall (s,t) \in \Delta, \ i = 1, \cdots, [p]$$

for some $K_5 \ge 0$ and $\theta > 1$, then $\widehat{X} = \widehat{Y}$.

Next we discuss integrals of 1-forms against multiplicative functionals. Let W be another separable Banach space. A map $\alpha\colon V \to \hom(V, W)$ is called a W-valued 1-form on V, where we use $\hom(B_1, B_2)$ to denote the Banach space of all bounded linear operators from B_1 to B_2. We will use α^j to denote the j-th differential if it exists, and we shall regard it as a map from V to $\hom(V^{\otimes(j+1)}, W)$. We say a 1-form α is $\mathrm{Lip}(\gamma)$, if $\alpha^1, \cdots, \alpha^{[p]}$ exist and

$$\alpha^j(X_t)(v) = \sum_{0\le i+j\le[p]}^{[p]} \alpha^{i+j}(X_s)(X_{st}^i \otimes v) + R_j(X_s, X_t), \quad \forall (s,t) \in \Delta,$$

$$|\alpha^j(x)| \le M, \quad |R(x,y)| \le M|x-y|^{\gamma-j}, \quad \forall x, y,$$

for any smooth path X in V, where

$$X_{st}^j = \int_{s<t_1<\cdots<t_j<t} dX_{t_1} \cdots dX_{t_j}, \quad \forall (s,t) \in \Delta,$$

and the right hand side is the conventional multi-integral, see (1). Note that it is an important fact that R_j does only depends on the path through its end points.

For simplicity, and to make the material more accessible to those interested in stochastic differential equations, we only consider the case where $2 \le p < 3$ and $n = 2$. One of the reasons to make such a restriction is, of course, for simplicity. A more important reason is that almost all Brownian paths are $\frac{1}{p}$-Hölder continuous on any bounded interval for any $p > 2$, so that our discussions here can be applied to continuous semimartingales. Suppose we are given a multiplicative functional valued in $T^{(2)}(V)$ with finite p-variation controlled by ω, and a 1-form $\alpha: V \to \hom(V, W)$ which is Lip(γ), $\gamma > 1$. Define $Y: \Delta \to T^{(2)}(W)$, $Y_{st} = (1, Y_{st}^1, Y_{st}^2)$ by

$$\begin{aligned} Y_{st}^1 &= \alpha(X_s)(X_{st}^1) + \alpha^1(X_s)(X_{st}^2), \\ Y_{st}^2 &= \alpha(X_s) \otimes \alpha(X_s)(X_{st}^2), \quad \forall (s,t) \in \Delta. \end{aligned}$$

If $\frac{\gamma+1}{p} > 1$, then it is shown in [7] that Y is an almost multiplicative functional with finite p-variation controlled by $K_6\omega$, where and in what follows, we will use K_i to denote constants depending only on p, $\max \omega$, T, and Lipschitz constant M. We denote the associated multiplicative functional $\widehat{Y}$ by $\int \alpha(X)\delta X$, and set

$$\int_s^t \alpha(X)\delta X \hat{=} \widehat{Y}_{st}, \quad \int_s^t \alpha(X)\delta X^i \hat{=} \left(\int \alpha(X)\delta X\right)_{st}^i, \quad \forall (s,t) \in \Delta, \ i = 1, 2.$$

By definition we have

$$\int_s^t \alpha(X)\delta X^1 = \lim_{m(D)\to 0} \sum_l \left\{\alpha(X_{t_{l-1}})(X_{t_{l-1}t_l}^1) + \alpha^1(X_{t_{l-1}})(X_{t_{l-1}t_l}^2)\right\},$$

and

$$\begin{aligned} \int_s^t \alpha(X)\delta X^2 &= \lim_{m(D)\to 0} \sum_l \Big\{\alpha(X_{t_{l-1}}) \otimes \alpha(X_{t_{l-1}})(X_{t_{l-1}t_l}^2) \\ &\quad + \int_s^{t_{l-1}} \alpha(X)\delta X^1 \otimes \int_{t_{l-1}}^{t_l} \alpha(X)\delta X^1\Big\}. \end{aligned}$$

Moreover, $X \to \int \alpha(X)\delta X$ is continuous, i.e. if X, Y are two multiplicative functionals with finite p-variation controlled by ω, and satisfy (4), then

$$\left| \int_s^t \alpha(X)\delta X^i - \int_s^t \alpha(Y)\delta Y^i \right| \le K_7 K(\varepsilon)\omega(s,t)^{\frac{i}{p}}, \quad \forall (s,t) \in \Delta.$$

for $i = 1, 2$.

The following Prop.1 shows how to construct integrals of 1-forms along an almost multiplicative functional, which will play a very important role in our next development.

Proposition 1. *Let $X: \Delta \to T^{(2)}(V)$ be an almost multiplicative functional with finite p-variation controlled by ω. Suppose X^1 is additive, i.e.*

$$X^1_{st} + X^1_{tu} = X^1_{su}, \quad \forall (s,t), (t,u) \in \Delta,$$

and $\alpha: V \to \hom(V, W)$ is a Lip(γ) 1-form with $\frac{\gamma+1}{p} > 1$. Then Z is an almost multiplicative functional, where $Z: \Delta \to T^{(2)}(W)$, $Z_{st} = (1, Z^1_{st}, Z^2_{st})$,

$$\begin{aligned} Z^1_{st} &= \alpha(X_s)(X^1_{st}) + \alpha^1(X_s)(X^2_{st}), \\ Z^2_{st} &= \alpha(X_s) \otimes \alpha(X_s)(X^2_{st}), \end{aligned}$$

and

$$\widehat{Z} = \int \alpha(\widehat{X})\delta \widehat{X}.$$

Proof. We first prove that Z is an almost multiplicative functional. We will use $\cong$ to mean that the difference of the two sides is bounded by $K_8\omega(s,u)^\theta$ for some constants K_8 and $\theta > 1$. Since α is Lip(γ), and X^1 is additive, we have

$$\begin{aligned} &\alpha(X_t) = \alpha(X_s) + \alpha^1(X_s)(X^1_{st}) + R_0(X_s, X_t), \\ &|\alpha^1(X_t) - \alpha^1(X_s)| \le M|X^1_{st}|^{\gamma-1}. \end{aligned}$$

Using the facts that $|R_0(X_s, X_t)| \le M|X^1_{st}|^\gamma$, and that $\frac{\gamma+1}{p} > 1$, we have

$$\begin{aligned} Z^1_{st} + Z^1_{tu} &= \alpha(X_s)(X^1_{st}) + \alpha^1(X_s)(X^2_{st}) \\ &\quad + \alpha(X_t)(X^1_{tu}) + \alpha^1(X_t)(X^2_{tu}) \\ &\cong \alpha(X_s)(X^1_{st} + X^1_{tu}) + \alpha^1(X_s)(X^1_{st} \otimes X^1_{tu}) \\ &\quad + \alpha^1(X_s)(X^2_{st}) + \alpha^1(X_s)(X^2_{tu}) \\ &= \alpha(X_s)(X^1_{su}) + \alpha^1(X_s)(X^2_{st} + X^2_{tu} + X^1_{st} \otimes X^1_{tu}), \end{aligned}$$

so that

$$|Z^1_{st} + Z^1_{tu} - Z^1_{su}| \le K_9\omega(s,t)^\theta.$$

Similarly we have

$$\begin{aligned}
&\left|(Z_{st}\otimes Z_{tu})^2 - Z^2_{su}\right| = Z^2_{st} + Z^2_{tu} + Z^1_{st}\otimes Z^1_{tu} - Z^2_{su}\\
=\ & \alpha(X_s)\otimes\alpha(X_s)(X^2_{st}) + \alpha(X_t)\otimes\alpha(X_t)(X^2_{tu})\\
& + \left(\alpha(X_s)(X^1_{st}) + \alpha^1(X_s)(X^2_{st})\right)\otimes\left(\alpha(X_s)(X^1_{tu}) + \alpha^1(X_t)(X^2_{tu})\right)\\
& -\alpha(X_s)\otimes\alpha(X_s)(X^2_{su})\\
\cong\ & \alpha(X_s)\otimes\alpha(X_s)(X^2_{st} + X^2_{tu} - X^2_{su})\\
& +\alpha(X_s)(X^1_{st})\otimes\alpha(X_t)(X_{tu})\\
\cong\ & \alpha(X_s)\otimes\alpha(X_s)(X^2_{st} + X^2_{tu} + X^1_{st}\otimes X^1_{tu} - X^2_{su}).
\end{aligned}$$

However X is an almost multiplicative functional, so that

$$\left|(Z_{st}\otimes Z_{tu})^2 - Z^2_{su}\right| \le K_{10}\omega(s,t)^\theta.$$

That is, Z is an almost multiplicative functional. Since X^1 is additive, so that $\widehat{X}^1 = X^1$. Let

$$\begin{aligned}
Y^1_{st} &= \alpha(\widehat{X}_s)(\widehat{X}^1_{st}) + \alpha^1(\widehat{X}_s)(\widehat{X}^2_{st}),\\
Y^2_{st} &= \alpha(\widehat{X}_s)\otimes\alpha(\widehat{X}_s)(\widehat{X}^2_{st}).
\end{aligned}$$

Then

$$\begin{aligned}
\left|Y^1_{st} - Z^1_{st}\right| &= \left|\alpha^1(X_s)(X^2_{st} - \widehat{X}^2_{st})\right|\\
&\le K_{11}\omega(s,t)^\theta,
\end{aligned}$$

and

$$\begin{aligned}
\left|Y^2_{st} - Z^2_{st}\right| &= \left|\alpha(X_s)\otimes\alpha(X_s)(X^2_{st} - \widehat{X}^2_{st})\right|\\
&\le K_{12}\omega(s,t)^\theta,
\end{aligned}$$

so that $\widehat{Z} = \widehat{Y}$. However $\widehat{Y} = \int\alpha(\widehat{X})\delta\widehat{X}$, so that $\widehat{Z} = \int\alpha(\widehat{X})\delta\widehat{X}$.

3. Itô's formula

Let A be the anti-symmetrization operator on $V^{\otimes 2}$, i.e.

$$A(v_1\otimes v_2) = \frac{1}{2}(v_1\otimes v_2 - v_2\otimes v_1), \quad \forall v_1\otimes v_2\in V^{\otimes 2},$$

which induces a natural map from $T^{(2)}(V)$ to $T^{(2)}(V)$. Given a multiplicative functional $X\colon\Delta\to T^{(2)}(V)$, $X_{st} = (1, X^1_{st}, X^2_{st})$, define a map $\widetilde{X}\colon\Delta\to T^{(2)}(V)$ by

$$\begin{aligned}
\widetilde{X}_{st} &= (1, X^1_{st}, \widetilde{X}^2_{st}),\\
\widetilde{X}^2_{st} &\triangleq A(X^2_{st}) + \frac{1}{2}(X_t - X_s)^{\otimes 2},
\end{aligned}$$

and we say $\widetilde{X}$ is the minimal multiplicative functional associated to X. A multiplicative functional X is called minimal if $\widetilde{X} = X$.

The definition of a Lip(γ) map is similar to that of 1-form, i.e. a map $f: V \to W$ is called Lip(γ), if the derivatives f^j, $j = 0, \cdots, [\gamma]$, exist, and

$$f^j(X_t)(v) = \sum_{0 \le i+j \le [\gamma]} f^{i+j}(X_s)(X^i_{st} \otimes v) + R_j(X_s, X_t)(v), \tag{6}$$

$$|f^j(x)| \le M, \quad |R_j(x,y)| \le M|x-y|^{\gamma - j},$$

for $j = 0, 1, \cdots, [\gamma]$, $(x, y) \in V$ and any smooth path (X_t), where

$$X^i_{st} \widehat{=} \int_{s<t_1<\cdots<t_i<t} dX_{t_1} \cdots dX_{t_i},$$

where the integral in the right side is the conventional multi-integral.

Note that we can replace X^i_{st} by its symmetric part in Eq.(6), so that we have the following

Proposition 2. *Let $X: \Delta \to T^{(2)}(V)$ be a minimal multiplicative functional and let $f: V \to W$ be of* Lip(γ), $2 \le \gamma < 3$. *Then*

$$f^j(X_t)(v) = \sum_{0 \le i+j \le 2} f^{i+j}(X_s)(X^i_{st} \otimes v) + R_j(X_s, X_t)(v), \quad \forall (s,t) \in \Delta,$$

for $j = 0, 1, 2$. Similar conclusion holds for 1-forms.

Theorem 3 (Change of variable formula). *Let $X: \Delta \to T^{(2)}(V)$ be a minimal multiplicative functional with finite p-variation controlled by ω, and let f be a* Lip(γ) *function on V with values in a separable Banach space W, $\frac{\gamma}{p} > 1$. Then*

$$f(X_t) - f(X_s) = \left(\int f^1(X)\delta X \right)^1_{st}, \quad \forall (s,t) \in \Delta.$$

Proof. By definition, $\int f^1(X)\delta X = \widehat{Y}$, where $Y_{st} = (1, Y^1_{st}, Y^2_{st})$,

$$\begin{aligned} Y^1_{st} &= f^1(X_s)(X^1_{st}) + f^2(X_s)(X^2_{st}), \\ Y^2_{st} &= f^1(X_s) \otimes f^1(X_s)(X^2_{st}). \end{aligned}$$

Now let $Z: \Delta \to T^{(2)}(V)$, $Z_{st} = (1, Z^1_{st}, Z^2_{st})$, defined by

$$\begin{aligned} Z^1_{st} &= f(X_t) - f(X_s), \\ Z^2_{st} &= Y^2_{st}. \end{aligned}$$

Since X is minimal, we have

$$f(X_t) - f(X_s) = f^1(X_s)(X^1_{st}) + f^2(X_s)(X^2_{st}) + R_0(X_s, X_t),$$

so that

$$\begin{aligned}|Z^1_{st} - Y^1_{st}| &= |R_0(X_s, X_t)| \\ &\le M|X^1_{st}|^\gamma \\ &\le K_1\omega(s,t)^{\frac{\gamma}{p}}.\end{aligned}$$

However, $\frac{\gamma}{p} > 1$, so that Z is an almost multiplicative functional, and

$$\left|Z^i_{st} - Y^i_{st}\right| \le M\omega(s,t)^{\frac{\gamma}{p}}, \quad \forall (s,t) \in \Delta,$$

so that $\widehat{Z} = \widehat{Y}$. In particular

$$\widehat{Z}^1_{st} = \int_s^t f^1(X)\delta X^1, \quad \forall (s,t) \in \Delta.$$

Since Z^1 is additive, we have $\widehat{Z}^1_{st} = Z^1_{st}$. Hence

$$f(X_t) - f(X_s) = \int_s^t f^1(X)\delta X^1, \quad \forall (s,t) \in \Delta.$$

Definition 1. *1). A multiplicative functional $X: \Delta \to T^{(n)}(V)$ is called a smooth multiplicative functional if $t \to X^1_{0t}$ is continuous and piecewisely smooth, and each X^k_{st} is the conventional multi-integral, i.e.*

$$X^k_{st} = \int_{s<t_1<\cdots<t_k<t} dX^1_{0t_1} \cdots dX^1_{0t_k}, \quad \forall (s,t) \in \Delta.$$

2). Let $X: \Delta \to T^{(n)}(V)$ be a multiplicative functional with finite p-variation controlled by ω, $n = [p]$. We say it is a geometric multiplicative functional if there is a smooth multiplicative functional $X(\varepsilon)$ for any $\varepsilon > 0$ such that

$$\left|X(\varepsilon)^i_{st} - X^i_{st}\right| \le \varepsilon\omega(s,t)^{\frac{i}{p}}, \quad \forall (s,t) \in \Delta,\ i = 1, \cdots, [p].$$

Note that if X is a geometric multiplicative functional, then the symmetric part of X^i_{st} is $\frac{1}{i!}(X^1_{st})^{\otimes i}$, so that Eq.(6) holds for a geometric multiplicative functional X and a Lip(γ) function. In particular, when $2 \le p < 3$, and $n = 2$, a geometric multiplicative functional is minimal.

Example. Let (W_t) be a d-dimensional Brownian motion on a probability space (Ω, P). Put $\widetilde{W}_{st} = (1, \widetilde{W}^1_{st}, \widetilde{W}^2_{st})$, where $\widetilde{W}^1_{st} = W_t - W_s$ and

$$\widetilde{W}^2_{st} = \int_{s<t_1<t_2<t} \circ dW_{t_1} \circ dW_{t_2},$$

where $\circ d$ denotes the Stratonovich's differential. More precisely,

$$\widetilde{W}^2_{st}(\xi \otimes \eta) = \int_{s<t_1<t_2<t} \circ d\left\langle W_{t_1}, \xi\right\rangle \circ d\left\langle W_{t_2}, \eta\right\rangle, \quad \forall \xi, \eta \in R^d.$$

We can decompose $\widetilde{W}^2_{st}$ into two parts, i.e.

$$\widetilde{W}^2_{st}(\xi \otimes \eta) = A_{st}(\xi \otimes \eta) + \frac{1}{2}\langle W_t - W_s, \xi\rangle \langle W_t - W_s, \eta\rangle ,$$

where A_{st} is the so-called area process defined by

$$A_{st}(\xi \otimes \eta) = \frac{1}{2}\int_s^t (\langle W_u, \xi\rangle d\langle W_u, \eta\rangle - \langle W_u, \eta\rangle d\langle W_u, \xi\rangle) .$$

In particular, if $|\xi| = |\eta| = 1$ and $\xi \perp \eta$, then $(\langle W_t, \xi\rangle, \langle W_t, \eta\rangle)$ is a standard 2-dimensional Brownian motion, and $A_{st}(\xi \otimes \eta)$ is the Lévy area enclosed by the 2-dimensional Brownian motion. We can show that for any p, $2 < p < 3$, and almost all $\sigma \in \Omega$, $\widetilde{W}(\sigma): \Delta \to T^{(2)}((\mathrm{R}^d)^{\otimes 2})$ is a geometric multiplicative functional with finite p-variation controlled by $K_\sigma(t-s)$, i.e.

$$\left|\widetilde{W}(\sigma)^i_{st}\right| \le K_\sigma(t-s)^{\frac{i}{p}}, \quad \forall (s,t) \in \Delta, \ i = 1, 2.$$

Remark. Consider the case of arbitrary $p > 1$. Let $X: \Delta \to T^{(n)}(V)$ be a *geometric multiplicative functional* with finite p-variation controlled by ω, and let $n = [p]$. Then we still can define the integral of the form $\int \alpha(X)\delta X$ as a multiplicative functional of degree $[p]$, although the construction of $(\int \alpha(X)\delta X)^i$, $i > 1$, can be more complicated. However,

$$\left(\int \alpha(X)\delta X\right)^1_{st} = \lim_{m(D)\to 0} \sum_l \sum_{k=1}^n \alpha^{k-1}(X_{t_{l-1}})(X^k_{t_{l-1}t_l}).$$

Let $f: V \to W$ be a Lip(γ) function such that $\frac{\gamma}{p} > 1$. Then one still has the fundamental theorem of calculus,

$$f(X_t) - f(X_s) = \left(\int f^1(X)\delta X\right)^1_{st}, \quad \forall (s,t) \in \Delta. \tag{7}$$

However, if $p > 3$ and X is a general multiplicative functional of degree $[p]$, we do not know how to define integrals of the form $\int \alpha(X)\delta X$.

It is easy to understand that a geometric multiplicative functional produces the deterministic equivalent of Stratonovich's integral. Next we consider "Itô's integral". We say $\psi: \Delta \to V^{\otimes 2}$ is an additive functional if

$$\psi_{st} + \psi_{tu} = \psi_{su}, \quad \forall (s,t), (t,u) \in \Delta.$$

Let $X: \Delta \to T^{(2)}(V)$, $X_{st} = (1, X^1_{st}, X^2_{st})$. Then we set $X_{st}(\psi): \Delta \to T^{(2)}(V)$, $X(\psi)_{st} = (1, X^1_{st}, X^2_{st} + \psi_{st})$.

Proposition 3. *Let $X: \Delta \to T^{(2)}(V)$ be a multiplicative functional, and let $\psi: \Delta \to V^{\otimes 2}$. Then $X(\psi)$ is a multiplicative functional if and only if ψ is additive. Moreover, if X is of finite p-variation controlled by ω, then $X(\psi)$ is of finite p-variation controlled by $K_2\omega$ if*

$$|\psi_{st}| \le K_3\omega(s,t)^{\frac{2}{p}}, \quad \forall (s,t) \in \Delta. \tag{8}$$

Proposition 4. *Let $\alpha: W \to \hom(V^{\otimes 2}, H)$ be γ-Hölder continuous, i.e.*

$$|\alpha(x) - \alpha(y)| \le M|x-y|^\gamma, \quad \forall x, y \in V^{\otimes 2},$$

$\frac{\gamma+2}{p} > 1$, *let $z: I \to W$ be a continuous path satisfying*

$$|z_s - z_t| \le K_4 \omega(s,t)^{\frac{1}{p}}, \quad \forall (s,t) \in \Delta,$$

and let $\psi: \Delta \to V^{\otimes 2}$ be an additive functional satisfying (8). Then the following limit exists,

$$\int_s^t \alpha(z_u) d\psi_u \widehat{=} \lim_{m(D)\to 0} \sum_l \alpha(z_{t_{l-1}})(\psi_{t_{l-1}t_l}).$$

Using these results we can show that if $X: \Delta \to T^{(2)}(V)$ is a multiplicative functional with finite p-variation controlled by ω, and let $\psi: \Delta \to V^{\otimes 2}$ be an additive functional satisfying (8), then

$$\int_s^t \alpha(X(\psi))\delta X(\psi)^1 = \int_s^t \alpha(X)\delta X^1 + \int_s^t \alpha^1(X_u) d\psi_u,$$

and

$$\begin{aligned} \int_s^t \alpha(X(\psi))\delta X(\psi)^2 &= \int_s^t \alpha(X)\delta X^2 + \int_s^t \alpha(X_u) \otimes \alpha(X_u) d\psi_u \\ &+ \int_{s<t_1<t_2<t} \alpha(X_{t_1}) dX^1_{t_1} \alpha^1(X_{t_2}) d\psi_{t_2} \\ &+ \int_{s<t_1<t_2<t} \alpha^1(X_{t_1}) d\psi_{t_1} \alpha(X_{t_2}) dX^1_{t_2}. \end{aligned}$$

The last three integrals appeared in the right hand side make sense as usual limits of Riemannian sums due to the fact that $\frac{2}{p} + \frac{1}{p} > 1$, see [8] and [5].

Theorem 4 (Itô's formula). *Let $X: \Delta \to T^{(2)}(V)$ be a multiplicative functional with finite p-variation controlled by ω, let $f: V \to W$ be a $Lip(\gamma)$ function, $\frac{\gamma}{p} > 1$, and let $\psi_{st} = X^2_{st} - \widetilde{X}^2_{st}$.*

1). For any $(s,t) \in \Delta$,

$$f(X_t) - f(X_s) = \int_s^t f^1(\widetilde{X})\delta \widetilde{X}^1. \tag{9}$$

2). ψ is an additive functional and satisfies (8).

3). For any $(s,t) \in \Delta$,

$$f(X_t) - f(X_s) = \int_s^t f^1(X)\delta X^1 - \int_s^t f^2(X_u) d\psi_u. \tag{10}$$

Proof. 1) and 2) are obvious, and (10) follows from (9).

4. Differential equations

In this section we consider the following differential equation,

$$\left\{ \begin{array}{rcl} dY_t & = & f(Y_t)(\delta X_t), \\ Y_0 & = & z, \end{array} \right. \tag{11}$$

where $X: \Delta \to T^{(2)}(V)$ is a minimal multiplicative functional with finite p-variation, and $f: W \to \hom(V, W)$ is a vector field. We shall assume that f is $\mathrm{Lip}(\gamma)$, $\frac{\gamma}{p} > 1$. To motivate a definition of solutions to the Eq.(11) , we first assume that X is a smooth path in V and look at multiplicative functional of degree 1. We can write Eq.(11) to be

$$\left\{ \begin{array}{rcl} Y^1_{st} & = & \int_s^t f(Y_u)dX_u, \\ Y_0 & = & z, \end{array} \right. \tag{12}$$

where $Y^1_{st} = Y_t - Y_s$. Set $X^1_{st} = X_t - X_s$, and put it together with Eq.(12), we get

$$\left\{ \begin{array}{rcl} X^1_{st} & = & \int_s^t dX_u, \\ Y^1_{st} & = & \int_s^t f(Y_u)dX_u, \\ Y_0 & = & z. \end{array} \right. \tag{13}$$

Let $Z^1_{st} = (X^1_{st}, Y^2_{st})$, and define a map $\alpha_f: V \oplus W \to \hom(V \oplus W, V \oplus W)$ by

$$\alpha_f(x, y)(\xi, \eta) = (\xi, f(z + y)\xi), \quad \forall (x, y), (\xi, \eta) \in V \oplus W.$$

Then we can rewrite Eq.(13) to be

$$Z_{st} = \int_s^t \alpha_f(Z)dZ. \tag{14}$$

Observe that Eq.(14) makes sense for multiplicative functionals of any degree.

Now we can give the definition of solutios to Eq.(11). Decompose the spaces $V \oplus W$ and $(V \oplus W)^{\otimes 2}$ into the direct sums,

$$\begin{array}{rcl} V \oplus W & = & H_1 \oplus H_2, \\ (V \oplus W)^{\otimes 2} & = & H_{20} \oplus H^1_{11} \oplus H^2_{11} \oplus H_{02}, \end{array} \tag{15}$$

where

$$\begin{array}{rcl} H_1 & = & \{(v, 0) \in V \oplus W : v \in V\} \cong V, \\ H_2 & = & \{(0, w) \in V \oplus W : w \in W\} \cong W, \end{array}$$

and

$$\begin{array}{rcl} H_{20} & = & \mathrm{span}\,\{(u, 0) \otimes (v, 0) : u, v \in V\} \\ H^1_{11} & = & \mathrm{span}\,\{(u, 0) \otimes (0, v) : u \in V, v \in W\} \\ H^2_{11} & = & \mathrm{span}\,\{(0, u) \otimes (v, 0) : u \in W, v \in V\} \\ H_{02} & = & \mathrm{span}\,\{(0, u) \otimes (0, v) : u, v \in W\} \end{array}$$

A multiplicative functional $Z: \Delta \to T^{(2)}(V \oplus W)$ is called a solution of differential equation (11) if

1). Z is of finite p-variation.

2). Let $Z_{st} = (1, Z_{st}^1, Z_{st}^2)$ and let

$$\begin{aligned} Z_{st}^1 &= (Z_{st}^{10}, Z_{st}^{01}), \\ Z_{st}^2 &= (Z_{st}^{20}, Z_{st}^{11}(1), Z_{st}^{11}(2), Z_{st}^{02}), \end{aligned} \tag{16}$$

be the decomposition corresponding to (15). Then

$$Z_{st}^{i0} = X_{st}^i, \quad i = 1, 2, \quad \forall (s,t) \in \Delta,$$

3). For any $(s,t) \in \Delta$,

$$Z_{st}^i = \int_s^t \alpha_f(Z) \delta Z^i, \quad \forall i = 1, 2.$$

We have the following (see Lyons [7])

Theorem 5. *Let $X : \Delta \to T^{(2)}(V)$ be a minimal multiplicative functional with finite p-variation controlled by ω, and let $f : W \to \mathrm{hom}(V, W)$ be a Lip(γ) vector field such that $\frac{\gamma}{p} > 1$. Then there is a unique solution Z to the differential equation (11) for any $z \in W$. Moreover Y is a multiplicative functional with finite p-variation, where $Y: \Delta \to T^{(2)}(W)$, $Y_{st} = (1, Y_{st}^1, Y_{st}^2)$, $Y_{st}^i \widehat{=} Z_{st}^{0i}$, $i = 1, 2$, (see (16)).*

We say Y is the multiplicative functional solution of Eq.(11), and $t \to Y_{0t}^1 + z$ is the path solution to Eq.(11). In this case we also use Y_t to denote $Y_{0t}^1 + z$.

The Itô's map is continuous, see [7]:

Theorem 6. *If $X(1), X(2)$ are two minimal multiplicative functionals with finite p-variation controlled by ω and*

$$|X_{st}^i(1) - X_{st}^i(2)| \le \varepsilon K_1 \omega(s,t)^{\frac{i}{p}}, \quad i = 1, 2, \quad \forall (s,t) \in \Delta,$$

then

$$|Z_{st}^i(1) - Z_{st}^i(2)| \le K(\varepsilon) K_2 \omega(s,t)^{\frac{i}{p}}, \quad i = 1, 2, \quad \forall (s,t) \in \Delta.$$

where $Z(k)$ is the solution of (11) with the driven multiplicative functional $X(k)$ and same initial $z \in W$.

We next study the perturbation of differential equation driven by a multiplicative functional. Let $Z: \Delta \to T^{(2)}(V \oplus W)$ be the solution of the Eq.(11). Then we use $\langle X, Y \rangle$ and $\langle Y, X \rangle$ to denote $Z^{11}(1)$ and $Z^{11}(2)$, respectively, so that $Z_{st} = (1, Z_{st}^1, Z_{st}^2)$,

$$\begin{aligned} Z_{st}^1 &= (X_{st}^1, Y_{st}^1), \\ Z_{st}^2 &= (X_{st}^2, \langle X, Y \rangle_{st}, \langle Y, X \rangle_{st}, Y_{st}^2). \end{aligned}$$

Let $K: \Delta \to T^{(2)}(V \oplus W)$, $K_{st} = (1, K^1_{st}, K^2_{st})$,

$$\begin{aligned} K^1_{st} &= (X^1_{st}, Y^1_{st}), \\ K^2_{st} &= \alpha_f(Z_s) \otimes \alpha_f(Z_s)(Z^2_{st}). \end{aligned}$$

Then K is an almost multiplicative functional, and $\int \alpha_f(Z)\delta Z = \widehat{K}$, so that $\widehat{K} = Z$, and $K^1_{st} = Z^1_{st}$. Using Prop.1, and the fact that

$$K^2_{st} = (X^2_{st}, 1 \otimes f(Y_s)(X^2_{st}), f(Y_s) \otimes 1(X^2_{st}), f(Y_s) \otimes f(Y_s)(X^2_{st})),$$

we obtain that

$$\begin{aligned} Y^1_{st} &= \lim_{m(D)\to 0} \textstyle\sum_l \Big\{ f(Y_{t_{l-1}})(X^1_{t_{l-1}t_l}) \\ &\quad + f^1(Y_{t_{l-1}})((1 \otimes f(Y_{t_{l-1}}))(X^2_{t_{l-1}t_l}))\Big\}. \end{aligned} \tag{17}$$

Assume we are given an additive functional $\psi: \Delta \to V^{\otimes 2}$ satisfying

$$|\psi_{st}| \le K_3 \omega(s,t)^{\frac{2}{p}}, \quad \forall (s,t) \in \Delta. \tag{18}$$

Then we also use ψ^1_{st} to denote ψ_{st}, and define a multiplicative functional with values in $T^{(2)}(V^{\otimes 2})$, also denoted by ψ, by $\psi_{st} = (1, \psi^1_{st}, \psi^2_{st})$,

$$\psi^2_{st} = \int_{s<t_1<t_2<t} d\psi_{0t_1} d\psi_{0t_2}, \quad \forall (s,t) \in \Delta,$$

where the double integral is understood in the conventional sense, i.e.

$$\psi^2_{st} = \lim_{m(D)\to 0} \sum_l \psi^1_{st_{l-1}} \otimes \psi^1_{t_{l-1}t_l}$$

which exists because $\frac{2}{p} + \frac{2}{p} > 1$. It is easily seen that ψ is of finite $\frac{p}{2}$-variation controlled by $K_4\omega$. Notice that $\frac{2}{p} + \frac{1}{p} > 1$. So we can define a multiplicative functional $U: \Delta \to T^{(2)}(V \oplus V^{\otimes 2})$ as follows. $U_{st} = (1, U^1_{st}, U^2_{st})$,

$$\begin{aligned} U^1_{st} &= (X^1_{st}, \psi^1_{st}), \\ U^2_{st} &= (X^2_{st}, \int_s^t X^1_{su} d\psi_{su}, \int_s^t \psi_{su} dX^1_{su}, \psi^2_{st}). \end{aligned}$$

Then it is easy to check that U is a geometric multiplicative functional with finite p-variation controlled by $K_5\omega$, and if $X(1), X(2): \Delta \to T^{(2)}(V)$ are two multiplicative functionals with finite p-variation controlled by ω, and

$$\left|X^i_{st}(1) - X^i_{st}(2)\right| \le \varepsilon \omega(s,t)^{\frac{i}{p}}, \quad \forall (s,t) \in \Delta,\ i = 1, 2,$$

then

$$\left|U^i_{st}(1) - U^i_{st}(2)\right| \le K_6 \varepsilon \omega(s,t)^{\frac{i}{p}}, \quad \forall (s,t) \in \Delta,\ i = 1, 2.$$

Now we further assume that ψ is anti-symmetric, and $X\colon \Delta \to T^{(2)}(V)$ is geometric, so that $X(\psi)$ is minimal. Denote by $Y(\psi)$ the unique multiplicative functional solution of the differential equation

$$\left\{\begin{array}{rcl} dY_t &=& f(Y_t)(\delta X(\psi)_t), \\ Y_0 &=& z. \end{array}\right. \tag{19}$$

If in addition X is a smooth multiplicative functional, then by (17) we have

$$\begin{aligned} Y^1_{st}(\psi) &= \lim_{m(D)\to 0} \sum_l \Big\{ f(Y(\psi)_{t_{l-1}})(X^1_{t_{l-1}t_l}) \\ &\quad + f^1(Y(\psi)_{t_{l-1}})(1\otimes f(Y(\psi)_{t_{l-1}})(X^2_{t_{l-1}t_l}))\Big\} \\ &\quad + \lim_{m(D)\to 0} \sum_l \{ f^1(Y(\psi)_{t_{l-1}})(1\otimes f(Y(\psi)_{t_{l-1}})(\psi_{t_{l-1}t_l}))\}, \\ &= \int_s^t f(Y_u(\psi))dX_u + \int_s^t f^1\circ(1\otimes f)(Y_u(\psi))d\psi_u, \end{aligned}$$

so that $Y^1(\psi)$ satisfies the following equation

$$Y^1_{st}(\psi) = \int_s^t f(Y_u(\psi))dX_u + \int_s^t f^1\circ(1\otimes f)(Y_u(\psi))d\psi_u.$$

Since

$$\begin{aligned} f(Y(\psi)_s)\otimes f(Y(\psi)_s)(X^2_{st}+\psi_{st}) &= f(Y(\psi)_s)\otimes f(Y(\psi)_s)(X^2_{st}) \\ &\quad + f(Y(\psi)_s)\otimes f(Y(\psi)_s)(\psi_{st}) \end{aligned}$$

so that

$$\begin{aligned} Y^2_{st}(\psi) &= \int_{s<t_1<t_2<t} dY_{t_1}(\psi)^1 dY_{t_2}(\psi)^2 \\ &\quad + \int_s^t f(Y(\psi)_u)\otimes f(Y(\psi)_u)d\psi_u. \end{aligned}$$

Define a vector field $\tilde{f}\colon W \to \hom(V\oplus V^{\otimes 2}, W)$ by

$$\tilde{f}(x)(\xi,\eta) = f(x)\xi + f^1(x)\circ(1\otimes f(x))(\eta), \quad \forall x\in W, \xi\in V, \eta\in V^{\otimes 2}.$$

If E is the multiplicative functional solution to the following differential equation,

$$\left\{\begin{array}{rcl} dE_t &=& \tilde{f}(E_t)(\delta U_t), \\ E_0 &=& z, \end{array}\right. \tag{20}$$

then $Y(\psi)^1_{st} = E^1_{st}$, and

$$Y(\psi)^2_{st} = E^2_{st} + \int^t f(E_u)\otimes f(E_u)(d\psi_u).$$

Theorem 7. *Let X be a geometric multiplicative functional with finite p-variation controlled by ω, $\psi: \Delta \to V^{\otimes 2}$ be an additive, anti-symmetric functional satisfying (18), i.e.*

$$|\psi_{st}| \leq K_3 \omega(s,t)^{\frac{2}{p}},$$

and let $f: W \to \hom(V, W)$ be of $Lip(\gamma)$ with $\frac{\gamma-1}{p} > 1$. Let E be the multiplicative functional solution of the following equation,

$$\begin{cases} dE_t &= \tilde{f}(E_t)(\delta U_t), \\ E_0 &= z. \end{cases} \tag{21}$$

Then

$$Y(\psi)^1_{st} = E^1_{st}, \tag{22}$$

$$Y(\psi)^2_{st} = E^2_{st} + \int_s^t f(E_u) \otimes f(E_u)(d\psi_u). \tag{23}$$

Proof. We have proved that if X is a smooth multiplicative functional, then we have (22) and (23). Now assume that X is geometric. Choosing a sequence of smooth multiplicative functionals $X(n): \Delta \to T^{(2)}(V)$ such that

$$\left|X(n)^i_{st} - X^i_{st}\right| \leq \frac{1}{n}\omega(s,t)^{\frac{i}{p}}, \quad \forall (s,t) \in \Delta, \quad i = 1, 2.$$

Denote by $E(n), E$ the multiplicative functional solutions of (20) with driven multiplicative functional $U(n)$, U, respectively. Then we have

$$\begin{aligned} \left|X(n)(\psi)^i_{st} - X^i_{st}(\psi)\right| &\leq \frac{1}{n}\omega(s,t)^{\frac{i}{p}}, \\ \left|U(n)^i_{st} - U^i_{st}\right| &\leq K_7 K\left(\frac{1}{n}\right)\omega(s,t)^{\frac{i}{p}}, \end{aligned}$$

where $K(\varepsilon)$ is a constant depending only on $\varepsilon > 0$, and $\lim_{\varepsilon \to 0} K(\varepsilon) = 0$, so that

$$\left|Y(n)(\psi)^i_{st} - Y^i_{st}(\psi)\right| \leq K_8 K\left(\frac{1}{n}\right)\omega(s,t)^{\frac{i}{p}},$$

and

$$\left|E(n)^i_{st} - E^i_{st}\right| \leq K_9 K\left(\frac{1}{n}\right)\omega(s,t)^{\frac{i}{p}}.$$

However $E(n)^1_{st} = Y(n)(\psi)^1_{st}$ and

$$E(n)^2_{st} + \int_s^t f(E(n)_u) \otimes f(E(n)_u)(d\psi_u) = Y(n)(\psi)^2_{st}$$

for any n, so that (22)and (23) hold.

Thus we can regard equation (21) as Stratonovich's form of the Itô's equation (19). By Th.7 we can make the following

Definition 2. *Let $X: \Delta \to T^{(2)}(V)$ be a geometric multiplicative functional with finite p-variation controlled by ω, ψ be an additive functional valued in $V^{\otimes 2}$ satisfying (18), and let $f: W \to \hom(V, W)$ be Lip(γ) with $\frac{\gamma-1}{p} > 1$. Let E be the multiplicative functional solution of Eq.(21). Then the multiplicative functional $Y(\psi)$ defined by (22) and (23) is called the multiplicative functional solution of the following equation,*

$$\begin{cases} dY_t &= f(Y_t)(\delta X(\psi)_t), \\ Y_0 &= z. \end{cases}$$

Acknowledgements. The first author acknowledges the support of the SERC via senior fellowship B/93/sf/445, EEC grants SC1-CT92-784 and SC1-0062. Both authors are grateful to the support of YYYY 908 SERC grant GR/J55946.

References

1. Chen, K.T.: Integration of paths, geometric invariants and generalized Baker-Hausdorff formula, Ann. of Math., 163-178, (1957).
2. Ikeda, N. & Watanabe, S.: Stochastic Differential Equations and Diffusion Processes, North-Holland Pub. Com., Amsterdam, Oxford, New York, Tokyo, (1981).
3. Ito, K.: Stochastic integral, Proc. Imp. Acad. Tokyo, 20, 519-524, (1944).
4. Ito, K.: On stochastic differential equations, Mem. Amer. Math. Soc., 4 (1951).
5. Lyons, T.: Differential equations driven by rough signals (I): An extension of an inequality of L.C.Young, Math. Research Letters 1, 451-464 (1994).
6. Lyons,T.: The interpretation and solution of ordinary differential equation driven by rough noise, Proc. of Symposia in Pure Math. vol.57, (1995).
7. Lyons,T.: Differential equations driven by rough signals, (1995).
8. Young, L.C.: An inequality of Hölder type, connected with Stieltjes integration, Acta Math. 67, 251-282, (1936).

A Martin boundary connected with the ∞-volume limit of the focussing cubic Schrödinger equation

Henry P. McKean*

Courant Institute of Mathematical Sciences, New York University, 251 Mercer Street, New York, NY 10012, USA

Summary. The existence of a change of phase in the micro-canonical ensemble for the focussing cubic Schrödinger system was suggested by Lebowitz-Rose-Speer [1989]. Chorin [1994] disputes their numerical evidence; his own is based on a more sophisticated approximation to the micro-canonical distribution and leads him to the opposite conclusion. Perhaps the source of this contradictory testimony is the fact, proved here, that *the ∞-volume limit does not exist at any temperature* $0 < T < \infty$ *or density* $0 < D < \infty$. This does not preclude a more or less dramatic change in the ensemble, from high to low temperatures, but it does guarantee the existence of several distinct ∞-volume Gibbs states. These are related to a sort of "boundary" of the type introduced by Martin [1941] for the description of classical harmonic functions in general 3-dimensional regions, as will be explained below.

1. Introduction

Martin [1941] was the first to extend Poisson's formula for classical harmonic functions from the ball to general domains $D \subset \mathbb{R}^d$. This beautiful work was ignored at the time; only after his sad early death was it appreciated as being the "right" way to think about transient diffusions in their behavior for $t \uparrow \infty$; see D. Williams [1979] for a spirited account of this. The purpose of the present note is to explain the relevance of Martin's idea to the cubic Schrödinger equation

$$\begin{aligned} \partial Q/\partial t &= -\partial^2 P/\partial x^2 \pm 4(Q^2+P^2)P \\ \partial P/\partial t &= +\partial^2 Q/\partial x^2 \mp 4(Q^2+P^2)Q \end{aligned}$$

in the *focussing* case when the *lower* signatures are taken. The (unsolved) problem posed here is to describe all the ∞-volume Gibbs states of this system, starting from the micro-canonical ensemble for a circle of large perimeter L and fixed "particle number" $N = \int_0^L (Q^2+P^2)$, and passing to the limit $L \uparrow \infty$ at fixed temperature T and fixed density $D = N/L$. I begin in art. 2 with the ∞-volume limit for the simpler defocussing case, in which the *upper* signatures are taken in the display. The focussing case was studied by Lebowitz-Rose-Speer [1989] in hopes of detecting some kind of change of

* This work was performed at the Courant Institute of Mathematical Sciences with the partial support of the National Science Foundation, under NSF Grant No. DMS–9112664 which is gratefully acknowledged.

phase; the existence of the micro-canonical ensemble is due to them. Their idea, that solitons/radiation should be favored at low/high temperatures, was supported by numerical evidence indicating a break in the "concentration" = the mean value of $\int[(Q^2+P^2)]^2/[\int(Q^2+P^2)]^2$. The validity of the evidence has been disputed by Chorin [1994] who uses a more sophisticated simulation of the micro-canonical ensemble; the outcome is moot. The connection with Martin-like boundaries is explained in art. 3. The fact is that, in the focussing case, the ∞-volume limit of the micro-canonical ensemble fails to exist at any temperature T or density D *whatever*; this may explain the inconsistency of the numerical evidence reported above. Be that as it may, it follows that, for every T and D, different (micro-canonical) Gibbs states can be produced by fixing $D = N/L$ and making $L \uparrow \infty$ in different ways. These states are members of a compact convex figure of which the extreme points form a Martin-like boundary for the partial differential operator[1]

$$\mathfrak{G} = \partial/\partial x + \frac{1}{2}(\partial^2/\partial Q^2 + \partial^2/\partial P^2) + (Q^2+P^2)^2 + (Q^2+P^2)\partial/\partial I.$$

The reason for the non-existence of the ∞-volume limit and the resulting multiplicity of Gibbs states is to be found in the competition between a) the factor $\exp[+\int(Q^2+P^2)^2 dx]$ figuring in the micro-canonical measure, favoring large displacements and b) the micro-canonical restriction $N/L = D$ which wants to rein them in. This can be resolved, for $L \uparrow \infty$, in different ways, and it is the Martin boundary that lists them. It is a challenging problem to describe all these ∞-volume Gibbs states.

2. Defocussing case

The upper signatures are taken. The flow is Hamiltonian $[Q^\bullet = \partial H/\partial P,\ P^\bullet = -\partial H/\partial Q]$ with

$$H = \frac{1}{2}\int_0^L [(Q')^2 + (P')^2]dx + \int_0^L (Q^2+P^2)^2 dx.$$

The associated petit ensemble is determined by the canonical distribution

$$e^{-H/T} d^\infty Q d^\infty P = \frac{e^{-\frac{1}{2T}\int_0^L (Q')^2}}{(2\pi 0+)^{\infty/2}} d^\infty Q \times \frac{e^{-\frac{1}{2T}\int_0^L (P')^2}}{(2\pi 0+)^{\infty/2}} d^\infty P \times e^{-\frac{1}{T}\int_0^L (Q^2+P^2)^2}$$

suitably interpreted: in detail, the first factor signifies that Q is a circular Brownian motion (CBM), i.e., it is standard Brownian motion, conditioned so that $Q(L) = Q(0)$, this common height h being distributed on the line according to the infinite measure dh. The second factor signifies

[1] The reader will excuse the use of x as a time-like parameter; it's natural to the problem.

that P is an independent copy of Q. The third factor is just a density; it spoils the independence of Q and P, but controls the total mass so that $\mathfrak{Z} = \int e^{-H/T} d^\infty Q d^\infty P < \infty$. The flow, suitably interpreted, exists in this petit ensemble and preserves it; see McKean [1994 (2)] and, for a better result, Bourgain [1994]. A more transparent expression of the petit canonical distribution is easily obtained. Take $T = 1$ for ease of writing (it plays no real rule so long as D is free), let $\wedge$ be the ground energy of the operator $\mathfrak{K} = -(1/2)(\partial^2/\partial Q^2 + \partial^2/\partial P^2) + (Q^2 + P^2)^2$, and let ψ be the associated ground state with $\int \psi^2 dQdP = 1$. Then, with $m = \text{grad}\, \psi$,

$$e^{-H} d^\infty Q d^\infty P = \text{CBM } \times \text{CBM } \times e^{\int_0^L m \bullet (dQ,dP) - \frac{1}{2}\int_0^L m^2 dx} \times e^{-\wedge L}.$$

Here, you recognize $e^{-H} d^\infty Q d^\infty P$, re-normalized to remove the nuisance factor $\exp(-\wedge L)$, as the law of the circular diffusion (Brownian motion with drift) regulated by the infinitesimal operator

$$\frac{1}{2}(\partial^2/\partial Q^2 + \partial^2/\partial P^2) + m \bullet [\text{grad } = (\partial/\partial Q, \partial/\partial P)],$$

and it comes as no surprise, nor is it difficult to prove, that as $L \uparrow \infty$ without micro-canonical restriction, this approximates the law of the stationary diffusion with free parameter $x \in R$, the same infinitesimal operator, and invariant density $\psi^2(Q)$; see McKean-Vaninsky [1994] for such matters. The imposition, by conditioning, of the micro-canonical restriction $N/L = D$ has only a small effect: $\mathfrak{K}$ is modified by the addition of a constant multiple of $(Q^2 + P^2)$, the constant being adjusted so that mean $(Q^2) = D$ for infinite volume, in conformity with Gibbs' principle of maximal entropy production; compare McKean [1994 (1)] for discussion and a model computation.

3. Focussing case

This is more difficult. The energy $H = \frac{1}{2}\int[(Q')^2 + (P')^2]$ **minus** $\int(Q^2 + P^2)^2$ is now indefinite, and the total petit canonical mass is $+\infty$. This prompted Lebowitz-Rose-Speer [1989] to impose the micro-canonical restriction $N/L = D$ with a fixed number D, independent of L and of T. The latter is fixed at $T = 1$, as before, and the micro-canonical mean value of a "short" function $F(QP)$, depending only upon $Q(x')$ and $P(x')$ for $0 \leq x' \leq$ some fixed $x < L$, is expressed as

$$M_{N/L=D}(F) \quad = \quad \frac{\int E_c \left[e^{\int_0^L Z^4} F(Z),\ Z(L) = c,\ \int_0^L Z^2 = N \right] dc}{\int E_c \left[e^{\int^L Z^4},\ Z(L) = c,\ \int_0^L Z^2 = N \right] dc}$$

$$= \int E_a \left[e^{\int_0^x Z^4} F(Z),\ Z(x) = b,\ \int_0^x Z^2 = b \right]$$
$$\times \mathfrak{Z}^{-1} P(L-x, b, a, N-I) da\, db\, dI$$

in which $E_\bullet$ is the expectation for the 2-dimensional Brownian motion starting at $\bullet$; Z is the pair QP; $Z^2 = Q^2 + P^2$; $E_a[F,\ Z = b,\ \int Z^2 = I]$ and the like represent densities such as $(\partial^3/\partial b_1 \partial b_2 \partial I) E_a[F,\ Q \le b_1,\ P \le b_2,\ \int Z^2 \le I]$; $P(x,a,b,I) = E_a \left[e^{\int_0^x Z^4},\ Z(x) = b,\ \int_0^x Z^2 = I \right]$ is the elementary solution of the $\partial p/\partial x = \frac{1}{2}\Delta p + Z^4 p - Z^2 \partial p/\partial I$ and $\mathfrak{Z} = \int p(L,c,c,N)dc$ is the bottom of the line before, normalizing things so that $M(1) = 1$. The existence of the smooth density $p > 0$ is guaranteed by the "hypo-elliptic" character[2] of the parabolic equation. The fact that $\mathfrak{Z} < \infty$ is the chief point here; see Lebowitz-Rose-Speer [1989] and, for a variant proof, McKean [1994 (2)]. Now the question is: What does this mean-value do as $L \uparrow \infty$ with fixed $N/L = D$? which is to say: How does the competition between $\exp \int Z^4$ and $\int Z^2 = N$ come out? The only thing which changes here is the ratio $\mathfrak{Z}^{-1} p(L-x, a, b, N-I)$ so that is what you have to understand.

The density $p(x,a,b,I)$ is symmetric in ab owing to the reversibility of standard Brownian paths and the fact that $\int Z^4$ and $\int Z^2$ are reversible, too. Next, you remark that the micro-canonical measure is invariant under translations of the circle $0 \le x < L$. Now it is easy to see that $\mathfrak{Z}^{-1} p(x,a,b,I) p(L-x,b,a,N-I) da\, db\, dI$ is tight for $L \uparrow \infty$ due to the micro-canonical fiat $N/L = D$: in fact, the tail of the measure is controlled by

$$\begin{aligned}
&\frac{1}{2}\mathfrak{Z}^{-1} \int p(x,a,b,I) p(L-x,b,a,N-I)(a^2+b^2) da\, db\, dI \\
&= \mathfrak{Z}^{-1} \int p(L,c,c,N) c^2 dc \\
&= M\, Z^2(0) \\
&= M\, L^{-1} \int_0^L Z^2 = N/L = D
\end{aligned}$$

and by

$$\begin{aligned}
&\mathfrak{Z}^{-1} p(x,a,b,I) p(L-x,b,a,N-I) I\, da\, db\, dI \\
&= \int_0^x M\left[Z^2(x')\right] dx' \\
&= xD,
\end{aligned}$$

and you conclude that, for nice short functions F which vanish at large values of $a = Z(0)$, $b = Z(x)$ and/or $I = \int_0^x Z^2$, and for *suitable* $L \uparrow \infty$, the mean value will have a limit of the form

[2] $[\partial/\partial Q, [\partial/\partial Q, Z^2 \partial/\partial I]] = 2\partial/\partial I$; see Krylov [1987] for such matters.

$$M_\infty(F) = \int E_a \left[e^{\int_0^x Z^4} F(Z),\ Z(x) = b,\ \int_0^x Z^2 = I \right] h(x,a,b,I) da\, db\, dI,$$

in which $h(x,a,b,I)$ is the formal density of the 5-dimensional measure determined by the weak limit of $\mathfrak{Z}^{-1}p(L-x,a,b,N-I)da\,db\,dI$ as $L \uparrow \infty$ in its special way. This function is now the object of study. It is a solution 1) $0 = \partial h/\partial x + \frac{1}{2}\Delta h + c^4 h + c^2 \partial h/\partial I \equiv \mathfrak{H}h$ relative to $c = a$ or b, and so also smooth in all its variables $x > 0$, $ab \in R^4$, $I > 0$, as per the last footnote. It is also 2) symmetric in ab like p, and 3) it is positive. It is only 3) that needs little proof.

$$p(L-x,a,b,N-I) = \int p(L-x-x',a,b',N-I-I')p(x',b',b,I')db'\,dI',$$

is divided by $\mathfrak{Z}$ and the limit $L \uparrow \infty$ is taken to produce

$$h(x,a,b,I) \geq \int h(x+x',a,b',I+I')p(x',b',b,I')db'\,dI',$$

first as formal densities and then in the naive sense, so that if $h(x_0 a_0, b_0, I_0)$ vanishes for some $x_0 > 0$, $a_0\, b_0 \in R^4$, and $I_0 \geq 0$, then $h(x,a_0,b,I) = 0$ for any $x \geq x_0$, $b \in R^2$, and $I \geq I_0$ in view of $p > 0$. But a_0 can be replaced by any $a \in R^2$, by the symmetry of h inherited from p and a reprise, so

$$xD = M\left[\int_0^x Z^2\right] = \int ph(x,a,b,I)I\, da\, db\, dI \leq I_0,$$

which is contradictory for $x \uparrow \infty$.

The *micro-canonical functions* for the operator $\mathfrak{H} = \partial/\partial x + (1/2)\Delta + c^4 + c^2\partial/\partial I$ are now declared to be those functions $h(x;a,b,I)$ defined for $x > 0$, $ab \in R^4$, and $I > 0$ which satisfy 1), 2), 3), and the natural normalization, 4) $\int ph\, da\, db\, dI = 1$. This function class is convex and (hopefully) compact in any reasonable sense you like;[3] its extreme points $\mathfrak{E}$ comprise the *micro-canonical boundary* for $\mathfrak{H}$. It has been proved that every ∞-volume micro-canonical Gibbs state arises from such a micro-canonical function h; naturally, there might be only one of these for a fixed value of D, but the next art. 4 will show that, in fact, they abound. I don't know if the ∞-volume Gibbs states account for them all.

4. Non-existence of the ∞-volume limit

This means that the full limit, as $L \uparrow \infty$, of $M_{N/L=D}$ does not exist. The proof is by contradiction: if the full limit did exist, then you would have

[3] Krylov [1987] covers this.

$h(x,a,b,I) \geq \exp(-\vartheta I)$ for $I \uparrow \infty$ with a constant ϑ depending upon x, a, b alone, contradicting 4): [4]

$$1 = \int ph\, da\, db\, dI \geq \int da\, db\, E_a\left[e^{\int_0^x Z^4} \times e^{-\vartheta \int_0^x Z^2}\right] = +\infty.$$

Proof. The assumption is that the full limit $h(x,a,b,I)da\,db\,dI = \lim_{L\uparrow\infty} 3^{-1}p(L-x,a,b,N-I)da\,db\,dI$ exists in the weak topology of measures for each $x > 0$, without restriction on the mode of increase of L to ∞. Now, in the language of formal densities,

$$\begin{aligned}
& h(x,a,b,I) \\
= \ & \lim_{L\uparrow\infty} 3^{-1}(L')p(L'-x,a,b,N'-I) \\
& \qquad\qquad \text{with } L' = L - x_0 \text{ and } N' = DL' = N - Dx_0 \\
= \ & \lim_{L\uparrow\infty} 3^{-1}(L)p(L-x-x_0,a,b,N-I-Dx_0) \times 3(L)3^{-1}(L-x_0) \\
= \ & h(x+x_0,a,b,I+Dx_0) \times \lim_{L\uparrow\infty} 3(L)3^{-1}(L-x_0),
\end{aligned}$$

in which a) the final limit $\equiv e(x_0)$ exists by itself, h being positive; b) $e(x_1)e(x_2) = e(x_1+x_2)$ so that $e(x_0)$ is an exponential $\exp(\vartheta x_0)$; and c) $h(x,a,b,I) = h(x+x_0,a,b,I+Dx_0)\exp(\vartheta x_0)$ provided $x + x_0$ and $I + Dx_0$ are positive. Let T be the "loop time" for the 2-dimensional Brownian motion $Z(x) : x \geq 0$ starting at $Z(0) = a$, to return to the "inner" circle with center at a and radius 1, *via* the "outer" circle of radius 2, and use c) and

$$p(L-x,a,b,N-I) \geq E_a\left[e^{\int_0^T Z^4} p\left(L-x-T, Z(T), b, N-I-\int_0^T Z^2\right)\right]$$

to estimate h from below: with $m(I)$ = the minimum of $h(x,\bullet,b,I)$ on the (solid) inner disc, $n \geq 3$, $n \leq I \leq n+1$, and $m_n = \min m(I)$ on that interval, you find

[4] The proof of the divergence is easy: if $B(x) : x \geq 0$ is the standard 1-dimensional Brownian motion, then $P_0\left[\int_0^1 B^2 > I\right]$ exceeds the probability that $B(x)$ hits $\sqrt{2I}+1$ at $x = x_0 \leq 1/2$ and does not return to $\sqrt{2I}$ before $x_0 + 1$, which is more than

$$\begin{aligned}
& \int_0^{1/2} (2\pi t^3)^{-1/2} e^{-I/t}dt \times \int_1^{\infty} (2\pi t^3)^{-1/2} e^{-1/2t}dt \\
& \geq \text{constant} \times e^{-\vartheta I}, \quad \text{and so forth.}
\end{aligned}$$

$$
\begin{aligned}
& h(x,a,b,I) \\
\geq\ & E_a\left[h(x+T,Z(T),b,I+\int_0^T Z^2)\right] \\
\geq\ & E_a\left[h(x,Z(T),b,I+\int_0^T Z^2-DT)e^{-\vartheta T}, DT<I+\int_0^T Z^2\right] \\
\geq\ & E_a\left[m(I+\int_0^T Z^2-DT)e^{-\vartheta T}, -2\leq\int_0^T Z^2-DT\leq -1\right] \\
\geq\ & E_a\left[e^{-\vartheta T}, -2\leq\int_0^T Z^2-DT\leq -1\right]\times \text{ the smaller of } m_{n-2} \text{ and } m_{n-1},
\end{aligned}
$$

so that m_n is underestimated by the nth power of a fixed so that m_n is underestimated by the nth power of a fixed positive constant. The rest will be plain.

5. Statistics in a Gibbs state

Let $h(x,a,b,I)$ be a micro-canonical function as in art. 3. The statistics of the associated ∞-volume Gibbs state are described by a shift-invariant probability measure on paths $Z: x\in R\to R^2$ with finite-dimensional densities

$$
\begin{aligned}
& P^h[Z(x_0)=a_0,\ Z(x_1)=a_1,\ldots,Z(x_n)=a_n] \\
&= \int p(x_1-x_0,a_0,a_1,I_1)p(x_2-x_1,a_1,a_2,I_2)\cdots p(x_n-x_{n-1},a_{n-1},a_n,I_n) \\
&\times h(x_n-x_0,a_0,a_n,I_1+I_2+\cdots+I_n)d^nI,
\end{aligned}
$$

the prescription being consistent because $M[Z(0)=a,\ Z(x)=b,\ \int_0^x Z^2=I]$ can be expressed, indifferently, either as $ph(x,a,b,I)$ or as

$$
p(x,a,b,I)\times\int p(x',b,b',I')h(x+x',a,b',I+I')db'\,dI',
$$

which is to say that the 2nd factor is the same as $h(x,a,b,I)$ almost everywhere relative to $da\,db\,dI$.

Remark 1. The process is reversible because p and h are symmetric in ab.

Remark 2. The pair $(Z,\int_0 Z^2)$ is *never Markovian*, contrary to expectation; compare art. 2.

Proof.

$$P^h\left[Z(x)=b,\ \int_0^x Z^2 = I|Z(0)=a,\ Z(-x_0),\ \int_{-x_0}^0 Z^2\right]$$
$$= p(x,a,b,I)\frac{h\left(x_0+x, Z(-x_0), b, \int_{-x_0}^0 Z^2+I\right)}{h\left(x_0, Z(-x_0), a, \int_{-x_0}^0 Z^2\right)}$$

and

$$P^h\left[Z(x)=b,\ \int_0^x Z^2 = I|Z(0)=a\right] = p(x,a,b,I)\frac{h(x,a,b,I)}{h(0,a,a,0)}$$[5]

cannot agree unless $h(x,a,b,I)$ is exponential in I (and more besides), and that cannot be if $\int ph\, dI < \infty$.

Remark 3. The process is *metrically transitive* if and only if h belongs to the micro-canonical boundary $\mathfrak{E}$.

Proof. Let $\mathfrak{F}(L)$ be the field of $Z(x)$ and $\int_0^x Z^2$ for $x \geq L$.

$$P^h\left[Z(x_0)=a,\ Z(x_0+x)=b,\ \int_{x_0}^{x+x_0} Z^2 = I|Z(0),\ \mathfrak{F}(L)\right]$$
$$= \frac{\int p(x_0, Z(0), a, L')p(x,a,b,I)p\left(L-x-x_0, b, Z(L), \int_0^L Z^2 - I - I'\right)dI'}{P\left(L, Z(0), Z(L), \int_0^L Z^2\right)}$$

is averaged over $0 \leq x_0 \leq L-x$ and L is taken to $+\infty$. The individual ergodic theorem and the martingale theorem, combined, show that the left side approximates

$$P^h\left[Z(0)=a,\ Z(x)=b,\ \int_0^x Z^2 = I|\mathfrak{J}\right] \equiv ph^*(x,a,b),$$

$\mathfrak{J}$ being the field of shift-invariant events, and inspection of the right side shows that h^* is a micro-canonical function. But the mean-value of ph^* is $P^h[Z(0)=a,\ Z(x)=b,\ \int_0^x Z^2 = I] = ph$, and now you realize that $h^* = h$ if the latter belongs to $\mathfrak{E}$, which is to say $Z(0)$, $Z(x)$, and $\int_0^x Z^2$ are independent of $\mathfrak{J}$. The proof is finished by a self-evident reprise.

[5] $h(0,a,a,0) = \int ph(x,a,b,I)db\, dI$ is the density for $Z(0)=a$.

6. The true Martin boundary

This can be viewed as a list $\mathfrak{M}$ of the *minimal* solutions $m(x,a,I) \geq 0$ of $\mathfrak{H}m = 0$. The adjective means that a positive solution $h(x,a,I)$ which is majorized by m is a multiple of m, and it is a fact that *any* positive solution is a center of gravity of $m \in \mathfrak{M} : h(x,a,I) = \int m(x,a,I)de(m)$ with a non-negative mass distribution $e \in C^*(\mathfrak{M})$.

Remark 1. The micro-canonical functions $h(x,a,b,I)$ are *never minimal* in either variable a or b: if $h(\bullet,\bullet,b,\bullet)$ is minimal for fixed $b \in R^2$, then

$$h(x,a,b,I) \geq \int h(x+x',a,b',I+I')p(x',b',b,I')db'\,dI'$$

implies $h(x+x',a,b,I+I') = k(x',I')h(x,a,b,I)$ so that $h(x,a,b,I)$ is an exponential function of I, contradicting $\int ph\,dI < \infty$.

Remark 2. Translation by x and by I act on $\mathfrak{M}$, prompting the conjecture that $\dim \mathfrak{M} = 2$, but I have no real evidence.

Nothing more is known to me.

References

1. Bourgain, J.: The Fourier transform restriction phenomena for certain lattice subsets and applications to non-linear evolution equations. IAS preprint (1994), to appear 1995.
2. Chorin, A.: Univ. CA, Berkeley preprint (1994).
3. Krylov, N. V., *Non-Linear Elliptic and Parabolic Equations of Second Order*, Riedel, Boston, Tokyo, 1987.
4. Lebowitz, J., H. Rose, and E. Speer : Statistical mechanics of the non-linear Schrödinger equation (2). *J. Stat. Phys.* **54** (1989), 17–56.
5. Martin, R.: Minimal positive harmonic functions. *TAMS* **49** (1941), 137–164.
6. McKean, H. P.: Brownian motion with restoring drift: the micro-canonical ensemble. *Comm. Math. Phys.* **160** (1994), 615–630.
7. McKean, H. P. and K. Vaninsky: Statistical mechanics of non-linear wave equations: the petit and micro-canonical ensembles. *Trends and Perspective Appl. Math.*, ed. L. Sirovich, Springer-Verlag, New York, 1994.
8. ———: Statistical mechanics of non-linear wave equations (4): cubic Schrödinger. *Comm. Math. Phys.* **168** (1995), 479–491.
9. Williams, D.: *Diffusions, Markov Processes, and Martingales.* J. Wiley & Sons, New York, 1979.

The true Martin boundary

[illegible]

Diffusion processes on an open time interval and their time reversal

Masao Nagasawa and Thomas Domenig

Institut für Mathematik, Universität Zürich, Winterthurerstrasse 190, CH-8057 Zürich, Switzerland

Summary. To discuss time reversal of (Schrödinger's) diffusion processes, which are in general time-inhomogeneous, they must be defined on a closed time interval $[a, b]$, $-\infty < a < b < \infty$, because prescribed initial and terminal distributions μ_a and μ_b at $t = a$, and b, respectively, are involved. If a diffusion process is given only on an open time interval (a, b), we must first consider the process on a closed time interval $[a', b']$, $a < a' < b' < b$, and then analyze the limiting behaviour of the process as $a' \downarrow a$, and $b' \uparrow b$. This requires closer analysis of stochastic differential equations in connection with time reversal. In this context, a Skorokhod problem with singular drift is discussed.

1. Diffusion processes on an open time interval

Let D be a space-time open domain in $(a, b) \times \mathbb{R}^d$. We denote by ∂D the spatial boundary of the domain. Let $p(s, x; t, y)$, $(s, x), (t, y) \in D \cup \partial D$, $s < t$, be a transition probability density defined on $D \cup \partial D$. In terms of the transition probability $P(s, x; t, dy) = p(s, x; t, y)\, dy$, we can define a (Schrödinger's)[1] diffusion process on a closed time interval $[a', b']$, $a < a' < b' < b$. However, it is not clear if the given transition probability determines well a diffusion process on the closed time interval $[a, b]$. In fact, the existence (and uniqueness) of

$$\lim_{(s,x)\to(a,z)\in \bar{D}} P(s, x; t, B) \tag{1.1}$$

is not evident in general.

We begin with a simple but typical example. Let $p(s, x; t, y)$ be the transition probability density of a one-dimensional Brownian motion, i.e.,

$$p(s, x; t, y) = \frac{1}{\sqrt{2\pi(t-s)}} \exp\left(-\frac{|y-x|^2}{2(t-s)}\right), \quad 0 \le s < t,$$

and let $\phi(t, x)$ be a space-time harmonic function given by

$$\phi(t, x) = \frac{1}{\sqrt{2\pi t}} \exp\left(\frac{x^2}{2t}\right), \quad t > 0,$$

which satisfies

[1] Schrödinger's processes have prescribed initial and terminal distributions, for details cf. Nagasawa [7]

$$\frac{\partial \phi}{\partial t} + \frac{1}{2}\frac{\partial^2 \phi}{\partial x^2} = 0.$$

In terms of the Brownian transition density $p(s,x;t,y)$ and the space-time harmonic function $\phi(t,x)$, we can construct a (Schrödinger's) diffusion process on a time interval $[\varepsilon, b]$, $0 < \varepsilon < b$, which has the transition probability density

$$q(s,x;t,y) = \frac{1}{\phi(s,x)} p(s,x;t,y)\phi(t,y) = \frac{\sqrt{s}}{\sqrt{2\pi(t-s)t}} \exp\left(-\frac{s|y - \frac{t}{s}x|^2}{2(t-s)t}\right).$$

This is the fundamental solution of a diffusion equation

$$\frac{\partial u}{\partial s} + \frac{1}{2}\frac{\partial^2 u}{\partial x^2} + \frac{x}{s}\frac{\partial u}{\partial x} = 0, \quad (s,x) \in [\varepsilon, b] \times \mathbb{R}. \tag{1.2}$$

By $Q(s,x;t,dy) = q(s,x;t,y)\,dy$ we define a transition probability:

$$Q(s,x;t,dy) = \frac{\sqrt{s}}{\sqrt{2\pi(t-s)t}} \exp\left(-\frac{s|y - \frac{t}{s}x|^2}{2(t-s)t}\right) dy. \tag{1.3}$$

Lemma 1.1. *The transition probability $Q(s,x;t,B)$ given in (1.3) cannot be extended to the closed time-interval $[0,b]$, namely the set of probability measures $\{\, Q(s,x;t,dy) : s \in (0,\varepsilon] \,\}$ is not tight, and hence*

$$\lim_{s \downarrow 0} Q(s,x;t,B) \tag{1.4}$$

does not exist.

Proof. For any non-negative continuous function f of compact support, we have

$$\int Q(s,x;t,dy) f(y) \le \frac{\sqrt{s}}{\sqrt{2\pi(t-s)t}} \int f(y)\,dy,$$

which vanishes as $s \downarrow 0$. Therefore, $\{\, Q(s,x;t,dy) : s \in (0,\varepsilon] \,\}$ is not tight, and hence the limit (1.4) does not exist. This completes the proof. □

As will be seen in Section 5, if there is a boundary, we can start from the origin (not in a unique way, in general), since $\{\, Q(s,x;t,dy) : s \in (0,\varepsilon] \,\}$ will turn out to be tight in this case.

On the other hand we can analyze the (Schrödinger's) diffusion process path-wise as a solution of a stochastic differential equation

$$X_t = x + B_t - B_s + \int_s^t \frac{X_r}{r}\,dr, \quad 0 < s < t, \tag{1.5}$$

where B_t is a one-dimensional Brownian motion. If we let $s \downarrow 0$ and $x \downarrow 0$ in (1.5), then we get a new stochastic differential equation

$$X_t = B_t + \int_0^t \frac{X_r}{r}\, dr. \tag{1.6}$$

Since the diffusion process with singular drift x/t (cf. equation (1.2)) cannot start from the origin, solutions of equation (1.6) describe something which is not a diffusion process. As a matter of fact, the right way of reading equation (1.6) is with time t reversed, running backward decreasing toward 0 ($t \downarrow 0$). To see this we look at time reversal of (Schrödinger's) diffusion processes.

2. Time-reversed (Schrödinger's) diffusion processes

We consider a (Schrödinger's) diffusion process $\{X_t, t \uparrow\in [0,b], Q\}$ in $\mathbb{R}^d$ with drift $\mathbf{a}(t,x)$, and its *time reversal* $\{X_t, t \downarrow\in [0,b], Q\}$ with drift $\hat{\mathbf{a}}(t,x)$, where $t \uparrow\in [0,b]$ indicates that time runs *normally* from 0 to b, while $t \downarrow\in [0,b]$ means that time runs *backward* from b to 0. We assume, for simplicity, the diffusion coefficient $\sigma^2 = 1$. We have then the duality relation of time reversal

$$\mathbf{a}(t,x) + \hat{\mathbf{a}}(t,x) = \nabla(\log \mu(t,x)), \tag{2.1}$$

where $\mu(t,x) = Q[X_t \in dx]/dx$ (cf. Nagasawa [7]). The (Schrödinger's) diffusion process can be obtained as a solution of Itô's stochastic differential equation

$$X_t = X_0 + B_t + \int_0^t \mathbf{a}(r, X_r)\, dr, \tag{2.2}$$

where B_t is a d-dimensional Brownian motion, and X_0 is an initial value which is independent of the Brownian motion B_t. Since stochastic differential equations of Itô are defined in normal time evolution, namely the time parameter must increase, we introduce a *new time-parameter reversed from (fixed)* t, to discuss the time reversal of a Schrödinger process in terms of a stochastic differential equation. Namely, we define the time-reversed process with increasing time parameter s by

$$\hat{X}_s = X_{t-s}, \quad (\text{with } s \uparrow). \tag{2.3}$$

Then $\hat{X}_s$ satisfies a stochastic differential equation

$$\hat{X}_s = \hat{X}_0 + \beta_{t-s} - \beta_t + \int_0^s \hat{\mathbf{a}}(t-u, \hat{X}_u)\, du, \tag{2.4}$$

where β_s is a Brownian motion, and $\hat{X}_0$ is independent of $\beta_{t-s} - \beta_t$, $\forall s \in [0,t]$. We should pay attention to that the Brownian motion β_s is different from B_t which appears in equation (2.2) and will be determined in Theorem 2.2 below.

Theorem 2.1. *The time reversal of the diffusion process X_s (with $s \downarrow$) can be represented as*

$$X_s = X_t + \beta_s - \beta_t + \int_s^t \hat{\mathbf{a}}(r, X_r)\, dr, \quad (\text{with } s \downarrow) \tag{2.5}$$

where s runs backward toward 0, and X_t is independent of $\beta_t - \beta_s$, $s \in [0, t]$. Moreover, equation (2.5) yields, with $s = 0$,

$$X_t = X_0 + \beta_t - \int_0^t \hat{\mathbf{a}}(r, X_r)\, dr, \quad (\text{with } t \downarrow)$$

where t runs backward.

Proof. Substituting (2.3) in equation (2.4), we have

$$X_{t-s} = X_t + \beta_{t-s} - \beta_t + \int_{t-s}^t \hat{\mathbf{a}}(r, X_r)\, dr, \quad (\text{with } s \uparrow) \tag{2.6}$$

and then replacing $t - s$ by the normal parameter $s \downarrow$ in (2.6), we have equation (2.5). This completes the proof. □

Theorem 2.1 is a special case of a theorem on the time reversal of diffusion processes. It is important to remark that we can look at a Schrödinger process $\{X_t, t \in [a, b], Q\}$ forward with increasing t, and backward with decreasing t (cf. [7, chapter III]). Then, with increasing (reversed) time parameter t, we define

$$\hat{X}_t = X_{a+b-t}. \tag{2.7}$$

Let us call the process $\{\hat{X}_t, t \uparrow\in [a, b], Q\}$ "the (first) time-reversed process" of $\{X_t, t \uparrow\in [a, b], Q\}$. Since t increases, we can express $\hat{X}_t$ in terms of a stochastic differential equation of Itô. Now we apply the "time reversal" introduced in (2.7) to $\hat{X}_t$ but with decreasing $t \downarrow\in [a, b]$. Then we have a diagram:

$$\begin{aligned} &X_t \ (\text{with } t, a \uparrow b) \ \rightarrow \hat{X}_t = X_{a+b-t} \ (\text{with } t, a \uparrow b) \\ &\hat{\hat{X}} = \hat{X}_{a+b-t} \ (\text{with } t, b \downarrow a) \leftarrow \hat{X}_t \ (\text{with } t, a \uparrow b) \end{aligned} \tag{2.8}$$

We call the process $\{\hat{\hat{X}}_t, t \downarrow\in [a, b], Q\}$ *the second time-reversed process of* X_t. Actually $\hat{\hat{X}}_t = X_t$, for each $t \in [a, b]$, but the time parameters of $\hat{\hat{X}}_t$ and X_t run in the opposite directions. In fact, (2.8) implies that we get the second time-reversed process $\hat{\hat{X}}_t$, if we trace the first time-reversed process $\hat{X}_t$ with the normal time parameter but backward; i.e. the second time-reversed process $\hat{\hat{X}}_t$ is the time-reversed Kolmogorov representation with $t, b \downarrow a$. Through this the stochastic differential equation satisifed by X_t (with $t \downarrow$) is obtained, which is a backward stochastic differential equation, and is not of Itô's one.

Applying the above argument, we shall prove a theorem on (second) time reversal. We denote $\mathbf{a}(t,x)\colon [a,b]\times\mathbb{R}^d \to \mathbb{R}^d$, $\sigma(t,x)\colon [a,b]\times\mathbb{R}^d \to \mathbb{R}^d\times\mathbb{R}^d$, and consider a diffusion process $\{X_t,\ t \uparrow\in [a,b],\ Q\}$ which is determined by $A = \Delta/2 + \mathbf{a}(t,x)\cdot\nabla$, with the Laplace–Beltrami operator

$$\Delta = \frac{1}{\sqrt{\sigma_2(t,x)}}\frac{\partial}{\partial x^i}\left(\sqrt{\sigma_2(t,x)}((\sigma\sigma^T(t,x))^{ij}\frac{\partial}{\partial x^j}\right),$$

where $\sigma_2 = |(\sigma\sigma^T)_{ij}|$. Then the time-reversed Kolmogorov representation has the drift field $\hat{\mathbf{a}}(t,x)$ which is determined by the duality relation

$$\mathbf{a}(t,x) + \hat{\mathbf{a}}(t,x) = \sigma\sigma^T\nabla(\log\mu(t,x)), \tag{2.9}$$

where $\mu(t,x)\sqrt{\sigma_2(t,x)}\,dx = Q[X_t \in dx]$ (cf. [7, chapter III]). Let us assume that the diffusion process $\{X_t,\ t\uparrow\in[a,b],\ Q\}$ is determined by a stochastic differential equation

$$X_t = X_a + \int_1^t \{\mathbf{a}(s,X_s) + \mathbf{a}_\sigma(s,X_s)\}\,ds + \int_a^t \sigma(s,X_s)\,dB_s, \tag{2.10}$$

with

$$\mathbf{a}_\sigma(t,x)^j = \frac{1}{2}\frac{1}{\sqrt{\sigma_2(t,x)}}\frac{\partial}{\partial x^i}\big(\sqrt{\sigma_2(t,x)}(\sigma\sigma^T(t,x))^{ij}\big).$$

where B_t, $t\in[a,b]$, is a d-dimensional Brownian motion with $B_a = 0$, and the initial value X_a is independent of the Brownian motion. The correction term $\mathbf{a}_\sigma(t,x)$ appears in equation (2.10), because we adopted the Laplace–Beltrami operator to obtain the duality relation (2.9).

Theorem 2.2.[2] *Let $\{X_t,\ t\uparrow\in[a,b],\ Q\}$ be a diffusion process on $\mathbb{R}^d$ in the Kolmogorov representation, which is determined by a stochastic differential equation* (2.10)*. Then we have the following.*

(i) *The diffusion process in the time-reversed Kolmogorov representation satisfies the backward stochastic differential equation*

$$X_t = X_b + \int_t^b \{\hat{\mathbf{a}}(s,X_s) + \mathbf{a}_\sigma(s,X_s)\}\,ds - \int_t^b \sigma(s,X_{s}+)\,d\beta_s, \tag{2.11}$$

where $\int_t^b \sigma(s,X_s+)\,d\beta_s = \lim \Sigma_k\sigma(s_k,X_{s_k})(\beta_{s_k}-\beta_{s_{k-1}})$ denotes the backward stochastic integral, t runs backward from b to a, β_t is a d-dimensional Brownian motion with $\beta_a = 0$, which will be given in (2.13)*, and the terminal value X_b is independent of the Brownian motion β_t. The drift field $\mathbf{a}(t,x)$ and the drift field $\hat{\mathbf{a}}(t,x)$ satisfy the duality relation* (2.9).

(ii) *Moreover, X_t satisfies another stochastic differential equation*

$$X_t = X_a - \int_a^t \{\hat{\mathbf{a}}(s,X_s) + \mathbf{a}_\sigma(s,X_s)\}\,ds + \int_a^t \sigma(s,X_{s}+)\,d\beta_s, \tag{2.12}$$

[2] Cf. Meyer (preprint)

where X_a is, in general, not independent of the Brownian motion β_t.

(iii) *The d-dimensional Brownian motion β_t in equations (2.11) and (2.12) is given, with $\mathbf{c} = \{c^i\}$, $c^i = \Sigma_{k,j}(\nabla_k \sigma^i_j)\sigma^k_j$, through*

$$\beta_t = B_t + \int_a^t \{\sigma^T \nabla \log \mu + \sigma^{-1}(2\mathbf{a}_\sigma - \mathbf{c})\}(s, X_s)\, ds, \tag{2.13}$$

where B_t is the d-dimensional Brownian motion in equation (2.10).

Proof. In terms of a d-dimensional Brownian motion β_t, we set $\hat{\beta}_t = \beta_b - \beta_{a+b-t}$. The duality relation (2.9) implies that the time-reversed process $\hat{X}_t$ is determined by a stochastic differential equation

$$\hat{X}_t = \hat{X}_a + \int_a^t \hat{\mathbf{a}}_0(a+b-r, \hat{X}_r)\, dr + \int_a^t \sigma(a+b-r, \hat{X}_r)\, d\hat{\beta}_r,$$

where $\hat{X}_a = X_b$ is independent of the Brownian motion $\{\hat{\beta}_t : t \in [a,b]\}$, and we denote $\hat{\mathbf{a}}_0(t,x) = \hat{\mathbf{a}}(t,x) + \mathbf{a}_\sigma(t,x)$. Since $\hat{X}_{a+b-s} = X_s$, and

$$\begin{aligned} \int_a^t \sigma(a+b-r, \hat{X}_r)\, d\hat{\beta}_r &= -\int_{a+b-t}^b \sigma(s, X_{s+})\, d\beta_s \\ &= -\int_{a+b-t}^b \{\sigma(s, X_s)\, d\beta_s + \mathbf{c}(s, X_s)\, ds\}, \end{aligned} \tag{2.14}$$

we have

$$\hat{X}_t = \hat{X}_a + \int_{a+b-t}^b \hat{\mathbf{a}}_0(s, X_s)\, ds - \int_{a+b-t}^b \sigma(s, X_{s+})\, d\beta_s. \tag{2.15}$$

Substituting $a+b-t$ in place of t in equation (2.15), we have equation (2.11), since $\hat{X}_a = X_b$ and $\hat{X}_t = \hat{X}_{a+b-t} = X_t$. Equation (2.12) follows immediately from equation (2.11), with a (resp. t) in place of t (resp. b). Finally substituting the duality relation (2.9) and formula (2.14) in equation (2.12), we have

$$\begin{aligned} X_t = X_a &+ \int_a^t \{\mathbf{a}(s, X_s) + \mathbf{a}_\sigma(s, X_s)\}\, ds \\ &+ \int_a^t \{\sigma(s, X_s)\, d\beta_s + (\mathbf{c} - 2\mathbf{a}_\sigma - \sigma\sigma^T \nabla \log \mu)(s, X_s)\, ds\}. \end{aligned} \tag{2.16}$$

A comparison of (2.16) with (2.10) yields (2.13). This completes the proof.□

As a corollary we have Theorem 2.1, namely,

Theorem 2.3. *If X_t is a solution of*

$$X_t = X_a + \sigma B_t + \int_a^t \mathbf{a}(s, X_s)\, ds,$$

where B_t is a d-dimensional Brownian motion and σ = constant, then it satisfies

$$X_t = X_a + \sigma\beta_t - \int_a^t \hat{\mathbf{a}}(s, X_s)\, ds, \tag{2.17}$$

where β_t is given by

$$\beta_t = B_t + \int_a^t \sigma\nabla \log \mu(s, X_s)\, ds.$$

Theorem 2.3 (or 2.1) applied to a Brownian motion yields

Theorem 2.4. *Let B_t be a d-dimensional Brownian motion. Then it solves a stochastic differential equation*

$$X_t = \beta_t + \int_0^t \frac{X_r}{r}\, dr, \tag{2.18}$$

where β_t is a d-dimensional Brownian motion given by

$$\beta_t = B_t - \int_0^t \frac{B_r}{r}\, dr. \tag{2.19}$$

Proof. Since $B_0 = 0$, $\mathbf{a}(t,x) \equiv 0$, and $\mu(t,x) = (2\pi t)^{-d/2} \exp(-|x|^2/2t)$, the duality formula (2.1) yields

$$\hat{\mathbf{a}}(t,x) = -\frac{x}{t}.$$

Therefore, with (2.17) applied to B_t, we have

$$B_t = \beta_t + \int_0^t \frac{B_r}{r}\, dr, \tag{2.20}$$

which shows that B_t solves (2.18), and β_t is given by (2.19). □

Remark. Let $\mathcal{F}_t^\beta$ and $\mathcal{F}_t^B$ be the natural filtrations of β_t and B_t, respectively. Then, the representation (2.19) shows that β_t is an $\mathcal{F}_t^B$-semimartingale, and B_t is its martingale part in the Doob–Meyer decomposition of β_t. Therefore, $\mathcal{F}_t^\beta$ is strictly smaller than $\mathcal{F}_t^B$, because of the uniqueness of the decomposition.[3] Therefore, B_t is not an adapted (Markov) solution of (2.18). All solutions of equation (2.18) will be determined in the next section.

[3] Cf. Meyer (preprint)

3. A Theorem of Jeulin–Yor

For solutions of equation (2.18) we can give another representation, in which the singular drift term disappears. We prepare a simple lemma.

Lemma 3.1. *Let Y_t and β_t be continuous functions on $[0,\infty)$ which vanish at $t=0$. If*

$$Y_t = -t\int_t^\infty \frac{d\beta_s}{s} + tY, \tag{3.1}$$

with a constant Y, then

$$\beta_t = Y_t - \int_0^t \frac{Y_s}{s}\,ds, \tag{3.2}$$

where the integrals are assumed to be well-defined. Conversely, equation (3.1) follows from equation (3.2) if there exists

$$Y = \lim_{r\to\infty} \frac{Y_r}{r}. \tag{3.3}$$

Remark. We define, if necessary, such an integral appeared in (3.1) by the right-hand side of

$$\int_t^r \frac{d\beta_s}{s} = \frac{\beta_r}{r} - \frac{\beta_t}{t} + \int_t^r \frac{\beta_s}{s^2}\,ds \tag{3.4}$$

through the integration by parts formula as for the Wiener integral, which coincides with Itô's stochastic integral for Brownian motions. We require

$$\lim_{r\to\infty} \int_t^r \frac{d\beta_s}{s}$$

is well-defined in (3.1), i.e.

$$\lim_{r\to\infty} \frac{\beta_r}{r}, \quad \text{and} \quad \lim_{r\to\infty} \int_t^r \frac{\beta_s}{s^2}\,ds$$

exist, and also in (3.2)

$$\lim_{r\downarrow 0} \int_r^t \frac{Y_s}{s}\,ds$$

exists, but the absolute integrability is not assumed.

Proof of Lemma 3.1. We remark first of all

$$\begin{aligned}\int_0^t ds \int_s^\infty \frac{d\beta_r}{r} &= \int_0^t ds \int_s^t \frac{d\beta_r}{r} + \int_0^t ds \int_t^\infty \frac{d\beta_r}{r} \\ &= \beta_t + t\int_t^\infty \frac{d\beta_r}{r}.\end{aligned} \tag{3.5}$$

Substituting (3.1) in the right-hand side of equation (3.2), we have

$$\begin{aligned} Y_t - \int_0^t \frac{Y_s}{s} ds &= -t\int_t^\infty \frac{d\beta_s}{s} + tY + \int_0^t ds \int_s^\infty \frac{d\beta_r}{r} - \int_0^t Y ds \\ &= \beta_t, \end{aligned}$$

because of (3.5). Thus we have (3.2). Conversely, assume (3.2). Then, through (3.4) applied to Y_t (integration by parts formula), we have

$$\int_t^\infty \frac{Y_s}{s^2} ds = \frac{Y_t}{t} - \lim_{r\to\infty} \frac{Y_r}{r} + \int_t^\infty \frac{dY_s}{s}, \tag{3.6}$$

where the integral on the right-hand side and

$$Y = \lim_{r\to\infty} \frac{Y_r}{r}$$

exist, which we require. Assuming equation (3.2), we have, because of formula (3.6),

$$\begin{aligned} -t\int_t^\infty \frac{d\beta_s}{s} &= -t\int_t^\infty \frac{dY_s}{s} + t\int_t^\infty \frac{Y_s}{s^2} ds \\ &= Y_t - tY. \end{aligned}$$

Thus we have equation (3.1). This completes the proof. □

Applying Lemma 3.1, we have a theorem of Jeulin–Yor [3], Yor [10].

Theorem 3.1. *Let β_t be a one-dimensional Brownian motion. Then, X_t solves*

$$X_t = \sigma\beta_t + \int_0^t \frac{X_s}{s} ds, \quad \sigma > 0, \tag{3.7}$$

if and only if

$$X_t = \sigma B_t + tZ, \tag{3.8}$$

where Z is a random variable and

$$B_t = -t\int_t^\infty \frac{d\beta_s}{s}, \tag{3.9}$$

which is a one-dimensional Brownian motion.

Proof. It remains to show that B_t in (3.9) is a Brownian motion. But it is clear because B_t is a Gaussian process with independent increments, $P[B_t] = 0$, and $P[B_t^2] = t$.[4] □

In the following sections some details of the results which have been announced in [1] will be given.

---

[4] Cf. Yor [10]

4. Two-sided Skorokhod type problem

Let $L(t)$ and $R(t)$ be continuous functions on $[0,\infty)$ such that $L(t) < R(t)$ for $\forall t \in [0,\infty)$. Then we consider a two-sided Skorokhod problem:

$$\xi_t = w_t + \Phi_t, \quad L(t) \le \xi_t \le R(t), \tag{4.1}$$

where w_t is a continuous function on $[0,\infty)$ such that $L(0) \le w_0 \le R(0)$, and

$$\begin{aligned} &\Phi_t \text{ is continuous in } t \ge 0,\ \Phi_0 = 0, \\ &\Phi_t = \Phi_t^{(-)} - \Phi_t^{(+)}, \text{ for } t \ge 0, \\ &\Phi_t^{(-)} \text{ increases only on } \{\, s : \xi_s = L(s)\,\}, \\ &\Phi_t^{(+)} \text{ increases only on } \{\, s : \xi_s = R(s)\,\}. \end{aligned} \tag{4.2}$$

We apply the same method to construct a reflecting Brownian motion, but what we handle now is a two-sided problem.

Lemma 4.1. (Domenig–Nagasawa [1]) *The two-sided Skorokhod problem* (4.1) *with* (4.2) *has a unique solution.*

Proof. We assume that w_t hits the lower boundary first, and define

$$L_1 = \inf\{\, t \ge 0 : w_t = L(t)\,\} \ (= \infty, \text{ if such } t \text{ does not exist}).$$

Denote

$$\xi_t^{0,0} = w_t,$$

and set

$$\xi_t^{1,0} = w_t + \Phi_t^1,$$

where

$$\Phi_t^1 = \sup_{s \le t}(L(s) - \xi_s^{0,0}) \vee 0.$$

With

$$R_1 = \inf\{\, t \ge L_1 : \xi_t^{1,0} = R(t)\,\} \ (= \infty, \text{ if such } t \text{ does not exist}),$$

we modify $\xi_t^{1,0}$ as

$$\xi_t^{1,0} = w_t + \Phi_{t \wedge R_1}^1, \tag{4.3}$$

where we use the same notation, but there will be no confusion. Further on we set

$$\xi_t^{1,1} = w_t + \Phi_{t \wedge R_1}^1 - \Psi_t^1, \quad \Psi_t^1 = \sup_{s \le t}(\xi_s^{1,0} - R(s)) \vee 0.$$

With

$$L_2 = \inf\{\, t \ge R_1 : \xi_t^{1,1} = L(t)\,\} \ (= \infty, \text{ if such } t \text{ does not exist}),$$

we modify $\xi_t^{1,1}$ as

$$\xi_t^{1,1} = w_t + \Phi_{t \wedge R_1}^1 - \Psi_{t \wedge L_2}^1. \tag{4.4}$$

It is then clear that

$$L(t) \le \xi_t^{1,1} \le R(t), \quad \text{for } t \le L_2.$$

We define inductively

$$\begin{aligned}
\Phi_t^n &= \sup_{s\le t}(L(s) - \xi_s^{n-1,n-1}) \vee 0, \quad \Psi_t^n = \sup_{s\le t}(\xi_s^{n,n-1} - R(s)) \vee 0, \\
R_n &= \inf\{\, t \ge L_n : \xi_t^{n,n-1} = R(t)\,\} \ (= \infty, \text{ if such } t \text{ does not exist}), \\
L_{n+1} &= \inf\{\, t \ge R_n : \xi_t^{n,n} = L(t)\,\} \ (= \infty, \text{ if such } t \text{ does not exist}),
\end{aligned}$$

and

$$\xi_t^{n+1,n+1} = w_t + \sum_{k=1}^n \Phi_{t\wedge R_k}^k - \sum_{k=1}^n \Psi_{t\wedge L_{k+1}}^k.$$

Then we have

$$L(t) \le \xi_t^{n,n} \le R(t), \quad \text{for } t \le L_{n+1}. \tag{4.5}$$

Finally we set

$$\xi_t^\infty = w_t + \Phi_t^{(-)} - \Phi_t^{(+)}, \qquad \Phi_t^{(-)} = \sum_{k=1}^\infty \Phi_{t\wedge R_k}^k, \qquad \Phi_t^{(+)} = \sum_{k=1}^\infty \Psi_{t\wedge L_{k+1}}^k. \tag{4.6}$$

Then $\Phi_t^{(-)}$ and $\Phi_t^{(+)}$ fulfill the requirement (4.2), and

$$L(t) \le \xi_t^\infty \le R(t), \quad \text{for } t < L_\infty. \tag{4.7}$$

It remains to prove $L_\infty = \lim_{n\to\infty} L_n = \infty$. Denote $\varepsilon = \inf_{t\le T}(R(t) - L(t)) > 0$ for an arbitrary but fixed $T < \infty$. Since w_t, $L(t)$, and $R(t)$ are uniformly continuous on $[0, T]$, there exists $\delta > 0$ such that $|w_t - w_s| < \varepsilon/3$, $|L(t) - L(s)| < \varepsilon/3$, and $|R(t) - R(s)| < \varepsilon/3$ if $|t - s| < \delta$. Therefore, the duration time of each crossing of $\xi_t^{n,n}$ (n is arbitrary) from the lower boundary to the upper one is greater than δ, and the total number of such crossings of $\xi_t^{n,n}$ in $[0, T]$ is bounded by T/δ. Hence, $T \le L_\infty$. Since T is arbitrary, we have $L_\infty = \infty$. The uniqueness of solutions is clear from Lemma 4.2 below. This completes the proof. □

Lemma 4.2. *Let $b(t, x)$ be continuous on $[0, \infty) \times \mathbb{R}$, and let $\xi_i(x)$, $\eta_i(x)$ and $\Phi_i(t)$, $i = 1, 2$, satisfy*

$$\xi_i(t) = \omega(t) + \int_0^t b(s, \eta_i(s))\, ds + \Phi_i(t), \tag{4.8}$$

where $L(t) \le \xi_i(t) \le R(t)$, and $\Phi_i(t)$ satisfies (4.2). Then

$$|\xi_1(t) - \xi_2(t)|^2 \le 2\int_0^t \{\xi_1(s) - \xi_2(s)\}\{b(s, \eta_1(s)) - b(s, \eta_2(s))\}\, ds. \tag{4.9}$$

Proof. Setting

$$\Psi_i(t) = \int_0^t b_i(s, \eta_i(s))\, ds + \Phi_i(t), \quad i = 1, 2, \quad \text{and} \quad \Psi(t) = \Phi_1(t) - \Phi_2(t),$$

we have

$$\begin{aligned} |\xi_1(t) - \xi_2(t)|^2 &= |\Psi(t)|^2 \\ &= 2\int_0^t \{\Phi_1(s) - \Phi_2(s)\}\{d\Phi_1(s) - d\Phi_2(s)\} \\ &= 2\int_0^t \{\xi_1(s) - \xi_2(s)\}\{b(s, \eta_1(s)) - b(s, \eta_2(s))\}\, ds \\ &\qquad +2\int_0^t \{\xi_1(s) - \xi_2(s)\}\, d(\Phi_1(s) - \Phi_2(s)). \end{aligned}$$

The last term is equal to

$$\begin{aligned} &2\int_0^t \{\xi_1(s) - \xi_2(s)\}\, d\Phi_1^{(-)}(s) - 2\int_0^t \{\xi_1(s) - \xi_2(s)\}\, d\Phi_1^{(+)}(s) \\ &\quad -2\int_0^t \{\xi_1(s) - \xi_2(s)\}\, d\Phi_2^{(-)}(s) + 2\int_0^t \{\xi_1(s) - \xi_2(s)\}\, d\Phi_2^{(+)}(s). \end{aligned}$$

Here the first line is non-positive, since $L(s) \le \xi_2(s) \le R(s)$, $\xi_1(s) = L(s)$ on $\mathrm{supp}(d\Phi_1^{(-)})$, and $\xi_1(s) = R(s)$ on $\mathrm{supp}(d\Phi_1^{(+)})$. Interchanging the roles of $\xi_1(s)$ and $\xi_2(s)$, we see that the second line is also non-positive, completing the proof. □

Theorem 4.1. *Let $b(t, x)$ satisfy a Lipschitz condition*

$$|b(t, x) - b(t, y)| \le K|x - y|, \quad \text{for } \forall t \ge 0, \forall x, y \ge 0. \tag{4.10}$$

Then, subject to (4.2), *there exists a unique solution of equation*

$$X(t) = w(t) + \int_0^t b(s, X(s))\, ds + \Phi(t), \quad L(t) \le X(t) \le R(t). \tag{4.11}$$

5. Skorokhod problem with singular drift

Let $R(t)$ be strictly increasing and continuous in $t \ge 0$ with $R(0) = 0$. In a space-time domain

$$D = \{(t, x); t \ge 0, x \in [-R(t), R(t)]\}, \tag{5.1}$$

we consider a diffusion equation and its formal adjoint

$$
\begin{aligned}
\frac{\partial u}{\partial t}+\frac{1}{2}\sigma^2\frac{\partial^2 u}{\partial x^2}+\frac{x}{t}\frac{\partial u}{\partial x} &= 0,\\
-\frac{\partial u}{\partial t}+\frac{1}{2}\sigma^2\frac{\partial^2 u}{\partial x^2}-\frac{\partial}{\partial x}\left(\frac{x}{t}\mu\right) &= 0,
\end{aligned}
\tag{5.2}
$$

with the reflecting boundary condition, where σ is a constant.

Since (5.2) is not defined at the origin $(0,0)$, it is not evident whether equation (5.2) determines well a diffusion process starting out from the origin, because of the singularity of the drift field $a(t,x)=x/t$ in equation (5.2).

Let $Q(s,x;t,dy)$, $s,t\in[a,b]$, $0<a<b<\infty$, be the transition probability of the diffusion process determined by equation (5.2). Then, because of (5.1), $\{\,Q(s,x;t,dy): s\in(0,\varepsilon]\,\}$ is tight, and we can chose a sequence $\xi(s)\downarrow 0$ so that

$$
Q^{\xi}(0,0;t,dy)=\lim_{s\downarrow 0}Q(s,\xi(s);t,dy) \tag{5.3}
$$

exists, but the limit $Q^{\xi}(0,0;t,dy)$ depends on sequences $\xi(s)\downarrow 0$, in general. Therefore, contrary to the case without boundary discussed as an example in Section 1, the uniqueness does not hold in general, although we can start from the origin, thanks to the moving reflecting boundary. The uniqueness depends on the shape of the boundary, i.e. $R(t)$, as we shall see in the following sections.

We will discuss the existence and uniqueness (resp. non-uniqueness) of solutions of equation (5.2) in terms of a singular Skorokhod problem. Namely, instead of equation (5.2), we consider a two-sided Skorokhod problem

$$
X_t=\sigma\beta_t+\int_0^t\frac{X_s}{s}\,ds+\Phi_t,\quad |X_t|\le R(t), \tag{5.4}
$$

where β_t denotes a one-dimensional Brownian motion, and

$$
\begin{gathered}
\Phi_t \text{ is continuous in } t\ge 0,\ \Phi_0=0,\\
\Phi_t \text{ is of bounded variation on } [\varepsilon,1/\varepsilon], \text{ for any } \varepsilon>0,\\
\Phi_t \text{ increases only on } \{\,s>0:\xi_s=-R(s)\,\},\\
\Phi_t \text{ decreases only on } \{\,s>0:\xi_s=R(s)\,\},\\
\Phi_t \text{ is constant otherwise.}
\end{gathered}
\tag{5.5}
$$

We will consider solutions of the Skorokhod problem (5.4), and show that the shape of the boundary of the domain D influences the uniqueness and non-uniqueness of solutions to the problem (5.4) subject to (5.5).

For solutions of equation (5.4) we can give another representation, in which the singular drift term disappears. We prepare a simple lemma.

Lemma 5.1. *Assume that X_t, β_t, and Φ_t are continuous functions on $[0,\infty)$ which vanish at $t=0$, and moreover that Φ_t is of bounded variation on $[\varepsilon,1/\varepsilon]$, for any $\varepsilon>0$. If*

$$
X_t=\sigma B_t-t\int_t^{\infty}\frac{d\Phi_s}{s}, \tag{5.6}
$$

then

$$X_t = \sigma\beta_t + \int_0^t \frac{X_s}{s}\,ds + \Phi_t, \tag{5.7}$$

with

$$\beta_t = B_t - \int_0^t \frac{B_s}{s}\,ds. \tag{5.8}$$

Conversely, if

$$\lim_{t\to\infty} \frac{X_t}{t} = 0, \tag{5.9}$$

then, (5.6) *follows from* (5.7) *with*

$$B_t = -t\int_t^\infty \frac{d\beta_s}{s}. \tag{5.10}$$

Proof. Remark on Lemma 3.1 applies also in Lemma 5.1. Assume that X_t satisfies equation (5.6). Because of (3.5) with Φ_t in place of β_t, we have

$$\int_0^t ds \int_0^\infty \frac{d\Phi_r}{r} = \Phi_t + t\int_t^\infty \frac{d\Phi_r}{r}.$$

Therefore,

$$\begin{aligned} X_t - \int_0^t \frac{X_s}{s}\,ds - \Phi_t & \\ = \; & \sigma B_t - t\int_t^\infty \frac{d\Phi_s}{s} - \int_t^\infty \frac{\sigma B_s}{s}\,ds + \int_t^\infty ds \int_s^\infty \frac{d\Phi_r}{r} - \Phi_t \\ = \; & \sigma\left\{B_t - \int_t^\infty \frac{B_s}{s}\,ds\right\} \\ = \; & \sigma\beta_t, \end{aligned}$$

where (5.8) has been applied at the last equality. Thus X_t satisfies equation (5.7). Conversely, let X_t satisfy equation (5.7). Then,

$$\sigma\beta_t = X_t - \sigma\int_0^t \frac{X_s}{s}\,ds - \Phi_t. \tag{5.11}$$

Substituting this in (5.10), we get

$$\begin{aligned} \sigma B_t &= -t\int_t^\infty \frac{d\sigma\beta_s}{s} \\ &= -t\int_t^\infty \frac{dX_s}{s} + t\int_t^\infty \frac{X_s}{s^2}\,ds + t\int_t^\infty \frac{d\Phi_s}{s} \\ &= X_t + t\int_t^\infty \frac{d\Phi_s}{s}, \end{aligned}$$

where we have applied (3.6) with X_t in place of Y_t under the assumption (5.9). Hence X_t satisfies equation (5.6). This completes the proof. □

Corresponding to Therorem 3.1 of Jeulin–Yor [3] without boundary, we have, in the case of the moving reflecting boundary,

Theorem 5.1. (Domenig–Nagasawa [1]) *Assume*

$$\lim_{t\to\infty} \frac{R(t)}{t} = 0. \tag{5.12}$$

Then X_t *satisfies equation* (5.4), *namely,*

$$X_t = \sigma\beta_t + \int_0^t \frac{X_s}{s}\, ds + \Phi_t, \quad |X_t| \le R(t), \tag{5.13}$$

subject to (5.5), *if and only if*

$$X_t = \sigma B_t - t\int_t^\infty \frac{d\Phi_s}{s}, \tag{5.14}$$

where B_t *is a one-dimensional Brownian motion given in* (5.10).

Proof. Since β_t is a Brownian motion and Φ_t satisfies (5.5), the integrals appearing in Lemma 5.1 are well-defined, and (5.9) is satisfied because of (5.12). Therefore, Lemma 5.1 can be applied to complete the proof. □

The representation (5.14) implies that if one looks at the process X_t backward in time, knowing the future, then the singular drift field in (5.13) disappears; namely, the process X_t starts from the origin as a Brownian motion, if it does not hit the boundary immediately.

6. The minimum and maximum solutions

Avoiding the singularity at the origin in the problem (5.4), we consider first of all the problem after $\varepsilon > 0$, starting from the lower boundary $-R(\varepsilon)$:

$$X_t = -R(\varepsilon) + \sigma(\beta_t - \beta_\varepsilon) + \int_\varepsilon^t \frac{X_s}{s}\, ds + \Phi_t, \quad |X_t| \le R(t), \quad t \ge \varepsilon, \tag{6.1}$$

subject to (4.2) with $t \ge \varepsilon$ and $\Phi_\varepsilon = 0$. Then we have

Lemma 6.1. (i) *There exists a unique solution* $X_t^{(-\varepsilon)}$ *to equation* (6.1) *subject to* (4.2) *with* $t \ge \varepsilon$. *If* $0 < \varepsilon' < \varepsilon$, *then*

$$X_t^{(-\varepsilon)} \le X_t^{(-\varepsilon')}, \quad \textit{for } t \ge \varepsilon, \tag{6.2}$$

namely $X_t^{(-\varepsilon)}$ *is monotone increasing as* $\varepsilon \downarrow 0$.

(ii) *There exists*

$$\underline{X}_t = \lim_{\varepsilon\downarrow 0} X_t^{(-\varepsilon)}, \tag{6.3}$$

which is a solution of the two-sided Skorokhod problem (5.4) *with* (5.5).

For a proof, cf. [1].

Remark. Since $X_t^{(-\varepsilon)}$ converges, so does $\Phi_t^{(-\varepsilon)} = \Phi_t^{(-\varepsilon,-)} - \Phi_t^{(-\varepsilon,+)}$ also, because of equation (6.1). However, $\Phi_t^{(-\varepsilon,-)}$ and $\Phi_t^{(-\varepsilon,+)}$ might diverge in general, when $\varepsilon \downarrow 0$. Hence, we cannot decompose like $\Phi_t = \Phi_t^{(-)} - \Phi_t^{(+)}$ as in (4.2).

We consider also a similar problem, but starting from the upper boundary $R(\varepsilon)$ at $\varepsilon > 0$:

$$X_t = R(\varepsilon) + \sigma(\beta_t - \beta_\varepsilon) + \int_\varepsilon^t \frac{X_s}{s} ds + \Phi_t, \quad |X_t| \leq R(t), \quad t \geq \varepsilon, \tag{6.4}$$

subject to (5.5) with $\Phi_\varepsilon = 0$. Then we have

Lemma 6.2. (i) *There exists a unique solution* $X_t^{(\varepsilon)}$ *to equation* (6.4) *subject to* (4.2) *with* $t \geq \varepsilon$. *Let* $0 < \varepsilon' < \varepsilon$. *Then*

$$X_t^{(\varepsilon)} \geq X_t^{(\varepsilon')}, \quad \text{for } t \geq \varepsilon, \tag{6.5}$$

namely $X_t^{(\varepsilon)}$ *is monotone decreasing as* $\varepsilon \downarrow 0$.
(ii) *There exists*

$$\bar{X}_t = \lim_{\varepsilon \downarrow 0} X_t^{(\varepsilon)}, \tag{6.6}$$

which is a solution of the two-sided Skorokhod problem (5.4) *with* (5.5).

For a proof, cf. [1].

Theorem 6.1. (Domenig–Nagasawa [1]) *Let* $\underline{X}_t$ *and* $\bar{X}_t$ *be defined by* (6.3) *and* (6.6), *respectively, and* X_t *be any solution of the two-sided Skorokhod problem* (5.4) *subject to* (5.5). *Then*

$$\underline{X}_t \leq X_t \leq \bar{X}_t, \quad \text{for } t \geq 0, \tag{6.7}$$

for almost all Brownian paths. Therefore, $\underline{X}_t$ *is the minimum solution (resp.* $\bar{X}_t$ *is the maximum one) and it reaches the lower- (resp. upper-)boundary immediately.*

7. The uniqueness and non-uniqueness of solutions

We shall analyze the behaviour of solutions near the origin, assuming

$$R(t) = (\alpha t)^\gamma, \quad 0 < \gamma < 1, \quad \text{for small } t, \tag{7.1}$$

where $\alpha > 0$.

Theorem 7.1. (Domenig–Nagasawa [1]) *Assume that $R(t)$ is given by* (7.1). *Then, there exist solutions of the two-sided problem* (5.4) *subject to* (5.5). *Solutions of equation* (5.4) *are not uniquely determined if* $0 < \gamma < 1/2$, *while the uniqueness of solutions holds if* $1/2 \leq \gamma < 1$.

Proof. First of all we remark that the law of iterated logarithm holds for a one-dimensional Brownian motion B_t:

$$\overline{\lim_{t \downarrow 0}} \frac{B_t}{\sqrt{2t \log\log t^{-1}}} = 1, \tag{7.2}$$

for almost all Brownian paths. Hence, if $1/2 \leq \gamma < 1$, any solution of equation (5.4) hits immediately the lower and upper boundaries $\{-R(t), R(t)\}$, as do the minimum and maximum solutions $\underline{X}_t$ and $\bar{X}_t$. Therefore, the uniqueness of solutions holds. On the other hand, if $0 < \gamma < 1/2$, the law of iterated logarithm (7.2) implies that a Brownian motion B_t does not immediately hit the boundary $\{-R(t), R(t)\}$, and hence processes $X_t = \sigma B_t + tZ$ also do not immediately hit the boundary. Let a Brownian motion B_t be given by

$$B_t = -t \int_t^\infty \frac{d\beta_s}{s},$$

in terms of a Brownian motion β_t in (5.4), and define a process by

$$X_t = \sigma B_t + tZ, \quad \text{for } t < \varepsilon, \tag{7.3}$$

where $\varepsilon > 0$ is the first hitting time to the boundary and Z is any random variable (cf. Theorem 3.1); and for $t \geq \varepsilon$ we define X_t by a solution of a Skorokhod problem

$$X_t = \sigma B_\varepsilon + \varepsilon Z + \sigma(\beta_t - \beta_\varepsilon) + \int_\varepsilon^t \frac{X_s}{s} ds + \Phi_t, \quad |X_t| \leq R(t), \tag{7.4}$$

subject to (5.5). Then the process X_t solves the two-sided Skorokhod problem (5.4). This completes the proof. □

Taking a solution with $Z \equiv 0$ in (7.3) and (7.4) when $0 < \gamma < 1/2$, we call it a "central solution" or "central process". It is central, in the sense that it starts from the origin as a Brownian motion. This means that in the representation (5.14) the integral term vanishes for sufficiently small t, namely, there is $\varepsilon > 0$ such that

$$t \int_t^\infty \frac{d\Phi_s}{s} = 0, \quad \text{for } t \leq \varepsilon. \tag{7.5}$$

Therefore, we have

Theorem 7.2. (Domenig–Nagasawa [1]) *Assume* $R(t) = (\alpha t)^{\gamma}$, $0 < \gamma < 1/2$, *for small* t, *and let* X_t^0 *be the central solution. If* $R(t) = (\alpha' t)^{\gamma_2}$ *with* $0 < \gamma_2 \le 1/2$ *for large* t, *then* $X_t - X_t^0 \to 0$ *in law as* $t \uparrow \infty$, *for any solution* X_t *of the Skorokhod problem* (5.4); *while* $\bar{X}_t - X_t^0$ *(resp.* $\underline{X}_t - X_t^0$*) does not converge in law as* $t \uparrow \infty$, *if* $1/2 < \gamma_2 < 1$ *for large* t, *where* $\bar{X}_t$ *(resp.* $\underline{X}_t$*) is the maximum (resp. minimum) solution of equation* (5.4).

Proof. Without loss of generality we can assume $\sigma = 1$ and $\alpha, \alpha' = 1$ in the proof. It is clear that X_t and X_t^0 hit each other before X_t^0 hits both boundaries:

$$T = \inf\{\, t > 0 \colon X_t = X_t^0 \,\} \le T_l(X^0) \vee T_r(X^0), \tag{7.6}$$

where

$$T_l(X^0) = \inf\{\, t > 0 \colon X_t^0 = -R(t) \,\}, \quad T_r(X^0) = \inf\{\, t > 0 \colon X_t^0 = R(t) \,\}.$$

However, since X_t^0 is the central solution, we have $X_t = B_t$ before the first hitting of the boundary. Therefore, $T_l(X^0) \wedge T_r(X^0) = T_l(B) \wedge T_r(B)$. Let $T_l(X^0) < T_r(X^0)$ (resp. $T_l(X^0) > T_r(X^0)$). Then

$$\int_0^\infty \frac{d\Phi_s}{s} = 0, \tag{7.7}$$

because of (7.5). Therefore, (5.14) together with (7.7) yields

$$X_t^0 = B_t + t \int_0^t \frac{\Phi_s^{(-)}}{s} ds \ge B_t, \quad \text{for } \forall t \in [0, T_r(X^0)),$$

and in the same way

$$X_t^0 = B_t - t \int_0^t \frac{\Phi_s^{(+)}}{s} ds \le B_t, \quad \text{for } \forall t \in [0, T_l(X^0)).$$

Hence, $T_r(X^0) \le T_r(B)$ and $T_l(X^0) \le T_l(B)$. Consequently,

$$T_l(X^0) \vee T_r(X^0) \le T_l(B) \vee T_r(B). \tag{7.8}$$

Therefore, if $R(t) = (\alpha' t)^{\gamma_2}$ with $0 < \gamma_2 \le 1/2$ for $t \ge t_0 > 1$, the law of iterated logarithm for large t

$$P\left[\overline{\lim_{t\uparrow\infty}} \frac{B_t}{\sqrt{2t \log(\log t)}} = 1, \quad \text{and} \quad \underline{\lim}_{t\uparrow\infty} \frac{B_t}{\sqrt{2t \log(\log t)}} = -1\right] = 1,$$

together with (7.6) and (7.8) yields

$$P[X_t = X_t^0 \text{ for some } t < \infty] = P[T < \infty] \ge P[T_l(B) \vee T_r(B) < \infty] = 1.$$

On the other hand, if $R(t) = (\alpha' t)^{\gamma_2}$ with $1/2 < \gamma_2 < 1$ for $t \ge t_0 > 1$, then the probability that a Brownian motion B_t does not hit the boundary is positive by the law of iterated logarithm for large t. Moreover, the central solution X_t^0 is equal to B_t by definition as long as X_t^0 doesn't hit the boundary. Taking κ, $1/2 < \kappa < \gamma_2$, we define an increasing function $R_\kappa(t) \leq R(t)$ such that

$$R_\kappa(t) = R(t), \quad \text{for } \forall t \leq 1, \quad R_\kappa(t) = (\alpha' t)^\kappa, \quad \text{for } \forall t \geq t_0.$$

We choose $\varepsilon_0 > 0$ so that the subset A of Ω, defined by

$$A = \{\, |B_t| < R_\kappa(t), \text{for } \forall t > 0, \text{ and } |B_t| + \varepsilon_0 t < R(t) \text{ for } \forall t \leq t_0 \,\},$$

has positive probability. For $\omega \in A$ and $t \geq 0$, we have $B_t = X_t^0$. Define $\varepsilon(\tau)$ by

$$\varepsilon(\tau) = \varepsilon_0 \wedge \frac{(\alpha'\tau)^{\gamma_2} - (\alpha'\tau)^\kappa}{\tau}. \tag{7.9}$$

Since $\varepsilon(\tau)$ is decreasing for $\tau \geq \tau_0$, where τ_0 depends on γ_2 and κ, we have

$$\varepsilon(\tau)t \leq (\alpha' t)^{\gamma_2} - (\alpha' t)^\kappa, \qquad \text{for } \tau_0 \leq t \leq \tau. \tag{7.10}$$

We can assume that $\tau_0 \leq t_0$. It follows that for any $\tau > 0$,

$$|B_t| + \varepsilon(\tau)t < R(t), \quad \text{for } \forall t \leq \tau, \text{ on } A. \tag{7.11}$$

In fact, by (7.9) and (7.10)

$$\begin{aligned} |B_t| + \varepsilon(\tau)t \leq |B_t| + \varepsilon_0 t < R(t), &\qquad \text{for } t \leq t_0, \\ |B_t| + \varepsilon(\tau)t \leq |B_t| + (\alpha' t)^{\gamma_2} - (\alpha' t)^\kappa < (\alpha' t)^{\gamma_2} = R(t), &\qquad \text{for } t_0 < t \leq \tau. \end{aligned}$$

Because of (7.11) and Theorem 3.1, we have

$$B_\tau(\omega) + \varepsilon(\tau)\tau \leq \bar{X}_\tau(\omega), \quad \text{for } \omega \in A.$$

Therefore,

$$\begin{aligned} \bar{X}_\tau(\omega) - \bar{X}_\tau^0(\omega) &\geq B_\tau(\omega) + \varepsilon(\tau)\tau - B_\tau(\omega) \\ &= \varepsilon_0\tau \wedge ((\alpha'\tau)^{\gamma_2} - (\alpha'\tau)^\kappa), \end{aligned}$$

where τ is arbitrary, and hence $\bar{X}_\tau(\omega) - \bar{X}_\tau^0(\omega)$ does not converge as τ tends to infinity for $\omega \in A$. This completes the proof. □

For an application of Theorems 7.1 and 7.2 to the origin of universes see [1].

Acknowledgement. We'd like to thank Hiroshi Tanaka, who pointed out an error.

References

1. T. Domenig and M. Nagasawa, *A Skorokhod problem with singular drift and its application to the origin of universes*, Proc. of Japan Acad. Science **70**, Ser. A, no. 4 (1994), 88–93.
2. K. Itô, *Stochastic differential equations*, Memoirs A.M.S. 4, 1951.
3. Th. Jeulin and M. Yor, *Filtration des ponts browniens et équations différentielles stochastiques linéaires*, Sem. de Probab. XXIV, 227–265. Lecture Notes in Math. **1426**, Springer-Verlag, 1988/89.
4. A. Kolmogoroff, *Zur Umkehrbarkeit der statistischen Naturgesetze*, Math. Ann. **113** (1937), 766–772.
5. P.A. Meyer, *Sur une transformation du mouvement brownien due á Jeulin et Yor*, (preprint), to appear in: Sém de Probab. XXVIII.
6. M. Nagasawa, *Time reversal of Markov processes*, Nagoya Math. Jour. **24** (1964), 177–204.
7. M. Nagasawa, *Schrödinger Equations and Diffusion Theory*, Birkhäuser Verlag, Basel · Boston · Berlin, 1993.
8. M. Nagasawa, *Mathematical Foundations of Quantum Mechanics*, Lecture Notes at the University of Zürich, 1994.
9. E. Schrödinger, *Über die Umkehrung der Naturgesetze*, Sitzungsberichte der preussischen Akad. der Wissenschaften Physicalisch Mathematische Klasse, 1931, 144–153.
10. M. Yor, *Some Aspects of Brownian Motion, Part 1: Some special Functionals*, Birkhäuser Verlag AG, Basel · Boston · Berlin, 1992.

On sensitive control and differential games in infinite dimensional spaces

Makiko Nisio

Faculty of Engineering, Osaka Electro-Communication University, Hatsu-cho, Neyagawa, Osaka 572, Japan

1. Introduction

We are considering a risk sensitive control for a system governed by a stochastic partial differential equation. The aim of this paper is to investigate the relationship between small noise asymptotics of the value function and a differential game.

A stochastic partial differential equation is a parabolic equation with random external force and can be viewed as a stochastic differential equation in infinite dimensional space [3],[6],[8]. Here, we consider a stochastic partial differential equation in a finite time interval $[0, T]$ as follows.

Let D be a bounded domain of R^n with smooth boundary. We put $H = L^2(D)$, $\| * \| =$ its norm and

$$A\zeta = \sum_{ij=1}^{n} \frac{\partial}{\partial x_i}(a^{ij}(x)\frac{\partial \zeta}{\partial x_j}) + \sum_{i=1}^{n} r^i(x)\frac{\partial \zeta}{\partial x_i} - c(x)\zeta.$$

Let W_k, $k = 1, 2, \cdots$, be independent 1-dimensional Brownian motions, defined on a probability space (Ω, F, P). F_t denotes the $\sigma - field$ generated by $\{W_k(s), s \leq t, k = 1, 2, \cdots\}$. For a given compact convex set Γ of R^q, an $F_t - progressively$ measurable process with values in $L^2(D, \Gamma)$, say γ, is called an admissible control. Assuming some regularity conditions for A (see $(A1) \sim (A3)$ in Sect.2), we will consider the following stochastic partial differential equation, for an admissible control U,

$$d\xi(t, x) = (A\xi(t, x) + b(x, \xi(t, x), U(t)(x)))dt + \sqrt{\varepsilon} dM(t, x),$$
$$x \in D, 0 < t < T, \quad (1)$$

$$\xi(t, x) = 0 \quad on \quad \partial D \quad and \quad \xi(0, *) = \eta \ (\in H),$$

where ε is a small positive parameter and a random force M is a colored noise of the form

$$M(t, x) = \sum_{k=1}^{\infty} \sqrt{m_k} e_k(x) W_k(t)$$

with $\sum m_k < \infty$ and a smooth orthonormal base $\{e_k, k = 1, 2, \cdots\}$ of H. Defining β ; $H \times \gamma \longmapsto H$ by $\beta(\zeta, u)(x) = b(x, \zeta(x), u(x))$, we can regard (1) as the following stochastic differential equation (2) in the Hilbert space H (cf.[6]),

$$d\xi(t) = (A\xi(t) + \beta(\xi(t), U(t)))dt + \sqrt{\varepsilon}dM(t), \qquad 0 < t < T, \tag{2}$$
$$\xi(0) = \eta.$$

We assume an exponential cost criterion J^ε expressed in the integral form

$$J^\varepsilon(t, \eta, U) = E\exp(\frac{1}{\varepsilon}\int_0^t h(\xi(s, \eta, U))ds).$$

Then, from the dynamic programming principle and Itô's formula, the value function $v^\varepsilon(t, \eta) = \inf_U J^\varepsilon(t, \eta, U)$ turns out to be a unique viscosity solution of the Hamilton-Jacobi-Bellman equation in H,

$$\begin{aligned} 0 &= \frac{\partial v}{\partial t}(t, \eta) - \langle A^* \partial v(t, \eta), \eta\rangle - \inf_{u\in\gamma}\langle \partial v(t, \eta), \beta(t, u)\rangle \\ &\quad -\frac{\varepsilon}{2}trace(S\partial^2 v(t, \eta)) - \frac{1}{\varepsilon}h(\eta)v(t, \eta), \qquad 0 < t < T, \end{aligned} \tag{3}$$
$$v(0, \eta) = 1,$$

where ∂ = Fréchet derivative, A^* = adjoint of A, $\langle\ ,\ \rangle$ = duality pair between $H^{-1}(D)$ and $H_0^1(D)$ and S ; $H \longmapsto H$ is defined by $Se_k = m_k e_k$, $k = 1, 2, \cdots$. Hence its logarithmic transformation

$$V^\varepsilon(t, \eta) = \varepsilon \log v^\varepsilon(t, \eta)$$

is a unique viscosity solution of the following equation

$$\begin{aligned} 0 &= \frac{\partial V}{\partial t}(t, \eta) - \langle A^* \partial V(t, \eta), \eta\rangle - \inf_{u\in\gamma}\langle \partial V(t, \eta), \beta(\eta, u)\rangle - h(\eta) \\ &\quad -\frac{\varepsilon}{2}trace(S\partial^2 V(t, \eta)) - \frac{1}{2}\langle S\partial V(t, \eta), \partial V(t, \eta)\rangle, \qquad 0 < t < T, \end{aligned} \tag{4}$$
$$V(0, \eta) = 0.$$

In [7], we have shown that the small noise limit V of V^ε exists and again becomes a unique viscosity solution of (4) with $\varepsilon = 0$, namely

$$\begin{aligned} 0 &= \frac{\partial V}{\partial t}(t, \eta) - \langle A^* \partial V(t, \eta), \eta\rangle - \inf_{u\in\gamma}\langle \partial V(t, \eta), \beta(\eta, u)\rangle \\ &\quad -h(\eta) - \frac{1}{2}\langle S\partial V(t, \eta), \partial V(t, \eta)\rangle, \qquad 0 < t < T, \end{aligned} \tag{5}$$
$$V(0, \eta) = 0.$$

Applying the Legendre transformation to the last term of (5), this equation turns out to be the Isaacs equation (6), related to a differential game in H,

$$\begin{aligned} 0 &= \frac{\partial V}{\partial t}(t,\eta) - \langle A^*\partial V(t,\eta),\eta\rangle - \inf_{u\in\gamma}\langle \partial V(t,\eta),\beta(\eta,u)\rangle - h(\eta) \\ &\quad - \sup_{\zeta\in H}(\langle S\zeta,\partial V(t,\eta)\rangle - \frac{1}{2}\langle S\zeta,\zeta\rangle), \qquad 0<t<T, \end{aligned} \tag{6}$$

$$V(0,\eta)=0.$$

This fact seems to characterize V as the value of game. However, the differential game, associated with (6), has the unbounded control region H . Consequently some difficulties arise, when we try to prove the continuity of the value of game. In finite dimensional spaces, this characterization has already been obtained by several authors (cf.[1],[4]), under some conditions. Among them, Bensoussan-Nagai [1] developed an interesting method to obtain continuity. Using their method, we can prove the continuity [Theorem 3.1] and characterize V by a differential game [Theorems 4.2 and 4.3].

In Section 2, we shall recall some results on sensitive control, based on [7]. Sections 3 and 4 are devoted to differential games associated with (6).

2. Sensitive control

Let us assume the following conditions,
$(A1)$. a^{ij} and r^i are in $C^3(\bar{D})$
$(A2)$. $n\times n$ matrix (a^{ij}) is uniformly positive definite
$(A3)$. c is nonnegative and continuous in $\bar{D}$
$(A4)$. $b\,;\,\bar{D}\times R^1\times\Gamma\longmapsto R^1$ is bounded and Lipschitz continuous
$(A5)$. $h\,;\,H\longmapsto R^1$ is bounded and Lipschitz continuous, say

$$|h|\le\bar{h} \qquad and \qquad |h(\xi)-h(\zeta)|\le\bar{l}\|\xi-\zeta\|.$$

Put $H^k=$ Sobolev space $H_0^k(D)$ and $\|*\|_k=$ its norm. A can be regarded as an operator from H^1 into H^{-1} satisfying the coercive condition,

$$\langle -A\zeta,\zeta\rangle+\bar{\mu}\|\zeta\|^2 \quad\ge\quad \bar{\lambda}\|\zeta\|_1^2 \qquad for \qquad \zeta\in H^1,$$

with constants $\bar{\mu}\ge 0$ and $\bar{\lambda}>0$. Hence,

$$\langle -A\zeta,\zeta\rangle+\quad\mu\|\zeta\|^2\ge 0 \tag{7}$$

holds with a constant $\mu \geq 0$. When $-A$ is a monotone operator, $\mu = 0$. From $(A4)$, β is bounded and Lipschitz continuous, say

$$\|\beta\| \leq \bar{\beta} \quad \textit{and} \quad \|\beta(\xi, u) - \beta(\zeta, v)\| \leq l(\|\xi - \zeta\| + \|u - v\|).$$

The operator B ; $H \longmapsto H^2$, defined by

$$B = [I - (A - \sum_{i=1}^{n} r^i \frac{\partial}{\partial x_i})]^{-1} \qquad \textit{with boundary value } 0,$$

is compact. We set $|\zeta|_B^2 = \langle B\zeta, \zeta\rangle$.

Denoting, by $M^2(0,T;H^1)$, the subset of $L^2([0,T] \times \Omega, H^1)$ consisting of F_t- *progressively* measurable processes, we will define a solution of (2).

Definition 2.1. $\xi \in M^2(0,T;H^1)$ is called a solution of (2), if $\xi \in C([0,T], H)$ a.s. and for any t and smooth function φ with support in D

$$\begin{aligned} &\langle \xi(t), \varphi\rangle \\ = \;& \langle \eta, \varphi\rangle + \int_0^t \langle A\xi(s), \varphi\rangle + \langle \beta(\xi(s), U(s)), \varphi\rangle ds + \sqrt{\varepsilon}\langle M(t), \varphi\rangle, \quad a.s(8) \end{aligned}$$

Proposition 2.1. There is a unique solution $\xi^\varepsilon(t, \eta, U)$ of (2) having the following properties,

$$E(\sup_{t\leq T} \|\xi^\varepsilon(t,\eta,U)\|^2 + \int_0^T \|\xi^\varepsilon(t,\eta,U)\|_1^2 dt) \leq K_1(1 + \|\eta\|^2) \tag{9}$$

$$\sup_{t\leq T} \|\xi^\varepsilon(t,\eta_1,U) - \xi^\varepsilon(t,\eta_2,U)\| \leq K_2\|\eta_1 - \eta_2\| \tag{10}$$

$$\int_0^T \|\xi^\varepsilon(t,\eta_1,U) - \xi^\varepsilon(t,\eta_2,U)\|^2 dt \leq K_3|\eta_1 - \eta_2|_B^2 \tag{11}$$

where K_i, $i = 1,2,3$, stand for constants independent of U, ε and ω.

Moreover, we will recall the following propositions.

Proposition 2.2. There is a constant K_4 independent of ε, such that

$$|V^\varepsilon(t,\eta)| \leq \bar{h}T \tag{12}$$

$$|V^\varepsilon(t_1,\eta_1) - V^\varepsilon(t_2,\eta_2)| \leq K_4(|t_1 - t_2| + |\eta_1 - \eta_2|_B) \tag{13}$$

Proposition 2.3. For any sequence of $\{V^\varepsilon, \varepsilon > 0\}$, there is a subsequence which converges uniformly on bounded sets of $[0,T] \times H$. The limit function also satisfies (12) and (13).

Hence, denoting by $BUC \cap C_w$ the space of bounded, uniformly continuous and weakly continuous functions, defined on $[0,T] \times H$, we see that V^ε and the limit function belong to $BUC \cap C_w$.

Now we shall define a viscosity solution for the nonlinear equation (14) below, according to Crandall and Lions [2],[9]. $\Phi \in C^{12}((0,T) \times H)$ is called a test function, if (i) Φ is weakly lower semicontinuous and bounded from below and (ii) $\partial\Phi(t,\eta) \in H^2$ and $A^*\partial\Phi$ is continuous. $g \in C^2(H)$ is called radial, if $g(\zeta) = \tilde{g}(\|\zeta\|)$ with $\tilde{g} \in C^2([0,\infty))$ strictly increasing from 0 to ∞. Let us consider the following equation

$$\begin{aligned} 0 &= \frac{\partial v}{\partial t}(t,\eta) - \langle A^*\partial v(t,\eta),\eta\rangle + F(t,\eta,v(t,\eta),\partial v(t,\eta)) \\ &\quad -\lambda trace(S\partial^2 v(t,\eta)), \quad 0 < t < T, \qquad (14) \\ & v(0,\eta) = \Psi(\eta) \end{aligned}$$

where $\lambda \geq 0$ and F ; $[0,T] \times H \times R^1 \times H \longmapsto R^1$ is uniformly continuous on bounded sets.

Definition 2.2. $v \in C([0,T]\times H)$ is called a subsolution (resp. supersolution) of (14), if $v(0,\eta) = \Psi(\eta)$ and the following condition (i) (resp.(ii)) holds for any test function Φ and radial function g,

(i) If v - Φ - g has a global maximum at $(\hat{t},\hat{\eta})$ with $\hat{t} \in (0,T)$, then

$$\frac{\partial \Phi}{\partial t}(\hat{t},\hat{\eta}) - \langle A^*\partial\Phi(\hat{t},\hat{\eta}),\hat{\eta}\rangle + F(\hat{t},\hat{\eta},v(\hat{t},\hat{\eta}),\partial(\Phi+g)(\hat{t},\hat{\eta}))$$

$$-\lambda trace(S\partial^2(\Phi+g)(\hat{t},\hat{\eta})) \leq \mu\tilde{g}'(\|\hat{\eta}\|)\|\hat{\eta}\|.$$

(ii) If $v + \Phi + g$ has a global maximum at $(\hat{t},\hat{\eta})$ with $\hat{t} \in (0,T)$, then

$$-\frac{\partial \Phi}{\partial t}(\hat{t},\hat{\eta}) + \langle A^*\partial\Phi(\hat{t},\hat{\eta}),\hat{\eta}\rangle + F(\hat{t},\hat{\eta},v(\hat{t},\hat{\eta}),-\partial(\Phi+g)(\hat{t},\hat{\eta}))$$

$$+\lambda trace(S\partial^2(\Phi+g)(\hat{t},\hat{\eta})) \geq -\mu\tilde{g}'(\|\hat{\eta}\|)\|\hat{\eta}\|.$$

where μ is the constant of (7). v is called a viscosity solution if it is both a subsolution and a supersolution.

For our equation, it follows that v^ε is a viscosity solution of (3), from the dynamic programming principle and Itô's formula. Moreover, applying the comparison theorem (Theorem 4.8 in [9]), we can see that v^ε is a unique viscosity solution in $BUC \cap C_w$ from Proposition 2.2. Therefore V^ε is also a unique viscosity solution of (4) in $BUC \cap C_w$. Noting Proposition 2.3, we can prove that its limit function V is a unique viscosity solution of (5) in $BUC \cap C_w$, by virtue of the stability property of viscosity solutions. Again recalling Proposition 2.2, we have

Theorem 2.1. As $\varepsilon \to 0$, V^ε converges to V uniformly on bounded sets of $[0,T] \times H$ and V is a unique viscosity solution of (5) in $BUC \cap C_w$.

3. Differential games

This section is devoted to the differential game associated with (6). Let us set

$\mathbf{Y} = L^2([0,T], H\ ), \mathbf{Z} = L^2([0,T], \gamma\ ),$

$\mathbf{M} = \{M\ ;\ \mathbf{Z} \longmapsto \mathbf{Y}$, non-anticipative$\}$, $\mathbf{N} = \{N\ ;\ \mathbf{Y} \longmapsto \mathbf{Z}$, non-anticipative$\}$
where M is said to be non-anticipative, if M is measurable and $M(Z) = M(Z')$ in $L^2[0,t]$ whenever $Z = Z'$ in $L^2[0,t]$ for any t. Y (resp.Z) is called a control for an opponent (resp. controller) and M (resp.N) a strategy for an opponent (resp. controller).

For Y and Z, we shall consider the system X governed by the following evolution equation,

$$\frac{\partial X}{\partial t}(t,x) = AX(t,x)+b(x,X(t,x),Z(t)(x))+SY(t)(x), \quad x \in D, 0<t<T,$$
$$X(t,x) = 0 \qquad for \quad x \in \partial D \quad and \quad X(0,x) = \eta(x). \tag{15}$$

Again, considering the solution as a function of t with value in H, we have a unique solution $X(*,\eta,Y,Z)$ in $C([0,T],H) \cap L^2([0,T],H^1)$ with properties (16) and (17),

$$\sup_{t\leq T} \|X(t,\eta_1,Y,Z) - X(t,\eta_2,Y,Z)\| \leq k_1\|\eta_1 - \eta_2\| \tag{16}$$

$$\int_0^T \|X(t_1,\eta_1,Y,Z) - X(t_2,\eta_2,Y,Z)\|^2 dt \leq k_2|\eta_1 - \eta_2|_B^2 \tag{17}$$

where k_1 and k_2 are independent of Y and Z . We assume the criterion J below,

$$\mathbf{J}(t,\eta,Y,Z) = \int_0^t h(X(\theta,\eta,Y,Z)) - \frac{1}{2}\langle SY(\theta),Y(\theta)\rangle d\theta \tag{18}$$

and put $\mathbf{J}(t,\eta,Y,N) = \mathbf{J}(t,\eta,Y,NY)$, $\mathbf{J}(t,\eta,M,Z) = \mathbf{J}(t,\eta,MZ,Z)$. According to Elliot-Kalton [4], the lower value ν^- (resp. upper value ν^+) is defined by (19) (resp. (20)),

$$\nu^-(t,\eta) = \inf_{N\in\mathbf{N}} \sup_{Y\in\mathbf{Y}} \mathbf{J}(t,\eta,Y,N) \tag{19}$$

$$\nu^+(t,\eta) = \sup_{M\in\mathbf{M}} \inf_{Z\in\mathbf{Z}} \mathbf{J}(t,\eta,M,Z). \tag{20}$$

ν is called the value of game, if it is the lower value and the upper value.

Hereafter we will deal with the lower value, since we can apply the same arguments to the upper value. First we can easily see

$$|\nu^-(t,\eta)| \leq \bar{h}t \tag{21}$$

and, with a constant k_3 independent of t,

$$|\nu^-(t,\eta_1) - \nu^-(t,\eta_2)| \leq k_3|\eta_1 - \eta_2|_B \tag{22}$$

For the continuity with respect to t, we will use the same method as in [1]. Let us put

$$\mathbf{Y}_t = \{Y \in \mathbf{Y}; \int_0^t \langle SY(\theta), Y(\theta)\rangle d\theta \leq 4\bar{h}t\}. \tag{23}$$

Then, we get, by (21)

$$\nu^-(t,\eta) = \inf_{N\in\mathbf{N}} \sup_{Y\in\mathbf{Y}_t} \mathbf{J}(t,\eta,Y,N). \tag{24}$$

Proposition 3.1.

$$|\nu^-(t,\eta) - \nu^-(s,\eta)| \leq \bar{h}|t-s|. \tag{25}$$

Proof. Using (24), we will prove

$$\nu^-(t,\eta) - \nu^-(s,\eta) \geq -\bar{h}(t-s), \qquad for \quad s < t. \tag{26}$$

For $\varepsilon > 0$, there is an ε optimal strategy $\tilde{N} = N(\varepsilon, t, \eta)$

$$\nu^-(t,\eta) \leq \sup_{Y\in\mathbf{Y}_t} \mathbf{J}(t,\eta,Y,\tilde{N}) < \nu^-(t,\eta) + \varepsilon. \tag{27}$$

Let us take $\tilde{Y} \in \mathbf{Y}_s$ such that

$$\mathbf{J}(s,\eta,\tilde{Y},\tilde{N}) > \sup_{Y\in\mathbf{Y}_s} \mathbf{J}(s,\eta,Y,\tilde{N}) - \varepsilon \geq \nu^-(s,\eta) - \varepsilon. \tag{28}$$

Then, putting

$$Y^*(\theta) = \begin{cases} \tilde{Y}(\theta) & for \quad \theta \in [0,s] \\ 0 & for \quad \theta \in (s,T] \end{cases} \tag{29}$$

we have $X(\theta,\eta,Y^*,\tilde{N}) = X(\theta,\eta,\tilde{Y},\tilde{N})$ for $\theta \in [0,s]$ and, by (27),

$$\mathbf{J}(t,\eta,Y^*,\tilde{N}) < \nu^-(t,\eta) + \varepsilon. \tag{30}$$

Now, combining (28) $\sim$ (30) together, we obtain

$$\nu^-(t,\eta) - \nu^-(s,\eta) \geq \mathbf{J}(t,\eta,Y^*,\tilde{N}) - \mathbf{J}(s,\eta,Y^*,\tilde{N}) - 2\varepsilon \geq -\bar{h}(t-s) - 2\varepsilon.$$

Next we will choose $\bar{N} \in \mathbf{N}$ and $\bar{Y} \in \mathbf{Y}_t$ such that

$$\sup_{Y\in\mathbf{Y}} \mathbf{J}(s,\eta,Y,\bar{N}) < \nu^-(s,\eta) + \varepsilon \tag{31}$$

and

$$\nu^-(t,\eta) \leq \sup_{Y \in \mathbf{Y}_t} \mathbf{J}(t,\eta,Y,\bar{N}) < \mathbf{J}(t,\eta,\bar{Y},\bar{N}) + \varepsilon. \tag{32}$$

Since [$\mathbf{J}(s,\eta,\bar{Y},\bar{N}) < \nu^-(s,\eta) + \varepsilon$] holds by (31), we have

$$\nu^-(t,\eta) - \nu^-(s,\eta) \leq \mathbf{J}(t,\eta,\bar{Y},\bar{N}) - \mathbf{J}(s,\eta,\bar{Y},\bar{N}) + 2\varepsilon \leq \bar{h}(t-s) + 2\varepsilon.$$

This completes the proof of Proposition 3.1.

Consequently, from (31) and (32), we see

Theorem 3.1. ν^- belongs to $BUC \cap C_w$ and satisfies the following,

$$|\nu^-(t,\eta)| \leq \bar{h}T \quad and \quad |\nu^-(t,\eta) - \nu^-(s,\zeta)| \leq k_4(|t-s| + |\eta - \zeta|_B).$$

4. Isaacs equation on H

First we will prove the dynamic programming principle for ν^-.

Theorem 4.1. For any $(s,\zeta) \in (0,T) \times H$,

$$\nu^-(s,\zeta) = \inf_{N \in \mathbf{N}} \sup_{Y \in \mathbf{Y}} [\mathbf{J}(\tau,\zeta,Y,N) + \nu^-(s-\tau, X(\tau,\zeta,Y,N))], \qquad for \quad \tau < s. \tag{33}$$

Proof. Let us denote the right hand side of (33) by $\omega(s,\zeta)$. Taking an ε optimal strategy $N^* \in \mathbf{N}$

$$\nu^-(s,\zeta) + \varepsilon > \sup_{Y \in \mathbf{Y}} \mathbf{J}(s,\zeta,Y,N^*), \tag{34}$$

we can choose $\hat{Y} \in \mathbf{Y}$, such that

$$\omega(s,\zeta) \leq \mathbf{J}(\tau,\zeta,\hat{Y},N^*) + \nu^-(s-\tau, X(\tau,\zeta,\hat{Y},N^*)) + \varepsilon. \tag{35}$$

For any $\tilde{Y} \in \mathbf{Y}$, we define $Y_1 = (\hat{Y},\tilde{Y})$ and $\tilde{N} \in \mathbf{N}$ as follows,

$$Y_1(\theta) = \begin{cases} \hat{Y}(\theta) & for \quad \theta \in [0,\tau] \\ \tilde{Y}_{-\tau}(\theta) & for \quad \theta \in (\tau,T] \end{cases}$$

where the shifted function $\tilde{Y}_t$ is defined by $\tilde{Y}_t(\theta) = \tilde{Y}(\theta + t)$, and

$$\tilde{N}\tilde{Y}(\theta) = N^* Y_1(\theta + \tau).$$

We denote the solution of (36) by X_1,

$$\dot{X}_1(t) = AX_1(t) + \beta(X_1(t), N^*Y_1(t)) + SY_1(t), \qquad \tau < t < T, \tag{36}$$

$$X_1(\tau) = X(\tau,\zeta,\hat{Y},N^*)$$

where $\dot{X}_1$ = derivative of X_1. Then

$$\mathbf{J}(t-\tau, X(\tau,\zeta,\hat{Y},N^*),\tilde{Y},\tilde{N}) = \int_\tau^t h(X_1(\theta)) - \frac{1}{2}\langle SY_1(\theta), Y_1(\theta)\rangle d\theta$$

holds. Let us choose a nearly optimal $\tilde{Y}_0$, such that

$$\nu^-(s-\tau, X(\tau,\zeta,\hat{Y},N^*)) \le \sup_{\tilde{Y}\in\mathbf{Y}} \int_\tau^s h(X_1(\theta)) - \frac{1}{2}\langle SY_1(\theta), Y_1(\theta)\rangle d\theta$$

$$\le \int_\tau^s h(X_0(\theta)) - \frac{1}{2}\langle SY_0(\theta), Y_0(\theta)\rangle d\theta, \tag{37}$$

where $Y_0 = (\hat{Y}, \tilde{Y}_0)$ and X_0 is the solution of (36) with Y_0 instead of Y_1. Now, combining (34) $\sim$ (37) together, we obtain

$$\omega(s,\zeta) \le \mathbf{J}(s,\zeta,Y,N^*) + 2\varepsilon \le \nu^-(s,\zeta) + 3\varepsilon.$$

Hence we get

$$\omega(s,\zeta) \le \nu^-(s,\zeta). \tag{38}$$

Next, we take a nearly optimal strategy $N(t,\eta)$ such that

$$\nu^-(t,\eta) + \varepsilon > \sup_{Y\in\mathbf{Y}} \mathbf{J}(t,\eta,Y,N(t,\eta)). \tag{39}$$

On the other hand, there exists N^0 with (40),

$$\omega(s,\zeta) + \varepsilon > \sup_{Y\in\mathbf{Y}} [\mathbf{J}(\tau,\zeta,Y,N^0) + \nu^-(s-\tau, X(\tau,\zeta,Y,N^0))]$$

$$\ge \mathbf{J}(\tau,\zeta,Y,N^0) + \nu^-(s-\tau, X(\tau,\zeta,Y,N^0), \quad for\ any \quad Y \in \mathbf{Y}. \tag{40}$$

Let us define N as follows

$$NY(\theta) = \begin{cases} N^0Y(\theta) & for \quad \theta \in [0,\tau] \\ N(s-\tau, X(\tau,\zeta,Y,N^0))Y_\tau(\theta-\tau) & for \quad \theta \in (\tau, T] \end{cases}$$

Then, from (39) and (40), it follows that for any $Y \in \mathbf{Y}$

$$\omega(s,\zeta)+\varepsilon > \mathbf{J}(\tau,\zeta,Y,N^0)+\mathbf{J}(s-\tau, X(\tau,\zeta,Y,N^0), Y_\tau, N(s-\tau, X(\tau,\zeta,Y,N^0))$$

$$= \mathbf{J}(s,\zeta,Y,N) \qquad for\ any \qquad Y \in \mathbf{Y}.$$

Therefore, we get the inverse inequality of (38).

Let us set

$$\mathbf{Y}(\tau) = \{Y \in \mathbf{Y}; \int_0^T \langle SY(\theta), Y(\theta)\rangle d\theta \le 4\bar{h}T\}.$$

Then, from (21), it follows that for any $Y \notin \mathbf{Y}(\tau)$

$$\mathbf{J}(\tau,\zeta,Y,N) < \bar{h}\tau - 2\bar{h}T < -\bar{h}\tau \qquad for \quad any \qquad N \in \mathbf{N}.$$

Remark. The dynamic programming principle (33) turns out to be the following form

$$\nu^-(s,\zeta) = \inf_{N\in\mathbf{N}} \sup_{Y\in\mathbf{Y}(\tau)} [\mathbf{J}(\tau,\zeta,Y,N) + \nu^-(s-\tau, X(\tau,\zeta,Y,N))]. \tag{41}$$

Noting, for any $Y \in \mathbf{Y}(\tau)$ and $t \in [0,\tau]$, $(\int_0^t \|SY(\theta)\|d\theta)^2 \le 4\bar{h}Tt(\max m_i)$, we have a constant k_5, independent of $N \in \mathbf{N}$ and $Y \in \mathbf{Y}(\tau)$, such that

$$\|X(t,\eta,Y,N) - \eta\| \le \|e^{tA}\eta - \eta\| + k_5\sqrt{t} \qquad for \quad t \le \tau. \tag{42}$$

Now appealing to (41) and the continuity properties (42) and (22), we will state our main theorems.

Theorem 4.2. ν^- is a viscosity solution of (5) in $BUC \cap C_w$.

This theorem implies $\nu^- = V$, since V is a unique viscosity solution in $BUC \cap C_w$. Moreover, we can show that the upper value ν^+ coincides with V, employing similar arguments.

Theorem 4.3. The differential game associated with (6) has the value, which coincides with the limit function V.

Proof of Theorem 4.2. First we will show that ν^- is a subsolution. Putting $\Psi = \Phi + g$, we suppose that ν^- - Ψ has a global maximum at $(\hat{t},\hat{\eta}) \in (0,T)\times H$. Then, the dynamic programming principle (41) implies the following evaluation,

$$\begin{aligned} 0 &\le \inf_{N\in\mathbf{N}} \sup_{Y\in\mathbf{Y}(\tau)} \mathbf{J}(\tau,\hat{\eta},Y,N) + \Psi(\hat{t}-\tau, X(\tau),Y,N) - \Psi(\hat{t},\hat{\eta}) \\ &= \inf_{N\in\mathbf{N}} \sup_{Y\in\mathbf{Y}(\tau)} \int_0^\tau \Big[h(X(\theta)) - \frac{\partial\Phi}{\partial t}(\hat{t}-\theta, X(\theta)) \\ &\quad + \langle A^*\Psi(\hat{t}-\theta, X(\theta)), X(\theta)\rangle + \langle \partial\Psi(\hat{t}-\theta, X(\theta)), \beta(X(\theta), NY(\theta))\rangle \\ &\quad + \langle \partial\Psi(\hat{t}-\theta, X(\theta)), SY(\theta)\rangle - \frac{1}{2}\langle SY(\theta), Y(\theta)\rangle \Big] d\theta, \end{aligned} \tag{43}$$

where $X(\theta)$ stands for $X(\theta,\hat{\eta},Y,N)$.

From (7), it follows that

$$\begin{aligned} \langle A^*\partial g(X(\theta)), X(\theta)\rangle &= \tilde{g}'(\|X(\theta)\|)\langle A^*X(\theta), X(\theta)\rangle \|X(\theta)\|^{-1} \\ &\le \mu\tilde{g}'(\|X(\theta)\|)\|X(\theta)\|. \end{aligned} \tag{44}$$

Moreover, noting

$$2\langle S\zeta,\xi\rangle - \langle S\zeta,\zeta\rangle = -\langle S(\zeta-\xi), \zeta-\xi\rangle + \langle S\xi,\xi\rangle \quad \le \quad \langle S\xi,\xi\rangle,$$

we get

$$\text{the integrand of right hand side of (43)}$$
$$\leq h(X(\theta)) - \frac{\partial \Phi}{\partial t}(\hat{t}-\theta, X(\theta)) + \langle A^* \partial \Phi(\hat{t}-\theta, X(\theta)), X(\theta)\rangle$$
$$+\mu \tilde{g}'(\|X(\theta)\|)\|X(\theta)\| + \langle \partial \Psi(\hat{t}-\theta, X(\theta)), \beta(X(\theta), NY(\theta))\rangle$$
$$+\frac{1}{2}\langle S\partial\Psi(\hat{t}-\theta, X(\theta)), \partial\Psi(\hat{t}-\theta, X(\theta))\rangle$$
$$\leq h(\hat{\eta}) - \frac{\partial \Phi}{\partial t}(\hat{t}, \hat{\eta}) + \langle A^*\partial\Phi(\hat{t},\hat{\eta}), \hat{\eta}\rangle + \mu\tilde{g}'(\|\hat{\eta}\|)\|\hat{\eta}\|$$
$$+ \langle \partial\Psi(\hat{t},\hat{\eta}), \beta(\hat{\eta}, NY(\theta))\rangle + \frac{1}{2}\langle S\partial\Psi(\hat{t},\hat{\eta}), \partial\Psi(\hat{t},\hat{\eta})\rangle + o(1), \tag{45}$$

where $o(1)$ is uniformly small in $Y \in \mathbf{Y}(\tau)$ and N, by virtue of (42). On the other hand, we have

$$\inf_{N\in\mathbf{N}} \sup_{Y\in\mathbf{Y}} \int_0^\tau \langle \xi, \beta(\hat{\eta}, NY(\theta)\rangle d\theta \leq \inf_{z\in\gamma}\langle \xi, \beta(\hat{\eta}, z)\rangle\tau, \tag{46}$$

because $\mathbf{N}$ contains any constant strategy N_z; $N_zY(\theta) = z$, for any Y and θ.

Consequently, from (43) $\sim$ (46), it follows that ν^- is a subsolution of (5).

Suppose that $\nu^- + \Psi$ has a global minimum at $(\hat{t},\hat{\eta}) \in (0,T) \times H$. Then, employing the same arguments as (43) $\sim$ (45), we get

$$0 \geq \inf_{N\in\mathbf{N}} \sup_{Y\in\mathbf{Y}(\tau)} [h(\hat{\eta}) + \frac{\partial\Phi}{\partial t}(\hat{t},\hat{\eta}) - \langle A^*\partial\Phi(\hat{t},\hat{\eta}), \hat{\eta}\rangle$$
$$-\mu\tilde{g}'(\|\hat{\eta}\|)\|\hat{\eta}\| + \frac{1}{2}\langle S\partial\Psi(\hat{t},\hat{\eta}), \partial\Psi(\hat{t},\hat{\eta})\rangle]\tau$$
$$+\int_0^\tau [-\frac{1}{2}\langle S(Y(\theta) - \partial\Psi(\hat{t}-\theta, X(\theta))), Y(\theta) - \partial\Psi(\hat{t}-\theta, X(\theta))\rangle$$
$$+ \langle \partial\Psi(\hat{t},\hat{\eta}), \beta(\hat{\eta}, NY(\theta))\rangle] d\theta + o(\tau). \tag{47}$$

Again noting $\langle S(\zeta-\xi), \zeta-\xi\rangle \leq 2\langle S(\zeta-\eta), \zeta-\eta\rangle + 2\langle S(\eta-\xi), \eta-\xi\rangle$ and $\langle \xi, \beta(\hat{\eta}, NY(\theta))\rangle \geq \inf_{z\in\gamma}\langle\xi, \beta(\hat{\eta}, z)\rangle$, we have

$$\textit{the integrand of } (47) \geq -\langle S(Y(\theta) - \partial\Psi(\hat{t},\hat{\eta})), Y(\theta) - \partial\Psi(\hat{t},\hat{\eta})\rangle$$
$$+ \inf_{z\in\gamma}\langle \partial\Psi(\hat{t},\hat{\eta}), \beta(\hat{\eta}, z)\rangle + o(1). \tag{48}$$

Let us set $\tilde{Y}(\theta) = \partial\Psi(\hat{t},\hat{\eta})\theta$. Then, $\tilde{Y} \in \mathbf{Y}(\tau)$ for small τ. Therefore from (47) and (48), it follows that ν^- is a supersolution of (5).

Application. When $\beta = 0$ and $m_j \neq 0$ for any j, Theorem 4.3 is related to the large deviation of Ventcel-Freidlin (cf. Chapter 12 in [3]). Since the solution $X(*, \eta, Y)$ of (15) belongs to $\mathbf{W}(\eta) = \{x \in W^{12}((0,T)\times R^n); x(0) = \eta\}$, $SY(t) = \dot{X}(t,\eta,Y) - AX(t,\eta,Y)$ holds. Therefore we get

$$\lim_{\varepsilon \to 0} \varepsilon \log E \exp(\frac{1}{\varepsilon} \int_0^T h(\xi^\varepsilon(t,\eta))dt)$$

$$= \sup_{x \in \mathbf{X}(\eta)} \int_0^T h(x(t)) - \frac{1}{2}\langle S^{-1}(\dot{x}(t) - Ax(t)), \dot{x}(t) - Ax(t)\rangle dt.$$

References

1. Bensoussan, A.,Nagai, H. : *Min* – max characterization of a small noise limit on risk sensitive control. preprint
2. Crandall, M.G., Lions, P.L. (1990): Hamilton-Jacobi equations in infinite dimensions, Part 4. J.Funct.Anal. 90, 237-283
3. Da Prato, G., Zabczyk, J., (1992): Stochastic equations in infinite dimensions. Cambridge Univ. Press, Encyclopedia of Math. Appl.
4. Elliot, R.J., Kalton, N.J., (1972): The existence of value in differential games. Mem. Amer. Math. Soc. 126
5. Fleming, W.H., McEneaney, W.M., (1992): Risk sensitive control and differential games. L. N. Contr. Inf. Sci. 184, 185-197
6. Itô, K., (1984): Foundation of stochastic differential equations in infinite dimensional spaces. SIAM Reg. Conf. Series 47
7. Nisio, M.,(1994): On sensitive control for stochastic partial differential equations. Pitman Res.Notes Math. 310, 231-241
8. Rozovskii, B.L., (1990): Stochastic evolution systems. Kluwer
9. Świech, A.,(1993): Viscosity solutions of fully nonlinear partial differential equations with unbounded terms in infinite dimensions. Ph.D.dissertation, Univ. of Calif.

Decomposition at the maximum for excursions and bridges of one-dimensional diffusions*

Jim Pitman[1] and Marc Yor[2]

[1] Department of Statistics, University of California, 367 Evans Hall # 3860, Berkeley, CA 94720-3860, USA
[2] Laboratoire de Probabilités, Université Pierre et Marie Curie, 4, Place Jussieu - Tour 56, 75252 Paris Cedex 05, France

1. Introduction

In his fundamental paper [25], Itô showed how to construct a Poisson point process of excursions of a strong Markov process X over time intervals when X is away from a recurrent point a of its statespace. The point process is parameterized by the local time process of X at a. Each point of the excursion process is a path in a suitable space of possible excursions of X, starting at a at time 0, and returning to a for the first time at some strictly positive time ζ, called the *lifetime* of the excursion. The intensity measure of the Poisson process of excursions is a σ-finite measure Λ on the space of excursions, known as Itô's excursion law. Accounts of Itô's theory of excursions can now be found in several textbooks [48, 46, 10]. His theory has also been generalized to excursions of Markov processes away from a set of states [34, 19, 10] and to excursions of stationary, not necessarily Markovian processes [38].

Itô's excursion theory has been applied to the study of the distribution of functionals of the trajectories of one-dimensional Brownian motion and Bessel processes [41, 9, 58, 39], and to the study of random trees [2, 3, 4, 5, 18, 1, 8] and measure valued diffusions [16]. In such studies, the following two descriptions of Itô's law Λ for excursions away from 0 of a reflecting Brownian motion X on $[0,\infty)$ have proved useful. Both involve BES(3), the 3-dimensional Bessel process. We recall that for positive integer δ a BES(δ) process can be defined as the radial part of a δ-dimensional Brownian motion, and that this definition can be extended using additivity properties of squares of Bessel processes to define a BES(δ) process for all real $\delta \geq 0$ [49]. The first description of Itô's law Λ is drawn from Itô's definition and observations of Lévy [33], Itô-McKean [26], and Williams [54]. The second description is due to Williams [56] and proved in Rogers [47].

> Description I: Conditioning on the lifetime: *First pick a lifetime t according to the σ-finite density $(2\pi)^{-1/2}t^{-3/2}dt$ on $(0,\infty)$; then given t, run a* BES(3) *bridge from* 0 *to* 0 *over time t.*

* Research supported in part by N.S.F. Grants MCS91-07531 and DMS-9404345

Description II: Conditioning on the maximum: *First pick a maximum value m according to the σ-finite density $m^{-2}dm$ on $(0,\infty)$; then given m, join back to back two independent* BES(3) *processes, each started at* 0 *and run till it first hits m.*

As explained in Biane-Yor [9] and Williams [57], the agreement between these two descriptions of Itô's excursion law, combined with Brownian scaling, implies an identity relating the distribution of the maximum of the standard Brownian excursion (or BES(3) bridge from 0 to 0 over time 1) and the distribution of the sum of two independent copies of the hitting time of 1 by BES(3). These authors show how this identity, expressed in terms of moments, is related to the functional equation for Riemann's zeta function. A central result of this paper is the following generalization of this identity from dimension $\delta = 3$ to arbitrary positive real δ:

Theorem 1.1. *For each $\delta > 0$, on the space of continuous non-negative paths with a finite lifetime, starting and ending at 0, there exists a σ-finite measure Λ_{00}^{δ} that is uniquely determined by either of the following descriptions:*

Description I: Conditioning on the lifetime: *First pick a lifetime t according to the σ-finite density $2^{-\frac{\delta}{2}}\Gamma(\frac{\delta}{2})^{-1}t^{-\frac{\delta}{2}}dt$ on $(0,\infty)$; then given t, run a* BES(δ) *bridge from* 0 *to* 0 *over time t.*

Description II: Conditioning on the maximum: *First pick a maximum value m according to the σ-finite density $m^{1-\delta}dm$ on $(0,\infty)$; then given m, join back to back two independent* BES(δ) *processes, each started at* 0 *and run till it first hits m.*

The measures Λ_{00}^{δ} defined by Description II for $\delta > 2$ were considered in [41] and further studied by Biane-Yor [9], who gave Description I in this case. It was shown in [41] that for $2 < \delta < 4$ the measure Λ_{00}^{δ} is Itô's excursion law for excursions of BES$(4-\delta)$ away from zero. For all $\delta \geq 2$ the measure Λ_{00}^{δ} concentrates on excursion paths starting at 0 and first returning to 0 at their lifetime. But the measure with density $t^{-\frac{\delta}{2}}dt$ on $(0,\infty)$ is a Lévy measure only for $2 < \delta < 4$. So for $\delta \leq 2$ or $\delta \geq 4$ the measure Λ_{00}^{δ} is not the excursion law of any Markov process. Nonetheless, these measures Λ_{00}^{δ} are well defined for all $\delta > 0$, and have some interesting properties and applications. As shown in [41], the measure $4\Lambda_{00}^{4}$ appears, due to the Ray-Knight description of Brownian local times, as the distribution of the square root of the total local time process of a path governed by the Itô's Brownian excursion law Λ_{00}^{3}. Consequently, Λ_{00}^{4} appears also in the Lévy-Khintchine representation of the infinitely divisible family of squares of Bessel processes and Bessel bridges [41, 39]. For $0 < \delta < 2$, the point 0 is a recurrent point for BES (δ), and the measure Λ_{00}^{δ} concentrates on paths which, unlike excursions, return many times to 0 before finally being killed at 0.

Here we establish Theorem 1.1 for all $\delta > 0$ using a general formulation of Williams' path decomposition at the maximum for one-dimensional diffusion bridges, presented in Section 2.. This formulation of Williams' decomposition, due to Fitzsimmons [15], contains an explicit factorization of the joint density of the time and place of the maximum of a one-dimensional diffusion bridge. For Brownian bridge this density factorization appears already in the work of Vincze [50] in 1957, and its extensions to Brownian excursion, Brownian meander and diffusion processes have been derived by several authors [13, 21, 23, 12]. As an application of this decomposition, in Section 3. we describe the law of the standard BES(δ) bridge by its density on path space relative to the law obtained by taking two independent BES(δ) processes started at 0 and run till they first hit 1, joining these processes back to back, and scaling the resultant process with a random lifetime and maximum 1 to have lifetime 1 and a random maximum.

Our approach to the family of measures $(\Lambda^{\delta}_{00}, \delta > 0)$ leads us to consideration of a σ-finite measure Λ_{xy} associated with a general one-dimensional diffusion process instead of BES(δ), for an arbitrary initial point x and final point y. Some instances of these measures were considered in [40]. Some of the results in this paper were announced in [42].

2. Williams' Decomposition for a One-dimensional diffusion

2.1 Decomposition at the maximum over a finite time interval

Let $X = (X_t, t \geq 0)$ be a regular one-dimensional diffusion on a sub-interval I of the real line. See [26] for background and precise definitions. To keep things simple, assume that I contains $[0, \infty)$, and that X has infinite lifetime. The infinitesimal generator A of X, restricted to smooth functions vanishing in some neighbourhoods of boundary points of I, is of the form

$$A = \frac{d}{dm}\frac{d}{ds}, \tag{2.1}$$

where $s = s(dx)$ and $m = m(dx)$ are the scale and speed measures of the diffusion. The semigroup of X admits a jointly continuous transition density relative to the speed measure

$$p(t, x, y) = P_x(X_t \in dy)/m(dy), \tag{2.2}$$

which is symmetric in (x, y). Here $P_x(\cdot) = P(\cdot\,|X_0 = x)$ defines the distribution on a suitable path space of the diffusion process started at $X_0 = x$. Let $P^t_{x,y}$ govern the *diffusion bridge* of length t from x to y:

$$P^t_{x,y}(\cdot) = P_x(\cdot\,|\,X_t = y) \tag{2.3}$$

Under $P^t_{x,y}$ the process $(X_s, 0 \leq s \leq t)$ is an inhomogeneous Markov process with continuous paths, starting at x at time 0 and ending at y at time t. The one-dimensional and transition probability densities of the diffusion bridge are derived from $p(t, x, y)$ in the obvious way via Bayes rule [14].

Let

$$M_t = \sup_{0 \leq s \leq t} X_s; \qquad \rho_t = \inf\{s : X_s = M_t\}. \tag{2.4}$$

For a diffusion X whose ultimate maximum M_∞ is a.s. finite, Williams [55] gave a path decomposition of X at the time ρ_∞ that X first attains this ultimate maximum value. Since this fundamental work of Williams variations of his idea have been developed and applied in a number of different contexts. See for instance Denisov [13], Millar [35, 36], Jeulin [27], Le Gall [17]. In particular, Fitzsimmons [15] gave the following decomposition at the maximum over a finite time interval, part (i) of which appears also in Csáki et al [12]. The density factorization (2.7) for Brownian bridge was found already by Vincze [50]. See also Imhof [21, 22] for related results, and Asmussen et al. [6] for an application to discretization errors in the simulation of reflecting Brownian motion. Let

$$f_{xz}(t) = P_x(T_z \in dt)/dt \tag{2.5}$$

where $T_z = \inf\{t : X_t = z\}$ is the first passage time to z. See [26],p.154, regarding the existence of continuous versions of such first passage densities. This allows rigorous construction of nice versions of the conditioned processes appearing in part (ii) of the following theorem, along the lines of [14].

Theorem 2.1. [55, 15, 12]
(i) *For $x, y \leq z < \infty, 0 \leq u \leq t$, the P_x joint distribution of M_t, ρ_t and X_t is given by*

$$P_x(M_t \in dz, \rho_t \in du, X_t \in dy) = f_{xz}(u) f_{yz}(t-u) s(dz) du\, m(dy). \tag{2.6}$$

Consequently the $P^t_{x,y}$ joint distribution of M_t and ρ_t is given by

$$P^t_{x,y}(M_t \in dz, \rho_t \in du) = \frac{f_{xz}(u) f_{yz}(t-u)}{p(t,x,y)} s(dz) du \tag{2.7}$$

(ii) *Under P_x conditionally given $M_t = z, \rho_t = u$ and $X_t = y$, that is to say under $P^t_{x,y}$ given $M_t = z$ and $\rho_t = u$, the path fragments*

$$(X_s, 0 \leq s \leq u) \text{ and } (X_{t-s}, 0 \leq s \leq t-u)$$

are independent, distributed respectively like

$$(X_s, 0 \leq s \leq T_z) \text{ under } P_x \text{ given } T_z = u$$

and

$$(X_s, 0 \leq s \leq T_z) \text{ under } P_y \text{ given } T_z = t-u$$

Integrating out u in formula (2.7) gives an expression of convolution type for the density at z of the maximum M_t of a diffusion bridge from x to y over time t. A second integration then yields

$$P^t_{x,y}(M_t \geq z)p(t,x,y) = \int_0^t du \int_z^\infty s(da) f_{xa}(u) f_{ya}(t-u) \tag{2.8}$$

$$= \int_0^t f_{xz}(u) p(t-u,z,y) du \tag{2.9}$$

Here the equality between (2.9) and the left side of (2.8) is clear directly by interpreting the latter as

$$P_x(M_t \geq z, X_t \in dy)/m(dy) = P_x(t \geq T_z, X_t \in dy)/m(dy)$$

and conditioning on T_z. Following the method used by Gikhman [20] in the case of Bessel processes, explicit formulae for the bridge probabilities $P^t_{x,y}(M_t \geq z)$ for particular diffusions can be computed using the Laplace transformed version of (2.9), which is

$$\int_0^\infty e^{-\alpha t} P^t_{x,y}(M_t \geq z)p(t,x,y)dt = \phi^\uparrow(\alpha,x)\phi^\uparrow(\alpha,y)\frac{\phi^\downarrow(\alpha,z)}{\phi^\uparrow(\alpha,z)} \tag{2.10}$$

where $\phi^\uparrow(\alpha,x)$ and $\phi^\downarrow(\alpha,x)$ are the increasing and decreasing solutions of $Au = \alpha u$, for $\alpha > 0$, normalized so that

$$\int_0^\infty e^{-\alpha t} p(t,x,y)dt = \phi^\uparrow(\alpha, x \wedge y)\phi^\downarrow(\alpha, x \vee y). \tag{2.11}$$

Then

$$E_x(e^{-\alpha T_z}) = \int_0^\infty e^{-\alpha t} f_{xz}(t)dt = \frac{\phi^\uparrow(\alpha,x)}{\phi^\uparrow(\alpha,z)} \text{ for } x \leq z \tag{2.12}$$

and the same holds with $\phi^\downarrow(\alpha,\cdot)$ instead of $\phi^\uparrow(\alpha,\cdot)$ for $x \geq z$. See Itô-McKean [26] for these formulae. In view of (2.11) and (2.12), the equality between the right sides of (2.8) and (2.9) reduces by Laplace transforms to the classical Wronskian identity

$$\phi^\downarrow(\alpha,x)\frac{d\phi^\uparrow(\alpha,x)}{s(dx)} - \phi^\uparrow(\alpha,x)\frac{d\phi^\downarrow(\alpha,x)}{s(dx)} = 1 \tag{2.13}$$

To see this, note from (2.11) and (2.12) that the Laplace transform of the righthand expression in (2.8) becomes

$$\phi^\uparrow(\alpha,x)\phi^\uparrow(\alpha,y)\int_z^\infty s(da)\frac{1}{[\phi^\uparrow(\alpha,a)]^2} \tag{2.14}$$

This equals the Laplace transform of the right-hand side of (2.9), which, as already remarked, is the expression in (2.10). Indeed, the Wronskian formula (2.13) makes

$$\frac{d}{s(dz)}\left(\frac{\phi^{\downarrow}(\alpha,z)}{\phi^{\uparrow}(\alpha,z)}\right)=\frac{-1}{[\phi^{\uparrow}(\alpha,z)]^2}$$

and each of the expressions in (2.10) and (2.14) vanishes as $z\uparrow\infty$, because the assumption of infinite lifetime implies $\phi^{\uparrow}(\alpha,z)\uparrow\infty$ as $z\uparrow\infty$. Csáki et al [12] used a variation of this argument to derive (2.6).

2.2 The agreement formula for a diffusion bridge

With one more simplifying assumption, Theorem 2.1 can be expressed as in the next corollary. This corollary is a generalization of Theorem 1.1 suggested by work of Williams [56], Pitman-Yor [41], Biane-Yor [9], Biane [7]. The notation is taken from Section 6 of [9], where more formal definitions can be found. For a distribution Q on path space, and a random time T, let Q^T be the distribution of the path obtained by killing at time T. Let $Q^\wedge$ be the image of Q by time reversal. For a second distribution of paths Q', let $Q\circ Q'$, the *concatenation* of Q and Q', be the distribution of the path obtained by first following a path distributed according to Q, then continuing independently according to Q'.

Corollary 2.1. Agreement Formula for Diffusion Bridges. *Assume that for all $x,y\in I$ with $x<y$, $P_x(T_y<\infty)=1$. Then for all $x,y\in I$ there is the following identity of measures on path space:*

$$\int_0^\infty dt\,p(t,x,y)P^t_{x,y}=\int_{x\vee y}^\infty s(dz)(P_x^{T_z})\circ(P_y^{T_z})^\wedge \tag{2.15}$$

Theorem 1.1 amounts to the following instance of this formula when the basic diffusion is BES(δ) and $x=y=0$:

$$\Lambda^\delta_{00}=\int_0^\infty dt\frac{P^t_{0,0}}{(2t)^{\frac{\delta}{2}}\Gamma(\frac{\delta}{2})}=\int_0^\infty\frac{dz(P_0^{T_z})\circ(P_0^{T_z})^\wedge}{z^{\delta-1}} \tag{2.16}$$

Definition 2.1. *For a one-dimensional diffusion subject to the conditions of Corollary 2.1, let Λ_{xy} denote the measure on path space defined by either side of the agreement formula* (2.15).

The measure Λ_{xy} is always σ-finite. Its total mass is the 0-potential density

$$\int_0^\infty p(t,x,y)dt=s(\infty)-s(x\vee y)$$

which may be either finite or infinite. Informally, the agreement formula states that each of the following two schemes derived from a basic diffusion process X can be used to generate Λ_{xy}:
(LHS) *Pick t according to $p(t,x,y)dt$ and then run an X bridge of length t from x to y*

(RHS) *Pick z according to the speed measure $s(dz)$ restricted to $(x \vee y, \infty)$, then join back to back a copy of X started at x run to T_z and a copy of X started at y run to T_z.*

The (LHS) amounts to conditioning on the lifetime of the path from x to y, while the (RHS) amounts to conditioning on the maximum.

Clearly, Λ_{xy} concentrates on paths starting at x and ending at y, and attaining a maximum value, M say, at a unique intermediate time. Note that $\Lambda_{xy}^{\wedge} = \Lambda_{yx}$. This is obvious from the right side of (2.15), and can be seen also on the left side, because $p(t,x,y) = p(t,y,x)$, and $(P_{x,y}^t)^{\wedge} = P_{y,x}^t$.

2.3 Relation to last exit times

We now consider the case when X is transient, i.e. $X_t \to \infty$ as $t \to \infty$. We can then choose s such that $s(\infty) = 0$. In this transient case, the measure Λ_{xy} is finite, and in fact is a multiple of the restriction of $P_x^{L_y}$ to $L_y > 0$, where $L_y = \sup\{t > 0 : X_t = y\}$ with the usual convention that $\sup(\emptyset) = 0$. To be precise, by formula (6.e) of [40],

$$P_x(L_y \in dt) = -s(y)^{-1} p(t,x,y) dt \tag{2.17}$$

where we have dropped a factor of 2 from the formula of [40] due to our definition of the speed measure m here using $A = \frac{d}{dm}\frac{d}{ds}$ rather than $A = \frac{1}{2}\frac{d}{dm}\frac{d}{ds}$ as in [40]. Furthermore, from [40] there is the formula

$$P_x^{L_y}(\,\cdot \mid L_y = t) = P_{x,y}^t \tag{2.18}$$

so for transient X the agreement formula (2.15) can be written

$$P_x^{L_y}(\,\cdot \cap (L_y > 0)\,) = -\frac{1}{s(y)} \int_{x \vee y}^{\infty} s(dz)(P_x^{T_z}) \circ (P_y^{T_z})^{\wedge} \tag{2.19}$$

When X is the BES(3) process on $[0,\infty)$, and $x = y = 0$, the σ -finite measure Λ_{00} appearing in (2.15) is Itô's excursion law. The LHS is the description of Itô's law for Brownian excursions due to Lévy [33] and Itô [25], while the RHS is Williams' [56] description. As noted in Biane-Yor [9] and Williams [57], the agreement between these two descriptions of Itô's law has interesting consequences related to the functional equation for the Riemann zeta function.

Corollary 2.1 allows the identity (2.8) to be lifted to an identity of measures on path space:

$$\text{the restriction of } \Lambda_{xy} \text{ to } (M > z) \text{ is } P_x^{T_z} \circ \Lambda_{zy} \tag{2.20}$$

We note also that integration with respect of $m(dy)$ yields the following version of the agreement formula for unconditioned diffusions:

$$\int_0^{\infty} dt\, P_x^t = \int_x^{\infty} s(dz)(P_x^{T_z}) \circ \left(\int_{-\infty}^{z} m(dy)(P_y^{T_z})^{\wedge}\right) \tag{2.21}$$

Similar representations of the left side of (2.21) for Brownian motion appear in [9] (see also [46], Ex. (4.18) of Ch. XII). These too can be formulated much as above for a general diffusion.

2.4 Relation to excursion laws

The connection between BES(3) and BM is that BES(3) is BM on $[0,\infty)$ *conditioned to approach* ∞ *before 0*, a concept made precise by Doob's theory of h-transforms. More generally, if 0 is a recurrent point of a regular diffusion Y on an interval I which contains $[0,\infty)$, and X is Y conditioned to approach ∞ before 0, then Λ_{00} derived from X admits a similar interpretation as Itô's law for excursions of Y above 0. See Section 3 of Pitman-Yor [41], where Williams' representation of Λ_{00} is given along with two other representations of the measure in this case, due to Itô and Williams. In view of (2.20), the second of these two other descriptions also identifies Λ_{x0} derived from X, for $x > 0$, as

$$\Lambda_{x0} = s(x,\infty)Q_x^{T_0}$$

where Q_x is the distribution of Y started at x. The description of Itô's excursion law for a general one-dimensional diffusion Y, via the LHS of (2.15) for X as above, is less well known. According to this description, the Lévy measure governing the duration of excursions of the recurrent diffusion Y above 0 has density at t identical to $p(t,0,0)$ for the diffusion X on $[0,\infty)$ obtained by conditioning Y to approach ∞ before hitting 0. See Knight [31], Kotani-Watanabe [32], concerning the problem of characterizing such Lévy densities.

The two other descriptions of an Itô excursion law, given in Section 3 of [41], do not make sense in the generality of Corollary 2.1, because they involve conditioning on sets which might have infinite mass. In particular, this is the case if X is recurrent, for example a standard Brownian motion. If y is a recurrent point for X, the measure Λ_{xy}, while σ-finite on path space, has finite dimensional distributions that are not σ-finite. This follows from the LHS of the agreement formula (2.15) combined with the fact that $\int_v^\infty p(t,x,y)dt = \infty$ for every $v > 0$. The measures Λ_{yy}, as defined by the LHS of the agreement formula for a recurrent point y, were considered in Pitman-Yor [40], and applied in case X is a Bessel process to establish complete monotonicity of some particular ratios of Bessel functions. As noted in [40], if $(\tau_\ell, \ell \geq 0)$ is the inverse of the local time process $(L_t, t \geq 0)$ at a recurrent point y, normalized so that $E_x(dL_t) = p(t,x,y)dt$, then there is the further identity

$$\int_0^\infty dt\, p(t,x,y)P_{x,y}^t = \int_0^\infty d\ell P_x^{\tau_\ell} \tag{2.22}$$

That is to say , for a recurrent point y a third description of the measure Λ_{xy} in (2.15) is obtained by first picking ℓ according to Lebesgue measure, then running the diffusion started at x until the time τ_ℓ that its local time at y first equals ℓ.

3. The Agreement Formula for Bessel Processes

3.1 Definition and basic properties of Bessel Processes

Let $R = (R_t, t \geq 0)$ be a BES(δ) process started at $R_0 = 0$. Here δ is a strictly positive real parameter. For $\delta = 1, 2, \ldots$, a BES(δ) diffusion R is obtained as the radial part of BM in $\mathbb{R}^\delta$. See Itô-McKean [26] Section 2.7. For positive integer parameters, this representation displays the *Pythagorean property* of Bessel processes: the sum of squares of independent BES(δ) and BES(δ') processes is the square of a BES($\delta + \delta'$) process. As shown by Shiga-Watanabe [49], the family of BES(δ) processes for real $\delta > 0$ is characterized by extension of this Pythagorean property to all positive real δ and δ'. Typical properties of Bessel processes are consequences of the Brownian representation for integer δ that admit natural extensions to all $\delta > 0$. See [46] for further background and proofs of the basic properties of BES(δ) now recalled.

The BES(δ) process is a diffusion on $[0, \infty)$ whose infinitesimal generator A_δ acts on smooth functions vanishing in a neighbourhood of 0 as

$$A_\delta = \frac{1}{2}\frac{d^2}{dx^2} + \frac{\delta - 1}{2x}\frac{d}{dx} = \frac{d}{dm_\delta}\frac{d}{ds_\delta}. \tag{3.1}$$

where the scale and speed measures s_δ and m_δ can be chosen to be

$$s_\delta(dx) = x^{1-\delta}dx, \qquad m_\delta(dx) = 2x^{\delta-1}dx. \tag{3.2}$$

For $0 < \delta < 2$, the definition of the generator is completed by specifying that the boundary point 0 acts as a simple instantaneously reflecting barrier.

The Pythagorean property implies easily that for all $\delta > 0$

$$\text{the law of } R_t^2/2t \text{ is gamma } (\tfrac{\delta}{2}).$$

That is to say

$$P(R_t \in dy) = 2^{1-\frac{1}{2}\delta}\Gamma(\delta/2)^{-1}t^{-\frac{\delta}{2}}y^{\delta-1}e^{-\frac{y^2}{2t}}dy = p_\delta(t,0,y)m_\delta(dy) \tag{3.3}$$

where

$$p_\delta(t,0,y) = (2t)^{-\frac{\delta}{2}}\Gamma(\tfrac{\delta}{2})^{-1}e^{-\frac{y^2}{2t}} \tag{3.4}$$

is the transition probability density relative to the speed measure. This is the simple special case $x = 0$ of the general formula for the transition probability function $p_\delta(t, x, y)$ of the Bessel diffusion, for which see Itô-McKean [26] Section 2.7, Molchanov and Ostrovski [37].

The BES(δ) process R for each real $\delta > 0$ inherits the familiar *Brownian scaling property* from integer dimensions: for every $c > 0$

$$(c^{-1/2}R_{ct}, t \geq 0) \stackrel{d}{=} (R_t, t \geq 0)$$

A *standard Bessel* (δ) *bridge* is a process

$$(R_u^{\mathrm{br}}, 0 \le u \le 1) \stackrel{d}{=} (R_u, 0 \le u \le 1 | R_1 = 0).$$

For all $\delta > 0$ a standard BES(δ) bridge R^{br} is conveniently constructed from the unconditioned BES(δ) process R as

$$R_u^{\mathrm{br}} = (1-u)R(u/(1-u)), 0 \le u < 1.$$

In particular, for positive integer δ, the square of the standard BES(δ) bridge is distributed as the sum of squares of δ independent standard one-dimensional Brownian bridges.

By Brownian scaling, for $t > 0, \delta > 0$, a BES(δ) bridge from 0 to 0 over time t can be represented in terms of the standard BES(δ) bridge R^{br} as

$$\sqrt{t} R_{s/t}^{\mathrm{br}}, 0 \le s \le t$$

We note in passing that an interesting continuum of processes, passing from the Bessel bridges to the free Bessel processes and including the Bessel meanders, is introduced and studied in [45].

3.2 Random Scaling Construction of the Standard Bessel Bridge.

The following theorem is an expression of the agreement formula (2.15) for Bessel processes. This is a probabilistic expression in terms of standard bridges of Theorem 1.1.

Theorem 3.1. *Let R and $\hat{R}$ be two independent $BES(\delta)$ processes starting at 0, T and $\hat{T}$ their first hitting times of 1. Define $\tilde{R}$ by connecting the paths of R on $[0, T]$ and $\hat{R}$ on $[0, \hat{T}]$ back to back:*

$$\tilde{R}_t = \begin{cases} R_t \text{ if } t \le T \\ \hat{R}_{T+\hat{T}-t} \text{ if } T \le t \le T + \hat{T}, \end{cases}$$

and let $\tilde{R}^{\mathrm{br}}$ be obtained by Brownian scaling of $\tilde{R}$ onto the time scale $[0,1]$:

$$\tilde{R}_u^{\mathrm{br}} = (T+\hat{T})^{-1/2} \tilde{R}_{u(T+\hat{T})}, \ 0 \le u \le 1.$$

Let R^{br} be a standard BES^δ bridge. Then for all positive or bounded measurable functions $F : C[0,1] \to \mathbb{R}$,

$$E[F(R^{\mathrm{br}})] = c_\delta E[F(\tilde{R}^{\mathrm{br}})(\tilde{M}^{\mathrm{br}})^{2-\delta}] \tag{3.5}$$

where

$$\tilde{M}^{\mathrm{br}} = \sup_{0 \le u \le 1} \tilde{R}_u^{\mathrm{br}} = (T+\hat{T})^{-1/2} \tag{3.6}$$

$$c_\delta = 2^{\frac{\delta}{2}-1} \Gamma(\tfrac{\delta}{2}). \tag{3.7}$$

Proof. Fix δ. Let f_z be the density of $T_z = \inf\{t : R_t = z\}$ for the unconditional BES(δ) diffusion R started at $X_0 = 0$. Applied to the standard BES(δ) bridge R^{br}, and using (3.2), (3.4) and the scaling property

$$f_z(t) = z^{-2} f_1(tz^{-2}),\ t > 0,\ z > 0,$$

formula (2.7) yields

$$\frac{P(M^{\text{br}} \in dz, \rho^{\text{br}} \in dt)}{dz\, dt} = 2c_\delta f_1\left(\frac{t}{z^2}\right) f_1\left(\frac{1-t}{z^2}\right) z^{-\delta-3}. \tag{3.8}$$

On the other hand, $\tilde{R}^{\text{br}}$ constructed as above has maximum value

$$\tilde{M}^{\text{br}} = (T + \hat{T})^{-1/2} \text{ attained at time } \tilde{\rho}^{\text{br}} = \frac{T}{T+\hat{T}} \tag{3.9}$$

where T and $\hat{T}$ are independent with density f_1. Thus by a change of variables

$$\frac{P(\tilde{M}^{\text{br}} \in dz, \tilde{\rho}^{\text{br}} \in dt)}{dz\, dt} = 2 f_1\left(\frac{t}{z^2}\right) f_1\left(\frac{1-t}{z^2}\right) z^{-5}. \tag{3.10}$$

Comparison of (3.8) and (3.10) shows that (3.5) holds for F a function of the maximum of the process and the time it is attained. To lift the formula from the above identity of joint laws for the time and level of the maximum, to the identity of laws on the path space $C[0,1]$, it only remains to be seen that the two laws on path space share a common family of conditional laws given the time and level of the maximum: for $y > 0,\ 0 < t < 1$,

$$P(R^{\text{br}} \in \cdot\, | M^{\text{br}} = y, \rho^{\text{br}} = t) = P(\tilde{R}^{\text{br}} \in \cdot\, | \tilde{M}^{\text{br}} = y, \tilde{\rho}^{\text{br}} = t).$$

But this follows immediately from Williams decomposition as stated in part (ii) of Theorem 2.1, and Brownian scaling. □

According to (3.5), the law of the standard Bessel bridge R^{br} on $C[0,1]$ is mutually absolutely continuous with respect to that of $\tilde{R}^{\text{br}}$, with density at $w \in C[0,1]$

$$\frac{P(R^{\text{br}} \in dw)}{P(\tilde{R}^{\text{br}} \in dw)} = c_\delta \left(\sup_{0 \le u \le 1} w_u\right)^{2-\delta}. \tag{3.11}$$

Our formulation of Theorems 1.1 and 3.1 was suggested by Section 3 of Biane-Yor [9], where some forms of these results are discussed for $\delta > 2$. The present development shows that everything works also for $0 < \delta \le 2$. Recall the well known fact that dimension $\delta = 2$ is the threshold between recurrence and transience of BES(δ) processes:

for $\delta > 2$, there are no recurrent points for the BES(δ) diffusion;
for $\delta = 2$, every $x > 0$ is a recurrent point, but 0 is only neighbourhood-recurrent, not point recurrent;
for $0 < \delta < 2$, every $x \ge 0$ is a recurrent point for BES(δ).

Dimension 2 plays a special role here, as the unique dimension that makes the density factor (3.11) identically equal to 1. Thus for $(\tilde{R}^{\mathrm{br}}_u, 0 \le u \le 1)$ defined as in Theorem 3.1 by pasting back to back two independent BES(δ) processes run till they first hit 1 then Brownian scaling the result to have lifetime 1, there is the following immediate consequence of Theorem 3.1:

Corollary 3.1. *The process $(\tilde{R}^{\mathrm{br}}_u, 0 \le u \le 1)$ is a standard BES(δ) bridge if and only if $\delta = 2$.*

Combined with the skew-product description of planar Brownian motion, (see e.g. [26] or [46]) this yields in turn:

Corollary 3.2. *Run each of two independent planar Brownian motions Z and $\hat{Z}$ starting at the origin until hitting the unit circle, at times T and $\hat{T}$ respectively. Rotate the entire path of $\hat{Z}$ over the time interval $[0, \hat{T}]$ to make the two paths meet when they first reach the unit circle at times T and $\hat{T}$. Define a path $Z^{\dagger}$ with lifetime $T + \hat{T}$ by first travelling out to the unit circle over time T via Z, then returning via the reversed and rotated path of $\hat{Z}$ over a following time interval of length $\hat{T}$. Finally, rescale $Z^{\dagger}$ to have lifetime 1 by Brownian scaling. Then the resultant process is a standard planar Brownian bridge.*

Some applications of this result have been made by Werner [52, 53] to study the shape of the small connected components of the complement of a 2-dimensional Brownian path. We note also the following *asymptotic representation* of the 2-dimensional Brownian bridge as the limit in distribution as $r \to 0$ of

$$\left(\frac{1}{\sqrt{T_r}} Z(uT_r); 0 \le u \le 1\right)$$

where $Z = (Z(t), t \ge 0)$ is a 2-dimensional Brownian motion started at $Z(0) \ne 0$, and T_r is the hitting time of $\{z : |z| = r\}$ by Z.

A construction like that in Corollary 3.2 can be made starting from δ dimensional Brownian motion for any $\delta = 1, 2, 3, \ldots$. But the result is the standard bridge only for $\delta = 2$. For other dimensions δ the result has distribution absolutely continuous with respect to that of the bridge, with density the function of the maximum of the radial part indicated by (3.11).

3.3 Applications

A subscript δ will now be used to indicate the dimension of the underlying Bessel process. So

$$\begin{aligned} T_\delta &= \text{hitting time of 1 for a BES}(\delta) \text{ started at 0} \\ \hat{T}_\delta &= \text{independent copy of } T_\delta \\ (M^{\mathrm{br}}_\delta, \rho^{\mathrm{br}}_\delta) &= \text{level and time of the maximum for a standard} \\ &\quad\ \text{BES}(\delta) \text{ bridge.} \end{aligned}$$

The distribution of T_δ is determined by its Laplace transform (Kent [29])

$$\varphi_\delta(\lambda) = E\exp(-\lambda T_\delta) = \frac{(2\lambda)^{\mu/2}}{c_\delta I_\mu(\sqrt{2\lambda})} \tag{3.12}$$

where $\mu = \frac{\delta}{2} - 1$ is the *index* corresponding to dimension δ, and $c_\delta = 2^{\frac{\delta}{2}-1}\Gamma(\frac{\delta}{2}) = 2^\mu \Gamma(\mu+1)$ as in (3.7). According to Ismail-Kelker ([24], Theorem 4.10) the corresponding density f_δ can be written as a series expansion involving the zeros of J_μ, the usual Bessel function of index μ.

3.3.1 Moment identities. Several consequences of (3.5), all of which are apparent at the level of the joint laws (3.8) and (3.10), were noted for $\delta > 2$, i.e. $\mu > 0$, as formulae (3.k), (3.l), (3.k'), (3.k") of [9]. According to the Theorem 3.1, these identities in fact hold for all $\delta > 0$: for all positive measurable functions f:

$$E[f(M_\delta^{\mathrm{br}})] = c_\delta E[f((T_\delta + \hat{T}_\delta)^{-1/2})(T_\delta + \hat{T}_\delta)^{\frac{\delta}{2}-1}] \tag{3.13}$$

$$E[f(\rho_\delta^{\mathrm{br}})] = c_\delta E\left[f\left(\frac{T_\delta}{T_\delta + \hat{T}_\delta}\right)(T_\delta + \hat{T}_\delta)^{\frac{\delta}{2}-1}\right]. \tag{3.14}$$

In particular,

$$E(M_\delta^{\mathrm{br}})^{\delta-2} = c_\delta \tag{3.15}$$

$$E(T_\delta + \hat{T}_\delta)^{\frac{\delta}{2}-1} = 1/c_\delta \tag{3.16}$$

3.3.2 Relation to Kiefer's formula. Let $f_\delta^{(2)}(t) = f_\delta * f_\delta(t)$ denote the density of $T_\delta + \hat{T}_\delta$, with Laplace transform $[\varphi_\delta(\lambda)]^2$. According to (3.13),

$$P[(M_\delta^{\mathrm{br}})^2 \in da] = c_\delta a^{-\frac{\delta}{2}-1} f_\delta^{(2)}(a^{-1})da. \tag{3.17}$$

For integer dimensions δ, Kiefer ([30], (3.21)) found a formula for the density of $(M_\delta^{\mathrm{br}})^2$ which also involves the zeros of J_μ. It appears that Kiefer's method and formula are valid also for arbitrary $\delta > 0$. Comparison of Kiefer's formula and (3.17) using the formula of Ismail-Kelker for f_δ leads to some tricky identities involving the zeros of J_μ. Kiefer [30],p.429 discusses the cases $\delta = 1$ and $\delta = 3$. The second case is of special interest because, as noted by Williams [54], the standard Brownian excursion is a $BES(3)$ bridge. Kiefer's formulae were rediscovered in the context of Brownian excursions by Kennedy [28] and Chung [11].

3.3.3 Moment identities for dimension 2. Differentiation of (3.16) with respect to δ at $\delta = 2$ yields

$$E[\log(T_2 + \hat{T}_2)] = -\log 2 - \Gamma'(1) \tag{3.18}$$

From (3.13) with $\delta = 2$ one also gets

$$2E[\log(M_2^{\mathrm{br}})] = \log 2 + \Gamma'(1) \tag{3.19}$$

Recently, formula (3.18) has been useful in checking the following asymptotic result, which is of interest in certain questions related to random environments: For $(B_s, s \geq 0)$ a one-dimensional BM

$$E\left[\log\left(\int_0^t \exp(B_s)ds\right)\right] - \sqrt{\frac{2t}{\pi}} \to \log 2 - \Gamma'(1) \text{ as } t \to \infty. \tag{3.20}$$

This follows from the consequence of Theorem 2.1 and the Ray-Knight description of Brownian local times that for $S_1 = \sup_{0 \leq s \leq 1} B_s$

$$t \int_0^1 \exp(-\sqrt{t}(S_1 - B_s))ds \xrightarrow{d} 4(T_2 + \hat{T}_2) \text{ as } t \to \infty \tag{3.21}$$

where $\xrightarrow{d}$ denotes convergence in distribution.

3.3.4 A check for dimensions less than 2. For $0 < \delta < 2$, corresponding to $-1 < \mu < 0$, we have a check that the evaluation of the constant $c_\delta = 2^\mu \Gamma(\mu + 1)$, in (3.5), (3.13) etc. is correct, starting from (3.12). For a r.v. $X \geq 0$ with Laplace transform $\varphi(\lambda) = Ee^{-\lambda X}$, there is the formula

$$EX^p = \frac{1}{\Gamma(-p)} \int_0^\infty \lambda^{-p-1} \varphi(\lambda) d\lambda, \quad p < 0 \tag{3.22}$$

Applied to $X = T_\delta + \hat{T}_\delta$, $p = \mu$, for $-1 < \mu < 0$, this yields

$$E(T_\delta + \hat{T}_\delta)^\mu = \frac{1}{c_\delta} \int_0^\infty \frac{d}{d\lambda} \frac{I_{-\mu}(\sqrt{2\lambda})}{I_\mu(\sqrt{2\lambda})} d\lambda \tag{3.23}$$

due to the standard formula for the Wronskian of I_μ and $I_{-\mu}$ (Watson [51], p.77) By standard asymptotics of I_μ, this confirms that (3.16) holds for $-1 < \mu < 0$, with $c_\delta = 2^\mu \Gamma(1 + \mu)$ as in (3.7).

3.3.5 A check for integer dimensions. In case $\mu = k$ is a positive integer, formula (3.16) can be checked using

$$E\left[(T_\delta + \hat{T}_\delta)^k\right] = (-1)^k \left.\frac{d^k}{d\lambda^k}\right|_{\lambda=0} [\phi_\delta(\lambda)^2] \tag{3.24}$$

Note also the easy equality

$$E\left[(T_\delta + \hat{T}_\delta)\right] = 2E(T_\delta) = \frac{2}{\delta} E\left[(R_\delta(T_\delta))^2\right] = \frac{2}{\delta} \tag{3.25}$$

which, in case $\delta = 4$, agrees with (3.16), since $c_4 = 2$.

3.3.6 Chung's identity. To further illustrate formula (3.13), we now show how it implies the remarkable identity

$$(M_1^{\rm br})^2 \stackrel{d}{=} \frac{\pi^2}{4} T_3 \tag{3.26}$$

which was discovered by Chung [11]. See Biane-Yor [9] and Pitman-Yor [42, 44, 43] for further discussion and related identities.

Take $f(x) = e^{-\frac{1}{2}\lambda^2 x^2}$, so $f(\frac{1}{\sqrt{t}}) = e^{-\lambda^2/2t}$ in (3.13):

$$E\exp\left(-\frac{\lambda^2}{2}(M_\delta^{\rm br})^2\right) = c_\delta E\left[(T_\delta + \hat{T}_\delta)^{\frac{\delta}{2}-1}\exp\left(-\frac{\lambda^2}{2(T_\delta + \hat{T}_\delta)}\right)\right].$$

For $\delta = 1$, this expression equals

$$\pi\, E\left[\frac{1}{\sqrt{2\pi}\sqrt{T_1 + \hat{T}_1}}\exp\left(-\frac{\lambda^2}{2(T_1 + \hat{T}_1)}\right)\right] = \pi P\left(N\sqrt{T_1 + \hat{T}_1} \in d\lambda\right)/d\lambda$$

where N is Normal $(0,1)$ independent of $\sqrt{T_1 + \hat{T}_1}$. But

$$E\exp\left(i\lambda N\sqrt{T_1 + \hat{T}_1}\right) = E\exp\left(-\frac{\lambda^2}{2}(T_1 + \hat{T}_1)\right) = \frac{1}{\cosh^2(\lambda)}$$

whence in this case by Fourier inversion

$$\begin{aligned} E\exp\left(-\frac{\lambda^2}{2}(M_1^{\rm br})^2\right) &= \pi\frac{1}{2\pi}\int_{-\infty}^{\infty}\frac{d\theta e^{-i\lambda\theta}}{\cosh^2(\theta)} = \int_{-\infty}^{\infty}\frac{dx e^{i\frac{\lambda}{2}x}}{2\cdot 2\cosh^2(\frac{x}{2})} \\ &= \frac{\pi\frac{\lambda}{2}}{\sinh(\pi\frac{\lambda}{2})} = E\exp\left(-\tfrac{\lambda^2}{2}\left(\tfrac{\pi^2}{4}\right)T_3\right) \end{aligned}$$

This proves (3.26).

References

1. D. Aldous and J. Pitman. Brownian bridge asymptotics for random mappings. *Random Structures and Algorithms*, 5:487–512, 1994.
2. D.J. Aldous. The continuum random tree I. *Ann. Probab.*, 19:1–28, 1991.
3. D.J. Aldous. The continuum random tree II: an overview. In M.T. Barlow and N.H. Bingham, editors, *Stochastic Analysis*, pages 23–70. Cambridge University Press, 1991.
4. D.J. Aldous. The continuum random tree III. *Ann. Probab.*, 21:248–289, 1993.
5. D.J. Aldous. Recursive self-similarity for random trees, random triangulations and Brownian excursion. *Ann. Probab.*, 22:527–545, 1994.
6. S. Asmussen, P. Glynn, and J. Pitman. Discretization error in simulation of one-dimensional reflecting Brownian motion. To appear in *Ann. Applied Prob.*, 1996.

7. Ph. Biane. Decompositions of Brownian trajectories and some applications. In A. Badrikian, P-A Meyer, and J-A Yan, editors, *Probability and Statistics; Rencontres Franco-Chinoises en Probabilités et Statistiques; Proceedings of the Wuhan meeting*, pages 51–76. World Scientific, 1993.
8. Ph. Biane. Some comments on the paper: "Brownian bridge asymptotics for random mappings" by D. J. Aldous and J. W. Pitman. *Random Structures and Algorithms*, 5:513–516, 1994.
9. Ph. Biane and M. Yor. Valeurs principales associées aux temps locaux Browniens. *Bull. Sci. Math. (2)*, 111:23–101, 1987.
10. R. M. Blumenthal. *Excursions of Markov processes*. Birkhäuser, 1992.
11. K.L. Chung. Excursions in Brownian motion. *Arkiv fur Matematik*, 14:155–177, 1976.
12. E. Csáki, A. Földes, and P. Salminen. On the joint distribution of the maximum and its location for a linear diffusion. *Annales de l'Institut Henri Poincaré, Section B*, 23:179 – 194, 1987.
13. I.V. Denisov. A random walk and a Wiener process near a maximum. *Theor. Prob. Appl.*, 28:821–824, 1984.
14. P. Fitzsimmons, J. Pitman, and M. Yor. Markovian bridges: construction, Palm interpretation, and splicing. In *Seminar on Stochastic Processes, 1992*, pages 101–134. Birkhäuser, Boston, 1993.
15. P.J. Fitzsimmons. Excursions above the minimum for diffusions. Unpublished manuscript, 1985.
16. J.-F. Le Gall. Brownian excursions, trees and measure-valued branching processes. *Ann. Probab.*, 19:1399–1439, 1991.
17. J.F. Le Gall. Une approche élémentaire des théorèmes de décomposition de Williams. In *Séminaire de Probabilités XX*, pages 447–464. Springer, 1986. Lecture Notes in Math. 1204.
18. J.F. Le Gall. The uniform random tree in a brownian excursion. *Probab. Th. Rel. Fields*, 96:369–383, 1993.
19. R.K. Getoor and M.J. Sharpe. Excursions of dual processes. *Advances in Mathematics*, 45:259–309, 1982.
20. I.I. Gikhman. On a nonparametric criterion of homogeneity for k samples. *Theory Probab. Appl.*, 2:369–373, 1957.
21. J.-P. Imhof. Density factorizations for Brownian motion, meander and the three-dimensional Bessel process, and applications. *Journal of Applied Probability*, 21:500 – 510, 1984.
22. J.-P. Imhof. On Brownian bridge and excursion. *Studia Sci. Math. Hungar.*, 20:1–10, 1985.
23. J. P. Imhof. On the range of Brownian motion and its inverse process. *Annals of Probability*, 13:1011 – 1017, 1985.
24. M. E. Ismail and D.H. Kelker. Special functions, Stieltjes transforms, and infinite divisibility. *SIAM J. Math. Anal.*, 10:884–901, 1979.
25. K. Itô. Poisson point processes attached to Markov processes. In *Proc. 6th Berk. Symp. Math. Stat. Prob.*, volume 3, pages 225–240, 1971.
26. K. Itô and H.P. McKean. *Diffusion Processes and their Sample Paths*. Springer, 1965.
27. Th. Jeulin. Temps local et théorie du grossissement: application de la théorie du grossissement à l'étude des temps locaux browniens. In *Grossissements de filtrations: exemples et applications. Séminaire de Calcul Stochastique, Paris 1982/83*, pages 197–304. Springer-Verlag, 1985. Lecture Notes in Math. 1118.
28. D.P. Kennedy. The distribution of the maximum Brownian excursion. *J. Appl. Prob.*, 13:371–376, 1976.

29. J. Kent. Some probabilistic properties of Bessel functions. *Annals of Probability*, 6:760 – 770, 1978.
30. J. Kiefer. K-sample analogues of the Kolmogorov-Smirnov and Cramér-von Mises tests. *Ann. Math. Stat.*, 30:420–447, 1959.
31. F.B. Knight. Characterization of the Lévy measure of inverse local times of gap diffusions. In *Seminar on Stochastic Processes, 1981*, pages 53–78. Birkhäuser, Boston, 1981.
32. S. Kotani and S. Watanabe. Krein's spectral theory of strings and generalized diffusion processes. In *Functional Analysis in Markov Processes*, pages 235–249. Springer, 1982. Lecture Notes in Math. 923.
33. P. Lévy. *Processus Stochastiques et Mouvement Brownien.* Gauthier-Villars, Paris, 1965. (first ed. 1948).
34. B. Maisonneuve. Exit systems. *Ann. of Probability*, 3:399 – 411, 1975.
35. P.W. Millar. Random times and decomposition theorems. In *Proc. of Symp. in Pure Mathematics*, volume 31, pages 91–103, 1977.
36. P.W. Millar. A path decomposition for Markov processes. *Ann. Probab.*, 6:345–348, 1978.
37. S. A. Molchanov and E. Ostrovski. Symmetric stable processes as traces of degenerate diffusion processes. *Theor. Prob. Appl.*, 14, No. 1:128–131, 1969.
38. J. Pitman. Stationary excursions. In *Séminaire de Probabilités XXI*, pages 289–302. Springer, 1986. Lecture Notes in Math. 1247.
39. J. Pitman. Cyclically stationary Brownian local time proceses. To appear in *Probability Theory and Related Fields*, 1996.
40. J. Pitman and M. Yor. Bessel processes and infinitely divisible laws. In *Stochastic Integrals*, pages 285–370. Springer, 1981. Lecture Notes in Math. 851.
41. J. Pitman and M. Yor. A decomposition of Bessel bridges. *Zeitschrift für Wahrscheinlichkeitstheorie und Verwandte Gebiete*, 59:425–457, 1982.
42. J. Pitman and M. Yor. Dilatations d'espace-temps, réarrangements des trajectoires browniennes, et quelques extensions d'une identité de Knight. *C.R. Acad. Sci. Paris*, t. 316, Série I:723–726, 1993.
43. J. Pitman and M. Yor. Some extensions of Knight's identity for Brownian motion. In preparation, 1995.
44. J. Pitman and M. Yor. Laws of homogeneous functionals of Brownian motion. In preparation, 1996.
45. J. Pitman and M. Yor. Quelques identités en loi pour les processus de Bessel. To appear in: *Hommage à P.A. Meyer et J. Neveu*, Astérisque, 1996.
46. D. Revuz and M. Yor. *Continuous martingales and Brownian motion.* Springer, Berlin-Heidelberg, 1994. 2nd edition.
47. L. C. G. Rogers. Williams' characterization of the Brownian excursion law: proof and applications. In *Séminaire de Probabilités XV*, pages 227–250. Springer-Verlag, 1981. Lecture Notes in Math. 850.
48. L.C.G. Rogers and D. Williams. *Diffusions, Markov Processes and Martingales.* Wiley, 1987.
49. T. Shiga and S. Watanabe. Bessel diffusions as a one-parameter family of diffusion processes. *Z. Wahrsch. Verw. Gebiete*, 27:37–46, 1973.
50. I. Vincze. Einige zweidimensionale Verteilungs- und Grenzverteilungssätze in der Theorie der georneten Stichproben. *Magyar Tud. Akad. Mat. Kutató Int. Kőzl.*, 2:183–209, 1957.
51. G.N. Watson. *A Treatise on the Theory of Bessel Functions, 2nd ed.* Cambridge University Press, 1944.
52. W. Werner. *Quelques propriétés du mouvement brownien plan.* Thèse de doctorat, Université Paris VI, March 1993.

53. W. Werner. Sur la forme des composantes connexes du complémentaire de la courbe brownienne plane. *Prob. Th. and Rel. Fields*, 98:307–337, 1994.
54. D. Williams. Decomposing the Brownian path. *Bull. Amer. Math. Soc.*, 76:871–873, 1970.
55. D. Williams. Path decomposition and continuity of local time for one dimensional diffusions I. *Proc. London Math. Soc. (3)*, 28:738–768, 1974.
56. D. Williams. *Diffusions, Markov Processes, and Martingales, Vol. 1: Foundations.* Wiley, Chichester, New York, 1979.
57. D. Williams. Brownian motion and the Riemann zeta-function. In *Disorder in Physical Systems*, pages 361–372. Clarendon Press, Oxford, 1990.
58. M. Yor. *Some Aspects of Brownian Motion.* Lectures in Math., ETH Zürich. Birkhaüser, 1992. Part I: Some Special Functionals.

Interacting diffusion systems over $\mathbf{Z}^d$

Tokuzo Shiga

Department of Applied Physics, Tokyo Institute of Technology, Oh-Okayama, Meguro-ku, Tokyo 152, Japan

Summary. Interacting diffusion systems are a class of diffusion processes taking values in an infinite product space of an interval, which models the variety of phenomena in physics and biology. This paper surveys the subject of interacting diffusion systems and related problems. Topics include:

1. Introduction.
2. Stationary distributions and ergodic theorems in transient case.
3. $\mathbf{Z}^d$-shift invariance of stationary distributions.
4. Local extinction in transient case.
5. Uniformity and local extinction in recurrent case.
6. Parabolic Anderson model and sample Lyapunov exponent.
7. Finite systems of interacting diffusions.
8. Approximation of infinite systems via finite systems.
9. Methods and some technicalities.

9.1 Duality.
9.2 Comparison.
9.3 Coupling.
9.4 Liouville property.
9.5 Random walk estimates.
9.6 Moments estimates.
9.7 Idea of the proof of the approximation result.

1. Introduction

Let I be a closed interval of $\mathbf{R}$ and let S be a countable set. Suppose that we are given an $S \times S$ real matrix $A = \{A_{ij}\}$ and a function $a(u) : I \mapsto \mathbf{R}$ satisfying the following condition.

Condition [A]

(A.1) $A = \{A_{ij}\}$ is a bounded infinitesimal matrix of a continuous time irreducible Markov chain with state space S, i.e.

$$A_{ij} \geq 0 \quad \text{for} \quad i \neq j, \quad \sum_{j\in S} A_{ij} = 0 \quad \text{and} \quad \sup_{i\in S} |A_{ii}| < \infty.$$

(A.2) $a(u)$ is a locally 1/2-Hölder continuous function defined on I, vanishing at each finite boundary point of I, and satisfies a linear growth condition

$$|a(u)| \leq \text{const.}(1 + |u|) \quad \text{for } u \in I \quad \text{if } I \text{ is unbounded.}$$

Let us consider the following stochastic differential equation (SDE):

$$dx_i(t) = \sum_{j \in S} A_{ij} x_j(t)dt + a(x_i(t))dB_i(t), \quad i \in S \tag{1}$$

where $\{B_i(t)\}_{i \in S}$ is an independent system of standard Brownian motions.

The diffusion process governed by the SDE (1) is called *an interacting diffusion system* (shortly, IDS). The IDS's are a class of stochastic interacting systems which have been extensively investigated especially for particle systems. (See [L].)

We here list several examples of interacting diffusion systems.

EXAMPLE 1. (Stepping stone diffusion model with random drift)

$$I = [0,1], \quad a(u) = \sqrt{u(1-u)}.$$

EXAMPLE 2. (Stepping stone diffusion model with random selection)

$$I = [0,1], \quad a(u) = u(1-u).$$

EXAMPLE 3. (Branching diffusion model)

$$I = [0,\infty), \quad a(u) = \sqrt{u}.$$

EXAMPLE 4. (Parabolic Anderson model)

$$S = \mathbf{Z}^d, \quad I = [0,\infty), \quad a(u) = u,$$

and for $\kappa > 0$,

$$A_{ij} = \begin{cases} \kappa & (\text{if} \quad |i-j| = 1) \\ -2d\kappa & (\text{if} \quad i = j) \\ 0 & (\text{otherwise}) \end{cases}$$

EXAMPLE 5 (Critical Ornstein-Uhlenbeck process)

$$I = (-\infty,\infty), \quad a(u) = \text{const}.$$

The stepping stone diffusion model of EXAMPLE 1 appears in population genetics, which is a diffusion approximation of a Markov chain model proposed by M. Kimura in [K], and was obtained complete description of stationary distributions and ergodic theorems in [S1], [S2]. EXAMPLE 2 is its modification incorporating random selection by Ohta and Kimura ([KO]), and the ergodic behaviors were studied in [NS]. EXAMPLE 3 is a diffusion model which is an diffusion approximation of population size processes in a branching random walk. If one takes a continuum approximation of both the population sizes and the space, it turns to a measure-valued branching diffusion, which was established by Watanabe [W] and the ergodic behaviors were investigated by Dawson [D]. Parabolic Anderson model is a time evolution of a Schrödinger operator on $\mathbf{Z}^d$ with time-dependent random potential, which might exhibit intermittency depending on the parameter $\kappa > 0$ and dimensionality.

To formulate the solutions of the SDE (1) we first choose a positive summable sequence $\gamma = \{\gamma_i\}_{i\in S}$ such that for some $C > 0$,

$$\sum_{i\in S} \gamma_i |A_{ij}| \le C\gamma_j, \quad j \in S.$$

The condition (A.1) guarantees existence of such a sequence $\gamma = \{\gamma_i\}_{i\in S}$. Let $\mathbf{E} = L^2(\gamma)$ be the totality of sequences $\mathbf{x} = \{x_i\}_{i\in S}$ satisfying $\|x\|^2 = \sum_{i\in S} x_i^2 < \infty$. Then under the condition [A], for every $\mathbf{x}(0) = \mathbf{x} \in E$, there exists a pathwise unique $\mathbf{E}$-valued solution, which induces a diffusion process $(\Omega, \mathcal{F}, \mathcal{F}_t, P^{\mathbf{x}}, \mathbf{x}(t))$ with state space $\mathbf{E}$, (cf. [SS]). The transition probability induces a Feller Markov semigroup T_t acting on $C_b(\mathbf{E})$ (the totality of bounded continuous functions on $\mathbf{E}$) such that

$$T_t f - f = \int_0^t T_s L f ds, \quad f \in C_0^2(\mathbf{E}) \tag{2}$$

where $C_0^m(\mathbf{E})$ stands for the totality of C^m-functions defined on $\mathbf{E}$ depending on finitely many components with f and Lf being bounded, and

$$Lf(\mathbf{x}) = \frac{1}{2}\sum_{i\in S} a(x_i)^2 \frac{\partial^2 f}{\partial x_i^2} + \sum_{i\in S}(\sum_{j\in S} A_{ij}x_j)\frac{\partial f}{\partial x_i}. \tag{3}$$

Let $\mathcal{P}(\mathbf{E})$ be the totality of probability measures on $\mathbf{E}$, equipped with the topology of weak convergence. T_t induces the dual semigroup T_t^* acting on $\mathcal{P}(\mathbf{E})$ by

$$\langle T_t^* \mu, f\rangle = \langle \mu, T_t f\rangle, \quad f \in C_b(\mathbf{E}),$$

where

$$\langle \mu, f\rangle = \int_{\mathbf{E}} f(\mathbf{x})\mu(d\mathbf{x}).$$

Let $\mathcal{S}$ be the totality of stationary distributions of the IDS $(\Omega, \mathcal{F}, \mathcal{F}_t, P^{\mathbf{x}}, \mathbf{x}(t))$, namely

$$\mathcal{S} = \{\mu \in \mathcal{P}(\mathbf{E}) | T_t^* \mu = \mu, \ (t > 0)\}.$$

We note that $\mathcal{S}$ is convex and simplex (cf. [Dy]), so that denote by $\mathcal{S}_{ext}$ the totality of extremal elements of $\mathcal{S}$.

The purpose of the present paper is to survey recent results on interacting diffusion systems with focus on the following two problems;
(i) Stationary distributions and ergodic theorems,
(ii) Relation between finite systems and infinite systems.

We consider these problems under the situation that $S = \mathbf{Z}^d$ and A is $\mathbf{Z}^d$-shift invariant. It is to be noted that the interacting diffusion systems may have many extremal stationary distributions, so that the first problem is to describe all extremal $\mathbf{Z}^d$-shift invariant stationary distributions in terms of some characteristics.

Now since we are assuming that the transition mechanism of the IDS is $\mathbf{Z}^d$-shift invariant, a natural problem occurs.

Is every stationary distribution $\mathbf{Z}^d$-shift invariant?

This problems has not been well developed even in the theory of interacting particle systems except the case that the process possesses a nice dual process as in the voter model and the stepping stone diffusion model.

For the second problem let us consider the situation in which the interacting diffusion system has many extremal stationary distributions, which are realized when the symmetrization of A-random walk is transient and the diffusion coefficient $a(u)$ has a mild growth rate. Even in such cases the corresponding finite systems have at most only some trivial stationary distributions. However by taking a critical time rescaling relevant to the system sizes of the finite systems, it would be possible to observe ergodic behaviors of the infinite system from the finite systems.

2. Stationary distributions and ergodic theorems in transient case

Recalling that $S = \mathbf{Z}^d$ and $A = \{A_{ij}\}$ is $\mathbf{Z}^d$-shift invariant, i.e. $A_{ij} = A_{0,j-i}$ $(i, j \in \mathbf{Z}^d)$, so that A is an infinitesimal matrix of a continuous time random walk on $\mathbf{Z}^d$, we impose a further assumption.

Condition [B]

(B.1) The symmetrized random walk generated by the infinitesimal matrix $\widehat{A} = \{\widehat{A}_{ij} = A_{ij}+A_{ji}\}$ is transient. (Hereafter we say it simply $\widehat{A}$ is transient.)

(B.2) If I is unbounded,

$$\limsup_{|u|\to\infty, u\in I} \frac{|a(u)|}{|u|} < \widehat{G}(0,0)^{1/2},$$

where $\widehat{G}(i,j)$ is the potential matrix of the $\widehat{A}$-random walk, i.e.

$$P_t = \exp tA, \quad \widehat{P}_t = P_t P_t^*, \quad \text{and} \quad \widehat{G} = \int_0^\infty \widehat{P}_t dt.$$

Noting that the state space $\mathbf{E}$ is $\mathbf{Z}^d$-shift invariant, denote by $\mathcal{T}(\mathbf{E})$ the totality of $\mathbf{Z}^d$-shift invariant probability measures on $\mathbf{E}$, and for $p > 0$ denote

$$\mathcal{T}_p(\mathbf{E}) = \{\mu \in \mathcal{T}(\mathbf{E}) | \langle \mu, |x_i|^p \rangle < \infty\}.$$

Since $\mathcal{S} \cap \mathcal{T}_p(\mathbf{E})$ also is convex and simplex, we denote by $(\mathcal{S} \cap \mathcal{T}_p(\mathbf{E}))_{ext}$ the totality of extremal elements of $\mathcal{S} \cap \mathcal{T}_p(\mathbf{E})$. Then we have the following result.

Theorem 1 **([CG2],[S3]).** *Assume the condision [A] and [B].*

1. *For each $\theta \in I$, there exists a unique $\nu = \nu_\theta \in (\mathcal{S} \cap \mathcal{T}_1(\mathbf{E}))_{ext}$ such that $\langle \nu, x_i \rangle = \theta$ for $i \in \mathbf{Z}^d$.*
2. $$(\mathcal{S} \cap \mathcal{T}_1(\mathbf{E}))_{ext} = \{\nu_\theta |\, \theta \in I\}.$$

   Moreover for every $\nu \in \mathcal{S} \cap \mathcal{T}_1(\mathbf{E})$ there exists a unique probability measure $m(d\theta)$ on I such that
   $$\nu = \int_I \nu_\theta dm(\theta).$$
3. *If $\mu \in \mathcal{T}_1(\mathbf{E})$ is $\mathbf{Z}^d$-shift ergodic, then*
   $$\lim_{t \to \infty} T_t^* \mu = \nu_\theta \quad \text{with} \quad \theta = \langle \mu, x_i \rangle.$$

   Accordingly for every $\mu \in \mathcal{T}_1(\mathbf{E})$, $\lim_{t\to\infty} T_t^*\mu$ exists in $\mathcal{S} \cap \mathcal{T}_1(\mathbf{E})$.

Remark 1. In the case $I = [0, \infty)$ Theorem 1 (b) is refined as follows,
$$(\mathcal{S} \cap \mathcal{T}(\mathbf{E}))_{ext} = \{\nu_\theta |\, \theta \in I\}.$$

Remark 2. (B.2) of the condition [B] would be far from the best possible one for Theorem 1. On the other hand, instead of (B.2) if the linear growth rate of $a(u)$ is sufficiently large, then there exists no longer any non-trivial stationary distribution, which will be discussed in the section 4.

3. $\mathbf{Z}^d$-shift invariance of stationary distributions

Since the transition mechanism of the IDS is $\mathbf{Z}^d$-shift invariant under the assumption [B], the following question arises naturally. Under what additional condition does it holds that every stationary distribution is $\mathbf{Z}^d$-shift invariant?

In the case $I = [0, 1]$ we have the following result.

Theorem 2 ([S4]). *Let $I = [0, 1]$. Assume that the $\widehat{A}$ random walk is transient. Then every stationary distribution of the IDS is $\mathbf{Z}^d$-shift invariant.*

On the other hand when I is unbounded, the problem would be more difficult. Very recently it was proved by Bramson, Cox and Greven in [BCG2] that for a super Brownian motion and a branching Brownian particles over $\mathbf{R}^d$ every stationary distribution is $\mathbf{R}^d$-shift invariant. We here present a partial solution for the IDS in the case $I = [0, \infty)$.

Let $\mathcal{S}_p$ $(p > 0)$ be the totality of stationary distributions with p-th finite moments, i.e.
$$\mathcal{S}_p = \{\nu \in \mathcal{S} |\, \langle \nu, |x_i|^p \rangle < \infty \text{ for } i \in \mathbf{Z}^d\}.$$

Theorem 3. *Let $I = [0, \infty)$. In addition to the condition [B], assume further that*

$$\sum_{i \in \mathbf{Z}^d} A_{0i} \cdot i = 0, \tag{4}$$

and for some $C > 0$,

$$|a(u)| \leq C(1 + \sqrt{x}). \tag{5}$$

Then every $\nu \in \mathcal{S}_2$ is $\mathbf{Z}^d$-shift invariant.

Remark 3. The condition (4) is necessary since it guarantees the Liouville property that every nonnegative A-harmonic function is constant. To the contrary if we consider a non-symmetric nearest neighbor random walk, there is a nonnegative and non-constant A-harmonic function. In such a case there exists a stationary distribution which is not $\mathbf{Z}^d$-shift invariant in higher dimension.

Theorem 4. *Let $I = [0, \infty)$. In addition to the condition [B], assume that $a(u)$ is linear, i.e. $a(u) = cu$. Then every $\nu \in \mathcal{S}_2$ is $\mathbf{Z}^d$-shift invariant.*

Remark 4. The assumption in Theorems 3 and 4 could be relaxed to the condition (B.2), which is an open problem.

4. Local extinction in transience case

As discussed in the previous sections, in the transient situation of $\widehat{A}$-random walk, if the linear growth rate of the coefficient $a(u)$ is modest, the IDS possesses a one-parameter family of stationary distributions. To the contrary one can show that if $I = [0, \infty)$ and the linear growth rate of $a(u)$ is sufficiently large there is only a trivial stationary state $\delta_{\mathbf{0}}$ and the distribution at time t converges to $\delta_{\mathbf{0}}$ as $t \to \infty$, where $\delta_{\mathbf{x}}$ stands for the Dirac measure at $\mathbf{x} \in \mathbf{E}$. This means that as far as we observe it from a local window (a finite region of $\mathbf{Z}^d$), the component processes $x_i(t)$ vanishes to zero in probability. Thus the local extinction occurs, which is a common picture in the theory of interacting particle systems, see [L].

Theorem 5 ([S3]). *Let $I = [0, \infty)$. Suppose that $\widehat{A}$ is transient and of finite range. Then there exists a constant $M > 0$ such that if*

$$|a(u)| > Mu \quad (u > 0), \tag{6}$$

then for every $\mu \in \mathcal{T}_1(\mathbf{E})$,

$$\lim_{t \to \infty} T_t^* \mu = \delta_{\mathbf{0}}.$$

Now we fix an A and a function $a(u)$ in the situation of Theorem 5, and set $a_\lambda(x) = \lambda a(x)$ for $\lambda > 0$. Then by Theorem 1 and Theorem 5 together with a comparison theorem due to [CFG] (see the section 9.2) there exists a critical parameter $0 < \lambda_c < \infty$ such that if $\lambda > \lambda_c$, the local extinction occurs, while if $\lambda < \lambda_c$, there is a one-parameter family of extremal stationary distributions.

5. Uniformity and local extinction in recurrent case

In the case that $\widehat{A}$-random walk is recurrent the ergodic behaviors of the IDS are simpler. In the case $I = [0,1]$ there are two extremal stationary distributions δ_1 and δ_0, which correspond to the genetically uniform states in population genetics. On the other hand when $I = [0,\infty)$, δ_0 is the unique stationary distribution.

Theorem 6 ([NS], [S4]). *Let $I = [0,1]$. Suppose that $\widehat{A}$-random walk is irreducible and recurrent. Then*

1.

$$\mathcal{S}_{ext} = \{\delta_0, \delta_1\}.$$

2. For a $\mu \in \mathcal{P}(\mathbf{E})$, $\lim_{t\to\infty} T_t^*\mu$ exists if and only if

$$\lim_{t\to\infty} \sum_{j\in\mathbf{Z}^d} P_t(i,j)\langle \mu, x_j\rangle$$

exists for any $i \in \mathbf{Z}^d$. In this case the limit is constant in $i \in \mathbf{Z}^d$, which we denote by θ, and

$$\lim_{t\to\infty} T_t^*\mu = \theta\delta_1 + (1-\theta)\delta_0.$$

Theorem 7 ([CFG]). *Let $I = [0,\infty)$. Suppose that $\widehat{A}$-random walk is irreducible and recurrent. Then for every $\mu \in \mathcal{T}_1(\mathbf{E})$,*

$$\lim_{t\to\infty} T_t^*\mu = \delta_0.$$

6. Parabolic Anderson model and sample Lyapunov exponent

Let $I = [0, \infty)$, and consider the following SDE:

$$dx_i(t) = \kappa \Delta x_i(t)dt + x_i(t)dB_i(t), \quad i \in \mathbf{Z}^d, \tag{7}$$

where $\kappa > 0$ is a constant and

$$\Delta x_i = \frac{1}{2d} \sum_{|j-i|=1} (x_j - x_i).$$

The r.h.s. of (7) is a Schrödinger operator with nonstationary Gaussian random potential, so that (7) is called *parabolic Anderson model*, which attracts much attention from the view point of intermittency. (cf.[ZMRS]) The parabolic Anderson model is a linear system for which one can define Lyapunov exponent for the sample path.

Theorem 8 ([S4]). *There exists a constant* $\lambda = \lambda(\kappa, d)$ *such that if* $\mathcal{L}(\mathbf{x}(0) = (x_i(0))) \in \mathcal{T}_1(\mathbf{E})$ *and* $\mathbf{x}(0) \neq \mathbf{0}$ *P-a.s. then the solution* $\mathbf{x}(t) = (x_i(t))$ *satisfies*

$$\lim_{t\to\infty} \frac{1}{t} \log x_i(t) = \lambda(\kappa, d) \quad in\ L^1(P). \tag{8}$$

$\lambda(\kappa, d)$ is called *the sample Lyapunov exponent.* It is very important to investigate its dependency on the parameters κ and d. By the comparison theorem by [CFG] there exists a critical $\kappa(d) \in (0, \infty]$ such that if $0 < \kappa < \kappa(d)$, $\lambda(\kappa, d) < 0$, while if $\kappa > \kappa(d)$, then $\lambda(\kappa, d) = 0$. When $d \geq 3$, if κ is large, the assumption of Theorem 1 is fulfilled, hence $\lambda(\kappa, d) = 0$. Therefore, $0 < \kappa(d) < \infty$ holds for $d \geq 3$. On the other hand it is conjectured that $\kappa(d) = \infty$ for $d = 1$ or 2, which is a challenging open problem.

One can also ask the behavior of the sample Lyapunov exponent $\lambda(\kappa, d)$ as $\kappa \to 0^+$. For this we have

Theorem 9. *There exists constants* $0 < c(d) < C(d) < \infty$ *and* $0 < \kappa_0 < 1$ *such that*

$$\frac{c(d)}{\log \frac{1}{\kappa}} \leq \lambda(\kappa, d) \leq \frac{C(d)}{\log \frac{1}{\kappa}} \quad for\ 0 < \kappa < \kappa_0. \tag{9}$$

The lower bound was obtained in [CM] extending an idea in [ZMRS]. Concerning the upper bound [S3] and [CM] obtained

$$\lambda(\kappa, d) \leq C(d) \frac{\log \log \frac{1}{\kappa}}{\log \frac{1}{\kappa}},$$

and very recently [CV] obtained the final conclusion.

7. Finite systems of interacting diffusions

Returning to the situation of the section 2, let A be the infinitesimal matrix of a continuous time irreducible random walk on $\mathbf{Z}^d$, and $a(u)$ be a function satisfying the condition [B]. Let us define finite systems which approximate the infinite system of the IDS. For each $N \geq 1$, let $\Lambda_N = (-N, N]^d \cap \mathbf{Z}^d$. Regarding it as a torus, let us introduce an infinitesimal matrix of a continuous time random walk on Λ_N induced by A:

$$A_{ij}^N = \sum_{k:k-j \in 2N\mathbf{Z}^d} A_{ik} \quad (i, j \in \Lambda_N). \tag{10}$$

A finite system is a diffusion process $(\Omega, \mathcal{F}, \mathcal{F}_t, P_N^{\mathbf{x}}, \mathbf{x}^N(t))$ with state space $\mathbf{E}_N = I^{\Lambda_N}$ which is governed by the following SDE:

$$dx_i^N(t) = \sum_{j \in \Lambda_N} A_{ij}^N x_j^N(t)dt + a(x_i^N(t))dB_i(t), \quad i \in \Lambda_N. \tag{11}$$

We use similar notation to the infinite systems; e.g.

$$\mathcal{P}(\mathbf{E}_N),\ \mathcal{T}(\mathbf{E}_N),\ \mathcal{T}_p(\mathbf{E}_N),\ T_t^N,\ T_t^{N*} \text{ etc.}$$

Then we have

Theorem 10. *Let $N \geq 1$ be fixed.*

1. *If $I = [0, 1]$, for every $\mu \in \mathcal{P}(\mathbf{E}_N)$*

$$\lim_{t \to \infty} T_t^{N*} \mu = \theta \delta_1 + (1 - \theta)\delta_0,$$

where

$$\theta = \frac{1}{|\Lambda_N|} \sum_{i \in \Lambda_N} \langle \mu, x_i \rangle.$$

2. *If $I = [0, \infty)$, for every $\mu \in \mathcal{P}(\mathbf{E}_N)$*

$$\lim_{t \to \infty} T_t^{N*} \mu = \delta_0.$$

3. *If $I = (-\infty, \infty)$, for every $\mu \in \mathcal{P}(\mathbf{E}_N)$*

$$\lim_{t \to \infty} T_t^{N*} \mu = 0,$$

in the sense of the vague convergence, more precisely,

$$\lim_{t \to \infty} \langle T_t^{N*} \mu, f \rangle = 0$$

for every continuous function f with compact support on $\mathbf{E}_N$.

8. Finite systems and infinite systems

As seen in the previous sections the finite systems and the infinite systems of the IDS exhibit quite different ergodic behaviors. However it would still be desirable to observe the ergodic behaviors of the infinite systems through the finite systems. Indeed it is possible by taking a suitable time scaling relevant to the system size of the finite systems.

For this purpose we introduce the following ingredients of the finite systems.

1. The time scale $\beta_N = |\Lambda_N| = (2N)^d$.
2. The empirical density $\Theta_N(t) = |\Lambda_N|^{-1} \sum_{i\in\Lambda_N} x_i^N(t)$.
3. The rescaled process of empirical densities $Z_N(t) = \Theta_N(t\beta_N)$.
4. The diffusion process $Z(t)$ on I defined by

$$dZ(t) = a^*(Z(t))dB(t), \quad Z(0) = \rho,$$

   where $B(t)$ is a standard Brownian motion and $a^*(\theta)$ is the function $a^*(\theta) = \sqrt{\langle \nu_\theta, a(x_0)^2\rangle}$. We denote by $Q_s(\rho, d\theta)$ the transition probability of $Z(t)$.
5. The empirical distribution of the finite systems

$$U_N(t) = |\Lambda_N|^{-1} \sum_{i\in\Lambda_N} \delta_{\sigma_i^N \mathbf{x}^N(t)},$$

   where σ_i^N is the shift by i on the torus Λ_N, i.e.

$$(\sigma_i^N \mathbf{x})_j = x_k, \ k = i + j \bmod (2N).$$

Let $\mu_N \in \mathcal{P}(\mathbf{E}_N)$, $N = 1, 2, \cdots$ and $\mu \in \mathcal{P}(\mathbf{E})$. To give the meaning of convergence $\mu_N \Longrightarrow \mu$ as $N \to \infty$, we introduce the periodic extension operators $\pi_N : \mathbf{E}_N \mapsto \mathbf{E}$, $(\pi_N \mathbf{x}^N)_j = x_i^N$ where $i \in \Lambda$, $i = j \bmod (2N)$. Let $\bar{\mu}_N$ be an induced probability measure on $\mathbf{E}$ of μ_N by π_N. If $\bar{\mu}_N$ converges to μ weakly, we write $\mu_N \Longrightarrow \mu$ and the same notation applies to $\mathbf{E}_N$-valued processes. Also we denote by $\mathcal{L}(X)$ and $\mathcal{L}(Y(\cdot))$ the distributions of a random variable X and a process $Y(\cdot)$.

Theorem 11 ([CGS1]). *Assume $d \geq 3$. Suppose that $\sup_N E\langle U_N(0), |x_0|^p\rangle < \infty$ for some $p > 2$, and for some random variable Z_0 satisfying that $\mathcal{L}(\Theta_N(0)) \Longrightarrow \mathcal{L}(Z_0)$ as $N \to \infty$. Then as $N \to \infty$,*

$$\mathcal{L}(Z_N(\cdot)) \Longrightarrow \mathcal{L}(Z(\cdot)), \quad Z(0) = Z_0,$$

and

$$\{U_N(t\beta_N); t > 0\} \stackrel{\text{fdd}}{\Longrightarrow} \{\nu_{Z(t)}; t > 0\},$$

where $\stackrel{\text{fdd}}{\Longrightarrow}$ stands for the convergence in the sense of finite dimensional distributions.

Theorem 12 ([CGS1]). *Assume $d \geq 3$. Suppose that for some $p > 2$, $\mathcal{L}(\mathbf{x}^N(0)) \in \mathcal{T}_p(\mathbf{E}_N)$, $\sup_N E|x_0^N(0)|^p < \infty$, and for some $\rho \in I$, $\Theta_N(0) \to \rho$ as $N \to \infty$ in probability. Let $t_N \uparrow \infty$ and $t_N/\beta_N \to s \in [0,\infty)$, and in the case $t_N/N^2 \not\to \infty$, assume also that $\mathcal{L}(\mathbf{x}^N(0)) \Longrightarrow$ some element of $\mathcal{T}_2(\mathbf{E})$. Then as $N \to \infty$,*

$$\mathcal{L}(\Theta_N(t_N)) \Longrightarrow \mathcal{L}(Z(s)), \quad Z(0) = \rho,$$

and

$$\mathcal{L}(\mathbf{x}^N(t_N)) \Longrightarrow \int_I Q_s(\rho, d\theta)\nu_\theta.$$

In particular if $t_N \nearrow \infty$ and $t_N = o(\beta_N)$, then

$$\mathcal{L}(\mathbf{x}^N(t_N)) \Longrightarrow \nu_\rho \quad \text{as } N \to \infty.$$

Remark 5. Even in the case that $\widehat{A}$-random walk is recurrent, it is possible to discuss a scaling limit of the empirical density process $\Theta_N(\cdot)$ like as for the two dimensional voter model in [CGri]. In fact, if $I = [0,1]$ and $d = 2$ it can be shown that the rescaled empirical density process $Z_N(\cdot) = \Theta_N(\cdot\beta_N)$ with $\beta_N = |\Lambda_N| \log \sqrt{\Lambda_N}$ converges to the Wright-Fisher diffusion. (cf. [CGS2])

9. Methods and some technicalities

In this section we would like to explain the methods and some technicalities which are exploited for the proof of the results on the IDS's.

9.1 Duality

If we specialize the situation to $I = [0,1]$ and $a(u) = \sqrt{u(1-u)}$, there is a nice dual process which is quite similar to the dual of the voter model.

For $\mathbf{n} = (n_i)_{i\in S} \in \mathbf{Z}_+^S$ with $|\mathbf{n}| = \sum_{i\in S} n_i < \infty$, set $f_\mathbf{n}(\mathbf{x}) = \prod_{i\in S} x_i^{n_i}$, and denote by $\mathbf{e}_i$ the unit vector supported by $i \in S$. From the form of the generator L it is easy to see

$$Lf_\mathbf{n}(\mathbf{x}) = \sum_\mathbf{m} Q(\mathbf{n},\mathbf{m}) f_\mathbf{m}(\mathbf{x}),$$

where

$$Q(\mathbf{n},\mathbf{m}) = \begin{cases} \frac{1}{2}n_i(n_i-1) & \text{if } \mathbf{m} = \mathbf{n} - \mathbf{e}_i \\ n_i A_{ij} & \text{if } \mathbf{m} = \mathbf{n} - \mathbf{e}_i + \mathbf{e}_j \\ 0 & \text{otherwise} \end{cases}$$

Note that $\mathbf{Q} = (Q(\mathbf{n},\mathbf{m}))$ generates a continuous time Markov chain, which we denote by $(\mathbf{n}(t), \mathbf{Q}_\mathbf{n})$. Then the following duality relation holds.

$$T_t f_{\mathbf{n}}(\mathbf{x}) = E^{\mathbf{Q_n}}(f_{\mathbf{n}(t)}(\mathbf{x})).$$

The Markov chain $(\mathbf{n}(t), \mathbf{Q_n})$ is called *a coalescing Markov chain with delay,* which has similar ergodic behaviors as the standard coalescing Markov chain and the detailed analysis of the Markov chain yields complete description of stationary distributions and ergodic theorems (cf. [S1], [S2]).

9.2 Comparison

To investigate ergodic behaviors of the IDS comparison arguments also are very useful.

(a) *Basic comparison*
Let $\mathbf{x}(t;\mathbf{x})$ and $\mathbf{x}(t;\mathbf{y})$ be the two strong solutions of the SDE (1) with common Brownian motions with the initial conditions $\mathbf{x} \in \mathbf{E}$ and $\mathbf{y} \in \mathbf{E}$. If $\mathbf{x} \leq \mathbf{y}$, i.e. $x_i \leq y_i$ for all $i \in S$, then it holds that

$$\mathbf{x}(t;\mathbf{x}) \leq \mathbf{x}(t;\mathbf{y}) \quad \text{for every } t \geq 0,\ P-a.s.$$

This comparison is due to the fact that the IDS is defined by a linear coupling of independent one-dimensional diffusions so that one-dimensional theory by Yamada-Watanabe is applicable (cf. [WY]).

(b) *New comparison*
Recently Cox, Fleischmann and Greven discovered another comparison for the IDS's. Let

$$\mathbf{F} = \{f \in C_0^2(\mathbf{E});\ \frac{\partial^2 f}{\partial x_i \partial x_j}(\mathbf{x}) \geq 0 \quad \text{for all } i, j \in S\}.$$

Then we have

Theorem 13 ([CFG]). *Let T_t^1 and T_t^2 be the transition semigroup of two IDS's on I associated with a common A and diffusion coefficients $a_1(u)$ and $a_2(u)$. Assume that*

$$|a_1(u)| \leq |a_2(u)| \quad \textit{for every } u \in I.$$

Then for $f \in \mathbf{F}$ it holds

$$T_t^1 f(\mathbf{x}) \leq T_t^2 f(\mathbf{x}) \quad \textit{for every } \mathbf{x} \in \mathbf{E},\ t > 0.$$

The CFG comparison is useful in discussing local extinction problem.

9.3 Coupling

Coupling is one of most important and applicable tools in the subject of interacting particle systems (cf. [L]), and it is true for the IDS's also.

Let $\mathbf{x}(t)$ and $\mathbf{y}(t)$ be two strong solutions of the SDE (1) with common Brownian motions. Then for $i \in \mathbf{Z}^d$, $|x_i - y_i|$ plays a role of a Lyapunov function. In fact one can show it by applying Ito formula to $|x_i - y_i|$ together with the $\mathbf{Z}^d$-shift invariance as follows.

Lemma 1. *Suppose that A and $\mathcal{L}(\mathbf{x}(0), \mathbf{y}(0))$ are $\mathbf{Z}^d$-shift invariant. Then, $E(|x_i(t) - y_i(t)|)$ is non-decreasing in $t \geq 0$ for $i \in \mathbf{Z}^d$.*

For the proof of Theorem 1 this lemma is essentially used as well as some routine coupling technique and second moment estimates as in [LS].

9.4 Liouville property

Recall that $A = \{A_{ij}\}$ is an infinitesimal matrix of a continuous time irreducible random walk on $\mathbf{Z}^d$. A function $f : \mathbf{Z}^d \mapsto \mathbf{R}$ is an A-harmonic function if Af is well-defined and $Af(i) = 0$ for $i \in \mathbf{Z}^d$. We say that A satisfies the Liouville property if every nonnegative A-harmonic function is constant. One can prove the Liouville property of A under an unbias condition on A.

Theorem 14. *Assume that*

$$\sum_{i \in \mathbf{Z}^d} i \cdot A_{0,i} = 0.$$

Then every nonnegative A-harmonic function is constant.

It is to be noted that the Liouville property plays a key role in the proof of Theorems 3 and 4.

9.5 Random walk estimates

In the proof of Theorems 11 and 12 the following random walk estimates play a crucial role, in which one can find a critical scaling order relevant to the size of the torus in the approximation problem of the infinite system via finite systems.

Let $Q_t^N = \{Q_t^N(i,j)\}$ and $\widehat{Q}_t^N = \{\widehat{Q}_t^N(i,j)\}$ be the transition matrix of the continuous time A^N-random walk and $\widehat{A}^N$-random walk on Λ^N respectively.

Lemma 2 ([C]). *1. If $t_N/N \to \infty$ as $N \to \infty$, then*

$$\lim_{N\to\infty} \sup_{t \geq t_N} \sup_{i,j \in \Lambda_N} |\Lambda_N||Q_t^N(i,j) - |\Lambda_N|^{-1}| = 0.$$

2. If $d \geq 3$ *and* $T_N/\beta_N \to s \in (0,\infty)$ *as* $N \to \infty$,

$$\lim_{N\to\infty} \int_0^{T_N} \widehat{Q}_t^N(i,j) = \widehat{G}(i,j) + s.$$

9.6 Moment estimates

Using the condition [B] and the random walk estimates one obtains the following moment estimates for the IDS's, which are used in the proof of Theorems 11 and 12.

Lemma 3. *1. If* $\mu \in \mathcal{P}(\mathbf{E})$ *satisfies* $\sup_{i\in\mathbf{Z}^d}\langle\mu, x_i^2\rangle < \infty$, *then*

$$\sup_{t\geq 0}\sup_{i\in\mathbf{Z}^d} E^{\mu}|x_i(t)|^2 < \infty.$$

2. For some $p > 2$,

$$\sup_{t\geq 0} E^{\mu}|x_0(t)|^p < \infty \quad \text{if } \mu \in \mathcal{T}_p(\mathbf{E}).$$

3. Assume $d \geq 3$. *If* $\mu^N \in \mathcal{T}_2(\mathbf{E}_N)$ *and* $\sup_{N\geq 1}\langle\mu,, |x_0^N|^2\rangle < \infty$, *then*

$$\sup_{N\geq 1}\sup_{0\leq t\leq T\beta_N} E^{\mu^N}|x_0^N(t)|^2 < \infty \quad \text{for any } T < \infty.$$

4. Assume $d \geq 3$. *If for some* $p > 2$, $\mu^N \in \mathcal{T}_p(\mathbf{E}_N)$ *and* $\sup_{N\geq 1}\langle\mu,, |x_0^N|^p\rangle < \infty$, *then*

$$\sup_{N\geq 1}\sup_{0\leq t\leq T\beta_N} E^{\mu^N}|x_0^N(t)|^p < \infty \quad \text{for any } T < \infty..$$

9.7 Idea of the proof of the approximation result

Now we try to explain some ideas for the proof of Theorem 11. First note that

$$\Theta_N(t) - \Theta_N(0) = |\Lambda_N|^{-1}\sum_{i\in\Lambda_N}\int_0^t a(x_i^N(s))dB_i(s), \tag{12}$$

so that $Z_N(t) = \Theta_N(t\beta_N)$ is a martingale with quadratic variation process

$$\langle Z_N\rangle(t) = \int_0^t \langle U_N(s\beta_N), a(x_0)^2\rangle ds. \tag{13}$$

The crucial step is to show that for $t > 0$ and $\phi \in C_0^1(\mathbf{E})$,

$$\lim_{N\to\infty} E|\langle U_N(t\beta_N),\phi\rangle - \langle \nu_{Z_N(t)},\phi\rangle| = 0. \tag{14}$$

Once we get (14), it follows from (13) that

$$\lim_{N\to\infty} E|\langle Z_N\rangle(t) - \int_0^t (a^*)^2(Z_N(s))ds| = 0, \tag{15}$$

where

$$a^*(\theta) = \sqrt{\langle \nu_\theta, a(x_0)^2\rangle}.$$

Accordingly (13), (14) and (15) together with some tightness arguments would yield that the rescaled density process $Z_N(\cdot)$ converges to the I-valued diffusion process $Z(\cdot)$ as $N \to \infty$.

To understand (14) recall Theorem 1, which claims that the limiting distribution of the infinite system depends merely upon the density of the initial distribution. This principle would be true even for finite systems, so that the large time behaviors of the finite systems would be determined by the density of the initial distribution. This idea is realized in the following way;

Take a sequence $\alpha_N = o(\beta_N) \nearrow \infty$ and set $t_N = \beta_N t - \alpha_N$. The empirical density process $\Theta_N(\cdot)$ attains almost the same level at time t_N and $t\beta_N$ since $\{Z_N(\cdot)\}$ forms a tight family. Accordingly, if one switches the process $\mathbf{x}^N(\cdot) = \{x_i^N(\cdot)\}$ at time t_N and define a new process $\mathbf{y}^N(\cdot)$ by the SDE (11) with initial condition $y_i^N(t_N) = \Theta_N(t_N)$ for all $i \in \Lambda_N$, its associated empirical distribution process

$$V_N(s) = |\Lambda_N|^{-1} \sum_{i\in\Lambda_N} \delta_{\sigma_i^N \mathbf{y}^N(s)} \quad \text{for } s \geq t_N$$

would well approximate the empirical distribution process $U_N(s)$ for $s \geq t_N$ since $\mathbf{x}^N(\cdot)$ and $\mathbf{y}^N(\cdot)$ have an identical density at t_N. At the same time $V_N(t\beta_N)$ would be identified with $\nu_{\Theta_N(t_N)}$. As a matter of fact this story is confirmed in [CGS1].

References

[BCG1] M. Bramson, J.T. Cox and A. Greven: *Ergodicity of critical spatial branching processes in low dimension*, Ann. Probab. **21**, 1946-1957 (1993).

[BCG2] M. Bramson, J.T. Cox and A. Greven: *Invariant measures of critical spatial branching processes in high dimensions*, to appear in Ann. Probab. (1996).

[C] J.T. Cox: *Coalescing random walks and voter model consensus times on the torus in* $\mathbf{Z}^d$. Ann. Probab. **17**, 1333-1366 (1989)

[CFG] J.T. Cox, K. Fleischmann and A. Greven: *Comparison of interacting diffusions and an application to their ergodic theory*, (preprint).

[CGre1] J.T. Cox and A. Greven: *On the long term behavior of some finite particle systems*, Probab. Th. Rel. Fields **85**, 195-237 (1990).

[CGre2] J.T. Cox and A. Greven: *Ergodic theorems for infinite systems of interacting diffusions*, Ann. Probab. **22**, 833-853 (1994).

[CGri] J.T. Cox and D. Griffeath: *Diffusive clustering in the two dimensional voter model*, Ann. Probab. **14**, 347-370 (1986).

[CGS1] J.T. Cox, A. Greven and T. Shiga: *Finite and infinite systems of interacting diffusions*, Probab. Th. Rel. Fields **103**, 165 - 197 (1995).

[CGS2] J.T. Cox, A. Greven and T. Shiga: *Finite and infinite systems of interacting diffusions, Part II*, (preprint).

[CM] R.A. Carmona and S.A. Molchanov: *Prabolic Anderson problem and intermittency*, AMS Memoir **108**, No.518 (1994).

[CV] R.A. Carmona and Viens: (personal communication).

[D] D.A. Dawson: *The critical measure diffusion process*, Z. Wahr. verw. Geb. **40**, 125-145 (1977).

[Deu1] J.-D. Deuschel: *Central limit theorem for an infinite lattice system of interacting diffusion processes*, Ann. Probab. **16**, 700-716 (1988)

[Deu2] J-N. Deuschel: *Algebraic L^2-decay of attractive critical process on the lattice*, Ann. Probab. **22**, 264-293 (1994).

[Dy] E.B. Dynkin: *Sufficient statistics and extreme points*, Ann. Probab. **6**, 705-730 (1978).

[DG] D.A. Dawson and A. Greven: *Multiple time scale analysis of interacting diffusions*, Prob. Th. Rel. Fields **95**, 467-508 (1993).

[K] M. Kimura: *"Stepping stone"model of population*, Ann. Rep. Nat. Inst. Gen. **3**, 62-63 (1953).

[KO] M. Kimura and T. Ohta: *Theoretical Aspects of Population Genetics*, Princeton University Press (1971).

[L] T.M. Liggett: *Interacting Particle Systems*, Springer Verlag (1985).

[LS] T.M. Liggett and F. Spitzer: *Ergodic theorems for coupled random walks and other systems with locally interacting components*, Z. Wahrsch. verw. Gebiete **56**, 443-468 (1981).

[NS] M. Notohara and T. Shiga: *Convergence to genetically uniform state in stepping stone models of population genetics*, J. Math. Biol. **10**, 281-294 (1980).

[Sa] K. Sato: *Limit diffusions of some stepping stone models*, J. Appl. Prob. **20**, 460-471 (1983).

[S1] T. Shiga: *An interacting system in population genetics*, J. Math. Kyoto Univ. **20**, 213-243 (1990).

[S2] T. Shiga: *An interacting system in population genetics II*, J. Math. Kyoto Univ. **20**, 723-733 (1990).

[S3] T. Shiga: *Ergodic theorems and exponential decay of sample paths for certain interacting diffusion systems*, Osaka J. Math. **29**, 789-807 (1992).

[S4] T. Shiga: *Stationary distribution problem for interacting diffusion systems*, CRM Proceeding and Lecture Notes **5**, 199-211 (1994).

[S5] T. Shiga: *A note on sample Lyapunov exponents of a class of SPDE*, (preprint).

[SS] T. Shiga and A. Shimizu: *Infinite-dimensional stochastic differential equations and their applications*, J. Math. Kyoto Univ. **20**, 395-416 (1980).

[SU] T. Shiga and K. Uchiyama: *Stationary states and their stability of the stepping stone model involving mutation and selection*, Prob. Th. Rel. Fields **73**, 87-117 (1986).

[Sp] F. Spitzer: *Principles of Random Walk*, Springer-Verlag (1976).

[W] S. Watanabe: *A limit theorem of branching processes and continuous branching processes*, J. Math. Kyoto Univ. **8**, 141-167 (1968).

[WY] S. Watanabe and T. Yamada: *On the uniqueness of solutions of stochastic differential equations*, J. Math. Kyoto Univ. **11**, 155-167 (1971).

[ZMRS] Ya.B. Zeldovich, S.A. Molchanov, A.A. Ruzmaikin and D.D. Sokoloff: *Intermittency, diffusion and generation in nonstationary random medium*, Soviet Sci. Rev. Math. Phys. **7**, 1-110 (1988).

A Kähler metric on a based loop group and a covariant differentiation

Ichiro Shigekawa[1] and Setsuo Taniguchi[2]

[1] Department of Mathematics, Graduate School of Science, Kyoto University, Kyoto 606-01, Japan
[2] Graduate School of Mathematics, Kyushu University, Hakozaki, Fukuoka 812, Japan

1. Introduction

Loop groups have been attracting many authors recently. In this paper, we are discussing a Kähler metric on a loop group. Let G be a d-dimensional compact Lie group and $\mathfrak{g}$ be its Lie algebra ($\equiv$ the space of *left* invariant vector fields). Then, $\mathfrak{g}$ admits an $\mathrm{Ad}(G)$-invariant inner product $(\cdot,\cdot)_{\mathfrak{g}}$ and we fix it through the paper. We denote the G-valued path space on $[0,1]$ by

$$PG := \{\gamma: [0,1] \to G\,; \text{continuous and } \gamma(0) = e\} \tag{1.1}$$

e being the unit element of G.

On the other hand, our interest is in the based loop group ΩG over G:

$$\Omega G := \{\gamma: [0,1] \to G\,; \text{continuous and } \gamma(0) = \gamma(1) = e\}. \tag{1.2}$$

We develop a differential geometry from an analytic point of view. In particular, we discuss several operators acting on, e.g., tensor fields. Usually, the following Cameron-Martin space H_0 is regarded as a tangent space:

$$H_0 = \left\{ \mathbf{h} : [0,1] \to \mathfrak{g} : \begin{array}{l} \mathbf{h} \text{ is absolutely continuous, } \mathbf{h}(0) = \mathbf{h}(1) = 0, \text{ and} \\ \text{the derivative } \dot{\mathbf{h}} \text{ satisfies that } \int_0^1 |\dot{\mathbf{h}}(t)|_{\mathfrak{g}}^2 dt < \infty \end{array} \right\},$$

where $|\cdot|_{\mathfrak{g}} = \sqrt{(\cdot,\cdot)_{\mathfrak{g}}}$. H_0 is a Hilbert space with the inner product

$$(\mathbf{h},\mathbf{k})_{H_0} = \int_0^1 (\dot{\mathbf{h}}(t), \dot{\mathbf{k}}(t))_{\mathfrak{g}} dt, \qquad \mathbf{h},\mathbf{k} \in H_0.$$

Using the left translation, this inner product defines a Riemannian metric.

Further, we can introduce an almost complex structure, denoted by J, (see, [11]), but under it, the above metric is not a Kähler metric. The following Kähler form S was introduced by Pressley [11]:

$$S(\mathbf{h},\mathbf{k}) = \int_0^1 (\dot{\mathbf{h}}(t), \mathbf{k}(t))_{\mathfrak{g}} dt.$$

The associated Riemannian metric is defined by $B(X, Y) = S(X, JY)$. This metric was discussed by [11, 2] in view of differential geometry. We will discuss it from a probabilistic point of view, in which the pinned Brownian motion measure plays an essential role.

The organization of this paper is as follows. In the section 2, we prepare the fundamental notions of differential geometry. We define vector fields, differential forms, exterior derivatives, etc. We also introduce an almost complex structure. Kähler metric is discussed in the section 3. We will calculate the Levi-Civita covariant derivative and the associated Riemannian curvature. The section 4 is devoted to showing the closability of operators. The Ricci curvature is computed in the section 5.

2. A based loop group and an almost complex structure

In this section, we introduce several notions in differential geometry. The Cameron-Martin space H_0 is a tangent space of ΩG. Defining a bracket by $[\mathbf{h}, \mathbf{k}](t) = [\mathbf{h}(t), \mathbf{k}(t)]_{\mathfrak{g}}$, $t \in [0, 1]$, H_0 becomes a Lie algebra.

Thinking of H_0 as $T_{\mathbf{e}}(\Omega G)$, the tangent space of ΩG at e, where $\mathbf{e}(s) \equiv e$, one may regard the product space $\Omega G \times H_0$ as the tangent bundle of ΩG. One then defines spaces of tensor fields on ΩG by

$$\mathcal{F}\Gamma_b^\infty(T_q^p(\Omega G)) = \left\{ u : \Omega G \to H_0^{\otimes p} \otimes (H_0^*)^{\otimes q} : \begin{array}{l} u = \sum_j^{\text{finite}} f_j e_j \text{ for some } f_j \in \mathcal{F}C_b^\infty(\Omega G) \\ \text{and } e_j \in H_0^{\otimes p} \otimes (H_0^*)^{\otimes q} \end{array} \right\},$$

where

$$\mathcal{F}C_b^\infty(\Omega G) = \left\{ u : \Omega G \to \mathbb{R} : \begin{array}{l} u(\gamma) = f(\gamma(t_1), \ldots, \gamma(t_n)) \text{ for some} \\ f \in C_b^\infty(G^n) \text{and } 0 \le t_1 < \ldots < t_n \le 1 \end{array} \right\}.$$

For $X \in \mathcal{F}\Gamma_b^\infty(T_0^1(\Omega G))$ and $u \in \mathcal{F}C_b^\infty(\Omega G)$, set

$$Xu(\gamma) = \lim_{\varepsilon \downarrow 0} \frac{1}{\varepsilon} \Big(f\big(\gamma e^{\varepsilon X(\gamma)}\big) - f(\gamma) \Big),$$

where for $\mathbf{h} \in H_0$, $\big(e^{\varepsilon \mathbf{h}}\big)(t) = e^{\varepsilon \mathbf{h}(t)}$, the position of the integral curve along $\mathbf{h}(t) \in \mathfrak{g}$ at time ε. As is easily seen, one has that

$$Xu(\gamma) = \sum_{i=1}^n \Big(\big(X(\gamma)\big)(t_i)^{(i)} f \Big)(\gamma(t_1), \ldots, \gamma(t_n)) \tag{2.1}$$

$$\text{for } u(\gamma) = f(\gamma(t_1), \ldots, \gamma(t_n)) \in \mathcal{F}C_b^\infty(\Omega G)$$

where, for $\xi \in \mathfrak{g}$,

$$\xi^{(i)}f(g_1,\ldots,g_n)=\lim_{\varepsilon\downarrow 0}\frac{1}{\varepsilon}\big(f(g_1,\ldots,g_{i-1},g_ie^{\varepsilon\xi},g_{i+1},\ldots,g_n)-f(g_1,\ldots,g_n)\big).$$

For $X,Y\in\mathcal{F}\Gamma_b^\infty(T_0^1(\Omega G))$, the Lie bracket $[X,Y]$ can be defined as a unique element of $\mathcal{F}\Gamma_b^\infty(T_0^1(\Omega G))$ so that

$$[X,Y]u=XYu-YXu,\quad X,Y\in\mathcal{F}\Gamma_b^\infty(T_0^1(\Omega G)),\ u\in\mathcal{F}C^\infty(\Omega G). \tag{2.2}$$

Defining constant vector field $X^{\mathbf{h}}\in\mathcal{F}\Gamma_b^\infty(T_0^1(\Omega G))$, $\mathbf{h}\in H_0$, by $X^{\mathbf{h}}(\gamma)=\mathbf{h}$, $\gamma\in\Omega G$. Then $X^{\mathbf{h}}$ is a *left* invariant vector field. Due to (2.1), one has that

$$[X^{\mathbf{h}},X^{\mathbf{k}}]=X^{[\mathbf{h},\mathbf{k}]},\quad \mathbf{h},\mathbf{k}\in H_0.$$

By virtue of (2.2), the Jacobi identity can be seen;

$$[X,[Y,Z]]+[Y,[Z,X]]+[Z,[X,Y]]=0,\quad X,Y,Z\in\mathcal{F}\Gamma_b^\infty(T_0^1(\Omega G)). \tag{2.3}$$

Let

$$\mathcal{F}\Gamma_b^\infty(\textstyle\bigwedge^pT^*(\Omega G))=\{u\in\mathcal{F}\Gamma_b^\infty(T_p^0(\Omega G)):u(\gamma)\text{ is anti-symmetric}\}.$$

The exterior derivative du can be defined, as in the finite dimensional case, for $u\in\mathcal{F}\Gamma_b^\infty(\bigwedge^pT^*(\Omega G))$

$$\begin{aligned}du(X_1,\ldots,X_{p+1})&\\=&\sum_{a=1}^{p+1}(-1)^{a-1}X_a\big(u(X_1,\ldots,\hat{X}_a,\ldots,X_{p+1})\big)\\&+\sum_{1\le a<b\le p+1}(-1)^{a+b}u([X_a,X_b],X_1,\ldots,\hat{X}_a,\ldots,\hat{X}_b,\ldots,X_{p+1}),\end{aligned} \tag{2.4}$$

for $X_1,\ldots,X_{p+1}\in\mathcal{F}\Gamma_b^\infty(T_0^1(\Omega G))$, where $\hat{X}_a$ means that X_a is omitted. As in the finite dimensional case, one has that

$$d^2=0. \tag{2.5}$$

Indeed, define $\mathcal{L}_X$ and $\iota(X)$ by

$$\begin{aligned}\mathcal{L}_Xu(X_1,\ldots,X_p)&=X\big(u(X_1,\ldots,X_p)\big)\\&\quad-\sum_{a=1}^{p}u(X_1,\ldots,X_{a-1},[X,X_a],X_{a+1},\ldots,X_p),\\ \iota(X)u(X_1,\ldots,X_{p-1})&=u(X,X_1,\ldots,X_{p-1}),\end{aligned}$$

for $X,X_1,\ldots,X_p\in\mathcal{F}\Gamma_b^\infty(T_0^1(\Omega G))$, $u\in\mathcal{F}\Gamma_b^\infty(\bigwedge^pT^*(\Omega G))$.

An elementary algebraic computation leads one to the identities

$$\begin{cases} d \circ \iota(X) + \iota(X) \circ d = \mathcal{L}_X, \\ d \circ \mathcal{L}_X = \mathcal{L}_X \circ d, \qquad X \in \mathcal{F}\Gamma_b^\infty(T_0^1(\Omega G)). \end{cases}$$

These yield that $d^2 \circ \iota(X) = \iota(X) \circ d^2$ from which (2.5) follows by induction on p.

We now introduce an almost complex structure on ΩG following [11]. To do this, put

$$e_n(t) = \frac{1}{2\pi\sqrt{-1}n}\left(e^{2\pi\sqrt{-1}nt} - 1\right), \qquad n \in \mathbb{Z} \setminus \{0\}$$

and we take an orthonormal basis $\{\xi\}_{i=1,\dots,d}$ in $\mathfrak{g}$. We fix it through the paper. We use the following convention. For $\alpha = (n, i)$, $n \in \{1, 2, \dots\}$, $i = 1, 2, \dots, d$, we define $\bar{\alpha} = (-n, i)$, and

$$\mathbf{e}_\alpha = e_n(t)\xi_i, \quad \mathbf{e}_{\bar{\alpha}} = e_{-n}(t)\xi_i. \tag{2.6}$$

Every $\mathbf{h} \in H_0$ can be expanded as

$$\mathbf{h} = \sum_{i=1}^d \sum_{n \neq 0} (\mathbf{h}, \mathbf{e}_{-n,i})_{H_0} \mathbf{e}_{n,i} \quad \text{in } H_0,$$

where $(\mathbf{h}, \mathbf{k}_1 + \sqrt{-1}\mathbf{k}_2)_{H_0} = (\mathbf{h}, \mathbf{k}_1)_{H_0} + \sqrt{-1}(\mathbf{h}, \mathbf{k}_2)_{H_0}$, $\mathbf{h}, \mathbf{k}_1, \mathbf{k}_2 \in H_0$. An almost complex structure $J : H_0 \to H_0$ on ΩG is defined by

$$J\mathbf{h} = \sqrt{-1}\sum_{i=1}^d \sum_{n>0} (\mathbf{h}, \mathbf{e}_{-n,i})_{H_0} \mathbf{e}_{n,i} - \sqrt{-1}\sum_{i=1}^d \sum_{n>0} (\mathbf{h}, \mathbf{e}_{n,i})_{H_0} \mathbf{e}_{-n,i}.$$

See [11]. As is easily seen, it holds that

$$J^2\mathbf{h} = -\mathbf{h} \quad \text{and} \quad (J\mathbf{h}, J\mathbf{k})_{H_0} = (\mathbf{h}, \mathbf{k})_{H_0}, \quad \mathbf{h}, \mathbf{k} \in H_0.$$

Put $H_0^{\mathbb{C}} = H_0 \oplus \sqrt{-1}H_0$ and

$$H_0^{(1,0)} = \{\eta \in H_0^{\mathbb{C}} : J\eta = \sqrt{-1}\eta\}, \quad H_0^{(0,1)} = \{\eta \in H_0^{\mathbb{C}} : J\eta = -\sqrt{-1}\eta\}.$$

Here we extend J to $H_0^{\mathbb{C}}$ by complex linearity. Obviously $H_0^{\mathbb{C}} = H_0^{(1,0)} \oplus H_0^{(0,1)}$. $J^*: H_0^* \to H_0^*$ satisfies the same property and therefore $H_0^{*(1,0)}$, $H_0^{*(0,1)}$ can be defined similarly. For $\mathbf{h} \in H_0$, define

$$\pi_+\mathbf{h} = \mathbf{h}^{(1,0)} = \tfrac{1}{2}(\mathbf{h} - \sqrt{-1}J\mathbf{h}), \tag{2.7}$$

$$\pi_-\mathbf{h} = \mathbf{h}^{(0,1)} = \tfrac{1}{2}(\mathbf{h} + \sqrt{-1}J\mathbf{h}). \tag{2.8}$$

Then $\mathbf{h}^{(1,0)} \in H_0^{(1,0)}$, $\mathbf{h}^{(0,1)} \in H_0^{(0,1)}$, and $\mathbf{h} = \mathbf{h}^{(1,0)} + \mathbf{h}^{(0,1)}$. By a straightforward computation one sees that

$$\mathbf{h}^{(1,0)} = \sum_{i=1}^{d}\sum_{n>0}(\mathbf{h}, \mathbf{e}_{-n,i})_{H_0}\mathbf{e}_{n,i}, \quad \mathbf{h}^{(0,1)} = \sum_{i=1}^{d}\sum_{n>0}(\mathbf{h}, \mathbf{e}_{n,i})_{H_0}\mathbf{e}_{-n,i}. \tag{2.9}$$

Since

$$[\mathbf{e}_{n,i}, \mathbf{e}_{m,j}](t) = \frac{1}{2\pi\sqrt{-1}mn}\{(m+n)e_{m+n}(t) - ne_n(t) - me_m(t)\}[\xi_i, \xi_j],$$

one can conclude from (2.9) that

$$[\mathbf{h}^{(1,0)}, \mathbf{k}^{(1,0)}] \in H_0^{(1,0)} \quad \text{and} \quad [\mathbf{h}^{(0,1)}, \mathbf{k}^{(0,1)}] \in H_0^{(0,1)} \quad \mathbf{h}, \mathbf{k} \in H_0. \tag{2.10}$$

Let us define the Newlander-Nirenberg tensor N as follows:

$$N(X,Y) = J[X,Y] - [JX,Y] - [X,JY] - J[JX,JY].$$

By (2.10), we can easily see $N = 0$ and in this sense, J is "integrable." Put

$$\bigwedge^{p,q} = \underbrace{H_0^{*(1,0)} \wedge \cdots \wedge H_0^{*(1,0)}}_{p\text{-times}} \wedge \underbrace{H_0^{*(0,1)} \wedge \cdots \wedge H_0^{*(0,1)}}_{q\text{-times}}$$

and define

$$\mathcal{F}\Gamma_b^\infty(\bigwedge^{p,q}T^*(\Omega G)) = \left\{ u : \Omega G \to \bigwedge^{p,q} : \begin{array}{l} u = \sum_i^{\text{finite}} (f_i + \sqrt{-1}g_i)e_i \\ f_i, g_i \in \mathcal{F}C_b^\infty(\Omega G), e_i \in \bigwedge^{p,q} \end{array} \right\}.$$

Observing that

$$\mathcal{F}\Gamma_b^\infty(\bigwedge^{p,q}T^*(\Omega G)) \subset \mathcal{F}\Gamma_b^\infty(T_{p+q}^0(\Omega G)) \oplus \sqrt{-1}\mathcal{F}\Gamma_b^\infty(T_{p+q}^0(\Omega G)),$$

we can extend the exterior derivative d in (2.1) to $\mathcal{F}\Gamma_b^\infty(\bigwedge^{p,q}(T^*(\Omega G)))$, and due to (2.4), obtain that

$$du(\gamma) \in \bigwedge^{p+1,q} \oplus \bigwedge^{p,q+1}, \ \gamma \in \Omega G, \quad \text{for } u \in \mathcal{F}\Gamma_b^\infty(\bigwedge^{p,q}T^*(\Omega G)).$$

Thus, operators

$$\partial : \mathcal{F}\Gamma_b^\infty(\bigwedge^{p,q}T^*(\Omega G)) \to \mathcal{F}\Gamma_b^\infty(\bigwedge^{p+1,q}T^*(\Omega G))$$
$$\overline{\partial} : \mathcal{F}\Gamma_b^\infty(\bigwedge^{p,q}T^*(\Omega G)) \to \mathcal{F}\Gamma_b^\infty(\bigwedge^{p,q+1}T^*(\Omega G))$$

can be defined so that for $u \in \mathcal{F}\Gamma_b^\infty(\bigwedge^{p,q}T^*(\Omega G))$, $\partial u(\gamma)$ (resp. $\overline{\partial} u(\gamma)$) is the projection of $du(\gamma)$ onto $\bigwedge^{p+1,q}$ (resp. $\bigwedge^{p,q+1}$). Obviously $d = \partial + \overline{\partial}$. Then, by virtue of (2.2), one obtains that

$$\partial^2 = 0, \quad \overline{\partial}^2 = 0, \quad \text{and} \quad \partial\overline{\partial} + \overline{\partial}\partial = 0 \tag{2.11}$$

on $\mathcal{F}\Gamma_b^\infty(\bigwedge^{p,q}T^*(\Omega G))$ for $p, q \geq 0$.

3. A Kähler metric

We now introduce a Kähler metric on ΩG following [11]. Define $S \in \mathcal{F}\Gamma_b^\infty(\bigwedge^2 T^*(\Omega G))$ by

$$S(X^{\mathbf{h}}, X^{\mathbf{k}}) = \int_0^1 (\dot{\mathbf{h}}(t), \mathbf{k}(t))_{\mathfrak{g}} dt, = -\int_0^1 (\mathbf{h}(t), \dot{\mathbf{k}}(t))_{\mathfrak{g}} dt, \quad \mathbf{h}, \mathbf{k} \in H_0.$$

Note that

$$|S(X^{\mathbf{h}}, X^{\mathbf{k}})| \leq |\mathbf{h}|_{H_0} |\mathbf{k}|_{H_0},$$

and hence that S is well-defined. By the integration by parts and the $\mathrm{Ad}(G)$-invariance of $(\cdot,\cdot)_{\mathfrak{g}}$, we obtain the following cyclic formula:

$$S([X^{\mathbf{h}}, X^{\mathbf{k}}], X^{\mathbf{l}}) + S([X^{\mathbf{k}}, X^{\mathbf{l}}], X^{\mathbf{h}}) + S([X^{\mathbf{l}}, X^{\mathbf{h}}], X^{\mathbf{k}}) = 0.$$

Combine this with (2.1), we can show that $dS = 0$. Define

$$B(X, Y) = S(X, JY), \quad X, Y \in \mathcal{F}\Gamma_b^\infty(T_0^1(\Omega G)).$$

We note that $S(X, Y) = B(JX, Y)$. Let $T : H_0^{\mathbb{C}} \to H_0^{\mathbb{C}}$ be a continuous linear operator so that $T\mathbf{e}_{n,i} = \frac{1}{2|n|\pi}\mathbf{e}_{n,i}$. Observe then that

$$S(X^{\mathbf{e}_{n,i}}, X^{\mathbf{e}_{m,j}}) = \frac{1}{2\pi m\sqrt{-1}}\delta_{n,-m}\delta_{i,j},$$

and hence that

$$B(X^{\mathbf{h}}, X^{\mathbf{k}}) = (T\mathbf{h}, \mathbf{k})_{H_0}. \tag{3.1}$$

In particular, for $X, Y \in H_0$,

$$B(X, X) \geq 0 \quad \text{and “} = 0\text{” if and only if } X = 0 \tag{3.2}$$

$$B(X, Y) = B(Y, X) \in \mathbb{R}. \tag{3.3}$$

Thus one have obtained the Kähler metric B on ΩG. We denote the completion of H_0 with respect to B by H_1. From now on, we regard H_0 as a tangent space. Moreover, by noting $JT = TJ$, J also defines an almost complex structure in H_1.

We now turn to the Levi-Civita covariant derivative. As in the finite dimensional case, the Levi-Civita covariant derivative is characterized by the following identity:

$$\begin{aligned} 2B(\nabla_X Y, Z) &= XB(Y, Z) + YB(X, Z) - ZB(X, Y) \\ &\quad + B([X, Y], Z) + B([Z, X], Y) + B(X, [Z, Y]). \end{aligned}$$

In particular, taking left invariant vector fields, we have

$$2B(\nabla_{X^{\mathbf{h}}} X^{\mathbf{k}}, X^{\mathbf{l}}) = B([X^{\mathbf{h}}, X^{\mathbf{k}}], X^{\mathbf{l}}) + B([X^{\mathbf{l}}, X^{\mathbf{h}}], X^{\mathbf{k}}) + B(X^{\mathbf{h}}, [X^{\mathbf{l}}, X^{\mathbf{k}}]).$$

Furthermore, due to the identity (see, e.g., [8, Proposition IX.4.2])

$$4B((\nabla_X J)Y, Z) = 6dS((X, JY, JY) - 6dS(X, Y, Z) + B(N(Y, Z), JX),$$

we have $\nabla J = 0$, i.e., the almost complex structure is parallel.

We easily see that

$$B(\nabla_X Y, Z) = 0 \quad \text{if } Y, Z \in H^{(1,0)} \text{ or } Y, Z \in H^{(0,1)}. \tag{3.4}$$

To see this, we note that the almost complex structure J is parallel. For example, if $Y, Z \in H^{(1,0)}$, then

$$\begin{aligned} B(\nabla_X Y, Z) &= -\sqrt{-1}B(\nabla_X JY, Z) \\ &= -\sqrt{-1}B(J\nabla_X Y, Z) \\ &= \sqrt{-1}B(\nabla_X Y, JZ) \\ &= -B(\nabla_X Y, Z). \end{aligned}$$

Thus we have (3.4).

Let us calculate the covariant derivative. From the definition,

$$\begin{aligned} &B(\nabla_{X^{\mathbf{h}}} X^{\mathbf{k}}, X^{\mathbf{l}}) \\ &= B([X^{\mathbf{h}}, X^{\mathbf{k}}], X^{\mathbf{l}}) + B([X^{\mathbf{l}}, X^{\mathbf{h}}], X^{\mathbf{k}}) + B(X^{\mathbf{h}}, [X^{\mathbf{l}}, X^{\mathbf{k}}]) \\ &= S(X^{\mathbf{l}}, J[X^{\mathbf{h}}, X^{\mathbf{k}}]) + S([X^{\mathbf{l}}, X^{\mathbf{h}}], JX^{\mathbf{k}}) + S([X^{\mathbf{l}}, X^{\mathbf{k}}], JX^{\mathbf{h}},) \\ &= \int_0^1 (\dot{\mathbf{l}}(t), J[\mathbf{h}, \mathbf{k}](t))_{\mathfrak{g}} dt + \int_0^1 (\frac{d}{dt}[\mathbf{l}(t), \mathbf{h}(t)], J\mathbf{k}(t))_{\mathfrak{g}} dt \\ &\quad + \int_0^1 (\frac{d}{dt}[\mathbf{l}(t), \mathbf{k}(t)], J\mathbf{h}(t))_{\mathfrak{g}} dt \\ &= \int_0^1 (\dot{\mathbf{l}}(t), J[\mathbf{h}, \mathbf{k}](t))_{\mathfrak{g}} dt + \int_0^1 ([\dot{\mathbf{l}}(t), \mathbf{h}(t)], J\mathbf{k}(t))_{\mathfrak{g}} dt \\ &\quad + \int_0^1 ([\mathbf{l}(t), \dot{\mathbf{h}}(t)], J\mathbf{k}(t))_{\mathfrak{g}} dt + \int_0^1 ([\dot{\mathbf{l}}(t), \mathbf{k}(t)], J\mathbf{h}(t))_{\mathfrak{g}} dt \\ &\quad + \int_0^1 ([\mathbf{l}(t), \dot{\mathbf{k}}(t)], J\mathbf{h}(t))_{\mathfrak{g}} dt \\ &= \int_0^1 (\dot{\mathbf{l}}(t), J[\mathbf{h}, \mathbf{k}](t))_{\mathfrak{g}} dt + \int_0^1 (\dot{\mathbf{l}}(t), [\mathbf{h}(t), J\mathbf{k}(t)])_{\mathfrak{g}} dt \\ &\quad + \int_0^1 (\mathbf{l}(t), [\dot{\mathbf{h}}(t), J\mathbf{k}(t)])_{\mathfrak{g}} dt + \int_0^1 (\dot{\mathbf{l}}(t), [\mathbf{k}(t), J\mathbf{h}(t)])_{\mathfrak{g}} dt \\ &\quad + \int_0^1 (\mathbf{l}(t), [\dot{\mathbf{k}}(t), J\mathbf{h}(t)])_{\mathfrak{g}} dt \\ &= \int_0^1 (\dot{\mathbf{l}}(t), J[\mathbf{h}, \mathbf{k}](t) + [\mathbf{h}(t), J\mathbf{k}(t)] + [\mathbf{k}(t), J\mathbf{h}(t)])_{\mathfrak{g}} dt \\ &\quad + \int_0^1 (\mathbf{l}(t), [\dot{\mathbf{h}}(t), J\mathbf{k}(t)] + [\dot{\mathbf{k}}(t), J\mathbf{h}(t)])_{\mathfrak{g}} dt. \end{aligned}$$

First we consider the case $X^{\mathbf{h}}, X^{\mathbf{k}} \in H^{(1,0)}$ or $X^{\mathbf{h}}, X^{\mathbf{k}} \in H^{(0,1)}$. We set $JX^{\mathbf{k}} = \varepsilon\sqrt{-1}X^{\mathbf{k}}$.

$$
\begin{aligned}
&2B(\nabla_{X^{\mathbf{h}}} X^{\mathbf{k}}, X^{\mathbf{l}}) \\
&= \int_0^1 (\dot{\mathbf{l}}(t), \varepsilon\sqrt{-1}[\mathbf{h},\mathbf{k}](t) + \varepsilon\sqrt{-1}[\mathbf{h}(t),\mathbf{k}(t)] + \varepsilon\sqrt{-1}[\mathbf{k}(t),\mathbf{h}(t)])_{\mathfrak{g}}dt \\
&\quad + \int_0^1 (\mathbf{l}(t), \varepsilon\sqrt{-1}[\dot{\mathbf{h}}(t),\mathbf{k}(t)] + \varepsilon\sqrt{-1}[\dot{\mathbf{k}}(t),\mathbf{h}(t)])_{\mathfrak{g}}dt \\
&= \int_0^1 (\dot{\mathbf{l}}(t), \varepsilon\sqrt{-1}[\mathbf{h},\mathbf{k}](t))_{\mathfrak{g}}dt \\
&\quad + \int_0^1 (\mathbf{l}(t), \varepsilon\sqrt{-1}\frac{d}{dt}[\mathbf{h}(t),\mathbf{k}(t)] - 2\varepsilon\sqrt{-1}[\mathbf{h}(t),\dot{\mathbf{k}}(t)])_{\mathfrak{g}}dt \\
&= 2\int_0^1 (\dot{\mathbf{l}}(t), \varepsilon\sqrt{-1}\int_0^t [\mathbf{h}(s),\dot{\mathbf{k}}(s)]ds)_{\mathfrak{g}}dt \\
&= 2S(\mathbf{l}, \varepsilon\sqrt{-1}\int_0^{\cdot} [\mathbf{h}(s),\dot{\mathbf{k}}(s)]ds)_{\mathfrak{g}} \\
&= -2B(\mathbf{l}, \varepsilon\sqrt{-1}J\int_0^{\cdot} [\mathbf{h}(s),\dot{\mathbf{k}}(s)]ds) \\
&= 2B(\mathbf{l}, \int_0^{\cdot} [\mathbf{h}(s),\dot{\mathbf{k}}(s)]ds).
\end{aligned}
$$

Here we used $J\int_0^{\cdot}[\mathbf{h}(s),\dot{\mathbf{k}}(s)]ds = \varepsilon\sqrt{-1}\int_0^{\cdot}[\mathbf{h}(s),\dot{\mathbf{k}}(s)]ds$ due to the expression of $\mathbf{e}_\alpha$. Setting

$$
A(\mathbf{h},\mathbf{k})(t) = \int_0^t [\mathbf{h}(s),\dot{\mathbf{k}}(s)]ds, \tag{3.5}
$$

we have

$$
\nabla_{X^{\mathbf{h}}} X^{\mathbf{k}} = A(\mathbf{h},\mathbf{k}).
$$

In the case that $X^{\mathbf{h}} \in H^{(1,0)}$ and $X^{\mathbf{k}} \in H^{(0,1)}$, or $X^{\mathbf{h}} \in H^{(0,1)}$ and $X^{\mathbf{k}} \in H^{(1,0)}$, we set $JX^{\mathbf{k}} = \varepsilon\sqrt{-1}X^{\mathbf{k}}$. Then

$$
\begin{aligned}
&2B(\nabla_{X^{\mathbf{h}}} X^{\mathbf{k}}, X^{\mathbf{l}}) \\
&= \int_0^1 (\dot{\mathbf{l}}(t), J[\mathbf{h},\mathbf{k}](t) + \varepsilon\sqrt{-1}[\mathbf{h}(t),\mathbf{k}(t)] - \varepsilon\sqrt{-1}[\mathbf{k}(t),\mathbf{h}(t)])_{\mathfrak{g}}dt \\
&\quad + \int_0^1 (\mathbf{l}(t), \varepsilon\sqrt{-1}[\dot{\mathbf{h}}(t),\mathbf{k}(t)] - \varepsilon\sqrt{-1}[\dot{\mathbf{k}}(t),\mathbf{h}(t)])_{\mathfrak{g}}dt \\
&= \int_0^1 (\dot{\mathbf{l}}(t), J[\mathbf{h},\mathbf{k}](t) + \varepsilon\sqrt{-1}[\mathbf{h}(t),\mathbf{k}(t)] - \varepsilon\sqrt{-1}[\mathbf{k}(t),\mathbf{h}(t)])_{\mathfrak{g}}dt \\
&\quad - \int_0^1 (\dot{\mathbf{l}}(t), \varepsilon\sqrt{-1}[\mathbf{h}(t),\mathbf{k}(t)])_{\mathfrak{g}}dt \\
&= S(\mathbf{l}, J[\mathbf{h},\mathbf{k}](t) + \varepsilon\sqrt{-1}[\mathbf{h},\mathbf{k}])
\end{aligned}
$$

$$
\begin{aligned}
&= -B(\mathbf{1}, J^2[\mathbf{h},\mathbf{k}](t) + \varepsilon\sqrt{-1}J[\mathbf{h},\mathbf{k}]) \\
&= B(\mathbf{1}, [\mathbf{h},\mathbf{k}] - \varepsilon\sqrt{-1}J[\mathbf{h},\mathbf{k}]).
\end{aligned}
$$

Thus we have (recall the definition of $\pi_\pm$ in (2.7), (2.8))

$$\nabla_{X^{\mathbf{h}}} X^{\mathbf{k}} = \frac{1}{2}\{[\mathbf{h},\mathbf{k}] - \varepsilon\sqrt{-1}J[\mathbf{h},\mathbf{k}]\} = \pi_\varepsilon[\mathbf{h},\mathbf{k}].$$

We sum up into a theorem.

Theorem 3.1. *The covariant derivatives are given as follows:*

$$\nabla_{X^{\mathbf{h}}} X^{\mathbf{k}} = \begin{cases} A(\mathbf{h},\mathbf{k}), & \text{if } \mathbf{h},\mathbf{k} \in H^{(1,0)} \text{ or } \mathbf{h},\mathbf{k} \in H^{(0,1)}, \\ \pi_+[\mathbf{h},\mathbf{k}], & \text{if } \mathbf{h} \in H^{(0,1)} \text{ and } \mathbf{k} \in H^{(1,0)}, \\ \pi_-[\mathbf{h},\mathbf{k}], & \text{if } \mathbf{h} \in H^{(1,0)} \text{ and } \mathbf{k} \in H^{(0,1)}. \end{cases}$$

Now we are ready to compute the Riemannian curvature. The Riemannian curvature is defined by

$$R(X,Y) := [\nabla_X, \nabla_Y] - \nabla_{[X,Y]}.$$

From the definition, it is easy to see that R satisfies

$$B(R(X,Y)Z,W) = -B(R(Y,X)Z,W) = B(R(Z,W)X,Y).$$

Accordingly, the non-trivial term is $B(R(X_\alpha, X_{\bar\beta})X_\gamma, X_{\bar\delta})$ where $X_\alpha = X^{\mathbf{e}_\alpha}$, $X_{\bar\beta} = X^{\mathbf{e}_{\bar\beta}}$. The Riemannian curvature is given as follows:

Theorem 3.2. *It holds that*

$$
\begin{aligned}
&B(R(X_\alpha, X_{\bar\beta})X_\gamma, X_{\bar\delta}) \\
&\quad = -\sqrt{-1}\int_0^1 (\pi_+[\mathbf{e}_{\bar\beta},\mathbf{e}_\gamma], [\dot{\mathbf{e}}_\alpha, \mathbf{e}_{\bar\delta}])dt - \sqrt{-1}\int_0^1 (\pi_+[\mathbf{e}_{\bar\beta},\mathbf{e}_\gamma], [\mathbf{e}_\alpha, \dot{\mathbf{e}}_{\bar\delta}])dt \\
&\qquad + \sqrt{-1}\int_0^1 (A(\mathbf{e}_\alpha,\mathbf{e}_\gamma), [\mathbf{e}_{\bar\beta}, \dot{\mathbf{e}}_{\bar\delta}])dt \\
&\qquad + \sqrt{-1}\int_0^1 (\pi_+[\mathbf{e}_\alpha,\mathbf{e}_{\bar\beta}], [\dot{\mathbf{e}}_\gamma, \mathbf{e}_{\bar\delta}])dt - \sqrt{-1}\int_0^1 (\pi_-[\mathbf{e}_\alpha,\mathbf{e}_{\bar\beta}], [\mathbf{e}_\gamma, \dot{\mathbf{e}}_{\bar\delta}])dt.
\end{aligned}
$$

Proof. By the definition,

$$
\begin{aligned}
&B(R(X_\alpha, X_{\bar\beta})X_\gamma, X_{\bar\delta}) \\
&\quad = B(\nabla_{X_\alpha}\nabla_{X_{\bar\beta}}X_\gamma - \nabla_{X_{\bar\beta}}\nabla_{X_\alpha}X_\gamma - \nabla_{[X_\alpha,X_{\bar\beta}]}X_\gamma, X_{\bar\delta}) \\
&\quad = B(\nabla_{X_\alpha}\pi_+[X_{\bar\beta},X_\gamma] - \nabla_{X_{\bar\beta}}A(X_\alpha,X_\gamma) - \nabla_{\pi_+[X_\alpha,X_{\bar\beta}]}X_\gamma \\
&\qquad - \nabla_{\pi_-[X_\alpha,X_{\bar\beta}]}X_\gamma, X_{\bar\delta}) \\
&\quad = B(A(X_\alpha, \pi_+[X_{\bar\beta},X_\gamma]) - \pi_+[X_{\bar\beta}, A(X_\alpha,X_\gamma)] - A(\pi_+[X_\alpha,X_{\bar\beta}],X_\gamma) \\
&\qquad - \pi_+[\pi_-[X_\alpha,X_{\bar\beta}],X_\gamma], X_{\bar\delta}) \\
&\quad = -\sqrt{-1}S(A(X_\alpha,\pi_+[X_{\bar\beta},X_\gamma]) - [X_{\bar\beta}, A(X_\alpha,X_\gamma)]
\end{aligned}
$$

$$
\begin{aligned}
&- A(\pi_+[X_\alpha, X_{\bar\beta}], X_\gamma) - [\pi_-[X_\alpha, X_{\bar\beta}], X_\gamma], X_{\bar\delta}) \\
=\ & \sqrt{-1}\int_0^1 (A(\mathbf{e}_\alpha, \pi_+[\mathbf{e}_{\bar\beta}, \mathbf{e}_\gamma]), \dot{\mathbf{e}}_{\bar\delta})dt - \sqrt{-1}\int_0^1 ([\mathbf{e}_{\bar\beta}, A(\mathbf{e}_\alpha, \mathbf{e}_\gamma)], \dot{\mathbf{e}}_{\bar\delta})dt \\
& -\sqrt{-1}\int_0^1 (A(\pi_+[\mathbf{e}_\alpha, \mathbf{e}_{\bar\beta}], \mathbf{e}_\gamma), \dot{\mathbf{e}}_{\bar\delta})dt - \sqrt{-1}\int_0^1 ([\pi_-[\mathbf{e}_\alpha, \mathbf{e}_{\bar\beta}], \mathbf{e}_\gamma], \dot{\mathbf{e}}_{\bar\delta})dt \\
=\ & -\sqrt{-1}\int_0^1 ([\mathbf{e}_\alpha, \frac{d}{dt}\pi_+[\mathbf{e}_{\bar\beta}, \mathbf{e}_\gamma]], \mathbf{e}_{\bar\delta})dt + \sqrt{-1}\int_0^1 (A(\mathbf{e}_\alpha, \mathbf{e}_\gamma), [\mathbf{e}_{\bar\beta}, \dot{\mathbf{e}}_{\bar\delta}])dt \\
& +\sqrt{-1}\int_0^1 ([\pi_+[\mathbf{e}_\alpha, \mathbf{e}_{\bar\beta}], \dot{\mathbf{e}}_\gamma], \mathbf{e}_{\bar\delta})dt - \sqrt{-1}\int_0^1 (\pi_-[\mathbf{e}_\alpha, \mathbf{e}_{\bar\beta}], [\mathbf{e}_\gamma, \dot{\mathbf{e}}_{\bar\delta}])dt \\
=\ & \sqrt{-1}\int_0^1 (\frac{d}{dt}\pi_+[\mathbf{e}_{\bar\beta}, \mathbf{e}_\gamma], [\mathbf{e}_\alpha, \mathbf{e}_{\bar\delta}])dt + \sqrt{-1}\int_0^1 (A(\mathbf{e}_\alpha, \mathbf{e}_\gamma), [\mathbf{e}_{\bar\beta}, \dot{\mathbf{e}}_{\bar\delta}])dt \\
& +\sqrt{-1}\int_0^1 (\pi_+[\mathbf{e}_\alpha, \mathbf{e}_{\bar\beta}], [\dot{\mathbf{e}}_\gamma, \mathbf{e}_{\bar\delta}])dt - \sqrt{-1}\int_0^1 (\pi_-[\mathbf{e}_\alpha, \mathbf{e}_{\bar\beta}], [\mathbf{e}_\gamma, \dot{\mathbf{e}}_{\bar\delta}])dt \\
=\ & -\sqrt{-1}\int_0^1 (\pi_+[\mathbf{e}_{\bar\beta}, \mathbf{e}_\gamma], [\dot{\mathbf{e}}_\alpha, \mathbf{e}_{\bar\delta}])dt - \sqrt{-1}\int_0^1 (\pi_+[\mathbf{e}_{\bar\beta}, \mathbf{e}_\gamma], [\mathbf{e}_\alpha, \dot{\mathbf{e}}_{\bar\delta}])dt \\
& +\sqrt{-1}\int_0^1 (A(\mathbf{e}_\alpha, \mathbf{e}_\gamma), [\mathbf{e}_{\bar\beta}, \dot{\mathbf{e}}_{\bar\delta}])dt \\
& +\sqrt{-1}\int_0^1 (\pi_+[\mathbf{e}_\alpha, \mathbf{e}_{\bar\beta}], [\dot{\mathbf{e}}_\gamma, \mathbf{e}_{\bar\delta}])dt - \sqrt{-1}\int_0^1 (\pi_-[\mathbf{e}_\alpha, \mathbf{e}_{\bar\beta}], [\mathbf{e}_\gamma, \dot{\mathbf{e}}_{\bar\delta}])dt
\end{aligned}
$$

which completes the proof. □

4. Closability

We will show the closability of operators that were introduced in the previous sections. Operators were considered in the framework of L^2 theory. To define an L^2 space, we need a measure. So we begin with introducing a measure on the path space PG. To do this, let us consider the following stochastic differential equation. We consider the following stochastic differential equation on G:

$$
\begin{cases} d\gamma_t &= \sum_{i=1}^{d} \xi_i(\gamma_t)\circ db_t^i, \\ \gamma_0 &= e \end{cases} \tag{4.1}
$$

where $(b_t^1, \ldots, b_t^d)_{t\in[0,T]}$ is a d-dimensional Brownian motion, and $\circ$ stands for the Stratonovich symmetric stochastic integral. Setting $b_t = \sum_i b_t^i \xi_i$, (b_t) is a Brownian motion on $\mathfrak{g}$. (b_t) induces a measure on the space $P\mathfrak{g}$ where $P\mathfrak{g}$ is the $\mathfrak{g}$-valued path space. The measure is called the Wiener measure and is denoted by P^W. There exists the unique strong solution to (4.1), i.e., there exists a measurable function $I: P\mathfrak{g} \to PG$ such that $\gamma = I(b)$ is the unique solution to (4.1). We call this map I the Itô map. We denote the image measure of P^W under I by μ. Then $(P\mathfrak{g}, P^W) \cong (PG, \mu)$ as measure spaces. We sometimes regard a function on (PG, μ) as a function on $(P\mathfrak{g}, P^W)$. We can restrict the measure to ΩG by taking a conditional probability. We set $m = E[\cdot\,|\gamma(1) = e]$. m is called a pinned measure or a bridge measure.

For function $u \in \mathcal{F}C_b^\infty(\Omega G)$, du is characterized by

$$\langle du, X\rangle = Xu, \quad \text{for } X \in \mathcal{F}\Gamma_b^\infty(T_0^1(\Omega G)).$$

In the section 2, X was an H_0-valued function. But we have replaced H_0 by H_1. Since $H_0 \subseteq H_1$, we have $H_1^* \subseteq H_0^*$ and du must be in H_1^*. Let us give an example of such a function. Set

$$u(\gamma) = \int_0^1 \varphi(t) f(\gamma(t))dt$$

where $\varphi \in C^\infty([0,1])$ and $f \in C^\infty(M)$. Then clearly

$$\langle du(\gamma), \mathbf{h}\rangle = \int_0^1 \varphi(t)\langle df(\gamma(t)), \mathbf{h}(t)\rangle dt.$$

It is now easy to see that $du(\gamma) \in H_1^*$. Since those functions are dense in $L^2(\Omega G)$, we can define the following pre-Dirichlet form:

$$\mathcal{E}(u,v) := \int_{\Omega G} (du, dv)_{H_1^*} m(d\gamma). \tag{4.2}$$

The closability of the pre-Dirichlet form is a fundamental problem. It is equivalent to the closability of the operator d. To show the closability, we are enough to show the existence of the dual operator.

Since $|\mathbf{h}|_{H_1} = |\sqrt{T}\mathbf{h}|_{H_0}$, we easily show that $\boldsymbol{\theta} \in H_1^*$ if and only if $\boldsymbol{\theta} \in \mathrm{Dom}(\sqrt{T^{*-1}})$ and

$$|\boldsymbol{\theta}|_{H_1^*} = |\sqrt{T^{*-1}}\boldsymbol{\theta}|_{H_0^*},$$

where $T^*\colon H_0^* \to H_0^*$ is the dual operator of T. We also notice that

$$(\mathbf{h}, \mathbf{k})_{H_1} = (T\mathbf{h}, \mathbf{k})_{H_0} \quad \text{for } \mathbf{h}, \mathbf{k} \in H_1$$

and similarly

$$(\boldsymbol{\theta}, \boldsymbol{\eta})_{H_1^*} = (T^{*-1}\boldsymbol{\theta}, \boldsymbol{\eta})_{H_0^*} \quad \text{for } \boldsymbol{\theta} \in \mathrm{Ran}(T^*),\ \boldsymbol{\eta} \in H_1^*.$$

Now we have

$$\int_{\Omega G} (du, \boldsymbol{\theta})_{H_1^*} dm = \int_{\Omega G} (du, T^{*-1}\boldsymbol{\theta})_{H_0^*} dm = \int_{\Omega G} u d'(T^{*-1}\boldsymbol{\theta}) dm$$

where d' is the dual operator with respect to the inner product $(\ ,\ )_{H_0}$. This implies that $d^* = d'T^{*-1}$.

We shall give a rather explicit expression of d^*. We recall the notion of divergence. For $X \in \mathcal{F}\Gamma_b^\infty(T_0^1(\Omega G))$, $\operatorname{div} X$ is characterized by the following identity:

$$\int_{\Omega G} X u dm = -\int_{\Omega G} (\operatorname{div} X) u dm$$

If X is left invariant, we can give an explicit formula of $\operatorname{div} X$:

$$\operatorname{div} X^{\mathbf{h}} = -\int_0^1 (\dot{\mathbf{h}}(s), db(s))_{\mathfrak{g}}.$$

Notice that this is valid only for $\mathbf{h} \in H_0$. Because we are given a Riemannian metric, a 1-form can be identified with a vector field, i.e., for any 1-form ω, there exists a unique vector field, which we denote by $\omega^\sharp$, such that

$$\langle \omega, X \rangle = B(\omega^\sharp, X).$$

We take $\boldsymbol{\theta} \in H_1^*$ (a constant 1-form). Then,

$$\int_{\Omega G} (du, \boldsymbol{\theta})_{H_1^*} dm = \int_{\Omega G} \langle du, \boldsymbol{\theta}^\sharp \rangle dm = \int_{\Omega G} \boldsymbol{\theta}^\sharp u dm = -\int_{\Omega G} (\operatorname{div} \boldsymbol{\theta}^\sharp) u dm$$

which implies $d^* \boldsymbol{\theta} = -\operatorname{div} \boldsymbol{\theta}^\sharp$. Of course, we have to assume that $\boldsymbol{\theta}^\sharp \in H_0$. Notice that $\boldsymbol{\theta}^\sharp \in H_0$ if and only if $\boldsymbol{\theta} \in T^*(H_0^*)$ and in this case $|\boldsymbol{\theta}^\sharp|_{H_0} = |T^{*-1}\boldsymbol{\theta}|_{H_0^*}$.

It is now easy to see that $\varphi = u\boldsymbol{\theta} \in \operatorname{Dom}(d^*)$ if $u \in \mathcal{F}C_b^\infty(\Omega G)$ and $\boldsymbol{\theta} \in \operatorname{Dom}(T^{*-1})$. Here and after, we identify an element of H_1^* with a constant 1-form (i.e., a left invariant 1-form) for notational simplicity. Hence $\operatorname{Dom}(d^*)$ is dense and thereby $\mathcal{E}$ is closable. It is easy to see that the closure of $\mathcal{E}$ is a Dirichlet form.

The closability of the exterior differentiation is valid for general p-forms. Of course, we have replaced H_0 by H_1 in the argument in the section 2. Hence p-form is a function on ΩG taking values in $\bigwedge^p = H_1^* \wedge \cdots \wedge H_1^*$. Now we notice that T^* can be naturally extended to the tensor product of H_1^* as follows:

$$\Gamma_k(T^*)(\boldsymbol{\theta}_1 \otimes \cdots \otimes \boldsymbol{\theta}_k) = (T^*\boldsymbol{\theta}_1) \otimes \cdots \otimes (T^*\boldsymbol{\theta}_k).$$

We also have

$$\begin{aligned} &(\boldsymbol{\theta}_1 \otimes \cdots \otimes \boldsymbol{\theta}_k, \boldsymbol{\eta}_1 \otimes \cdots \otimes \boldsymbol{\eta}_k)_{H_1^{*\otimes k}} \\ &= (\Gamma_k(T^{*-1})\boldsymbol{\theta}_1 \otimes \cdots \otimes \boldsymbol{\theta}_k, \boldsymbol{\eta}_1 \otimes \cdots \otimes \boldsymbol{\eta}_k)_{H_0^{*\otimes k}} \end{aligned}$$

for $\boldsymbol{\theta}_i \in \operatorname{Ran}(T^*)$, $\boldsymbol{\eta}_i \in H_1^*$.

Proposition 4.1. *The dual operator of d in $\bigwedge^k$ is given by*

$$d^* = \Gamma_{k-1}(T^*) d' \Gamma_k(T^{*-1}).$$

Proof. It is easy to see that

$$\begin{aligned}\int_{\Omega G}\frac{1}{k!}(\omega,d\varphi)_{H_1^{*\otimes k}}dm &= \int_{\Omega G}\frac{1}{k!}(\Gamma_k(T^{*-1})\omega,d\varphi)_{H_0^{*\otimes k}}dm\\ &= \int_{\Omega G}\frac{1}{(k-1)!}(d'\Gamma_k(T^{*-1})\omega,\varphi)_{H_0^{*\otimes k}}dm\\ &= \int_{\Omega G}\frac{1}{(k-1)!}(\Gamma_{k-1}(T^*)d'\Gamma_k(T^{*-1})\omega,\varphi)_{H_1^{*\otimes k}}dm.\end{aligned}$$

This implies the assertion. □

The closability of ∂, $\bar\partial$ can be shown in a similar manner.

Lastly we consider the covariant differentiation. We show the closability of covariant differentiation for 1-forms. Let ω be a 1-form, i.e., H_1^*-valued function on ΩG. Recall that the cotangent bundle $T^*(\Omega G)$ is identified with $\Omega G\times H_1^*$. Hence a section of $T^*(\Omega G)$ is nothing but a H_1^*-valued function. We define the covariant derivative of 1-form ω by

$$\langle\nabla_X\omega,Y\rangle = X\langle\omega,Y\rangle-\langle\omega,\nabla_XY\rangle.$$

Setting $\nabla\omega(X,Y)=\langle\nabla_X\omega,Y\rangle$, ∇ is a differential operator from $\Gamma(T_1^0(\Omega G))$ to $\Gamma(T_2^0(\Omega G))$. Let $(\mathbf{e}_\alpha,\mathbf{e}_{\bar\alpha})$ be a basis in H_0 defined by (2.6). If $\boldsymbol{\theta}\in H_1^{*(1,0)}$, then, by Theorem 3.1, we have $\nabla\boldsymbol{\theta}(\cdot,\mathbf{e}_{\bar\beta})=0$. Further, we have

$$\begin{aligned}\nabla\boldsymbol{\theta}(\mathbf{e}_\alpha,\mathbf{e}_\beta) &= -\boldsymbol{\theta}(A(\mathbf{e}_\alpha,\mathbf{e}_\beta))\\ \nabla\boldsymbol{\theta}(\mathbf{e}_{\bar\alpha},\mathbf{e}_\beta) &= -\boldsymbol{\theta}([\mathbf{e}_{\bar\alpha},\mathbf{e}_\beta]).\end{aligned}$$

Similar formula holds for $\boldsymbol{\theta}\in H_1^{*(0,1)}$ (just take a complex conjugate).

We will obtain the dual operator of ∇ acting on 1-forms. To do this, we recall the definition of the divergence operator, which is essentially same as the dual operator of d. Now, taking 1-forms $\boldsymbol{\theta}$, $\boldsymbol{\eta}$ and $\boldsymbol{\xi}\in H_1^*$ (constant 1-forms), we have

$$\begin{aligned}\int_{\Omega G}(\boldsymbol{\theta}\otimes\boldsymbol{\eta},\nabla\boldsymbol{\xi})_{H_1^*\otimes H_1^*}dm &= \int_{\Omega G}(\boldsymbol{\eta},\nabla_{\boldsymbol{\theta}^\sharp}\boldsymbol{\xi})_{H_1^*}dm\\ &= \int_{\Omega G}\{\boldsymbol{\theta}^\sharp(\boldsymbol{\eta},\boldsymbol{\xi})_{H_1^*}-(\nabla_{\boldsymbol{\theta}^\sharp}\boldsymbol{\eta},\boldsymbol{\xi})_{H_1^*}\}dm\\ &= \int_{\Omega G}(-(\operatorname{div}\boldsymbol{\theta}^\sharp)\boldsymbol{\eta}-\nabla_{\boldsymbol{\theta}^\sharp}\boldsymbol{\eta},\boldsymbol{\xi})_{H_1^*}dm.\end{aligned}$$

Thus we have

$$\nabla^*(\boldsymbol{\theta}\otimes\boldsymbol{\eta}) = -(\operatorname{div}\boldsymbol{\theta}^\sharp)\boldsymbol{\eta}-\nabla_{\boldsymbol{\theta}^\sharp}\boldsymbol{\eta}.$$

Now we can see that the domain of ∇^* is dense.

5. The Ricci curvature

We have obtained the Riemannian curvature. In this section we will get the Ricci curvature. We adopt the following definition of Ricci curvature.

$$dd^* + d^*d = \nabla^*\nabla + \mathrm{Ric} \tag{5.1}$$

To compute the Ricci curvature, we recall the following fact. For $\mathbf{h} \in H_0$, set

$$u = \int_0^1 (\dot{\mathbf{h}}(t), db(t))_{\mathfrak{g}}.$$

Then we have (see, e.g., Gross [5, Lemma 3.5])

$$\begin{aligned} \langle du, \mathbf{k}\rangle &= \int_0^1 (\dot{\mathbf{h}}(t), \dot{\mathbf{k}}(t))_{\mathfrak{g}} dt + \int_0^1 ([\dot{\mathbf{h}}(t), \mathbf{k}(t)], db(t))_{\mathfrak{g}} \\ &= (T^{-1}\mathbf{h}, \mathbf{k})_{H_1} + \int_0^1 ([\dot{\mathbf{h}}(t), \mathbf{k}(t)], db(t))_{\mathfrak{g}}. \end{aligned} \tag{5.2}$$

We prepare a proposition for later use.

Proposition 5.1. *Set Define Q and $\tilde{Q}$ as follows.*

$$Q(\mathbf{h}, \mathbf{k}) := \sqrt{-1} \sum_\beta \int_0^1 ([\mathbf{h}, \dot{\mathbf{e}}_\beta],_{\pi+} [\mathbf{k}, T^{-1}\mathbf{e}_{\bar{\beta}}])_{\mathfrak{g}} dt$$

for $\mathbf{h} \in H_0^{(0,1)}$ and $\mathbf{k} \in H_0^{(1,0)}$, and

$$\tilde{Q}(\mathbf{h}, \mathbf{k}) = \sqrt{-1} \sum_\beta \int_0^1 ([\mathbf{e}_\beta, \dot{\mathbf{h}}], A(T^{-1}\mathbf{e}_{\bar{\beta}}, \mathbf{k}))_{\mathfrak{g}} dt$$

for $\mathbf{h}, \mathbf{k} \in H_0^{(0,1)}$. Then it holds that

$$Q(\mathbf{h}, \mathbf{k}) = \frac{1}{2\pi} \int_0^1 K(J\mathbf{h}, \dot{\mathbf{k}}) dt, \quad \text{for } \mathbf{h} \in H_0^{(0,1)} \text{ and } \mathbf{k} \in H_0^{(1,0)}, \tag{5.3}$$

$$\tilde{Q}(\mathbf{h}, \mathbf{k}) = 0, \quad \text{for } \mathbf{h}, \mathbf{k} \in H_0^{(0,1)} \tag{5.4}$$

where K is the Killing form: $K(\xi, \eta) = \mathrm{tr}_{N(S)}(\mathrm{ad}\xi \mathrm{ad}\eta)$.

Proof. We compute Q by using a basis $(\mathbf{e}_\alpha, \mathbf{e}_{\bar{\alpha}})$.

$$
\begin{aligned}
&Q(\mathbf{e}_{-m,j},\mathbf{e}_{l,k}) \\
&= \sqrt{-1}\sum_{i=1}^{d}\sum_{n}\int_0^1([\mathbf{e}_{-m,j},,\dot{\mathbf{e}}_{n,i}],\pi_+[\mathbf{e}_{l,k},T^{-1}\mathbf{e}_{-n,i}])_{\mathfrak{g}}dt \\
&= \sqrt{-1}\sum_{i=1}^{d}\sum_{n}\int_0^1(e_{-m}\dot{e}_n[\xi_j,\xi_i],(2\pi n)\pi_+\{e_l e_{-n}[\xi_k,\xi_i]\})_{\mathfrak{g}}dt \\
&= \sqrt{-1}\sum_{i=1}^{d}\sum_{n}\int_0^1(\frac{1}{-2\pi m\sqrt{-1}}(e^{-2\pi\sqrt{-1}mt}-1)e^{2\pi\sqrt{-1}nt}[\xi_j,\xi_i], \\
&\quad\times(2\pi n)\frac{1}{(2\pi\sqrt{-1}l)(-2\pi\sqrt{-1}n)} \\
&\quad\times\pi_+\{(e^{2\pi\sqrt{-1}(l-n)t}-1)-(e^{2\pi\sqrt{-1}l}-1)-(e^{-2\pi\sqrt{-1}n}-1)\}[\xi_k,\xi_i])_{\mathfrak{g}}dt \\
&= \sum_n\frac{1}{4\pi^2 ml}\int_0^1(e^{-2\pi\sqrt{-1}mt}-1)e^{2\pi\sqrt{-1}nt} \\
&\quad\times\{1_{\{l>n\}}(e^{2\pi\sqrt{-1}(l-n)t}-1)-(e^{2\pi\sqrt{-1}l}-1)\}K(\xi_j,\xi_k)dt \\
&= \sum_n\frac{1}{4\pi^2 ml}\int_0^1\{1_{\{l>n\}}(e^{2\pi\sqrt{-1}(l-m)t}-e^{2\pi\sqrt{-1}(n-m)t}) \\
&\quad-(e^{2\pi\sqrt{-1}(n-m+l)}-e^{2\pi\sqrt{-1}(n-m)t})\}K(\xi_j,\xi_k)dt \\
&= \left\{\frac{1}{4\pi^2 ml}(l-1)\delta_{l,m}-\frac{1}{4\pi^2 ml}1_{\{l>m\}}\right. \\
&\quad\left.-\frac{1}{4\pi^2 ml}1_{\{m>l\}}+\frac{1}{4\pi^2 ml}\right\}K(\xi_i,\xi_j) \\
&= \left\{\frac{1}{4\pi^2 ml}(l-1)\delta_{l,m}+\frac{1}{4\pi^2 ml}\delta_{l,m}\right\}K(\xi_i,\xi_j) \\
&= \frac{1}{4\pi^2 m}\delta_{l,m}K(\xi_i,\xi_j).
\end{aligned}
$$

On the other hand,

$$
\begin{aligned}
&\frac{1}{2\pi}\int_0^1 K(J\mathbf{e}_{-m,j},\dot{\mathbf{e}}_{l,k})dt \\
&\quad= \frac{-\sqrt{-1}}{2\pi}\int_0^1\frac{1}{-2\pi\sqrt{-1}m}(e^{-2\pi\sqrt{-1}mt}-1)e^{2\pi\sqrt{-1}lt}K(\xi_j,\xi_k)dt \\
&\quad= \frac{1}{4\pi^2 m}\delta_{l,m}K(\xi_i,\xi_j).
\end{aligned}
$$

Now we have (5.3). Next we show (5.4)

$$
\begin{aligned}
&\tilde{Q}(\mathbf{e}_{-m,j},\mathbf{e}_{-l,k}) \\
&= \sqrt{-1}\sum_{n,i}\int_0^1([e_n(t)\xi_i,\dot{e}_{-m}(t)\xi_j],\int_0^t 2\pi n[e_{-n}(s)\xi_i,\dot{e}_{-l}(s)\xi_k]ds)_{\mathfrak{g}}dt
\end{aligned}
$$

$$
\begin{aligned}
&= \sqrt{-1}\sum_{n,i}\int_0^1 \Big(\frac{1}{2\pi\sqrt{-1}n}(e^{2\pi\sqrt{-1}nt}-1)e^{-2\pi\sqrt{-1}mt}[\xi_i,\xi_j],\\
&\qquad \Big\{\int_0^t (2\pi n)\frac{1}{-2\pi\sqrt{-1}n}(e^{-2\pi\sqrt{-1}ns}-1)e^{-2\pi\sqrt{-1}ls}[\xi_i,\xi_k]ds\Big\}\Big)_{\mathfrak{g}}dt\\
&= -\sqrt{-1}\sum_n K(\xi_j,\xi_k)\int_0^1 \frac{1}{2\pi n}(e^{2\pi\sqrt{-1}nt}-1)e^{-2\pi\sqrt{-1}mt}\\
&\qquad \times\Big\{\frac{-1}{2\pi\sqrt{-1}(n+l)}(e^{-2\pi\sqrt{-1}(n+l)t}-1)+\frac{1}{2\pi\sqrt{-1}l}(e^{-2\pi\sqrt{-1}lt}-1)\Big\}dt\\
&= -\sqrt{-1}\sum_n K(\xi_j,\xi_k)\\
&\qquad \times\int_0^1\Big\{\frac{-1}{(2\pi n)(2\pi\sqrt{-1}(n+l))}(e^{2\pi\sqrt{-1}(n-m-n-l)t}-e^{2\pi\sqrt{-1}(n-m)t})\\
&\qquad +\frac{1}{(2\pi n)(2\pi\sqrt{-1}l)}(e^{2\pi\sqrt{-1}(n-m-l)t}-e^{2\pi\sqrt{-1}(n-m)t})\Big\}dt\\
&= \sum_n K(\xi_j,\xi_k)\Big\{\frac{-\delta_{n,m}}{4\pi^2 n(n+l)}-\frac{\delta_{n,m+l}}{4\pi^2 nl}+\frac{\delta_{n,m}}{4\pi^2 nl}\Big\}\\
&= K(\xi_j,\xi_k)\Big\{\frac{-1}{4\pi^2 m(m+l)}-\frac{1}{4\pi^2(m+l)l}+\frac{1}{4\pi^2 ml}\Big\}\\
&= 0.
\end{aligned}
$$

This completes the proof. □

The Ricci curvature is given by the following theorem.

Theorem 5.1. *Set*

$$
\begin{aligned}
\mathrm{Ric}(\mathbf{h},\mathbf{k}) &= (T^{-1}\mathbf{h},\mathbf{k})_{H_1}+\frac{1}{2\pi}\int_0^1 K(J\mathbf{h},\dot{\mathbf{k}})dt+\Big(\int_0^1[\dot{\mathbf{h}}(u),\mathbf{k}(u)]du,b(1)\Big)_{\mathfrak{g}}\\
&\quad +\int_0^1(\rho_+[\dot{\mathbf{h}},\mathbf{k}],db(t))_{\mathfrak{g}}-\int_0^1(\rho_-[\mathbf{h},\dot{\mathbf{k}}],db(t))_{\mathfrak{g}} \qquad (5.5)
\end{aligned}
$$

for $\mathbf{h}\in H_0^{(0,1)}$ *and* $\mathbf{k}\in H_0^{(1,0)}$, *and*

$$\mathrm{Ric}(\mathbf{h},\mathbf{k})=0 \qquad (5.6)$$

for $\mathbf{h},\mathbf{k}\in H_0^{(0,1)}$. *If* $\mathbf{h}\in H_0^{(1,0)}$, *we define* Ric *by complex conjugation. Then identity (5.1) holds. Here* $\rho_\pm$ *in (5.5) is defined by*

$$\rho_+\mathbf{l} = \sum_\beta\Big\{\int_0^1 \mathbf{l}(t)\dot{\mathbf{e}}_{\bar\beta}(s)ds\Big\}\dot{\mathbf{e}}_\beta \qquad (5.7)$$

$$\rho_-\mathbf{l} = \sum_\beta\Big\{\int_0^1 \mathbf{l}(t)\dot{\mathbf{e}}_\beta(s)ds\Big\}\dot{\mathbf{e}}_{\bar\beta}. \qquad (5.8)$$

Proof. We first note that $d\eta(X,Y) = \nabla\eta(X,Y) - \nabla\eta(Y,X)$. So we set $\check{\nabla}\eta(X,Y) = \nabla\eta(Y,X)$. Since $d^* = \nabla^*$, we have

$$\begin{aligned} dd^*\eta + d^*d\eta - \nabla^*\nabla\eta &= dd^*\eta + \nabla^*(\nabla\eta - \check{\nabla}\eta) - \nabla^*\nabla\eta \\ &= dd^*\eta - \nabla^*\check{\nabla}\eta. \end{aligned}$$

Set $\eta^\sharp = \mathbf{h}$, $u = \int_0^1 (\dot{\mathbf{h}}(s), db(s))_{\mathfrak{g}}$. Note that $\mathbf{h} \in H_0^{(0,1)}$ since $\eta \in H_1^{*(1,0)}$ (we have assumed $\mathbf{h} \in H_0^{\mathbb{C}}$). As we saw in the previous section, $d^*\eta$ is given by

$$d^*\eta = -\operatorname{div}\eta^\sharp = u$$

and hence

$$dd^*\eta = du.$$

Let $(\theta^\alpha, \theta^{\bar{\alpha}})$ be a dual basis of $(\mathbf{e}_\alpha, \mathbf{e}_{\bar{\alpha}})$, i.e., $\langle\theta^\alpha, \mathbf{e}_\beta\rangle = \delta^\alpha_\beta$, $\langle\theta^\alpha, \mathbf{e}_{\bar{\beta}}\rangle = 0$, etc. By noting $\nabla\eta(\cdot, \mathbf{e}_{\bar{\beta}}) = 0$, we can expand $\nabla\eta$ as

$$\nabla\eta = \sum_{\alpha,\beta}\{\nabla\eta(\mathbf{e}_\alpha, \mathbf{e}_\beta)\theta^\alpha \otimes \theta^\beta + \nabla\eta(\mathbf{e}_{\bar{\alpha}}, \mathbf{e}_\beta)\theta^{\bar{\alpha}} \otimes \theta^\beta\}.$$

Reversing the order, we have

$$\check{\nabla}\eta = \sum_{\alpha,\beta}\{-\langle\eta, A(\mathbf{e}_\alpha, \mathbf{e}_\beta)\rangle\theta^\beta \otimes \theta^\alpha - \langle\eta, \pi_+[\mathbf{e}_{\bar{\alpha}}, \mathbf{e}_\beta]\rangle\theta^\beta \otimes \theta^{\bar{\alpha}}\}.$$

Since $(\theta^\alpha)^\sharp = T^{-1}\mathbf{e}_{\bar{\alpha}}$, we have

$$\nabla^*(\theta^\beta \otimes \theta^\alpha) = -(\operatorname{div} T^{-1}\mathbf{e}_{\bar{\beta}})\theta^\alpha - \nabla_{T^{-1}\mathbf{e}_{\bar{\beta}}}\theta^\alpha.$$

Therefore

$$\begin{aligned} \nabla^*\check{\nabla}\eta &= \sum_{\alpha,\beta}\{(\operatorname{div} T^{-1}\mathbf{e}_{\bar{\beta}})\theta^\alpha + \nabla_{T^{-1}\mathbf{e}_{\bar{\beta}}}\theta^\alpha\}\langle\eta, A(\mathbf{e}_\alpha, \mathbf{e}_\beta)\rangle \\ &\quad + \sum_{\alpha,\beta}\{(\operatorname{div} T^{-1}\mathbf{e}_{\bar{\beta}})\theta^{\bar{\alpha}} + \nabla_{T^{-1}\mathbf{e}_{\bar{\beta}}}\theta^{\bar{\alpha}}\}\langle\eta, \pi_+[\mathbf{e}_{\bar{\alpha}}, \mathbf{e}_\beta]\rangle. \end{aligned}$$

Hence, if $\mathbf{k} \in H_0^{(1,0)}$,

$$\begin{aligned} &\langle dd^*\eta - \nabla^*\check{\nabla}\eta, \mathbf{k}\rangle \\ &= \langle du - \nabla^*\check{\nabla}\eta, \mathbf{k}\rangle \\ &= (\mathbf{h}, \mathbf{k})_{H_0} + \int_0^1 ([\dot{\mathbf{h}}, \mathbf{k}], db(t))_{\mathfrak{g}} - \sum_{\alpha,\beta}\{(\operatorname{div} T^{-1}\mathbf{e}_{\bar{\beta}})\langle\theta^\alpha, \mathbf{k}\rangle\langle\eta, A(\mathbf{e}_\alpha, \mathbf{e}_\beta)\rangle \\ &\quad - \sum_{\alpha,\beta}\langle\nabla_{T^{-1}\mathbf{e}_{\bar{\beta}}}\theta^\alpha, \mathbf{k}\rangle\langle\eta, A(\mathbf{e}_\alpha, \mathbf{e}_\beta)\rangle\} \\ &= (T^{-1}(\eta^\sharp), \mathbf{k})_{H_1} + \int_0^1 ([\dot{\mathbf{h}}, \mathbf{k}], db(t))_{\mathfrak{g}} - \sum_\beta(\operatorname{div} T^{-1}\mathbf{e}_{\bar{\beta}})\langle\eta, A(\mathbf{k}, \mathbf{e}_\beta)\rangle \end{aligned}$$

$$
\begin{aligned}
&+\sum_{\alpha,\beta}\langle\theta^\alpha,\pi_+[T^{-1}\mathbf{e}_{\bar\beta},\mathbf{k}]\rangle\langle\eta,A(\mathbf{e}_\alpha,\mathbf{e}_\beta)\rangle\\
=\;&\langle T^{*-1}\eta,\mathbf{k}\rangle+\int_0^1([\dot{\mathbf{h}},\mathbf{k}],db(t))_{\mathfrak g}\\
&-\sum_\beta(\operatorname{div}T^{-1}\mathbf{e}_{\bar\beta})\langle\eta,A(\mathbf{k},\mathbf{e}_\beta)\rangle+\sum_\beta\langle\eta,A(\pi_+[T^{-1}\mathbf{e}_{\bar\beta},\mathbf{k}],\mathbf{e}_\beta)\rangle\}.
\end{aligned}
$$

We first compute the last term.

$$
\begin{aligned}
\langle\eta,A(\pi_+[T^{-1}\mathbf{e}_{\bar\beta},\mathbf{k}],\mathbf{e}_\beta)\rangle &= (\mathbf{h},A(\pi_+[T^{-1}\mathbf{e}_{\bar\beta},\mathbf{k}],\mathbf{e}_\beta))_{H_1}\\
&= \sqrt{-1}(J\mathbf{h},A(\pi_+[T^{-1}\mathbf{e}_{\bar\beta},\mathbf{k}],\mathbf{e}_\beta))_{H_1}\\
&= \sqrt{-1}S(\mathbf{h},A(\pi_+[T^{-1}\mathbf{e}_{\bar\beta},\mathbf{k}],\mathbf{e}_\beta))\\
&= -\sqrt{-1}\int_0^1(\mathbf{h},\frac{d}{dt}A(\pi_+[T^{-1}\mathbf{e}_{\bar\beta},\mathbf{k}],\mathbf{e}_\beta))_{\mathfrak g}dt\\
&= -\sqrt{-1}\int_0^1(\mathbf{h},[\pi_+[T^{-1}\mathbf{e}_{\bar\beta},\mathbf{k}],\dot{\mathbf{e}}_\beta])_{\mathfrak g}dt\\
&= \sqrt{-1}\int_0^1([\mathbf{h},\dot{\mathbf{e}}_\beta],\pi_+[\mathbf{k},T^{-1}\mathbf{e}_{\bar\beta}])_{\mathfrak g}dt.
\end{aligned}
$$

By virtue of Proposition 5.1, we can get

$$
\sum_\beta\langle\eta,A(\pi_+[T^{-1}\mathbf{e}_{\bar\beta},\mathbf{k}],\mathbf{e}_\beta)\rangle=\frac{1}{2\pi}\int_0^1 K(J\mathbf{h},\dot{\mathbf{k}})dt.
$$

Now we turn to the stochastic integral.

$$
\begin{aligned}
(\operatorname{div}T^{-1}\mathbf{e}_{\bar\beta})\langle\eta,A(\mathbf{k},\mathbf{e}_\beta)\rangle &= (\operatorname{div}T^{-1}\mathbf{e}_{\bar\beta})(\mathbf{h},A(\mathbf{k},\mathbf{e}_\beta))_{H_1}\\
&= \sqrt{-1}(\operatorname{div}T^{-1}\mathbf{e}_{\bar\beta})(J\mathbf{h},A(\mathbf{k},\mathbf{e}_\beta))_{H_1}\\
&= \sqrt{-1}(\operatorname{div}T^{-1}\mathbf{e}_{\bar\beta})S(\mathbf{h},A(\mathbf{k},\mathbf{e}_\beta))\\
&= -\sqrt{-1}(\operatorname{div}T^{-1}\mathbf{e}_{\bar\beta})\int_0^1(\mathbf{h},[\mathbf{k},\dot{\mathbf{e}}_\beta])_{\mathfrak g}dt\\
&= -\sqrt{-1}(\operatorname{div}T^{-1}\mathbf{e}_{\bar\beta})\int_0^1([\mathbf{h},\mathbf{k}],\dot{\mathbf{e}}_\beta)_{\mathfrak g}dt\\
&= -\sqrt{-1}(\operatorname{div}T^{-1}\mathbf{e}_{\bar\beta})S(\mathbf{e}_\beta,[\mathbf{h},\mathbf{k}])\\
&= -\sqrt{-1}(\operatorname{div}T^{-1}\mathbf{e}_{\bar\beta})(J\mathbf{e}_\beta,[\mathbf{h},\mathbf{k}])_{H_1}\\
&= (\operatorname{div}\mathbf{e}_{\bar\beta})(T^{-1}\mathbf{e}_\beta,[\mathbf{h},\mathbf{k}])_{H_1}\\
&= (\operatorname{div}\mathbf{e}_{\bar\beta})(\mathbf{e}_\beta,[\mathbf{h},\mathbf{k}])_{H_0}\\
&= (\operatorname{div}\mathbf{e}_{\bar\beta})\int_0^1([\dot{\mathbf{h}},\mathbf{k}]+[\mathbf{h},\dot{\mathbf{k}}],\dot{\mathbf{e}}_\beta)_{\mathfrak g}dt
\end{aligned}
$$

Further

$$
\begin{aligned}
&\int_0^1 ([\dot{\mathbf{h}},\mathbf{k}],db(t))_{\mathfrak{g}} - \sum_\beta (\operatorname{div} T^{-1}\mathbf{e}_{\bar\beta})\langle \eta, A(\mathbf{k},\mathbf{e}_\beta)\rangle \\
&= \int_0^1 (\rho_+[\dot{\mathbf{h}},\mathbf{k}],db(t))_{\mathfrak{g}} + \int_0^1 (\rho_-[\dot{\mathbf{h}},\mathbf{k}],db(t))_{\mathfrak{g}} \\
&\quad + \int_0^1 \Big(\int_0^1 [\dot{\mathbf{h}}(u),\mathbf{k}(u)]du, db(t)\Big)_{\mathfrak{g}} - \int_0^1 (\rho_-[\dot{\mathbf{h}},\mathbf{k}] + \rho_-[\mathbf{h},\dot{\mathbf{k}}],db(t))_{\mathfrak{g}} \\
&= \int_0^1 (\rho_+[\dot{\mathbf{h}},\mathbf{k}],db(t))_{\mathfrak{g}} + \Big(\int_0^1 [\dot{\mathbf{h}}(u),\mathbf{k}(u)]du, b(1)\Big)_{\mathfrak{g}} - \int_0^1 (\rho_-[\mathbf{h},\dot{\mathbf{k}}],db(t))_{\mathfrak{g}}.
\end{aligned}
$$

Next we consider the case $\mathbf{k} \in H_1^{(0,1)}$.

$$
\begin{aligned}
&\langle dd^*\eta - \nabla^*\check{\nabla}\eta, \mathbf{k}\rangle \\
&= \langle du - \nabla^*\check{\nabla}\eta, \mathbf{k}\rangle \\
&= \langle T^{*-1}\eta, \mathbf{k}\rangle + \int_0^1 ([\dot{\mathbf{h}},\mathbf{k}],db(t))_{\mathfrak{g}} \\
&\quad - \sum_{\alpha,\beta} \{(\operatorname{div} T^{-1}\mathbf{e}_{\bar\beta})\langle \theta^{\bar\alpha},\mathbf{k}\rangle\langle \eta, \pi_+[\mathbf{e}_{\bar\alpha},\mathbf{e}_\beta]\rangle + \langle \nabla_{T^{-1}\mathbf{e}_{\bar\beta}}\theta^\alpha,\mathbf{k}\rangle \eta(\pi_+[\mathbf{e}_{\bar\alpha},\mathbf{e}_\beta])\} \\
&= \int_0^1 ([\dot{\mathbf{h}},\mathbf{k}],db(t))_{\mathfrak{g}} \\
&\quad - \sum_\beta (\operatorname{div} T^{-1}\mathbf{e}_{\bar\beta})\langle \eta, [\mathbf{k},\mathbf{e}_\beta]\rangle + \sum_\beta \langle \eta, [A(T^{-1}\mathbf{e}_{\bar\beta},\mathbf{k}),\mathbf{e}_\beta]\rangle.
\end{aligned}
$$

Further,

$$
\begin{aligned}
\langle \eta, [A(T^{-1}\mathbf{e}_{\bar\beta},\mathbf{k}),\mathbf{e}_\beta]\rangle &= (\mathbf{h}, [A(T^{-1}\mathbf{e}_{\bar\beta},\mathbf{k}),\mathbf{e}_\beta])_{H_1} \\
&= \sqrt{-1}(J\mathbf{h}, [A(T^{-1}\mathbf{e}_{\bar\beta},\mathbf{k}),\mathbf{e}_\beta])_{H_1} \\
&= \sqrt{-1}S(\mathbf{h}, [A(T^{-1}\mathbf{e}_{\bar\beta},\mathbf{k}),\mathbf{e}_\beta]) \\
&= \sqrt{-1}\int_0^1 (\dot{\mathbf{h}}, [A(T^{-1}\mathbf{e}_{\bar\beta},\mathbf{k}),\mathbf{e}_\beta])_{\mathfrak{g}} dt \\
&= \sqrt{-1}\int_0^1 ([\mathbf{e}_\beta,\dot{\mathbf{h}}], A(T^{-1}\mathbf{e}_{\bar\beta},\mathbf{k}))_{\mathfrak{g}} dt.
\end{aligned}
$$

The stochastic integral vanishes because, for $\beta = (n,i)$,

$$
\begin{aligned}
&(\operatorname{div} T^{-1}\mathbf{e}_{\bar\beta})\langle \eta, [\mathbf{k},\mathbf{e}_\beta]\rangle \\
&= (\operatorname{div} T^{-1}\mathbf{e}_{\bar\beta})(\mathbf{h}, [\mathbf{k},\mathbf{e}_\beta])_{H_1} \\
&= \sqrt{-1}(\operatorname{div} T^{-1}\mathbf{e}_{\bar\beta})(J\mathbf{h}, [\mathbf{k},\mathbf{e}_\beta])_{H_1} \\
&= \sqrt{-1}(\operatorname{div} T^{-1}\mathbf{e}_{\bar\beta})S(\mathbf{h}, [\mathbf{k},\mathbf{e}_\beta]) \\
&= \sqrt{-1}(\operatorname{div} T^{-1}\mathbf{e}_{\bar\beta})\int_0^1 (\dot{\mathbf{h}}, [\mathbf{k},\mathbf{e}_\beta])_{\mathfrak{g}} dt
\end{aligned}
$$

$$
\begin{aligned}
&= \sqrt{-1}(\operatorname{div} T^{-1}\mathbf{e}_{\bar\beta}) \int_0^1 ([\dot{\mathbf{h}}, \mathbf{k}], \mathbf{e}_\beta])_{\mathfrak{g}} dt \\
&= \sqrt{-1}(\operatorname{div} \mathbf{e}_{\bar\beta})(2\pi n) \int_0^1 ([\dot{\mathbf{h}}, \mathbf{k}], \frac{1}{2\pi\sqrt{-1}n}(e^{2\pi\sqrt{-1}nt} - 1)\xi_j)_{\mathfrak{g}} dt \\
&= (\operatorname{div} \mathbf{e}_{\bar\beta}) \int_0^1 ([\dot{\mathbf{h}}, \mathbf{k}], e^{2\pi\sqrt{-1}nt}\xi_j)_{\mathfrak{g}} dt \\
&= (\operatorname{div} \mathbf{e}_{\bar\beta}) \int_0^1 ([\dot{\mathbf{h}}, \mathbf{k}], \dot{\mathbf{e}}_\beta)_{\mathfrak{g}} dt.
\end{aligned}
$$

Combining this with $\int_0^1 ([\dot{\mathbf{h}}, \mathbf{k}], \dot{\mathbf{e}}_{\bar\beta})_{\mathfrak{g}} dt = 0$, we have

$$
\sum_\beta (\operatorname{div} T^{-1}\mathbf{e}_{\bar\beta}) \langle \eta, [\mathbf{k}, \mathbf{e}_\beta] \rangle = \int_0^1 ([\dot{\mathbf{h}}, \mathbf{k}], db(t))_{\mathfrak{g}}.
$$

This completes the proof. □

References

1. S. Aida, Sobolev Spaces over Loop Groups, *J. Funct. Anal.*, **127** (1995), 155–172.
2. D. S. Freed, The geometry of loop groups, *J. Diff. Geom.*, **28** (1988), 223–276.
3. E. Getzler, Dirichlet forms on loop space, *Bull. Sci. Math.*, **113** (1989), 151–174.
4. L. Gross, Logarithmic Sobolev inequalities on loop groups, *J. Funct. Anal.*, **102** (1991), 268–313.
5. L. Gross, Uniqueness of ground states for Schrödinger operators over loop groups, *J. Funct. Anal.*, **112** (1993), 373–441.
6. T. Kazumi and I. Shigekawa, Differential calculus on a submanifold of an abstract Wiener space, I. Covariant derivative, in "*Stochastic analysis on infinite dimensional spaces*," ed. by H. Kunita and H.-H. Kuo, pp. 117–140, Longman, Harlow, 1994.
7. T. Kazumi and I. Shigekawa, Differential calculus on a submanifold of an abstract Wiener space, II. Weitzenböck formula, in "*Dirichlet forms and stochastic processes*," Proceedings of the International Conference held in Beijing, China, October 25-31, 1993, ed. by Z. M. Ma, M. Röckner and J. A. Yan, pp. 235–251, Walter de Gruyter, Berlin-New York, 1995.
8. S. Kobayashi and K. Nomizu, "*Foundations of differential geometry*," I, II, Interscience Publishers, New York-London, 1963, 1969.
9. S. Kusuoka and S. Taniguchi, Pseudoconvex domains in almost complex abstract Wiener spaces, *J. Funct. Anal.*, **117** (1993), 62–117.
10. M.-P. Malliavin and P. Malliavin, Integration on loop groups. I. Quasi Invariant measures, *J. Funct. Anal.*, **93** (1990), 207–237.
11. A. N. Pressley, The energy flow on the loop space of a compact Lie group, *J. London Math. Soc.* (2) **26** (1982), 557–566.
12. A. Pressley and G. Segal, "*Loop groups*," Oxford University Press, New York, 1986.
13. I. Shigekawa, Differential calculus on a based loop group, preprint.

Burgers system driven by a periodic stochastic flow

Ya. G. Sinai

Department of Mathematics, Princeton University, Fine Hall, Washington Road, Princeton, NJ 08544, USA and Landau Institute for Theoretical Physics, ul. Kosygina 2, Moscow 117940, GSP-1, V-334, Russia

Burgers System (BS) is Navier-Stokes system without pressure and incompressibility. It is one of the most popular models of hydrodynamics and has a lot of applications. In this paper, we consider BS driven by a potential force whose potential is a periodic stochastic flow. If $x = (x_1, \dots x_n)$ is the vector of coordinates and $u = (u_1, \dots, u_n)$ is the velocity vector then the n-dimensional BS takes the form

$$\frac{\partial u}{\partial t} + (u, \nabla) u = \mu \Delta u + \nabla \dot{B}(x,t) \tag{1}$$

or in the coordinate form

$$\frac{\partial u_i}{\partial t} + \sum_{k=1}^{n} \frac{\partial u_i}{\partial x_k} \cdot u_k = \mu \Delta u_i + \frac{\partial}{\partial x_i} \dot{B}(x,t). \tag{2}$$

Here and later ∇ means gradient with respect to space variables, dot means the differentiation with respect to time, ∂ is the differential with respect to time variable, μ is the viscosity. We impose periodic boundary conditions with period 1 for each x_i. The potential $\dot{B}(x,t)$ is the generalized derivative of a periodic stochastic flow. The flow itself can be represented in the form of Itô stochastic differential

$$\partial B(x,t) = \sum_{k \in \mathbb{Z}^n} c_k \, e^{2\pi i (x,k)} \, dB_k(t) \tag{3}$$

In the last expression $B_k(t)$ are standard complex Wiener processes, $B_{-k}(t) = \bar{B}_k(t)$ and independent otherwise. About the coefficients c_k we assume that $c_0 = 0$, $c_{-k} = \bar{c}_k$ and c_k decay so fast that the moment

$$\sum |c_k| \cdot \| \, k \, \|^4 < \infty \tag{4}$$

These conditions imply in particular that $B(x,t)$ is a.e. a C^2-smooth real-valued periodic function of x. When physicists write about BS with a potential force which is "white noise" in time they actually keep in mind forces generated by potentials which we just described.

An advantage (or disadvantage) of BS is due to the fact that it leaves invariant the space of gradient solutions. In the non-random case this means that if initially $u(x;0)$ is gradient of some function then $u(x;t)$ is also gradient for all $t > 0$. The resulting function satisfies the heat equation whose potential is the potential of the force. The derivation is based on the so-called Hopf-Cole substitution $u = -2\mu \frac{1}{\varphi} \nabla \varphi$ (see [C], [H]). However, in the case of random forces the deviation requires some changes which are due to the well-known rules in dealing with Itô stochastic differentials. In the first part of this paper, we discuss these changes. Our arguments are formal but they can be done easily rigorous.

Begin with stochastic partial differential equation

$$\partial\varphi = \mu \Delta\varphi \cdot dt + \left(-\frac{1}{2\mu}\partial B + c\, dt\right)\varphi \tag{5}$$

which is the heat equation with random potential. The value of the constant c will be chosen later. Denote by $\mathcal{F}_{t_1}^{t_2}$ the smallest σ-algebra generated by all increments $B_k(t') - B_k(t'')$, $t_1 \le t'' < t' \le t_2$, $k \in \mathbb{Z}^n$. Under solution (5) we mean an $\mathcal{F}_0^t$-measurable function $\varphi(x,t)$ for which

$$\varphi(x,t) = \varphi(x,0) + \mu\int_0^t \Delta\varphi(x,\tau)dt + c\int_0^t \varphi(x,\tau)dt - \frac{1}{2\mu}\int_0^t \varphi(x,\tau)\,\partial B(x,\tau) \tag{6}$$

The last expression is the usual Itô stochastic integral, x is considered as a parameter. The existence and smoothness of solutions of (6) for bounded periodic initial data follow from the general theory of stochastic flows (see [E], [Ku]). Put $v = -2\mu\, \ell n\varphi$. By Itô formula

$$\partial v = -2\mu\frac{\partial\varphi}{\varphi} + \mu\frac{(\partial\varphi)^2}{\varphi^2} = -2\mu^2\frac{\Delta\varphi}{\varphi}dt + \partial B - c\, dt + \frac{1}{4\mu^2}a\, dt.$$

where $a\, dt = E\,\partial B^2 = \sum_k |c_k|^2\, dt$. Choose $c = -\frac{a}{4\mu}$. Then

$$\partial v = -2\mu^2\frac{\Delta\varphi}{\varphi}dt + \partial B.$$

We shall show that $u = \nabla v$ is a solution of (1). We have

$$\partial u = \partial\nabla v = \nabla\partial v = 2\mu^2\frac{1}{\varphi^2}\Delta\varphi \cdot \nabla\varphi - 2\mu^2\frac{1}{\varphi}\nabla\Delta\varphi + \nabla\partial B.$$

From the other side $u_i = -2\mu\frac{1}{\varphi}\frac{\partial\varphi}{\partial x_i}$

and

$$\frac{\partial u_i}{\partial x_k} = 2\mu\frac{1}{\varphi^2}\frac{\partial\varphi}{\partial x_k}\cdot\frac{\partial\varphi}{\partial x_i} - 2\mu\frac{1}{\varphi}\frac{\partial^2\varphi}{\partial x_i\partial x_k},$$

$$((u,\nabla)u)_i = \sum_{k=1}^{n} \frac{\partial u_i}{\partial x_k}\cdot u_k = -4\mu^2\cdot\frac{1}{\varphi^3}(\nabla\varphi,\nabla\varphi)\frac{\partial\varphi}{\partial x_i} + 2\mu^2\frac{1}{\varphi^2}\frac{\partial}{\partial x_i}(\nabla\varphi,\nabla\varphi),$$

$$\begin{aligned}\frac{\partial^2 u_i}{\partial x_k^2} &= -4\mu\frac{1}{\varphi^3}\left(\frac{\partial\varphi}{\partial x_k}\right)^2\frac{\partial\varphi}{\partial x_i} + 2\mu\frac{1}{\varphi^2}\frac{\partial^2\varphi}{\partial x_k^2}\frac{\partial\varphi}{\partial x_i}\\ &\quad + 4\mu\frac{1}{\varphi^2}\frac{\partial\varphi}{\partial x_k}\cdot\frac{\partial^2\varphi}{\partial x_i\,\partial x_k} - 2\mu\frac{1}{\varphi}\frac{\partial^3\varphi}{\partial x_i\,\partial x_k^2}\end{aligned}$$

and

$$\begin{aligned}\mu\Delta u &= -4\mu^2\frac{1}{\varphi^3}\cdot(\nabla\varphi,\nabla\varphi)\,\nabla\varphi + 2\mu^2\frac{1}{\varphi^2}\Delta\varphi\cdot\nabla\varphi\\ &\quad + 2\mu^2\frac{1}{\varphi^2}\nabla(\nabla\varphi,\nabla\varphi) - 2\mu^2\frac{1}{\varphi}\nabla\Delta\varphi\end{aligned}$$

Therefore

$$\mu\Delta u - (u,\nabla)u = 2\mu^2\cdot\frac{1}{\varphi^2}\Delta\varphi\cdot\nabla\varphi - 2\mu^2\frac{1}{\varphi}\nabla\Delta\varphi = \partial u - \nabla\partial B,$$

or

$$\partial u + (u,\nabla)u = \mu\Delta u + \nabla\partial B.$$

In other words, u is a solution of BS.

Now we discuss the Feynman-Kac formula for solutions of (5). Assume that we are given an initial datum $\varphi(x,0)$ which is a bounded periodic function of x. A solution $\varphi(x,t)$ of (5) or (6) is an $\mathcal{F}_0^t$-measurable function which is a C^2-periodic function of x for which (6) holds. The Feynman-Kac formula takes the form

$$\varphi(x,t) = \int_{R^n} dy\,\varphi(y,0)\int e^{-\frac{1}{2\mu}\int_0^t \partial B(\omega(\tau),\tau)}\, d\sqcap_{(x,t)}^{(y,0)}(\omega) \qquad (7)$$

In this expression $d\sqcap_{(x,t)}^{(y,0)}$ is the notation for the differential of the (non-normed) Wiener measure with the diffusion constant 2μ defined on the Borel σ-algebra of the space of continuous Brownian trajectories ω going out of (x,t) and coming to $(y,0)$ (let us stress that the direction of time in ω is opposite to the one in (1)). Certainly, this Wiener measure is independent of the distribution of our stochastic flow. The integral $\int_0^t \partial B(\omega(\tau),\tau)$ is a stochastic integral which is understood as l^2-limit of integral sums

$$\sum_{r=0}^{R-1} e^{2\pi i(\omega(\frac{r}{R}t),k)} c_k \left(B_k\left(\frac{r+1}{R}t\right) - B_k\left(\frac{r}{R}t\right)\right)$$

as $R \longrightarrow \infty$. Under our assumptions the limit exists for a.e. ω and the inner integral in (7) is finite. It can be proven by the usual methods that (7) gives the solution of (5) satisfying given initial conditions. Let us pay attention to the term $c\,dt$ in (5) which is absent in the non-random case. Its appearance is due to the Itô formula for stochastic differentials.

We formulate the result of all previous considerations in the form of the following theorem.

Theorem 1. *The functional integral (7) gives the solution of the equation (6) or stochastic differential equation (5). The vector* $u = -2\mu \frac{\nabla\varphi}{\varphi}$ *is the solution of BS (2) if the initial datum for (1) is* $u(x,0) = \nabla\psi(x)$. *In this case* $\varphi(y,0) = e^{-\frac{1}{2\mu}\psi(y)}$.

The main purpose of this paper is to study invariant measures for BS. As far as we know, the problem was considered rigorously for the first time in [S] for a very restrictive class of forces. In this paper we extend the result of [S] for the whole class of stochastic flows considered in this paper.

Let us give a more detailed description of the problem. Assume that we are given a probability distribution Q_0 on the Borel σ-algebra of the space of C^2-periodic gradient functions $u(x;0)$. Then the probability distribution for stochastic flow $B(x,t)$ generates the probability measure P on the space of solutions of (1) or (2). The basic question is whether it is possible to find Q_0 such that the random process $u(x,t)$ is stationary in time and for each t has distribution Q_0. We prove below the following theorem.

Theorem 2. *Let* $u(x,0) = \nabla\psi(x,0)$ *where* ψ *is a* C^2*-periodic function. Then the probability distribution* Q_t *for the solutions* $u(x,t)$ *converges weakly to a limit* Q *which does not depend on* $\psi(x,0)$ *as* $t \longrightarrow \infty$.

Proof. The basic idea is to show that for large t, any solution $u(x,t)$ can be well approximated by random variables $u_s(x,t)$ measurable with respect to σ-algebras $\mathcal{F}_{t-s}^t$ and stationary in the natural sense. This means in particular, that solutions $u(x,t)$ exhibit some loss of dependence on initial data or "loss of memory."

We denote B the space of elementary events corresponding to our stochastic flow and all probabilities related to B will be denoted by P_B. Introduce the following operator $K(x,t;y,t') = \int e^{-\frac{1}{2\mu}\int_{t'}^{t} \partial B(\omega(\tau),\tau)} d\Pi_{(x,t)}^{(y,t')}(\omega)$, $t' < t$. Let us stress that the integration goes over Brownian trajectories which start from x at the moment t and come to y at the moment t'. The operators K are random as functions of B and their distribution depends on $t - t'$. They are periodic in x, y since B is periodic, i.e. $K(x+m,\ t;\ y+m,\ t') = K(x,t;\ y,t')$ for any $m \in \mathbb{Z}^n$.

Let us write

$$\begin{aligned}\varphi(x,t) &= \int e^{-\frac{1}{2\mu}\psi(y)}\,dy \int e^{-\frac{1}{2\mu}\int_0^t \partial B(\omega),\tau)}\, d\,\Pi_{(x,t)}^{(y,0)}(\omega) \\ &= \int_{R^n}\dots\int_{R^n} K(x,t;\ x_1,t-1)K(x_1,\ t-1;\ x_2,\ t-2)\,\cdot \\ &\dots\ \cdot K(x_{[t]-2},\ t-[t]+2;\ x_{[t]-1},\ t-[t]+1) \\ &\quad \cdot K\left(x_{[t]-1},\ t-[t]+1; y, 0\right)\cdot e^{-\frac{1}{2\mu}\psi(y)} \prod_{i=1}^{[t]-1} d\,x_i\,dy\end{aligned}$$

The n-dimensional torus T^n can be identified with the subset $\Delta^{(n)} = \{x\,|\,0 \le x_i < 1,\ 1 \le i \le n\} \subset R^n$. From the periodicity of K

$$\varphi(x,t) = \int_{\Delta^{(n)}}\dots\int_{\Delta^{(n)}} \sum_{m_1, m_2, \dots, m_{[t]} \in \mathbb{Z}^n} K(x,t;\ x_1 + m_1,\ t-1)$$

$$\prod_{j=1}^{[t]-2} K(x_j + m_j,\ t-j;\ x_{j+1} + m_{j+1},\ t-j-1)\cdot K(x_{[t]-1} + m_{[t]-1},\ t-[t]+1;$$

$$x_{[t]} + m_{[t]},\ 0)\, e^{-\frac{1}{2\mu}\psi(x_{[t]})} \prod_{j=1}^{[t]} d\,x_j = \int_{\Delta^{(n)}}\dots\int_{\Delta^{(n)}} \sum_{m_1\in\mathbb{Z}^n} K(x,t;\ x_1 + m_1,\ t-1)\,\cdot$$

$$\sum_{m_2,\dots,m_{[t]}\in\mathbb{Z}^n} \Pi\, K(x_j, t-j;\ x_{j-1} + m_{j-1} - m_j,\ t-j-1)\cdot K(x_{[t]-1},\ t-[t]+1;$$

$$x_{[t]} + m_{[t]} - m_{[t]-1}, 0)\cdot e^{-\frac{1}{2\mu}\psi(x_{[t]})} \prod_{j=1}^{[t]} d\,x_j$$

Introduce transfer-matrices

$$T(x,t;\ x',t') = \sum_{m\in\mathbb{Z}^n} K(x,t;\ x' + m,\ t'),\ x \in \Delta^{(n)},\ x' \in \Delta^{(n)}, t' < t.$$

and related probabilities as functions of m

$$q(x,t;\ x' + m,\ t-1) = K(x,t;\ x' + m, t-1)\cdot (T(x,t;\ x',t-1))^{-1}.$$

Then

$$\varphi(x,t) = \int_{\Delta^{(n)}}\dots\int_{\Delta^{(n)}} T(x,t;\ x_1,\ t-1)\cdot T(x_1,\ t-1;\ x_2,\ t-2)\cdot\ \dots\ \cdot$$

$$T(x_{[t]-1}, t-[t]+1; x_{[t]}, 0)\, e^{-\frac{1}{2\mu}\psi(x_{[t]},0)}\, dx_1\, dx_2 \ldots dx_{[t]-1}\, dx_{[t]}$$

$$\sum_{m_1, m_2, \ldots, m_{[t]} \in \mathbb{Z}^n} q(x,t;\, x_1+m_1,\, t-1) \cdot q(x_1,\, t-1;\, x_2+m_2-m_1,\, t-2) \cdot$$

$$\cdot \ldots \cdot q(x_{[t]-1},\, t-[t]+1;\, x_{[t]}+m_{[t]}-m_{[t]-1},\, 0)\, e^{-\frac{1}{2\mu}\psi(x_{[t]})}$$

The last sum equals obviously to 1 and therefore

$$\varphi(x,t) = \int_{\Delta^{(n)}} \cdots \int_{\Delta^{(n)}} T(x,t;\, x_1,\, t-1) \prod_{j=1}^{[t]-2} T(x_j,\, t_j;\, x_{j+1},\, t_j-1) \cdot$$

$$T(x_{[t]-1},\, t-[t]+1;\, x_{[t]},\, 0)\, e^{-\frac{1}{2\mu}\psi(x_{[t]})} \prod_{j=1}^{[t]} dx_j$$

The solution of BS

$$u = -2\mu \nabla \ln \varphi = -2\mu \frac{1}{\varphi} \nabla \varphi =$$

$$-2\mu \frac{\displaystyle\int_{\Delta^{(n)}} \cdots \int_{\Delta^{(n)}} \nabla_x T(x,t;\, x_1,\, t-1) \prod_{j=1}^{[t]-2} T(x_j,\, t_j;\, x_{j+1},\, t_j-1) \cdot}{\displaystyle\int_{\Delta^{(n)}} \cdots \int_{\Delta^{(n)}} T(x,t;\, x_1,\, t-1) \prod_{j=1}^{[t]-2} T(x_j,\, t_j;\, x_{j+1},\, t_j-1) \cdot}$$

$$\frac{T(x_{[t]-1},\, t-[t]+1;\, x_{[t]},\, 0)\, e^{-\frac{1}{2\mu}\psi(x_{[t]})} \prod_{j=1}^{[t]} dx_j}{T(x_{[t]-1},\, t-[t]+1;\, x_{[t]},\, 0)\, e^{-\frac{1}{2\mu}\psi(x_{[t]})} \prod_{j=1}^{[t]} dxj} \tag{8}$$

Consider the probability distribution P on the Borel σ-algebra of the space of sequences $\{x_1, x_2, \ldots, x_{[t]}\}$, $x_j \in \Delta^{(n)}$, $1 \le j \le n$, whose density with respect to $\prod_{j=1}^{[t]} dx_j$ is

$$\prod_{j=1}^{[t]-2} T(x_j,\, t_j;\, x_{j+1},\, t_j-1)\, T(x_{[t]-1},\, t-[t]+1;\, x_{[t]},\, 0)\, e^{-\frac{1}{2\mu}\psi(x_{[t]})}.$$

Then (8) becomes

$$u(x,t) = -2\mu \frac{E \nabla_x T(x,t;\, x_1,\, t-1)}{E\, T(x,\, t;\, x_1,\, t-1)}$$

where E is the expectation with respect to P. It is easy to see that P is a non-homogeneous Markov chain with the compact phase space $\Delta^{(n)}$. Conditional densities $\frac{dP}{dx}\{x_s \mid x_{s+1}, \ldots, x_{[t]}\}$ depend only on x_{s+1}. They are random since they are measurable with respect to $\mathcal{F}_{t-s}^{t}$ and with positive P_B-probability the corresponding transition operator is a contraction. Therefore, $u(x,t)$ can be represented as a limit as $s \longrightarrow \infty$ functions $u_s(x,t)$ where

$$u_s(x,t) = -2\mu \frac{E(\nabla_x T(x, t; x_1, t-1) \,|\, x_s)}{E(T(x, t; x_1, t-1) \,|\, x_s)}$$

Since the Markov chain is ergodic the convergence is exponential but not uniform in the sense that the value of s after which the exponential asymptotics holds depends on B. Thus, the limiting distribution of $u(x,t)$ as $t \longrightarrow \infty$ is the distribution of the limit $\lim_{s \to \infty} u_s(x,t)$.

Theorem is proven.

Concluding remarks.

1. It is not clear whether Theorem 2 remains true for more slowly decaying sequences c_k.
2. There are some doubts about the existence of invariant measures for BS when the force is an arbitrary stochastic flow in R^n with quickly decaying space correlations.

Recent results of Polyakov and Yakhot (see [P], [P-Y]) about the properties of the limit of the invariant measures for BS in R^n for $\mu \to 0$ are related to the limit of the measure constructed in this paper as $\mu \to 0$.

The financial supports from NSF, grant DMS-0404437 and FFR of Russia, grant N93-01-16090 are highly acknowledged.

References

[C] Cole, J., *Quart. Appl. Math.* **9**, 225 (1951).

[E] Elworthy, K. D., "Stochastic differential equations on manifolds," London Math. Society, Lect. Notes Series, **70**, Cambridge University Press, (1982).

[H] Hopf, E., *Comm. Pure Appl. Math.* **3**, 201 (1950).

[Ki] Kifer, Yu., "The Burgers Equation with a Random Force and a General Model for Directed Polymers in Random Environments," Preprint, Hebrew University, Jerusalem, June, 1995.

[Ku] Kunita, H., "Stochastic Flows and Stochastic Differential Equations," Cambridge University Press, (1990), p.346.

[P] Polyakov, A., "Turbulence without Pressure," *Physical Review E* **52**, (6) p.6183 (1995).

[P-Y] Polyakov, A. & Yakhot, V., Private communication.

[S] Sinai, Ya.G. *Journal of Stat. Physics* **64**, 1/2, 1 (1991).

An estimate on the Hessian of the heat kernel

Daniel W. Stroock

Department of Mathematics, Room 2-272, Massachusetts Institute of Technology, 77 Massachusetts Ave., Cambridge, MA 02139, USA

Summary. Let M be a compact, connected Riemannian manifold, and let $p_t(x,y)$ denote the fundamental solution to Cauchy initial value problem for the heat equation $\frac{\partial u}{\partial t} = \frac{1}{2}\Delta u$, where Δ is the Levi–Civita Laplacian. The purpose of this note is to show that the Hessian of $\log p_t(\,\cdot\,,y)$ at x is bounded above by a constant times $\frac{1}{t} + \frac{\mathrm{dist}(x,y)^2}{t^2}$ for $t \in (0,1]$.

0. Introduction

Let M be a compact, connected, d-dimensional Riemannian manifold, denote by $\mathcal{O}(M)$ with fiber map $\pi : \mathcal{O}(M) \longrightarrow M$ the associated bundle of orthonormal frames $\mathfrak{e}$, and use the Levi–Civita connection to determine the horizontal subspace $H_{\mathfrak{e}}(\mathcal{O}(M))$ at each $\mathfrak{e} \in \mathcal{O}(M)$. Next, given $\mathbf{v} \in \mathbb{R}^d$, let $\mathfrak{E}(\mathbf{v})$ be the *basic vector field* on $\mathcal{O}(M)$ determined by properties that

$$\mathfrak{E}(\mathbf{v})_{\mathfrak{e}} \in H_{\mathfrak{e}}(\mathcal{O}(M)) \quad \text{and} \quad d\pi\mathfrak{E}(\mathbf{v})_{\mathfrak{e}} = \mathfrak{e}\mathbf{v} \quad \text{for all } \mathfrak{e} \in \mathcal{O}(M).$$

(Here, and whenever convenient, we think of $\mathfrak{e}$ as a isometry from $\mathbb{R}^d$ onto $T_{\pi(\mathfrak{e})}(M)$.) In particular, if $\{\mathbf{e}_1, \dots, \mathbf{e}_d\}$ is the standard orthonormal basis in $\mathbb{R}^d$, then we set $\mathfrak{E}_k(\mathfrak{e}) = \mathfrak{E}(\mathbf{e}_k)_{\mathfrak{e}}$. If, for $\mathcal{O} \in O(d)$ (the orthogonal group on $\mathbb{R}^d$) $R_{\mathcal{O}} : \mathcal{O}(M) \longrightarrow \mathcal{O}(M)$ is defined so that

$$R_{\mathcal{O}}\mathfrak{e}\,\mathbf{v} = \mathfrak{e}\,\mathcal{O}\mathbf{v}, \quad \mathfrak{e} \in \mathcal{O}(M) \text{ and } \mathbf{v} \in \mathbb{R}^d,$$

then it easy to check that

$$dR_{\mathcal{O}}\mathfrak{E}(\mathbf{v})_{\mathfrak{e}} = \mathfrak{E}(\mathcal{O}^{\top}\mathbf{v})_{R_{\mathcal{O}}\mathfrak{e}}, \quad \mathfrak{e} \in \mathcal{O}(M) \text{ and } \mathbf{v} \in \mathbb{R}^d. \tag{0.1}$$

Given a smooth function F on $\mathcal{O}(M)$, we define $\nabla F : \mathcal{O}(M) \longrightarrow \mathbb{R}^d$, $\mathrm{Hess}\,(F) : \mathcal{O}(M) \longrightarrow \mathbb{R}^d \otimes \mathbb{R}^d$, and $\Delta F : \mathcal{O}(M) \longrightarrow \mathbb{R}$ by

$$\nabla F = \sum_1^d \mathfrak{E}_k F\, \mathbf{e}_k, \quad \mathrm{Hess}\,(F) = \big((\mathfrak{E}_k \circ \mathfrak{E}_\ell F)\big)_{1 \le k,\ell \le d}$$
$$\text{and} \quad \Delta F = \sum_1^d \mathfrak{E}_k^2 F. \tag{0.2}$$

Support was provided, in part, by NSF grant 9302709–DMS

In particular, when f is a smooth function on M, we set

$$\nabla f \equiv \nabla(f \circ \pi), \quad \text{Hess}\,(f) \equiv \text{Hess}\,(f \circ \pi), \quad \text{and} \quad \Delta f \equiv \Delta(f \circ \pi).$$

Starting from (0.1), it is an easy matter to check that

$$(\nabla f) \circ R_{\mathcal{O}} = \mathcal{O}^{\top} \nabla f, \quad (\text{Hess}\,(f)) \circ R_{\mathcal{O}} = \mathcal{O}^{\top} \text{Hess}\,(f)\, \mathcal{O}, \\ \text{and} \quad (\Delta f) \circ R_{\mathcal{O}} = \Delta f.$$

Hence, $|\nabla f|$, $\|\text{Hess}\,(f)\|_{\text{H.S.}}$ (the Hilbert–Schmidt norm), and Δf are all well-defined on M. In fact, Δf is precisely the action of the Levi–Civita Laplacian on f.

Now consider Cauchy initial value for the heat equation

$$\frac{\partial u}{\partial t} = \tfrac{1}{2} \Delta u, \quad t \in (0, \infty) \quad \text{with} \quad \lim_{t \searrow 0} u(t,x) = f(x), \quad x \in M.$$

By standard elliptic regularity theory, one knows that there is a unique, smooth function $(t, x, y) \in (0,\infty) \times M \times M \longmapsto p_t(x,y) \in (0, \infty)$ such that

$$u(t,x) = \int_M f(y)\, p_t(x,y)\, \lambda_M(dy), \quad (t,x) \in (0,\infty) \times M \text{ and } f \in C(M;\mathbb{R}),$$

where λ_M denotes the normalized Riemann measure on M. Moreover, because Δ is essentially self-adjoint in $L^2(\lambda_M)$, $p_t(x,y) = p_t(y,x)$. Our goal is to prove the existence of a constant $C \in (0, \infty)$ for which[1]

$$\begin{aligned} &\big|[\nabla \log p_T(\,\cdot\,, y)]\big|(x)^2 \vee \big\|[\text{Hess}\,(\log p_T(\,\cdot\,, y))]\big\|_{\text{H.S.}}(x) \\ &\qquad \le C \left(\frac{\text{dist}(x,y)^2}{T^2} + \frac{1}{T} \right) \quad \text{for all } (T,x,y) \in (0,1] \times M \times M. \end{aligned} \tag{0.3}$$

Our proof of (0.3) will rely on known estimates on $p_T(x,y)$ itself combined with estimates in which derivatives of $\log p_T(\,\cdot\,, y)$ are bounded in terms of $p_T(\,\cdot\,, y)$. We will simply barrow the known estimates from geometric analysis and will concentrate our attention on the latter estimates, which we will derive by a method which involves perturbation of Brownian paths.

1. Perturbation of Brownian Paths on a Manifold

In this section, we recall a procedure, introduced in [ES1], for perturbing Brownian paths on a manifold.

[1] Recently, W. Fleming pointed out to me that the estimate given below is closely related to and undoubted derivable from results obtained by S.–T. Sheu in [S].

Let $\mathfrak{W}$ be the separable Banach space of continuous paths $\mathbf{w} : [0,\infty) \longrightarrow \mathbb{R}^d$ satisfying

$$\mathbf{w}(0) = 0 = \lim_{t\to\infty} t^{-1}\mathbf{w}(t) \quad \text{with norm } \|\mathbf{w}\|_{\mathfrak{W}} \equiv \sup_{t\geq 0}(1+t)^{-1}|\mathbf{w}(t)|,$$

use $\mathcal{B}_{\mathfrak{W}}$ to denote the Borel field over $\mathfrak{W}$, and denote by μ the standard Wiener measure on $(\mathfrak{W}, \mathcal{B}_{\mathfrak{W}})$. In addition, for each $t \in [0,\infty)$, $\mathcal{B}_t$ will denote the σ-algebra generated $\mathbf{w} \in \mathfrak{W} \longmapsto \mathbf{w}(\tau) \in \mathbb{R}^d$ as τ varies over $[0,t]$. Next, given a frame $\mathfrak{e} \in \mathcal{O}(M)$, define $\mathfrak{F}_\mathfrak{e} : [0,\infty) \times \mathfrak{W} \longrightarrow \mathcal{O}(M)$ to be the μ-almost surely unique, progressively measurable (relative to $\{\mathcal{B}_t : t \geq 0\}$) solution to the Stratonovich stochastic differential equation

$$d\mathfrak{F}_\mathfrak{e}(t,\mathbf{w}) = \sum_{k=1}^{d} \mathfrak{E}_k\left(\mathfrak{F}_\mathfrak{e}(t,\mathbf{w})\right) \circ d\mathbf{w}(t)_k \quad \text{with } \mathfrak{F}_\mathfrak{e}(0,\mathbf{w}) = \mathfrak{e}.$$

As an easy application of Itô's formula and (0.2), one sees that, for any τ, $t \in [0,\infty)$ and $f \in C(M;\mathbb{R})$,

$$\text{(1.1)} \quad \mathbb{E}^\mu\Big[(f\circ\pi)\big(\mathfrak{F}_\mathfrak{e}(\tau+t)\big)\Big|\mathcal{B}_\tau\Big] = \int_M f(y)\,p_t\Big(\pi\big(\mathfrak{F}_\mathfrak{e}(\tau)\big),y\Big)\,\lambda_M(dy) \quad (\text{a.s.},\mu).$$

Hence, the distribution of $\{\pi\circ\mathfrak{F}_\mathfrak{e}(t) : t \in [0,\infty)\}$ under μ is that of Brownian motion on M starting at $\pi(\mathfrak{e})$. In particular, by taking $\tau = 0$ in (1.1), we get

$$\text{(1.2)} \quad \mathbb{E}^\mu\Big[(f\circ\pi)\big(\mathfrak{F}_\mathfrak{e}(t)\big)\Big] = [P_t f]\,(\pi(\mathfrak{e})) \equiv \int_M f(y)\,p_t(\pi(\mathfrak{e}),y)\,\lambda_M(dy).$$

Starting from (1.2), we will derive (0.3) by an application of a perturbation procedure alluded to above. Unfortunately, in order to describe this procedure, we will need some additional notation.

The *solder form* $\omega : T(\mathcal{O}(M)) \longrightarrow \mathbb{R}^d$ is the 1-form defined so that, for each $\mathfrak{e} \in \mathcal{O}(M)$ and $X_\mathfrak{e} \in T_\mathfrak{e}(\mathcal{O}(M))$, $d\pi X_\mathfrak{e} = \mathfrak{e}\omega(X_\mathfrak{e})$. Thus, the vertical subspace at $\mathfrak{e}$ is precisely the null space of $\omega \restriction T_\mathfrak{e}(\mathcal{O}(M))$. Next, let $\mathfrak{o}(d)$ stand for the Lie algebra of skew symmetric $d \times d$-matrices, remember that $\mathfrak{o}(d)$ can be identified with the Lie algebra of left-invariant vector fields on $O(d)$, and let λ be the map of $\mathfrak{o}(d)$ into the $T(\mathcal{O}(M))$ given by

$$\lambda(A)_\mathfrak{e} = \frac{d}{dt} R_{e^{tA}}\mathfrak{e}\Big|_{t=0}, \quad A \in \mathfrak{o}(d) \text{ and } \mathfrak{e} \in \mathcal{O}(M).$$

Clearly, $A \in \mathfrak{o}(d) \longmapsto \lambda(A)_\mathfrak{e} \in T_\mathfrak{e}(\mathcal{O}(M))$ provides an isomorphism between $\mathfrak{o}(d)$ and the vertical subspace at $\mathfrak{e}$. Thus, we can define the *connection* 1-*form* $\phi : T(\mathcal{O}(M)) \longrightarrow \mathfrak{o}(d)$ so that, for each $\mathfrak{e} \in \mathcal{O}(M)$ and $X_\mathfrak{e} \in T_\mathfrak{e}(\mathcal{O}(M))$,

$$X_\mathfrak{e} - \lambda(\phi(X_\mathfrak{e})) = \sum_{k=1}^{d} \omega(X_\mathfrak{e})_k \mathfrak{E}_k(\mathfrak{e}) \quad \text{is the horizontal part of } X_\mathfrak{e}.$$

Equivalently, $\lambda(\phi(X_{\mathfrak{e}}))$ is the vertical part of $X_{\mathfrak{e}}$. Finally, the Riemann curvature 2-form $\Phi : T(\mathcal{O}(M))^2 \longrightarrow \mathfrak{o}(d)$ is the horizontal part of the exterior derivative $d\phi$ of ϕ. We set

$$\Phi(\mathfrak{e})_{\ell,\ell'} = \Phi(\mathfrak{E}_\ell(\mathfrak{e}), \mathfrak{E}_{\ell'}(\mathfrak{e})), \quad \mathfrak{e} \in \mathcal{O}(M) \text{ and } 1 \le \ell, \ell' \le d, \tag{1.3}$$

and define the Ricci curvature matrix $\mathrm{Ric} : \mathcal{O}(M) \longrightarrow \mathbb{R}^d \otimes \mathbb{R}^d$ by

$$\left(\mathbf{v}, \mathrm{Ric}(\mathfrak{e})\mathbf{v}'\right)_{\mathbb{R}^d} = \sum_{k=1}^{d} \left(\Phi(\mathfrak{E}(\mathbf{v})_{\mathfrak{e}}, \mathfrak{E}(\mathbf{e}_k)_{\mathfrak{e}} \mathbf{e}_k, \mathbf{v}'\right)_{\mathbb{R}^d}, \quad \mathbf{v}, \mathbf{v}' \in \mathbb{R}^d.$$

Next, for each $m \ge 1$, let $W_2^{(m)}(\mathbb{R}; \mathcal{O}(M))$ be the Sobolev space of $\mathfrak{f} \in C(\mathbb{R}; \mathcal{O}(M))$ with m square integrable derivatives. That is, $\mathfrak{f} \in C(\mathbb{R}; \mathcal{O}(M))$ is an element of $W_2^{(m)}(\mathbb{R}; \mathcal{O}(M))$ if[2]

$$s \in \mathbb{R} \longmapsto \omega(\mathfrak{f}'(s)) \in \mathbb{R}^d \quad \text{and} \quad s \in \mathbb{R} \longmapsto \phi(\mathfrak{f}'(s)) \in \mathfrak{o}(d)$$

exist as elements of the Sobolev spaces $W_2^{(m-1)}(\mathbb{R}; \mathbb{R}^d)$ and $W_2^{(m-1)}(\mathbb{R}; \mathfrak{o}(d))$, respectively. Clearly, $W_2^{(m)}(\mathbb{R}; \mathcal{O}(M))$ becomes a Polish space when we use the metric

$$\rho^{(m)}(\mathfrak{f}, \mathfrak{g}) = \mathrm{dist}(\mathfrak{f}(0), \mathfrak{g}(0)) + \left\|\omega(\mathfrak{f}') - \omega(\mathfrak{g}')\right\|_{W_2^{(m-1)}(\mathbb{R};\mathbb{R}^d)} + \left\|\phi(\mathfrak{f}') - \phi(\mathfrak{g}')\right\|_{W_2^{(m-1)}(\mathbb{R};\mathfrak{o}(d))}.$$

Moreover, we can determine a continuous map $\mathfrak{f} \in W_2^{(1)}(\mathbb{R}; \mathcal{O}(M)) \longmapsto \mathcal{O}(\mathfrak{f}) \in C(\mathbb{R}; O(d))$ by the integral equation

$$[\mathcal{O}(\mathfrak{f})](s) = \mathbf{I} - \int_0^s \phi(\mathfrak{f}'(\sigma)) \, [\mathcal{O}(\mathfrak{f})](\sigma) \, d\sigma, \quad \sigma \in \mathbb{R}.$$

The following statement is a straight-forward generalization of results proved in §2 of [ES1]. In particular, it relies on the approximation procedure discussed in the Appendix of that article in order to reduce the conclusions drawn in (1.7) and (1.8) to the case of smooth paths, when they follow by the reasoning in Lemma 2.5 of [ES1]. Also, the relation between Stratonovich and Itô integrals in (1.9) is a consequence of computation done in Lemma 2.12 of [ES1] (cf. Lemma 2.12 in [ES2] for a slightly different derivation).

Theorem 1.4. *Suppose that*

$$V : [0,\infty) \times C([0,\infty); W_2^{(1)}(\mathbb{R}; \mathcal{O}(M))) \longrightarrow W_2^{(1)}(\mathbb{R}; \mathbb{R}^d)$$

---

[2] Throughout, "prime" will be used to indicate derivatives with respect to s, and "dot" will be reserved for derivatives with respect to t.

is a measurable function with the property that, for each $T \in [0,\infty)$, there is a $C(T) < \infty$ such that:

$$\int_0^t \left\|V(\tau,\mathfrak{F}_1) - V(\tau,\mathfrak{F}_2)\right\|^2_{W^{(1)}(\mathbb{R};\mathbb{R}^d)}\,d\tau \le C(T)\int_0^t \rho^{(1)}\big(\mathfrak{F}_1(\tau),\mathfrak{F}_2(\tau)\big)^2\,d\tau$$
$$\text{for all } t \in [0,T] \text{ and } \mathfrak{F}_1,\mathfrak{F}_2 \in C\Big([0,\infty);W_2^{(1)}\big(\mathbb{R};\mathcal{O}(M)\big)\Big).$$

Then, for each $\mathfrak{f} \in W_2^{(1)}\big(\mathbb{R};\mathcal{O}(M)\big)$, there is a μ-almost surely unique, progressively measurable pair of maps

$$\mathfrak{F}_{\mathfrak{f},V} : [0,\infty) \times \mathfrak{W} \longmapsto W_2^{(1)}\big(\mathbb{R};\mathcal{O}(M)\big)$$
$$\Theta_{\mathfrak{f},V} : [0,\infty) \times \mathfrak{W} \longmapsto W_2^{(1)}\big(\mathbb{R};\mathbb{R}^d\big)$$

such that

(1.5)
$$d_t\big[\mathfrak{F}_{\mathfrak{f},V}(t,\mathbf{w})\big](s) = \sum_{k=1}^d \mathfrak{E}_k\Big(\big[\mathfrak{F}_{\mathfrak{f},V}(t,\mathbf{w})\big](s)\Big) \circ d_t\big[\Theta_{\mathfrak{f},V}(t,\mathbf{w})\big](s)_k$$
$$\text{with } \big[\mathfrak{F}_{\mathfrak{f},V}(0,\mathbf{w})\big](s) = \mathfrak{f}(s),$$

and

(1.6)
$$d_t\big[\Theta_{\mathfrak{f},V}(t,\mathbf{w})\big](s) = \big[\mathcal{O}(\mathfrak{F}_{\mathfrak{f},V}(t,\mathbf{w})\big](s) \circ \Big(d\mathbf{w}(t) + \big[\mathbf{b}_{\mathfrak{f},V}(t,\mathbf{w})\big](s)\,dt\Big)$$
$$\text{with } \big[\Theta_{\mathfrak{f},V}(0,\mathbf{w})\big](s) = 0,$$

where

$$\big[\mathbf{b}_{\mathfrak{f},V}(t,\mathbf{w})\big](s) \equiv \int_0^s \big[\mathcal{O}(\mathfrak{F}_{\mathfrak{f},V}(t,\mathbf{w})\big](\sigma)^\top \Big[V\big(t,\mathfrak{F}_{\mathfrak{f},V}(\,\cdot\,,\mathbf{w})\big)\Big](\sigma)\,d\sigma.$$

Moreover, for μ-almost every $\mathbf{w} \in \mathfrak{W}$:

(1.7)
$$\frac{d}{dt}\omega\Big(\big[\mathfrak{F}_{\mathfrak{f},V}(t,\mathbf{w})\big]'(s)\Big) = \Big[V\big(t,\mathfrak{F}_{\mathfrak{f},V}(\,\cdot\,,\mathbf{w})\big)\Big](s)$$

and (cf. (1.3))

(1.8)
$$\begin{aligned}\phi\Big(\big[\mathfrak{F}_{\mathfrak{f},V}(t,\mathbf{w})\big]'(s)\Big) &- \phi\big(\mathfrak{f}'(s)\big)\\ &= \sum_{\ell,\ell'=1}^d \int_0^t \omega\Big(\big[\mathfrak{F}_{\mathfrak{f},V}(\tau,\mathbf{w})\big]'(s)\Big)_{\ell'}\\ &\qquad\times \Phi_{\ell,\ell'}\Big(\big[\mathfrak{F}_{\mathfrak{f},V}(\tau,\mathbf{w})\big](s)\Big) \circ d_\tau\big[\Theta_{\mathfrak{f},V}(\tau,\mathbf{w})\big](s)_\ell.\end{aligned}$$

Finally,

$$\begin{aligned}
&[\Theta_{\mathfrak{f},V}(t,\mathbf{w})](s) - [B_{\mathfrak{f},V}(t,\mathbf{w})](s) \\
(1.9)\quad &= \int_0^t [\mathcal{O}(\mathfrak{F}_{\mathfrak{f},V}(\tau,\mathbf{w}))](s)\Big([\mathbf{b}_{\mathfrak{f},V}(\tau,\mathbf{w})](s) + \tfrac{1}{2}[\Xi(\mathfrak{F}_{\mathfrak{f},V}(\tau,\mathbf{w}))](s)\Big)\,d\tau,
\end{aligned}$$

where

$$[B_{\mathfrak{f},V}(t,\mathbf{w})](s) \equiv \int_0^t [\mathcal{O}(\mathfrak{F}_{\mathfrak{f},V}(\tau,\mathbf{w}))](s)\,d\mathbf{w}(\tau)$$

and $\Xi : W_2^{(1)}(\mathbb{R};\mathcal{O}(M)) \longrightarrow W^{(1)}(\mathbb{R};\mathbb{R}^d\otimes\mathbb{R}^d)$ *is given by*

$$[\Xi(\mathfrak{g})](s) \equiv \int_0^s [\mathcal{O}(\mathfrak{g})](\sigma)^\top \mathrm{Ric}(\mathfrak{g}(\sigma))\,\omega(\mathfrak{g}'(\sigma))\,d\sigma.$$

In order to describe our applications of Theorem 1.4, we need to introduce the Ricci flow $A : [0,\infty)\times C([0,\infty);\mathcal{O}(M)) \longrightarrow \mathbb{R}^d\otimes\mathbb{R}^d$ determined by

$$(1.10)\quad \begin{aligned} &A(t,\mathfrak{p}) + \frac{1}{2}\int_0^t \mathrm{Ric}(\mathfrak{p}(\tau))\,A(\tau,\mathfrak{p})\,d\tau = \mathbf{I}, \\ &\qquad t\in[0,\infty) \text{ and } \mathfrak{p}\in C(([0,\infty);\mathcal{O}(M)). \end{aligned}$$

Next, given $1 \le i \le d$, set

$$\begin{aligned} &[\overleftarrow{V}_i(t,\mathfrak{F})](s) = \dot{A}\big(t,[\mathfrak{F}(\cdot)](s)\big)\mathbf{e}_i, \\ &\quad s\in\mathbb{R} \text{ and } (t,\mathfrak{F})\in[0,\infty)\times C\big([0,\infty);W_2^{(1)}(\mathbb{R};\mathcal{O}(M))\big), \end{aligned}$$

and, for $\mathfrak{e}\in\mathcal{O}(M)$, determine $\mathfrak{f}_{\mathfrak{e},i} : \mathbb{R} \longrightarrow \mathcal{O}(M)$ by

$$\mathfrak{f}_{\mathfrak{e},i}'(s) = \mathfrak{E}_i(\mathfrak{f}_{\mathfrak{e},i}(s)) \quad \text{with } \mathfrak{f}_{\mathfrak{e},i}(0) = \mathfrak{e}.$$

Clearly, $\overleftarrow{V}_i$ satisfies the hypotheses of Theorem 1.4, and so there exist progressively measurable functions $\overleftarrow{\mathfrak{F}}_{\mathfrak{e},i}$ and $\overleftarrow{\Theta}_{\mathfrak{e},i}$ satisfying (1.5) and (1.6) with $\mathfrak{f} = \mathfrak{f}_{\mathfrak{e},i}$ and $V = \overleftarrow{V}_i$. In particular, (1.7) yields

$$(1.11)\quad \omega\Big([\overleftarrow{\mathfrak{F}}_{\mathfrak{e},i}(t,\mathbf{w})]'(s)\Big) = A\Big(t,[\overleftarrow{\mathfrak{F}}_{\mathfrak{e},i}(\cdot,\mathbf{w})](s)\Big)\mathbf{e}_i.$$

Moreover, because

$$\dot{A}(t,\mathfrak{p}) + \tfrac{1}{2}\mathrm{Ric}(\mathfrak{p}(t))\,A(t,\mathfrak{p}) = 0,$$

one finds (cf. (1.9)) that

$$\overleftarrow{\Theta}_{\mathfrak{e},i} = \overleftarrow{B}_{\mathfrak{e},i} \equiv B_{\mathfrak{f}_{\mathfrak{e},i},\overleftarrow{V}_i}.$$

Hence, because $\mathbf{w} \rightsquigarrow \left[\overleftarrow{B}_{\mathfrak{e},i}(\,\cdot\,,\mathbf{w})\right](s)$ is again an $\mathbb{R}^d$-Brownian motion under μ for each $s \in \mathbb{R}$, we see that

$$
\begin{aligned}
&\left(\left[\overleftarrow{B}_{\mathfrak{e},i}(\,\cdot\,,\mathbf{w})\right](s), \left[\overleftarrow{\mathfrak{F}}_{\mathfrak{e},i}(\,\cdot\,,\mathbf{w})\right](s)\right) \quad \text{under } \mu \\
&\qquad \overset{\text{law}}{=} \quad \left(\mathbf{w}, \mathfrak{F}_{\mathfrak{f}_{\mathfrak{e},i}(s)}(\,\cdot\,,\mathbf{w})\right) \quad \text{under } \mu.
\end{aligned}
\tag{1.12}
$$

The importance of the process $\overleftarrow{\mathfrak{F}}_{\mathfrak{e},i}$ for us is that allows the transfer of derivatives from the backward to the forward variables. That is, because of (1.12), we have that

$$
\begin{aligned}
\left[\mathfrak{E}_i U\right](\mathfrak{e}) &= \frac{d}{ds}\mathbb{E}^{\mu}\left[F\left(\left[\overleftarrow{B}_{\mathfrak{e},i}(\,\cdot\,,\mathbf{w})\right](s), \left[\overleftarrow{\mathfrak{F}}_{\mathfrak{e},i}(\,\cdot\,,\mathbf{w})\right](s)\right)\right]\Bigg|_{s=0} \\
&\text{when } U(\mathfrak{e}) = \mathbb{E}\Big[F\big(\mathbf{w}, \mathfrak{F}_{\mathfrak{e}}(\,\cdot\,,\mathbf{w})\big)\Big].
\end{aligned}
\tag{1.13}
$$

We next want to introduce a process which will enable us to integrate by parts, and thereby remove derivatives with respect to the forward variables.

Given $1 \le i \le d$, set (cf. (1.10))

$$
\left[\overrightarrow{V}_i(t,\mathfrak{F})\right](s) = \left(t\dot{A}\left(t, [\mathfrak{F}(\,\cdot\,)](s)\right) + A\left(t, [\mathfrak{F}(\,\cdot\,)](s)\right)\right)\mathbf{e}_i,
$$

and take

$$
\overrightarrow{\mathfrak{F}}_{\mathfrak{e},i} = \mathfrak{F}_{\overrightarrow{\mathfrak{f}}_{\mathfrak{e}},\overrightarrow{V}_i} \quad \text{and} \quad \overrightarrow{\Theta}_{\mathfrak{e},i} = \Theta_{\overrightarrow{\mathfrak{f}}_{\mathfrak{e}},\overrightarrow{V}_i},
$$

where $\overrightarrow{\mathfrak{f}}_{\mathfrak{e}}(s) = \mathfrak{e}$ for all $s \in \mathbb{R}$. Then, by (1.7),

$$
\omega\left(\left[\overrightarrow{\mathfrak{F}}_{\mathfrak{e},i}(t,\mathbf{w})\right]'(s)\right) = tA\left(t, \left[\overrightarrow{\mathfrak{F}}_{\mathfrak{e},i}(\,\cdot\,,\mathbf{w})\right](s)\right)\mathbf{e}_i.
\tag{1.14}
$$

Also, by (1.6) and (1.9),

$$
\begin{aligned}
d_t\left[\overrightarrow{\Theta}_{\mathfrak{e},i}(t,\mathbf{w})\right](s) &= d_t\left[\overrightarrow{B}_{\mathfrak{e},i}(t,\mathbf{w})\right](s) \\
&\quad + \left[\mathcal{O}\left(\overrightarrow{\mathfrak{F}}_{\mathfrak{e},i}(t,\mathbf{w})\right)\right](s)\left[\overrightarrow{\beta}_{\mathfrak{e},i}(t,\mathbf{w})\right](s)\,dt,
\end{aligned}
\tag{1.15}
$$

where $\overrightarrow{B}_{\mathfrak{e},i} \equiv B_{\overrightarrow{\mathfrak{f}}_{\mathfrak{e}},\overrightarrow{V}_i}$ and

$$
\begin{aligned}
&\left[\overrightarrow{\beta}_{\mathfrak{e},i}(t,\mathbf{w})\right](s) \\
&\quad = \int_0^s \left[\mathcal{O}\left(\overrightarrow{\mathfrak{F}}_{\mathfrak{e},i}(t,\mathbf{w})\right)\right](\sigma)^{\top} A\left(t, \left[\overrightarrow{\mathfrak{F}}_{\mathfrak{e},i}(\,\cdot\,,\mathbf{w})\right](\sigma)\right)\mathbf{e}_i\,d\sigma.
\end{aligned}
\tag{1.16}
$$

In particular, if

$$
\begin{aligned}
\left[\overrightarrow{R}_{\mathfrak{e},i}(T,\mathbf{w})\right](s) = \exp\Bigg[&-\int_0^T \left(\left[\overrightarrow{\beta}_{\mathfrak{e},i}(t,\mathbf{w})\right](s), d\mathbf{w}(t)\right)_{\mathbb{R}^d} \\
&- \frac{1}{2}\int_0^T \left|\left[\overrightarrow{\beta}_{\mathfrak{e},i}(,\mathbf{w})\right](s)\right|^2 dt\Bigg],
\end{aligned}
$$

then, by the standard Cameron–Martin–Girsanov theory,

$$\begin{aligned}(1.17)\qquad &\Big(\big[\overrightarrow{\Theta}_{\epsilon,i}(\,\cdot\,,\mathbf{w})\big](s),\big[\overrightarrow{\mathfrak{F}}_{\epsilon,i}(\,\cdot\,,\mathbf{w})\big](s)\Big) \quad \text{under } \overrightarrow{\mu}_{s,\epsilon,i}\\ &\overset{\text{law}}{=} \Big(\mathbf{w},\mathfrak{F}_\epsilon(\,\cdot\,,\mathbf{w})\Big) \quad \text{under } \mu,\end{aligned}$$

where $\overrightarrow{\mu}_{s,\epsilon,i}$ is the unique probability measure on $\mathfrak{W}$ such that, for each $T \in (0,\infty)$,

$$\overrightarrow{\mu}_{s,\epsilon,i}(\Gamma) = \int_\Gamma \big[\overrightarrow{R}_{\epsilon,i}(T,\mathbf{w})\big](s)\,\mu(d\mathbf{w}), \quad \Gamma \in \mathcal{B}_T.$$

Hence, if $F : C([0,T];\mathbb{R}^d) \times C([0,T];\mathcal{O}(M)) \longrightarrow \mathbb{R}$ is a measurable function with the property that

$$s \in \mathbb{R} \longmapsto F\Big(\big[\overrightarrow{\Theta}_{\epsilon,i}(\,\cdot \wedge T)\big](s),\big[\overrightarrow{\mathfrak{F}}_{\epsilon,i}(\,\cdot \wedge T)\big](s)\Big) \in L^2(\mu) \quad \text{is continuous,}$$

then (1.17) says that

$$\frac{d}{ds}\mathbb{E}^\mu\Big[F\Big(\big[\overrightarrow{\Theta}_{\epsilon,i}(\,\cdot \wedge T,\mathbf{w})\big](s),\big[\overrightarrow{\mathfrak{F}}_{\epsilon,i}(\,\cdot \wedge T,\mathbf{w})\big](s)\Big)\big[\overrightarrow{R}_{\epsilon,i}(T,\mathbf{w})\big](s)\Big] = 0$$

and, therefore, that

$$\begin{aligned}(1.18)\qquad &\frac{d}{ds}\mathbb{E}^\mu\Big[F\Big(\big[\overrightarrow{\Theta}_{\epsilon,i}(\,\cdot \wedge T,\mathbf{w})\big](s),\big[\overrightarrow{\mathfrak{F}}_{\epsilon,i}(\,\cdot \wedge T,\mathbf{w})\big](s)\Big)\Big]\Big|_{s=0}\\ &= \mathbb{E}^\mu\left[\int_0^T \Big(A\big(t,\mathfrak{F}_\epsilon(\,\cdot\,,\mathbf{w})\big)\mathbf{e}_i, d\mathbf{w}(t)\Big)_{\mathbb{R}^d} F\big(\mathbf{w},\mathfrak{F}_\epsilon(\,\cdot \wedge T,\mathbf{w})\big)\right].\end{aligned}$$

2. Derivation of the Estimate

Our first application of these considerations is a little ridiculous.[3] Namely, as consequence of (1.13), we know that, for any $f \in C^1(M;\mathbb{R})$ and $T \in (0,\infty)$,

$$\big[\mathfrak{E}_i(P_T f)\circ\pi\big](\mathfrak{e}) = \frac{d}{ds}\mathbb{E}^\mu\Big[f\circ\pi\Big(\big[\overleftarrow{\mathfrak{F}}_{\epsilon,i}(T)\big](s)\Big)\Big]\Big|_{s=0},$$

and therefore, by (1.11),

$$(2.1)\qquad \big[\mathfrak{E}_i(P_T f)\circ\pi\big](\mathfrak{e}) = \mathbb{E}^\mu\Big[\Big(\mathfrak{F}_\epsilon(T)A\big(T,\mathfrak{F}_\epsilon(\,\cdot\,)\big)\mathbf{e}_i\Big)\, f\Big].$$

[3]The reason why the application is somewhat silly is that it can be seen as a more or less immediate consequence of Bochner's identity. This observation was made originally by Bismut in [B] and more recently in [SZ].

Next, we use (1.14) to write

$$T\mathbb{E}^{\mu}\Big[\Big(\mathfrak{F}_{\mathfrak{e}}(T)A(T,\mathfrak{F}_{\mathfrak{e}}(\cdot))\mathbf{e}_i\Big)f\Big] = \frac{d}{ds}\mathbb{E}^{\mu}\Big[f\circ\pi\Big(\big[\overrightarrow{\mathfrak{F}}_{\mathfrak{e},i}(T)\big](s)\Big)\Big]\Big|_{s=0}$$

and then combine this with (2.1) and (1.18) to arrive at

$$\begin{aligned}&T\big[\mathfrak{E}_i(P_T f\circ\pi)\big](\mathfrak{e})\\ &= \mathbb{E}^{\mu}\left[\int_0^T \Big(A(t,\mathfrak{F}_{\mathfrak{e}}(\cdot,\mathbf{w}))\mathbf{e}_i, d\mathbf{w}(t)\Big)_{\mathbb{R}^d} f\circ\pi\big(\mathfrak{F}_{\mathfrak{e}}(T)\big)\right],\end{aligned} \tag{2.2}$$

which is one of the many equivalent forms in which to write Bismut's beautiful formula (cf. (2.71) in [B] or [SZ]) for the logarithmic gradient of $p_t(\cdot,y)$.

Starting from (2.2), it is an easy matter to get the first part of estimate in (0.3). To see this, we review the line of reasoning introduced to pass from (18) to (20) in [SZ]. Namely, an elementary application of Jensen's inequality (cf. Lemma 3.2.13 in [DS]) shows that, for all non-negative $\varphi \in L^1(\mu)$ with integral 1 and all measurable ψ for which $\psi\varphi \in L^1(\mu)$:

$$\int \psi\varphi\,d\mu \le \int \varphi\log\varphi\,d\mu + \log\left[\int e^{\psi}\,d\mu\right]. \tag{2.3}$$

Thus, if $\lambda \in \mathbb{R}$, $f \in C\big(M;(0,\infty)\big)$, and we take

$$\varphi(\mathbf{w}) = \frac{f\circ\pi\big(\mathfrak{F}_{\mathfrak{e}}(T,\mathbf{w})\big)}{P_T f\big(\pi(\mathfrak{e})\big)} \text{ and } \psi(\mathbf{w}) = \lambda\int_0^T \Big(A(t,\mathfrak{F}_{\mathfrak{e}}(\cdot,\mathbf{w}))\mathbf{e}_i, d\mathbf{w}(t)\Big)_{\mathbb{R}^d},$$

we see, from (2.2), that

$$\begin{aligned}&\lambda T\,\frac{\big[\mathfrak{E}_i(P_T f)\circ\pi\big](\mathfrak{e})}{P_T f\big(\pi(\mathfrak{e})\big)}\\ &\le h_T\big(\pi(\mathfrak{e});f\big) + \log\mathbb{E}\left[\exp\left(\lambda\int_0^T \Big(A(t,\mathfrak{F}_{\mathfrak{e}}(\cdot,\mathbf{w}))\mathbf{e}_i, d\mathbf{w}(t)\Big)_{\mathbb{R}^d}\right)\right],\end{aligned}$$

where

$$\begin{aligned}h_T(x;f) &\equiv \int_M \frac{f(\xi)}{P_T f(x)}\log\frac{f(\xi)}{P_T f(x)}\,p_T(x,\xi)\,\lambda_M(d\xi)\\ &\le \max_{\xi\in M}\log\frac{f(\xi)}{P_T f(x)}.\end{aligned} \tag{2.4}$$

But, if

$$\alpha \equiv \inf_{\substack{\mathfrak{e}\in\mathcal{O}(M)\\ |\mathbf{v}|=1}} \Big(\mathbf{v},\mathrm{Ric}(\mathfrak{e})\mathbf{v}\Big)_{\mathbb{R}^d}, \tag{2.5}$$

then

$$\int_0^T \|A(t,\mathfrak{e})\|_{\mathrm{op}}^2\, dt \le E_\alpha(T) \equiv \int_0^T e^{-\alpha t}\, dt, \tag{2.6}$$

and so, by standard estimates for Brownian stochastic integrals,

$$\log \mathbb{E}\left[\exp\left(\lambda \int_0^T \Big(A\big(t,\mathfrak{F}_\mathfrak{e}(\,\cdot\,,\mathbf{w})\big)\mathbf{e}_i, d\mathbf{w}(t)\Big)_{\mathbb{R}^d}\right)\right] \le \frac{\lambda^2}{2}E_\alpha(T).$$

Hence, after minimizing with respect to λ, we find that

$$\frac{\left|\mathfrak{E}_i(P_T f)\circ\pi(\mathfrak{e})\right|}{P_T f(x)} \le T^{-1}\sqrt{2E_\alpha(T)h_T(x;f)} \quad \text{for } T \in (0,\infty) \text{ and } \varphi \in C\big(M;(0,\infty)\big). \tag{2.7}$$

Finally, because (cf. [LY] or [D]) there exists a $\theta \in (0,1)$ such that

$$\frac{\theta}{T^{\frac{d}{2}}}\exp\left[-\frac{\operatorname{dist}(x,y)^2}{\theta T}\right] \le p_T(x,y) \le \frac{1}{\theta T^{\frac{d}{2}}}, \quad \text{for } (T,x,y) \in (0,1] \times M \times M, \tag{2.8}$$

(2.7) with $f = p_T(\,\cdot\,,y)$ yields an estimate like the first one in (0.3), only with $2T$ in the place of T.

Because

$$\big[\mathfrak{E}_i\mathfrak{E}_j \log(f\circ\pi)\big](\mathfrak{e}) = \frac{\big[\mathfrak{E}_i\mathfrak{E}_j f\circ\pi\big](\mathfrak{e})}{f\circ\pi(\mathfrak{e})} - \frac{\big[\mathfrak{E}_i f\circ\pi\big](\mathfrak{e})\big[\mathfrak{E}_j f\circ\pi\big](\mathfrak{e})}{f\circ\pi(\mathfrak{e})^2} \tag{2.9}$$

the second estimate in (0.3) now reduces to showing that

$$\frac{\left|\big[\mathfrak{E}_i\mathfrak{E}_j p_T(\,\cdot\,,y)\big](\mathfrak{e})\right|}{p_T(x,y)} \le C\left(\frac{1}{T} + \frac{\operatorname{dist}(x,y)^2}{T^2}\right) \quad \text{for } T \in (0,1] \text{ and } x = \pi(\mathfrak{e}). \tag{2.10}$$

To this end, set

$$A_\mathfrak{e}(t,\mathbf{w}) = A\big(t,\mathfrak{F}_\mathfrak{e}(t,\mathbf{w})\big), \quad \mathbf{v}_{\mathfrak{e},i}(t,\mathbf{w}) = A_\mathfrak{e}(t,\mathbf{w})\mathbf{e}_i,$$
$$\overleftarrow{\phi}_{\mathfrak{e},i}(t,\mathbf{w}) = \phi\Big(\big[\overleftarrow{\mathfrak{F}}_{\mathfrak{e},i}(t,\mathbf{w})\big]'(0)\Big),$$
$$\text{and } \overleftarrow{A}'_{\mathfrak{e},i}(t,\mathbf{w}) = \frac{d}{ds}A\Big(t,\big[\overleftarrow{\mathfrak{F}}_{\mathfrak{e},i}(\,\cdot\,,\mathbf{w})\big](s)\Big)\Big|_{s=0}.$$

Then, by (2.2) (with $i = j$) and (1.13), we see that

$$\begin{aligned}
&T\big[\mathfrak{E}_i\mathfrak{E}_j(P_T f\circ\pi)\big](\mathfrak{e})\\
&= \frac{d}{ds}\mathbb{E}^{\mu}\Bigg[\int_0^T\Big(A\Big(t,[\overleftarrow{\mathfrak{F}}_{\mathfrak{e},i}(\cdot)](s)\Big)\mathbf{e}_j, d_t[\overleftarrow{B}_{\mathfrak{e},i}(t)](s)\Big)_{\mathbb{R}^d}\\
&\qquad\qquad\times f\Big(\pi\big([\overleftarrow{\mathfrak{F}}_{\mathfrak{e},i}(T)](s)\big)\Big)\Bigg]\Bigg|_{s=0}\\
&= \mathbb{E}^{\mu}\Big[\overleftarrow{G}_{\mathfrak{e},i,j}(T)\, f\circ\pi\big(\mathfrak{F}_{\mathfrak{e}}(T)\big)\Big]\\
&\qquad + \mathbb{E}^{\mu}\left[\int_0^T\Big(\mathbf{v}_{\mathfrak{e},j}(t,\mathbf{w}), d\mathbf{w}(t)\Big)_{\mathbb{R}^d}\Big(\mathfrak{F}_{\mathfrak{e}}(T,\mathbf{w})\big(\mathbf{v}_{\mathfrak{e},i}(T,\mathbf{w})\big)\Big)f\right],
\end{aligned}$$

where

$$\begin{aligned}
\overleftarrow{G}_{\mathfrak{e},i,j}(T,\mathbf{w}) &\equiv \frac{d}{ds}\int_0^T\Big(A(t,[\overleftarrow{\mathfrak{F}}_{\mathfrak{e},i}(\cdot,\mathbf{w})](s))\mathbf{e}_j, d_t[\overleftarrow{B}_{\mathfrak{e},i}(t,\mathbf{w})](s)\Big)_{\mathbb{R}^d}\Big|_{s=0}\\
&= \int_0^T\Big(\overleftarrow{A}'_{\mathfrak{e},i}(t,\mathbf{w})\mathbf{e}_j - \overleftarrow{\phi}_{\mathfrak{e},i}(t,\mathbf{w})\mathbf{v}_{\mathfrak{e},j}(t,\mathbf{w}), d\mathbf{w}(t)\Big)_{\mathbb{R}^d}.
\end{aligned}$$

Finally, set

$$\begin{aligned}
\overrightarrow{\phi}_{\mathfrak{e},i}(t,\mathbf{w}) &= \phi\Big([\overrightarrow{\mathfrak{F}}_{\mathfrak{e},i}(t,\mathbf{w})]'(0)\Big)\\
\text{and } \overrightarrow{A}'_{\mathfrak{e},i}(t,\mathbf{w}) &= \frac{d}{ds}A\Big(t,[\overrightarrow{\mathfrak{F}}_{\mathfrak{e},i}(\cdot,\mathbf{w})](s)\Big)\Big|_{s=0}.
\end{aligned}$$

Then, after another application of (1.18), we see (cf. (1.15) and (1.16)) that

$$\begin{aligned}
&T\mathbb{E}^{\mu}\left[\int_0^T\Big(\mathbf{v}_{\mathfrak{e},j}(t,\mathbf{w}), d\mathbf{w}(t)\Big)_{\mathbb{R}^d}\Big(\mathfrak{F}_{\mathfrak{e}}(T,\mathbf{w})\big(\mathbf{v}_{\mathfrak{e},i}(T,\mathbf{w})\big)\Big)f\right]\\
&\qquad = \mathbb{E}^{\mu}\Big[\Big(-\overrightarrow{G}_{\mathfrak{e},i,j}(T) + \overrightarrow{H}_{\mathfrak{e},i,j}(T)\Big)\, f\circ\pi\big(\mathfrak{F}_{\mathfrak{e}}(T)\big)\Big],
\end{aligned} \tag{2.11}$$

where

$$\begin{aligned}
\overrightarrow{G}_{\mathfrak{e},i,j}(T,\mathbf{w}) &\equiv \int_0^T\Big(\overrightarrow{A}'_{\mathfrak{e},i}(t,\mathbf{w})\mathbf{e}_j + \overrightarrow{\phi}_{\mathfrak{e},i}(t,\mathbf{w})\mathbf{v}_{\mathfrak{e},j}(t,\mathbf{w}), d\mathbf{w}(t)\Big)_{\mathbb{R}^d}\\
\overrightarrow{H}_{\mathfrak{e},i,j}(T,\mathbf{w}) &\equiv \int_0^T\Big(\mathbf{v}_{\mathfrak{e},i}(t,\mathbf{w}), d\mathbf{w}(t)\Big)_{\mathbb{R}^d}\int_0^T\Big(\mathbf{v}_{\mathfrak{e},j}(t,\mathbf{w}), d\mathbf{w}(t)\Big)_{\mathbb{R}^d}\\
&\qquad - \int_0^T\Big(\mathbf{v}_{\mathfrak{e},i}(t,\mathbf{w}), \mathbf{v}_{\mathfrak{e},j}(t,\mathbf{w})\Big)_{\mathbb{R}^d}\, dt.
\end{aligned}$$

Hence, when we combine (2.11) with the preceding, we find that

$$\begin{aligned}(2.12)\qquad &T^2\left[\mathfrak{E}_i\mathfrak{E}_j\left(P_T f\circ\pi\right)\right](\mathfrak{e})\\ &= \mathbb{E}^{\mu}\Big[\Big(T\overleftarrow{G}_{\mathfrak{e},i,j}(T)-\overrightarrow{G}_{\mathfrak{e},i,j}(T)+\overrightarrow{H}_{\mathfrak{e},i,j}(T)\Big)\, f\circ\pi\big(\mathfrak{F}_{\mathfrak{e}}(T)\big)\Big].\end{aligned}$$

In order to pass from (2.12) to the estimate in (2.10), we will need the completely standard results contained in the following lemma.

Lemma 2.13. *Let $\alpha : [0,\infty)\times\mathfrak{W}\longrightarrow\mathbb{R}^d$ be a progressively measurable function satisfying*

$$S(T,\mathbf{w})\equiv\int_0^T |\alpha(t)|^2\,dt<\infty,\quad (T,\mathbf{w})\in[0,\infty)\times\mathfrak{W},$$

and set

$$M(T,\mathbf{w})=\int_0^T\Big(\alpha(t,\mathbf{w}),d\mathbf{w}(t)\Big)_{\mathbb{R}^d}.$$

Then, for all $T\in[0,\infty)$,

$$(2.14)\qquad \mathbb{E}^{\mu}\Big[\exp\big(\lambda M(T)\big)\Big]\le\mathbb{E}^{\mu}\Big[\exp\big(2\lambda^2 S(T)\big)\Big]^{\frac{1}{2}},\quad \lambda\in\mathbb{R}.$$

Moreover, for any $\epsilon>0$,

$$\begin{aligned}(2.15)\qquad &4\epsilon\big\|S(T)\big\|_{L^\infty(\mu)}\le 1\implies\\ &\qquad\mathbb{E}^{\mu}\left[\exp\left(\epsilon\sup_{t\in[0,T]}M(t)^2\right)\right]\le 1+8\epsilon\big\|S(T)\big\|_{L^\infty(\mu)}.\end{aligned}$$

Proof. Without loss in generality, we assume that α is uniformly bounded, in which case

$$\left(\exp\Big(\lambda M(t)-\tfrac{\lambda^2}{2}S(t)\Big),\mathcal{B}_t,\mu\right)$$

is a continuous positive martingale with mean value 1. In particular, by Schwarz's inequality,

$$\begin{aligned}\mathbb{E}^{\mu}\Big[\exp\big(\lambda M(T)\big)\Big]^2&=\mathbb{E}^{\mu}\Big[\exp\big(\lambda M(T)-\lambda^2S(T)\big)\exp\big(\lambda^2S(T)\big]^2\\ &\le\mathbb{E}^{\mu}\Big[\exp\big(2\lambda^2S(T)\big)\Big].\end{aligned}$$

In addition, by Doob's inequality, for each $\lambda>0$,

$$\mu\left(\sup_{t\in[0,T]}M(t)\ge R\right)\le\exp\Big(-\lambda R+\tfrac{\lambda^2T}{2}\big\|S(T)\big\|_{L^\infty(\mu)}\Big),$$

and so, after minimizing with respect to λ and then repeating the argument for $-M(\cdot)$, one gets

$$\mu\left(\sup_{t\in[0,T]}|M(t)|\geq R\right)\leq 2\exp\left(-\frac{R^2}{2\|S(T)\|_{L^\infty(\mu)}}\right),\quad R\in(0,\infty).$$

Thus, (2.15) follows immediately from

$$\mathbb{E}^\mu\Big[\exp(\epsilon X^2)\Big]=1+\epsilon\int_0^\infty e^{\epsilon t}\mu(X^2>t)\,dt.\quad\square$$

As a consequence of (2.14), we see that, for any $\lambda\in\mathbb{R}$,

$$\begin{aligned}&\mathbb{E}^\mu\Big[\exp\Big(\lambda\big(T\overleftarrow{G}_{\mathfrak{e},i,j}(T)-\overrightarrow{G}_{\mathfrak{e},i,j}(T)\big)\Big)\Big]\\ &\qquad\leq\mathbb{E}^\mu\left[\exp\left(2\lambda^2\int_0^T\left|T\overleftarrow{g}_{\mathfrak{e},i,j}(t)-\overrightarrow{g}_{\mathfrak{e},i,j}(t)\right|^2dt\right)\right]^{\frac12},\end{aligned}\tag{2.16}$$

where

$$\begin{aligned}\overleftarrow{g}_{\mathfrak{e},i,j}(t,\mathbf{w})&=\overleftarrow{A}'_{\mathfrak{e},i}(t,\mathbf{w})\mathbf{e}_j+\overleftarrow{\phi}_{\mathfrak{e},i}(t,\mathbf{w})A_{\mathfrak{e}}(t,\mathbf{w})\mathbf{e}_j\\ \overrightarrow{g}_{\mathfrak{e},i,j}(t,\mathbf{w})&=\overrightarrow{A}'_{\mathfrak{e},i}(t,\mathbf{w})\mathbf{e}_j+\overrightarrow{\phi}_{\mathfrak{e},i}(t,\mathbf{w})A_{\mathfrak{e}}(t,\mathbf{w})\mathbf{e}_j.\end{aligned}$$

At the same time, from (1.8), (1.11), and (1.14),

$$\begin{aligned}\overleftarrow{\phi}_{\mathfrak{e},i}(t,\mathbf{w})=&\sum_{\ell,\ell'=1}^d\int_0^t A_{\mathfrak{e}}(\tau,\mathbf{w})_{\ell',i}\Phi_{\ell,\ell'}\big(\mathfrak{F}_{\mathfrak{e}}(\tau,\mathbf{w})\big)\,d\mathbf{w}(\tau)_\ell\\ &+\frac12\sum_{\ell,\ell'=1}^d\int_0^t A_{\mathfrak{e}}(\tau,\mathbf{w})_{\ell',i}\mathfrak{E}_\ell\Phi_{\ell,\ell'}\big(\mathfrak{F}_{\mathfrak{e}}(\tau,\mathbf{w})\big)\,d\tau\end{aligned}$$

and

$$\begin{aligned}\overrightarrow{\phi}_{\mathfrak{e},i}(t,\mathbf{w})=&\sum_{\ell,\ell'=1}^d\int_0^t \tau A_{\mathfrak{e}}(\tau,\mathbf{w})_{\ell',i}\Phi_{\ell,\ell'}\big(\mathfrak{F}_{\mathfrak{e}}(\tau,\mathbf{w})\big)\,d\mathbf{w}(\tau)_\ell\\ &+\frac12\sum_{\ell,\ell'=1}^d\int_0^t \tau A_{\mathfrak{e}}(\tau,\mathbf{w})_{\ell',i}\mathfrak{E}_\ell\Phi_{\ell,\ell'}\big(\mathfrak{F}_{\mathfrak{e}}(\tau,\mathbf{w})\big)\,d\tau\end{aligned}$$

Finally, starting from (1.10), one finds that

$$\begin{aligned}-2\frac{d}{dt}\overleftarrow{A}'_{\mathfrak{e},i}(t,\mathbf{w})&=\frac{d}{ds}\Big(\mathrm{Ric}\Big(\big[\overleftarrow{\mathfrak{F}}_{\mathfrak{e},i}(t,\mathbf{w})\big](s)\Big)A\Big(t,\big[\overleftarrow{\mathfrak{F}}_{\mathfrak{e},i}(\cdot,\mathbf{w})\big](s)\Big)\Big)\Big|_{s=0}\\ &=\Big(\mathfrak{E}\big(\mathbf{v}_{\mathfrak{e},i}(t,\mathbf{w})\big)\mathrm{Ric}\big(\mathfrak{F}_{\mathfrak{e}}(t,\mathbf{w})\big)+\Big[\overleftarrow{\phi}_{\mathfrak{e},i}(t,\mathbf{w}),\mathrm{Ric}\big(\mathfrak{F}_{\mathfrak{e}}(t,\mathbf{w})\big)\Big]\Big)A_{\mathfrak{e}}(t,\mathbf{w})\\ &\qquad+\mathrm{Ric}\big(\mathfrak{F}_{\mathfrak{e}}(t,\mathbf{w})\big)\overleftarrow{A}'_{\mathfrak{e},i}(t,\mathbf{w}),\end{aligned}$$

and therefore

$$\overleftarrow{A}'_{\mathfrak{e},i}(t,\mathbf{w}) = -\frac{1}{2}A_{\mathfrak{e}}(t,\mathbf{w})\int_0^t A_{\mathfrak{e}}(\tau,\mathbf{w})^{-1}\Big(\mathfrak{E}\big(\mathbf{v}_{\mathfrak{e},i}(\tau,\mathbf{w})\big)\mathrm{Ric}\big(\mathfrak{F}_{\mathfrak{e}}(\tau,\mathbf{w})\big) + \Big[\overleftarrow{\phi}_{\mathfrak{e},i}(\tau,\mathbf{w}),\mathrm{Ric}\big(\mathfrak{F}_{\mathfrak{e}}(\tau,\mathbf{w})\big)\Big]\Big)A_{\mathfrak{e}}(\tau,\mathbf{w})\,d\tau.$$

Similarly,

$$\overrightarrow{A}'_{\mathfrak{e},i}(t,\mathbf{w}) = -\frac{1}{2}A_{\mathfrak{e}}(t,\mathbf{w})\int_0^t A_{\mathfrak{e}}(\tau,\mathbf{w})^{-1}\Big(\tau\mathfrak{E}\big(\mathbf{v}_{\mathfrak{e},i}(\tau,\mathbf{w})\big)\mathrm{Ric}\big(\mathfrak{F}_{\mathfrak{e}}(\tau,\mathbf{w})\big) + \Big[\overrightarrow{\phi}_{\mathfrak{e},i}(\tau,\mathbf{w}),\mathrm{Ric}\big(\mathfrak{F}_{\mathfrak{e}}(\tau,\mathbf{w})\big)\Big]\Big)A_{\mathfrak{e}}(\tau,\mathbf{w})\,d\tau.$$

Hence, after combining these, using (2.16), and applying (2.15), we conclude that there exists an $\epsilon > 0$ such that

$$\mathbb{E}^{\mu}\Big[\exp\Big(\lambda\big(T\overleftarrow{g}_{\mathfrak{e},i,j}(t) - \overrightarrow{g}_{\mathfrak{e},i,j}(t)\big)\Big)\Big] \le 1 + \lambda^2T^4 \quad \text{for } 1 \le i,j \le d,\ 0 \le t \le T, \text{ and } |\lambda| \le \epsilon T^{-2}$$

Now let $f \in C^{\infty}\big(M;(0,\infty)\big)$ be given. By (2.3) with

$$\varphi(\mathbf{w}) = \frac{f\circ\pi\big(\mathfrak{F}_{\mathfrak{e}}(T,\mathbf{w})\big)}{P_Tf\big(\pi(\mathfrak{e})\big)} \text{ and } \psi(\mathbf{w}) = \lambda\Big(T\overleftarrow{G}_{\mathfrak{e},i,j}(T,\mathbf{w}) - \overrightarrow{G}_{\mathfrak{e},i,j}(T,\mathbf{w})\Big),$$

we can find an $\epsilon > 0$ for which (cf. (2.4))

$$\lambda\frac{\mathbb{E}^{\mu}\Big[\Big(T\overleftarrow{G}_{\mathfrak{e},i,j}(T) - \overrightarrow{G}_{\mathfrak{e},i,j}(T)\Big)f\circ\pi\big(\mathfrak{F}_{\mathfrak{e}}(T)\big)\Big]}{P_Tf\big(\pi(\mathfrak{e})\big)} \le h_T\big(\pi(\mathfrak{e});f\big) + \lambda^2T^4$$

when $T \in (0,1]$ and $|\lambda| \le \epsilon T^{-2}$. Hence, after taking $\lambda = \pm\epsilon T^{-2}$ and using (2.12), we have now shown that there is a $C \in (0,\infty)$ such that

$$\frac{\Big|\big[\mathfrak{E}_i\mathfrak{E}_j(P_Tf\circ\pi)\big](\mathfrak{e}) - T^{-2}\mathbb{E}^{\mu}\Big[H_{\mathfrak{e},i,j}(T)f\circ\pi\big(\mathfrak{F}_{\mathfrak{e}}(T)\big)\Big]\Big|}{P_Tf\big(\pi(\mathfrak{e})\big)} \le C\Big(1 + h_T\big(\pi(\mathfrak{e});f\big)\Big) \quad \text{for } 1 \le i,j \le d \text{ and } (T,\mathfrak{e}) \in (0,1]\times\mathcal{O}(M). \tag{2.17}$$

Next, define $\mathbf{Q}_Tf : \mathcal{O}(M) \longrightarrow \mathbb{R}^d\otimes\mathbb{R}^d$ by

$$\big[\mathbf{Q}_Tf\big](\mathfrak{e})_{i,j} = \frac{\mathbb{E}^{\mu}\Big[H_{\mathfrak{e},i,j}(T)f\circ\pi\big(\mathfrak{F}_{\mathfrak{e}}(T)\big)\Big]}{P_Tf\big(\pi(\mathfrak{e})\big)}.$$

Since, for $\mathbf{v} \in \mathbb{R}^d$, Itô's formula says that

$$\sum_{i,j=1}^{d} H_{\mathfrak{e},i,j}(T,\mathbf{w})\mathbf{v}_i\mathbf{v}_j = \left(\int_0^T \left(A_{\mathfrak{e}}(t,\mathbf{w})\mathbf{v}, d\mathbf{w}(t)\right)_{\mathbb{R}^d}\right)^2 - \int_0^T \left|A_{\mathfrak{e}}(t,\mathbf{w})\mathbf{v}\right|^2 dt$$
$$= 2\int_T \left(M(t,\mathbf{w})A_{\mathfrak{e}}(t,\mathbf{w})\mathbf{v}, d\mathbf{w}(t)\right)_{\mathbb{R}^d},$$

where

$$M(t,\mathbf{w}) \equiv \int_0^t \left(A_{\mathfrak{e}}(\tau,\mathbf{w}), d\mathbf{w}(\tau)\right)_{\mathbb{R}^d},$$

we see that

$$\left(\mathbf{v}, [\mathbf{Q}_T f](\mathfrak{e})\mathbf{v}\right)_{\mathbb{R}^d} = \frac{2\mathbb{E}^\mu\left[\int_0^T \left(M(t,\mathbf{w})A_{\mathfrak{e}}(t,\mathbf{w})\mathbf{v}, d\mathbf{w}(t)\right)_{\mathbb{R}^d} f\circ\pi\left(\mathfrak{F}_{\mathfrak{e}}(T)\right)\right]}{P_T f(\pi(\mathfrak{e}))}.$$

Hence, after an application of (2.3) and then (2.14) followed by (2.15), we can find $C \in (0,\infty)$ such that

$$\left|\left(\mathbf{v}, [\mathbf{Q}_T f](\mathfrak{e})\mathbf{v}\right)_{\mathbb{R}^d}\right| \le CT|\mathbf{v}|^2\left(1 + h_T(\pi(\mathfrak{e}); f)\right), \quad T \in (0,1] \tag{2.18}$$

Finally, by combining (2.12), (2.17), and (2.18), we arrive at first

$$\frac{\left|[\mathfrak{E}_i\mathfrak{E}_j(P_T f\circ\pi)](\mathfrak{e})\right|}{P_T f(\pi(\mathfrak{e}))} \le \frac{C}{T}\left(1 + h_T(\pi(\mathfrak{e}); f)\right)$$
$$\text{for } f \in C\big(M;(0,\infty)\big) \text{ and } (T,\mathfrak{e}) \in (0,1]\times\mathcal{O}(M).$$

and then, after taking $f = p_T(\,\cdot\,, y)$, using (2.8), and re-adjusting T, the estimate in (2.10).

3. Concluding Remarks

Many of the ideas employed in this article stem from an abortive attempt by the author and Ofer Zeitouni to find a stochastic route to the basic estimates in [LY]. In particular, we wanted to prove the Li–Yau estimate which says that, for any $\epsilon > 0$ and $f \in C\big([0,\infty); M\big)$,

$$\frac{|\nabla P_T f|^2}{(P_T f)^2} - (1+\epsilon)\frac{\Delta P_T f}{P_T f} \le d(1+\epsilon)^2\left(\frac{1}{T} + \frac{\alpha^- T}{4\epsilon}\right),$$

where α is the number in (2.5) and $\alpha^- = -(\alpha \wedge 0)$ is the negative part of α. Since (cf. (2.9))

$$\frac{|\nabla P_T f|^2}{(P_T f)^2} - \frac{\Delta P_T f}{P_T f} = -\mathrm{Trace}\Big(\mathrm{Hess}(\log P_T f)\Big),$$

their estimate is closely related to a lower bound on $\mathrm{Hess}(\log P_T f)$ as a symmetric matrix.

As the astute reader may have noticed, we can get a better lower bound on $\mathrm{Hess}(\log P_T f)$ than the one which follows trivially from the second part of (0.3). Namely, by (2.2), (2.9), (2.12), (2.17), and Schwarz's inequality:

$$\begin{aligned}
T^2\Big(\mathbf{v}, [\mathrm{Hess}(\log P_T f)](\mathfrak{e})\mathbf{v}\Big) &+ \frac{\mathbb{E}^{\mu}\left[\int_0^T \left|A_{\mathfrak{e}}(t,\mathbf{w})\mathbf{v}\right|^2 dt\, f(\pi(\mathfrak{F}_{\mathfrak{e}}(T))\right]}{P_T f(\pi(\mathfrak{e}))} \\
\geq \frac{\mathbb{E}^{\mu}\left[\left(\int_0^T \Big(A_{\mathfrak{e}}(t,\mathbf{w})\mathbf{v}, d\mathbf{w}(t)\Big)_{\mathbb{R}^d}\right)^2 f(\pi(\mathfrak{F}_{\mathfrak{e}}(T))\right]}{P_T f(\pi(\mathfrak{e}))} & \\
-\left(\frac{\mathbb{E}^{\mu}\left[\int_0^T \Big(A_{\mathfrak{e}}(t,\mathbf{w})\mathbf{v}, d\mathbf{w}(t)\Big)_{\mathbb{R}^d} f(\pi(\mathfrak{F}_{\mathfrak{e}}(T))\right]}{P_T f(\pi(\mathfrak{e}))}\right)^2 & \\
- CT^2\Big(1 + h_T(\pi(\mathfrak{e}); f)\Big)|\mathbf{v}|^2 & \\
\geq -CT^2\Big(1 + h_T(\pi(\mathfrak{e}); f)\Big)|\mathbf{v}|^2 &
\end{aligned}$$

for $(T, \mathfrak{e}) \in (0,1] \times \mathcal{O}(M)$. Hence (cf. (2.6))

$$[\mathrm{Hess}(\log P_T f)](\mathfrak{e}) \geq -\left(\frac{E_{\alpha}(T)}{T^2} + C\Big(1 + h_T(\pi(\mathfrak{e}); f)\Big)\right)\mathbf{I} \tag{3.1}$$

$$\text{for } f \in C\big(M; (0,\infty)\big) \text{ and } (T, \mathfrak{e}) \in (0,1] \times \mathcal{O}(M).$$

The estimate in (3.1) is closely related to one given by R. Hamilton in [H]. In fact, Hamilton's basic estimate (cf. Theorem 4.3 in [H] and take into account the difference in time scale on which his heat equation is running)

$$[\mathrm{Hess}(\log P_T f)](\mathfrak{e}) \geq -\frac{1}{T} - C\Big(1 - \log(T P_T f(\pi(\mathfrak{e}))\Big)\mathbf{I} \tag{3.2}$$

$$\text{for } f \in C\big(M; (0,\infty)\big) \text{ with } \int_M f\, d\lambda_M \leq 1 \text{ and } (T, \mathfrak{e}) \in (0,1] \times \mathcal{O}(M),$$

is an easy corollary of (3.1) and elementary estimates on the operator $\{P_t : t > 0\}$. To be precise, as is well-known and easily derived, for any $f \in C(M;(0,\infty))$, $\int_M P_t f \, d\lambda_M = \int_M f \, d\lambda_M$ for every $t \in (0,\infty)$ and

$$P_t f \le B t^{-\frac{d}{2}} \int_M f \, d\lambda_M, \quad t \in (0,1] \quad \text{and } f \in C(M;(0,\infty))$$

for some $B < \infty$. Hence, by taking $f_t = P_t f$ and using $f_T = P_{T-t} f$, (3.1) says that, for each $t \in (0,T)$ and $f \in C(M;(0,\infty))$ with $\int f \, d\lambda_M \le 1$,

$$\left[\text{Hess}(\log P_T f)\right](\mathfrak{e}) \ge -\frac{E_\alpha(T-t)}{(T-t)^2} - C\Big(1 - \log\big(B t^{\frac{d}{2}} P_T f(\pi(\mathfrak{e}))\big)\Big)\mathbf{I}.$$

Finally, take $t = \frac{T^2}{2}$ and make some elementary estimation to conclude that, with an appropriate choice of $C < \infty$, (3.2) follows.

It is significant that the lower bound in (3.2) does not have a complementary upper bound. In fact, as pointed out to the author by P. Malliavin, problems will occur at the cut locus. To wit, consider the unit length circle, for which

$$p_T(x,y) = (2\pi T)^{-\frac{1}{2}} \sum_{m\in\mathbb{Z}} \exp\left[-\frac{(y-x-m)^2}{2T}\right], \quad x,y \in \left(-\tfrac{1}{2}, \tfrac{1}{2}\right].$$

Then, after a simple computation, one discovers that

$$\lim_{T \searrow 0} T^2 \frac{\partial^2 p_T\left(\eta, \frac{1}{2}\right)}{\partial \eta^2}(0) = \frac{1}{4}.$$

As one might expect, the $\frac{1}{T}$ estimate does hold outside the cut locus, but to see this one needs a somewhat different line of reasoning (cf. [MS]).

References

[B] Bismut, J.M., *Large Deviations and the Malliavin Calculus*, Progress in Math. #45, Birkhäuser, Cambridge USA, 1984.

[D] Davies, E.B., *Heat Kernels and Spectral Theory*, Cambridge Tracts in Math. #92, Cambridge Univ. Press, Cambridge U.K. & NY, USA, 1989.

[ES1] Enchev, O. & Stroock, D., *Towards a Riemannian geometry on the path space over a Riemannian manifold* (to appear in J. Fnal. Anal.).

[ES2] ———, *Pinned Brownian motion and its perturbations* (to appear in Advances in Math.).

[DS] Deuschel, J.D. & Stroock, D., *Large Deviations*, Pure and Appl. Math. #137, Academic Press, San Diego USA, 1989.

[H] Hamilton, R., *A matrix Harnack estimate for the heat equation*, Comm. Anal. & Geom. **1** #1 (1993), 113–126.

[LY] Li, P. & Yau, S.T., *On the parabolic kernel of the Schrödinger operator*, Acta Math. **156** (1986), 153–201.

[MS] Malliavin, P. & Stroock, D., *Short time behavior of the heat kernel and its logarithmic derivatives* (to appear).

[S] Sheu, S.-T., *Estimates of the transition density of a non-degenerate diffusion Markov process*, Ann. Prob. **19** #2 (1991), 538–561.

[SZ] Stroock, D. and Zeitouni, O., *Variations on a theme by Bismut* (to appear).

Environment-wise central limit theorem for a diffusion in a Brownian environment with large drift

Hiroshi Tanaka

Department of Mathematics, Faculty of Science and Technology, Keio University, Hiyoshi, Yokohama, Kanagawa 223, Japan

Introduction

General theory of one-dimensional diffusion process was established by Itô and McKean ([1]) more than thirty years ago. In this paper we discuss the central limit theorem concerning a diffusion process in one-dimensional Brownian environment with large drift. Many of our methods benefit from the theory by Itô and McKean.

By a diffusion process in a Brownian environment with drift we mean a process $\mathbf{X}^{x_0} = \{\omega(t), t \geq 0, \mathcal{P}^{x_0}\}$ which is defined in the following way.
(i) W is the space of continuous functions on $\mathbb{R}$ vanishing at 0 and is endowed with the Wiener measure P. For an element w of W, w_κ denotes an element of W defined by $w_\kappa(x) = w(x) - \kappa x/2$ where κ is a given positive constant.
(ii) For $w \in W$ and $x_0 \in \mathbb{R}$, $P_w^{x_0}$ denotes the probability measure on $\Omega = C[0,\infty)$ such that $\mathbf{X}_w^{x_0} = \{\omega(t), t \geq 0, P_w^{x_0}\}$ is a diffusion process with generator

$$\mathcal{L}_w = \frac{1}{2} e^{w_\kappa(x)} \frac{d}{dx}\left(e^{-w_\kappa(x)} \frac{d}{dx}\right)$$

starting from x_0; $\omega(t)$ denotes the value of a trajectory $\omega(\in \Omega)$ at time t.
(iii) $\{\omega(t), t \geq 0\}$ is regarded as a process defined on the probability space $(W \times \Omega, \mathcal{P}^{x_0})$ where $\mathcal{P}^{x_0}(dwd\omega) = P(dw)P_w^{x_0}(d\omega)$. This process is what we denote by $\mathbf{X}^{x_0} = \{\omega(t), t \geq 0, \mathcal{P}^{x_0}\}$; x_0 indicates the starting point of our process.

The process $\mathbf{X}^{x_0}$ is a diffusion analogue of what was considered by Kesten-Kozlov-Spitzer [4] and Solomon [6] as a random walk in a random environment. Asymptotic behavior of the process $\mathbf{X}^0$ as t becomes large was investigated by Kawazu-Tanaka [3]. The results are analogous to those of [4][6] and vary with κ. In particular, when $\kappa > 1$, it was proved that

$$\lim_{x\to\infty} T_x/x = m, \quad \lim_{t\to\infty} \omega(t)/t = 1/m, \qquad \mathcal{P}^0\text{-a.s.},$$

where $m = 4(\kappa - 1)^{-1}$ and T_x is the hitting time: $T_x = \inf\{t > 0 : \omega(t) = x\}$.

The next problem is then to study a fluctuation. In this paper we assume $\kappa > 2$ and discuss central limit theorems concerning T_x and $\omega(t)$. We put

$$M_x = 2\int_0^x dy \int_{-\infty}^y e^{w_\kappa(y)-w_\kappa(z)}dz,$$

$$\mu(t) = \text{ the inverse function of } M_x,$$

$$\overline{\omega}(t) = \max\{\omega(s) : 0 \le s \le t\},\ \underline{\omega}(t) = \inf\{\omega(s) : s \ge t\},$$

$$A = 64(\kappa - 1)^{-2}(\kappa - 2)^{-1}.$$

Then it will be seen that $E_w^0\{T_x\} = M_x$, $E\{M_x\} = mx$ and $E\{\mathrm{Var}_w^0(T_x)\} = Ax$ for $x \ge 0$ where E_w^0 and Var_w^0 denotes the expectation and variance under P_w^0, respectively.

Our results are the following where $\kappa > 2$ is assumed.

Environment-wise central limit theorem

(I) *For almost all w with respect to P the process*

$$\left\{\frac{T_{\lambda x} - M_{\lambda x}}{\sqrt{A\lambda}}, x \ge 0, P_w^0\right\}$$

converges in law to a Brownian motion as $\lambda \to \infty$ (in the sense of convergence of probability measures on the Skorohod space).

(II) *For almost all w the process*

$$\left\{\frac{\omega(\lambda t) - \mu(\lambda t)}{\sqrt{m^{-3}A\lambda}}, t \ge 0, P_w^0\right\}$$

converges in law to a Brownian motion as $\lambda \to \infty$. The same is true when $\omega(\lambda t)$ is replaced by either of $\overline{\omega}(\lambda t)$ and $\underline{\omega}(\lambda t)$.

Central limit theorem in random environment, namely CLT under $\mathcal{P}^0$ will be discussed in another occasion.

§1. Some preliminaries

1.1. We compute the first and the second moments of various quantities related to hitting times for $\mathbf{X}^0$. Let $T_{a,b} = \inf\{t > 0 : \omega(t) = a \text{ or } b\}, b < a$. Then it is well-known that

$$u(x) = E_w^x\{\exp(-\lambda \int_0^{T_{a,b}} f(\omega(t))dt)\}$$

satisfies (e.g. see [1])

$$(1.1) \qquad \begin{cases} (\lambda f - \mathcal{L}_w)u = 0 & \text{in } (b,a), \\ u(a) = u(b) = 1. \end{cases}$$

For $n \ge 0$ we put $u_n(x) = E_w^x\{(\int_0^{T_{a,b}} f(\omega(t))dt)^n\}$. Then $u(x) = \sum_{n=0}^\infty u_n(x) \times (-\lambda)^n/n!$ and (1.1) imply $\sum_{n=0}^\infty (\lambda f - \mathcal{L}_w)u_n(-\lambda)^n/n! = 0$. Therefore we have $u_0 \equiv 1$ and for $n \ge 1$

$$(1.2)\qquad \begin{cases} \mathcal{L}_w u_n = -n u_{n-1} f & \text{in } (b,a), \\ u_n(a) = u_n(b) = 0. \end{cases}$$

Let $S_w(x) = \int_0^x \exp\{w_\kappa(y)\}dy$, the canonical scale of $\mathcal{L}_w$. Then

$$(1.3)\qquad S_w(-\infty) = -\infty, \quad P\text{-a.s., if } \kappa > 0.$$

The following formulas (1.4) and (1.5), which hold for almost all w, can be proved under the assumption $\kappa > 0$ by making use of (1.2) and (1.3): For $x < a$,

$$(1.4)\qquad E_w^x\{T_a\} = 2\int_x^a e^{w_\kappa(y)}dy \int_{-\infty}^y e^{-w_\kappa(z)}dz;$$

$$(1.5)\quad E_w^x\{T_a^2\} = 8\int_x^a e^{w_\kappa(y)}dy \int_{-\infty}^y e^{-w_\kappa(z)}dz \int_z^a e^{w_\kappa(u)}du \int_{-\infty}^u e^{-w_\kappa(v)}dv.$$

Making use of the above formulas we can compute $\mathrm{Var}_w^0\{T_x\}$ for $x > 0$; the result is

$$(1.6)\quad \mathrm{Var}_w^0\{T_x\} = 8\int_0^x dy \int_{-\infty}^y e^{w_\kappa(y)-w_\kappa(z)}dz \int_z^y dy \int_{-\infty}^u e^{w_\kappa(u)-w_\kappa(v)}dv.$$

It will be useful for our later discussions to introduce a one-parameter family of measure preserving transformations $\theta_t, t \in \mathbb{R}$, on (W,P) defined by $(\theta_t w)(x) = w(x+t) - w(t), x \in \mathbb{R}$. It is easy to see that $\theta_t\theta_s = \theta_{t+s}$ and $\{\theta_t\}$ is ergodic. If $\kappa > 0$, then

$$(1.7)\qquad f_0(w) = \int_{-\infty}^0 e^{-w_\kappa(t)}dt$$

is finite (P-a.s.) and $\theta_t f_0 \equiv f_0(\theta_t w) = \int_{-\infty}^t e^{w_\kappa(t)-w_\kappa(s)}ds$. From (1.4) and (1.6) we have for $x > 0$

$$(1.8)\qquad E_w^0\{T_x\} = M_x = 2\int_0^x \theta_y f_0 dy,$$

$$(1.9)\quad \begin{aligned} \mathrm{Var}_w^0\{T_x\} &= 8\int_0^x dy \int_{-\infty}^y e^{w_\kappa(y)-w_\kappa(z)}(\theta_z f_0)^2 dz \\ &= 8\int_0^x \theta_y g dy, \quad \left(g(w) = \int_{-\infty}^0 e^{-w_\kappa(t)}(\theta_t f_0)^2 dt\right). \end{aligned}$$

Making use of $E\{\exp(w_\kappa(x) - w_\kappa(y))\} = \exp\{-\gamma(x-y)\}$ for $x > y$ where

$$\gamma = (\kappa - 1)/2,$$

we have $E\{f_0\} = 1/\gamma$ and hence $E\{M_x\} = mx$, $x \geq 0$, if $\kappa > 1$ (note that $E\{f_0\} < \infty$ iff $\kappa > 1$ and $E\{{f_0}^2\} < \infty$ iff $\kappa > 2$). From now on we assume $\kappa > 2$. The constant A given in the introduction is expressed as $A = 16\gamma^{-2}(2\gamma-1)^{-1}$. We put $f = f_0 - \gamma^{-1}$. Then we have the following:

$$E\{f_0{}^2\} = 2\gamma^{-1}(2\gamma-1)^{-1}; \tag{1.10}$$

$$E\{f\theta_x f\} = \gamma^{-2}(2\gamma-1)^{-1}e^{-\gamma x} \quad \text{for } x \geq 0; \tag{1.11}$$

$$E\{\mathrm{Var}_w^0(T_x)\} = Ax \quad \text{for } x \geq 0. \tag{1.12}$$

1.2. The process $\{\theta_x f_0, x \in \mathbb{R}, P\}$ will often play an important role in our discussions. We make use of habitual notation t (instead of x) to indicate time. By an application of Itô's formula we have

$$d\theta_t f_0 = \theta_t f_0 dw(t) - (\gamma\theta_t f_0 - 1)dt, \tag{1.13}$$

and hence $\theta_t f_0$ is a stationary diffusion process obtained as the unique stationary positive solution of the stochastic differential equation. It can be also written as

$$\theta_t f_0 - f_0 = \int_0^t \theta_s f_0 dw(s) - \gamma \int_0^t \theta_s f ds. \tag{1.14}$$

From now on we assume $\kappa > 2$. Then we can easily prove that $E\{f_0{}^{2\delta}\} < \infty$ if $0 \leq \delta < \kappa/2$. Therefore by the Burkholder-Davis-Gundy inequalities we have

$$E\left\{\max_{0\leq s\leq t}\left|\int_0^s \theta_s f_0 dw(s)\right|^{2\delta}\right\} \leq \text{const.}t^\delta, \quad (t \geq 0, 1 \leq \delta < \kappa/2), \tag{1.15}$$

where const. means a constant that may depend on δ but not on t; such const. will also appear in later discussions and may vary from place to place.

Lemma 1. *$t^{-1/2}\max\{\theta_s f_0 : |s| \leq t\} \to 0$ as $t \to \infty$, P-a.s.*

Proof. Since the stationary diffusion $\theta_t f_0$ is reversible, it is enough to prove that $t^{-1/2}\max\{\theta_s f_0 : 0 \leq s \leq t\} \to 0$ as $t \to \infty$, P-a.s. Define stopping times σ_n, $n \geq 0$, by $\sigma_0 = \inf\{t > 0 : \theta_t f_0 = 2\}$ and $\sigma_n =$ the time of first return of $\theta_t f_0$ to 2 visiting 1 after σ_{n-1} $(n \geq 1)$ and then consider the random variables $X_n = \max\{\theta_t f_0 : \sigma_{n-1} \leq t \leq \sigma_n\}, n \geq 1$. Then $X_n, n \geq 1$, are i.i.d. random variables. Since $\sigma_n/n \to$ const.> 0 (a.s.),

$$\begin{aligned}\varlimsup_{t\to\infty} t^{-1/2}\max_{0\leq s\leq t}\theta_s f_0 &= \varlimsup_{n\to\infty}\sigma_n{}^{-1/2}\max_{1\leq k\leq n}X_k \\ &= \text{const.}\varlimsup_{n\to\infty} n^{-1/2}\max_{1\leq k\leq n}X_k, \quad \text{a.s.,}\end{aligned}$$

and the rightmost hand of the above equals to 0, a.s., because

$$P\{X_1 > x\} = \{S(2) - S(1)\}\{S(x) - S(1)\}^{-1} \sim \text{const.}x^{-\kappa}, x \to \infty,$$

(here $S(x)$ is the canonical scale of the diffusion $\theta_t f_0$; it is given by $S'(x) = x^{\kappa-1}\exp(2/x), x > 0$). This proves the lemma.

Lemma 2. *For any* $c_1 > 0$

$$M_{t+u} - M_t = mu(1 + o(1)) + o(\sqrt{\lambda}), \quad |t| \le c_1\lambda, \; u \in \mathbb{R},$$

where $o(1)$ *represents a general term that tends to* 0 *as* $\lambda \to \infty$ *uniformly in* (t,u) *such that* $|t| \le c_1\lambda$ *and* $u \in \mathbb{R}$, *for almost all* w; $o(\sqrt{\lambda})$ *is a term that can be expressed as* $o(1)\sqrt{\lambda}$.

Proof. Step1. For any positive constants c_1 and c_2

$$\sup \left\{ \left| \frac{M_{t+u} - M_t}{u} - m \right| : |t| \le c_1\lambda, |u| \ge c_2\sqrt{\lambda} \right\} \to 0 \text{ as } \lambda \to \infty, \; P\text{-a.s.} \tag{1.16}$$

We give a proof in the case $c_1 = c_2 = 1$ since the case of general c_1 and c_2 can be treated without any essential change in the proof. We use the notation $\sup_{(\lambda)}$ to denote supremum taken over all t, u satisfying $|t| \le \lambda$ and $u \ge \sqrt{\lambda}$. Since $M_t'/2 = \theta_t f_0$ is a stationary reversible process, it is enough to prove that

$$\sup_{(\lambda)} \left| \frac{1}{u} \int_t^{t+u} \theta_s f \, ds \right| \to 0 \text{ as } \lambda \to \infty, \text{ a.s.} \tag{1.17}$$

By Lemma 1 we have $\sup_{(\lambda)} u^{-1} |\theta_t f_0 - \theta_{t+u} f_0| \to 0$ as $\lambda \to \infty$, a.s. Combining this fact with (1.14) we see that, for the proof of (1.17), it is enough to show that

$$\sup_{(\lambda)} \left| \frac{1}{u} \int_t^{t+u} \theta_s f_0 dw(s) \right| \to 0 \text{ as } \lambda \to \infty, \text{ a.s.} \tag{1.18}$$

To prove (1.18) we take δ and α such that

$$1 < \delta < \kappa/2, \quad 0 < \alpha < 1/2, \quad \delta/2 + \alpha - 1 > 0, \tag{1.19}$$

and prepare the estimate

$$E\{R(\lambda)^{2\delta}\} \le \text{const.} \lambda^{-\delta_1}, \tag{1.20}$$

where $\delta_1 = \delta/2 + \alpha - 1$ and

$$R(t,u) = \frac{1}{u} \int_t^{t+u} \theta_s f_0 dw(s), \quad R(\lambda) = \sup_{(\lambda)} |R(t,u)| .$$

Putting $I_k = [k\lambda^\alpha, (k+1)\lambda^\alpha]$ for an integer k and denoting by $\sup_{(k,\lambda)}$ the supremum over all $t \in I_k$ and $u \ge \sqrt{\lambda}$, we write

$$\begin{aligned} &\sup_{(k,\lambda)} |R(t,u) - R(k\lambda^\alpha, u)| \\ = \; &\sup_{(k,\lambda)} \frac{1}{u} \left| - \int_{k\lambda^\alpha}^{t} \theta_s f_0 dw(s) + \int_{k\lambda^\alpha + u}^{t+u} \theta_s f_0 dw(s) \right| \le U + V, \end{aligned}$$

where

$$U = \lambda^{-1/2} \max_{0\le r\le \lambda^\alpha} \left| \int_{k\lambda^\alpha}^{k\lambda^\alpha + r} \theta_s f_0 dw(s) \right|,$$

$$V = \sup_{u\ge\sqrt{\lambda}} \max_{0\le r\le \lambda^\alpha} \left| \frac{1}{u} \int_{k\lambda^\alpha+u}^{k\lambda^\alpha + u + r} \theta_s f_0 dw(s) \right|.$$

Making use of (1.15) we have

(1.21) $$E\{U^{2\delta}\} \le \text{const.}\lambda^{-(1-\alpha)\delta},$$

and $E\{V^{2\delta}\}$ is dominated by

$$\sum_{n=1}^{\infty} E\left\{ \sup_{\sqrt{\lambda}n\le u\le\sqrt{\lambda}(n+1)} \max_{0\le r\le\lambda^\alpha} \left| \frac{1}{u} \int_{k\lambda^\alpha+u}^{k\lambda^\alpha+u+r} \theta_s f_0 dw(s) \right|^{2\delta} \right\}$$
$$\le \lambda^{-\delta} \sum_{n=1}^{\infty} n^{-2\delta} E\{R_\lambda^{2\delta}\} = \text{const.}\lambda^{-\delta} E\{R_\lambda^{2\delta}\},$$

where

$$R_\lambda = \sup_{0\le u\le\sqrt{\lambda}} \max_{0\le r\le\lambda^\alpha} \left| \int_u^{u+r} \theta_s f_0 dw(s) \right|.$$

If $0 \le u \le \sqrt{\lambda}$ and $0 \le r \le \lambda^\alpha$, then u and $u+r$ are simultaneously contained in one of the intervals $[l\lambda^\alpha, l\lambda^\alpha + 2\lambda^\alpha]$, $l = 0, 1, \dots, l_\lambda$, where $l_\lambda = [\lambda^{\frac{1}{2}-\alpha}]$. Therefore

$$R_\lambda \le 2 \max_{0\le l\le l_\lambda} \max_{0\le r\le 2\lambda^\alpha} \left| \int_{l\lambda^\alpha}^{l\lambda^\alpha+r} \theta_s f_0 dw(s) \right|,$$

and hence $E\{V^{2\delta}\}$ is dominated by

$$\text{const.}\lambda^{-\delta}(l_\lambda+1) E\left\{ \max_{0\le r\le 2\lambda^\alpha} \left| \int_0^r \theta_s f_0 dw(s) \right|^{2\delta} \right\}$$
$$\le \text{const.}\lambda^{-(1-\alpha)(\delta-1)-\frac{1}{2}} \text{ (use(1.15))}.$$

This combined with (1.21) implies

(1.22) $$E\left\{ \sup_{(k,\lambda)} |R(t,u) - R(k\lambda^\alpha, u)|^{2\delta} \right\}$$
$$\le \text{const.}\lambda^{-(1-\alpha)(\delta-1)-\frac{1}{2}},$$

because $(1-\alpha)\delta > (1-\alpha)(\delta-1) + \frac{1}{2}$. Let $k(t)$ be the integer determined from t by $k(t)\lambda^\alpha < t \le (k(t)+1)\lambda^\alpha$ and put $K_\lambda = \{k(t) : |t| \le \lambda\}$. Then $\#K_\lambda \le \text{const.}\lambda^{1-\alpha}$ and (1.22) implies

(1.23) $$E\left\{ \sup_{(\lambda)} |R(t,u) - R(k(t)\lambda^\alpha, u)|^{2\delta} \right\}$$
$$\le \text{const.}\#K_\lambda \cdot \lambda^{-(1-\alpha)(\delta-1)-\frac{1}{2}} \le \text{const.}\lambda^{-\delta_1},$$

because $(1-\alpha)(\delta-2)+\frac{1}{2}>\delta_1$. By a similar method we can prove that

$$E\left\{\sup_{(\lambda)}|R(k(t)\lambda^{\alpha},u)|^{2\delta}\right\}\leq \text{const.}\lambda^{-\delta_1} \tag{1.24}$$

and hence (1.20) follows from (1.23) and (1.24).

We finally prove (1.18). If β is a constant such that $\beta\delta_1>1$, then $E\{R(n^\beta)^{2\delta}\}\leq \text{const.}n^{-\beta\delta_1}$ by (1.20) so by using the Chebyschev inequality and the Borel-Cantelli lemma we see that $R(n^\beta)\to 0$ as $n\to\infty$, a.s. Likewise

$$R_c(n^\beta)\equiv \sup\left\{|R(t,u)|:|t|\leq cn^\beta, u\geq n^{\beta/2}\right\}\to 0,\ \text{a.s.},$$

for any constant $c>0$. The assertion (1.18) now follows from $R(\lambda)\leq R_c(n(\lambda)^\beta)$ with $c=2^\beta$ where $n(\lambda)$ is the integer satisfying $n(\lambda)^\beta<\lambda\leq (n(\lambda)+1)^\beta$.

Step 2 is to complete the proof of the lemma. The result (1.16) implies that

$$M_{t+u}-M_t=(m+o(1))u,\quad \text{if } |t|\leq c_1\lambda \text{ and } |u|\geq\sqrt{\lambda}/2. \tag{1.25}$$

When t and u are restricted to $|t|\leq c_1\lambda$ and $|u|<\sqrt{\lambda}/2$, we write $M_{t+u}-M_t=M'+M''$ where

$$M'=M_{t-\sqrt{\lambda}+(\sqrt{\lambda}+u)}-M_{t-\sqrt{\lambda}},\quad M''=M_{t-\sqrt{\lambda}}-M_t.$$

Then applying (1.16) we have $M'=(m+o(1))(\sqrt{\lambda}+u)$ and $M''=-(m+o(1))\sqrt{\lambda}$. Therefore $M_{t+u}-M_t=mu(1+o(1))+o(\sqrt{\lambda})$ if $|t|\leq c_1\lambda$ and $|u|<\sqrt{\lambda}/2$. This combined with (1.25) proves the lemma.

§2. Environment-wise central limit theorem

2.1. Proof of (I). We put $\tau_k=T_k-T_{k-1}, k\geq 1$. Then $\tau_k, k\geq 1$, are independent under the probability measure P_w^0. Since $T_n=\sum_1^n\tau_k$, the central limit theorem for T_n with fixed w can be obtained by verifying the Lindeberg condition. We have

$$\mathrm{Var}_w^0\{T_n\}=\sum_{k=1}^n\mathrm{Var}_w^0\{\tau_k\}=\sum_{k=1}^n V(\theta_{k-1}w)$$

where $V(w)=\mathrm{Var}_w^0\{\tau_1\}$. We now assume $\kappa>2$ to ensure the existence of $E\{V\}$, which is equal to A by (1.3). Since $\{\theta_t\}$ is ergodic, we have

$$\lim_{n\to\infty}\frac{\mathrm{Var}_w^0\{T_n\}}{n}=A,\ P\text{-a.s.} \tag{2.1}$$

Putting $\tilde{\tau}_k = \tau_k - E_w^0\{\tau_k\}$ and $V_N(w) = E_w^0\{|\tilde{\tau}_1|^2; |\tilde{\tau}_1| > N\}$ (the integral of $|\tilde{\tau}_1|^2$ over $\{|\tilde{\tau}_1| > N\}$ with respect to P_w^0), we have for any $\varepsilon > 0$ and for almost all w

$$
\begin{aligned}
(2.2) \qquad & \frac{1}{\mathrm{Var}_w^0\{T_n\}} \sum_{k=1}^{n} E_w^0 \left\{ |\tilde{\tau}_k|^2; |\tilde{\tau}_k| > \varepsilon\sqrt{\mathrm{Var}_w^0\{T_n\}} \right\} \\
\le & \frac{\text{const.}}{n} \sum_{k=1}^{n} E_w^0 \{|\tilde{\tau}_k|^2; |\tilde{\tau}_k| > N\} \quad \text{(by (2.1))} \\
= & \frac{\text{const.}}{n} \sum_{k=1}^{n} V_N(\theta_{k-1} w) \to \text{const.} E\{V_N\}, \quad n \to \infty.
\end{aligned}
$$

Since $E\{V_N\} \downarrow 0$ as $N \uparrow \infty$ because of $E\{V\} < \infty$, the left hand side of (2.2) tends to 0 as $n \uparrow \infty$ (P-a.s.), which shows that the Lindeberg condition is satisfied for $\{\tilde{\tau}_k\}$. Therefore for almost all w the central limit theorem holds for T_n with respect to P_w^0. By (1.4) $E_w^0\{T_n\} = M_n$ and by (2.1) $\mathrm{Var}_w^0\{T_n\} \sim An, n \to \infty$ (P-a.s.). Therefore the Lindeberg condition together with an application of Theorem 3.1 of [5] implies the assertion of (I). Further details will be given in a joint paper with K. Kawazu.

2.2. Proof of (II). The probability measure we consider in this proof is P_w^0 where w is taken from some subset of W that has P-measure 1. Thus events we consider here can be regarded as subsets of $\Omega_0 = \{\omega \in \Omega : \omega(0) = 0\}$. For any $s > 0$, $\rho \in \mathbb{R}$ and $\nu > 0$ we have, with convention $\underline{\omega}(\infty) = \infty$,

$$
\begin{aligned}
(2.3) \qquad \{T_{\rho^+} < s\} &= \{\overline{\omega}(s) > \rho\} \supset \{\omega(s) > \rho\} \supset \{\underline{\omega}(s) > \rho\} \\
&\supset \{T_{(\rho+\nu)^+} < s\} \cap \{\underline{\omega}(T_{\rho+\nu}) > \rho\},
\end{aligned}
$$

where $\rho^+ = \max\{\rho, 0\}$ and T_x denotes the hitting time to $x \in \mathbb{R}$. Similarly

$$
\begin{aligned}
(2.4) \qquad \{T_{\rho^+} > s\} &= \{\overline{\omega}(s) < \rho\} \subset \{\omega(s) < \rho\} \subset \{\underline{\omega}(s) < \rho\} \\
&\subset \{T_{(\rho+\nu)^+} > s\} \cup \{\underline{\omega}(T_{\rho+\nu}) < \rho\}.
\end{aligned}
$$

Lemma 3. *For any fixed $t_1 > 0$ let (φ, ψ) be a pair of continuous functions on $[0, t_1]$ such that*

$$
(2.5) \qquad \begin{array}{l} \varphi \text{ and } \psi \text{ are absolutely continuous, } \varphi', \psi' \in L^2[0, t_1], \\ \varphi(t) < \psi(t) \text{ for all } t \in [0, t_1] \text{ and } \varphi(0) \neq 0, \psi(0) \neq 0. \end{array}
$$

Then for almost all w

$$
\begin{aligned}
(2.6) \quad \lim_{\lambda\to\infty} \; & P_w^0 \left\{ \varphi(t) < \frac{\omega(\lambda t) - \mu(\lambda t)}{\sqrt{m^{-3} A \lambda}} < \psi(t) \text{ for all } t \in [t_0, t_1] \right\} \\
= \; & P\{\varphi(t) < w(t) < \psi(t) \text{ for all } t \in [t_0, t_1]\}.
\end{aligned}
$$

Proof. For the proof it is enough to consider the case $\varphi(0) < 0 < \psi(0)$. We put $\rho(t) = \mu(\lambda t) + \varphi(t)\sqrt{m^{-3}A\lambda}$ and $\rho_1(t) = \mu(\lambda t) + \psi(t)\sqrt{m^{-3}A\lambda}$. Then $\rho_1(t) > 0$ for all $t \in [0, t_1]$ if λ is sufficiently large, but $\rho(t) < 0$ if t is sufficiently small for each fixed λ. Since $\{\varphi(t) < (m^{-3}A\lambda)^{-1/2}(\omega(\lambda t) - \mu(\lambda t)) < \psi(t)\} = \{\rho(t) < \omega(\lambda t) < \rho_1(t)\}$, an application of (2.3) and (2.4) yields

$$\begin{aligned}&\{T_{(\rho(t)+\nu)^+} < \lambda t, T_{\rho_1(t)} > \lambda t, \underline{\omega}(T_{\rho(t)+\nu}) > \rho(t)\}\\ &\subset \{\varphi(t) < (m^{-3}A\lambda)^{-1/2}(\omega(\lambda t) - \mu(\lambda t)) < \psi(t)\}\\ &\subset \{T_{\rho(t)^+} < \lambda t\} \cap [\{T_{\rho_1(t)+\nu} > \lambda t\} \cup \{\underline{\omega}(T_{\rho_1(t)+\nu}) < \rho_1(t)\}].\end{aligned}$$

Therefore if we put

$$\begin{aligned}\Gamma_\lambda &= \{\varphi(t) < (m^{-3}A\lambda)^{-1/2}(\omega(\lambda t) - \mu(\lambda t)) < \psi(t) \text{ for all } t \in [0, t_1]\},\\ \Gamma_\lambda^- &= \{T_{(\rho(t)+\nu)^+} < \lambda t, T_{\rho_1(t)} > \lambda t \text{ for all } t \in [0, t_1]\},\\ \Gamma_\lambda^+ &= \{T_{\rho(t)^+} < \lambda t, T_{\rho_1(t)+\nu} > \lambda t \text{ for all } t \in [0, t_1]\},\\ A_\lambda &= \{\underline{\omega}(T_{\rho(t)+\nu}) > \rho(t) \text{ for all } t \in [0, t_1]\},\\ B_\lambda &= \{\underline{\omega}(T_{\rho_1(t)+\nu}) < \rho_1(t) \text{ for some } t \in [0, t_1]\},\end{aligned}$$

then

$$\Gamma_\lambda^- \cap A_\lambda \subset \Gamma_\lambda \subset \Gamma_\lambda^+ \cup B_\lambda. \tag{2.7}$$

Put $\nu = 6 \log \lambda$ and let us prove

$$\lim_{\lambda\to\infty} P_w^0\{A_\lambda^c\} = \lim_{\lambda\to\infty} P_w^0\{B_\lambda\} = 0, \quad P\text{-a.s.} \tag{2.8}$$

Since $\{\underline{\omega}(T_{\rho(t)+\nu}) \le \rho(t)\} \subset \{\underline{\omega}(T_{[\rho(t)+\nu]}) < [\rho(t) + \nu] - (\nu - 1)\}$, we have

$$A_\lambda^c \subset \bigcup_{k=\rho_\lambda^-}^{\rho_\lambda^+} \{\underline{\omega}(T_k) < k - (\nu - 1)\},$$

where $\rho_\lambda^- = \min\{[\rho(t)+\nu] : 0 \le t \le t_1\}$ and $\rho_\lambda^+ = \max\{[\rho(t)+\nu] : 0 \le t \le t_1\}$. Therefore

$$P_w^0\{A_\lambda^c\} \le \sum_{k=\rho_\lambda^-}^{\rho_\lambda^+} P_w^0\{\underline{\omega}(T_k) < k - (\nu - 1)\}.$$

By an ergodic theorem $M_x/x \to m$ as $x \to \infty$ (P-a.s.) and hence $\mu(t)/t \to m^{-1}$ as $t \to \infty$ (P-a.s.), which implies $\rho_\lambda^+ \sim \lambda m^{-1} t_1$ and $\rho_\lambda^- = o(\lambda)$ as $\lambda \to \infty$ (P-a.s.). Therefore for the proof of (2.8) it is enough to show that for any constant $c > 0$

$$\lim_{\lambda\to\infty} \sum_{|k|<c\lambda} P_w^0\{\underline{\omega}(T_k) < k - (\nu - 1)\} = 0, \quad P\text{-a.s.}$$

Taking the expectation we have

$$E\left\{\sum_{|k|<c\lambda} P_w^0\left\{\underline{\omega}(T_k) < k-(\nu-1)\right\}\right\}$$

$$= \sum_{|k|<c\lambda} \mathcal{P}^0\left\{\underline{\omega}(0) < -(\nu-1)\right\}$$

$$< \text{const.}\lambda \exp\left\{-(\kappa-1)(\nu-1)/2\right\} \le \text{const.}\lambda^{-2},$$

where we used the result of [2]. The above estimate holds for $\nu = 6\log\lambda + O(1)$. Therefore an application of Borel-Cantelli lemma implies

$$\lim_{n\to\infty} \sum_{|k|<cn} P_w^0\left\{\underline{\omega}(T_k) < k-(\nu_n-1)\right\} = 0, \quad P\text{-a.s.},$$

for any given sequence $\{\nu_n\}$ such that $\nu_n = 6\log n + O(1)$ as $n\to\infty$. We thus have

$$\sum_{|k|<c\lambda} P_w^0\left\{\underline{\omega}(T_k) < k-(\nu-1)\right\}$$

$$\le \sum_{|k|<c([\lambda]+1)} P_w^0\left\{\underline{\omega}(T_k) < k-(\nu-1)\right\}$$

$$\to 0 \quad \text{as } \lambda\to 0 \ (P\text{-a.s.}),$$

which implies the assertion (2.8) concerning A_λ^c. The assertion for B_λ can be proved similarly.

To proceed let $\xi = \lambda m^{-1}$ and put

$$\alpha(t) = \xi^{-1}\rho(t) = \lambda^{-1}m\left\{\mu(\lambda t) + \varphi(t)\sqrt{m^{-3}A\lambda}\right\},$$

$$\beta(t) = \xi^{-1}\rho_1(t) = \lambda^{-1}m\left\{\mu(\lambda t) + \psi(t)\sqrt{m^{-3}A\lambda}\right\}.$$

Then for almost all w both $\alpha(t)$ and $\beta(t)$ tend to t as $\lambda\to\infty$, the convergence being uniform on each finite t-interval. From now on we write $T(x)$ and $M(x)$ for T_x and M_x, respectively. Since $\mu(\lambda t) = \xi\alpha(t) - \varphi(t)\sqrt{m^{-3}A\lambda}$, an application of Lemma 2 yields

$$(2.9)\quad \lambda t = M(\xi\alpha(t) - \varphi(t)\sqrt{m^{-3}A\lambda}) = M(\xi\alpha(t)) - \varphi(t)\sqrt{m^{-1}A\lambda} + o(\sqrt{\lambda}),$$

where $o(\sqrt{\lambda})$ is a term that, when divided by $\sqrt{\lambda}$, tends to 0 uniformly in $t\in[0,t_1]$ as $\lambda\to\infty$ for almost all w. Similarly

$$(2.10)\qquad \lambda t = M(\xi\beta(t)) - \psi(t)\sqrt{m^{-1}A\lambda} + o(\sqrt{\lambda}).$$

In what follows the notation $o(1)$ represents a term that tends to 0 uniformly in $t\in[0,t_1]$ as $\lambda\to\infty$ for almost all w. Using (2.9) and (2.10) we can obtain the following (2.11) $\sim$ (2.14) that hold for almost all w:

$$(2.11)\quad T(\rho(t)^+) < \lambda t \Leftrightarrow \frac{T(\xi\alpha(t)^+) - M(\xi\alpha(t)^+)}{\sqrt{A\xi}} < -\varphi(t) + o(1);$$

$$(2.12)\quad T(\rho_1(t)) > \lambda t \Leftrightarrow \frac{T(\xi\beta(t)) - M(\xi\beta(t))}{\sqrt{A\xi}} > -\psi(t) + o(1);$$

$$\begin{aligned}(2.13)\ T\left((\rho(t)+\nu)^+\right) \ &< \ \lambda t \Leftrightarrow T\left((\rho(t)+\nu)^+\right) - M(\xi\alpha(t)) \\ &\qquad < -\varphi(t)\sqrt{m^{-1}A\lambda} + o(\sqrt{\lambda}) \\ &\Leftrightarrow \ \frac{T\left(\xi(\alpha(t)+\frac{\nu}{\xi})^+\right) - M\left(\xi(\alpha(t)+\frac{\nu}{\xi})^+\right)}{\sqrt{A\xi}} \\ &\qquad < -\varphi(t) + o(1);\end{aligned}$$

$$\begin{aligned}(2.14)\qquad &T\left((\rho_1(t)+\nu)\right) > \lambda t \Leftrightarrow \\ &\frac{T\left(\xi(\beta(t)+\frac{\nu}{\xi})\right) - M\left(\xi(\beta(t)+\frac{\nu}{\xi})\right)}{\sqrt{A\xi}} > -\psi(t) + o(1).\end{aligned}$$

In deriving the second equivalence in (2.13) as well as the equivalence (2.14) we again used Lemma 2. From (2.11) $\sim$ (2.14) we have the following: For *almost all* w, $P_w^0\{\Gamma_\lambda^+\}$ is equal to $p_w(\xi)$ where $p_w(\xi)$ is the probability, evaluated by P_w^0, of the event

$$\left\{\begin{array}{l}(A\xi)^{-1/2}\left\{T(\xi\alpha(t)^+) - M(\xi\alpha(t))^+\right\} < -\varphi(t) + o(1), \\ (A\xi)^{-1/2}\left\{T(\xi(\beta(t)+\nu\xi^{-1})) - M(\xi(\beta(t)+\nu\xi^{-1}))\right\} > -\psi(t) + o(1), \\ \qquad\qquad \text{for all } t \in [0,t_1]\end{array}\right\}.$$

But by (I), for almost all w, $p_w(\xi)$ tends to

$$\begin{aligned}&P\{-\psi(t) < w(t) < -\varphi(t) \text{ for all } t \in [0,t_1]\} \\ = \ &P\{\varphi(t) < w(t) < \psi(t) \text{ for all } t \in [0,t_1]\}\end{aligned}$$

as $\xi \to \infty$ (here we used the assumption (2.5)). We also have a similar statement for $P_w^0\{\Gamma_\lambda^-\}$. Therefore from (2.7) and (2.8) we finally obtain Lemma 3. It is also to be noted that, by virtue of (2.3) and (2.4), Lemma 3 remains valid when $\omega(\cdot)$ is replaced by either of $\overline{\omega}(\cdot)$ and $\underline{\omega}(\cdot)$.

For fixed $t_1 > 0$ we denote by $\mathbb{F}$ the set of all pairs (φ, ψ) of functions on $[0, t_1]$ of the form $\varphi = p_1 \vee p_2 \vee \cdots \vee p_n$(maximum) and $\psi = q_1 \wedge q_2 \wedge \cdots \wedge q_n$ (minimum) where p_j and q_j are polynomials with rational coefficients satisfying $p_j(t) < q_j(t)$ for all $t \in [0,t_1], p_j(0) \neq 0, q_j(0) \neq 0 (1 \leq j \leq n)$. Since $\mathbb{F}$ is countable, Lemma 3 implies that for almost all w (2.6) holds for all $(\varphi,\psi) \in \mathbb{F}$. Fixing an arbitrary w for which (2.6) holds for all $(\varphi,\psi) \in \mathbb{F}$, we denote by Q_λ the probability law of the process $\{(m^{-3}A\lambda)^{-1/2}(\omega(\lambda t) - \mu(\lambda t)), 0 \leq t \leq t_1, P_w^0\}$; we also denote by Q the Wiener measure on $C[0,t_1]$ with $Q\{\tilde{w}(0) = 0\} = 1$. A Borel set in the Banach space $C[0,t_1]$ is said to be admissible if $Q_\lambda(A) \to Q(A)$ as $\lambda \to \infty$. Then for each pair $(\varphi, \psi) \in \mathbb{F}$ the open set $U(\varphi, \psi) \equiv \{\tilde{w} \in C[0, t_1] : \varphi(t) < \tilde{w}(t) < \psi(t) \text{ for all } t \in [0, t_1]\}$ is admissible. On the other hand any open set G in $C[0, t_1]$ can be expressed as $\cup_{n=1}^{\infty} U_n$ where each U_n is of the form $U(\varphi, \psi)$ with $(\varphi, \psi) \in \mathbb{F}$. The intersection of any finite number of U_k's is admissible because it is still of the form $U(\varphi, \psi)$, so $G_n \equiv \cup_{k=1}^{n} U_k$ is admissible because of the inclusion-exclusion formula

$$Q_\lambda(G_n) = \sum_{k=1}^{n} (-1)^{k-1} \sum_{1 \le i_1 < \cdots < i_k \le n} Q_\lambda(U_{i_1} \cap U_{i_2} \cap \cdots \cap U_{i_k}).$$

Since $G_n \uparrow G$ we have $\underline{\lim}_{\lambda \to \infty} Q_\lambda(G) \ge Q(G)$, which implies that $Q_\lambda \to Q$ as $\lambda \to \infty$. Since $t_1 > 0$ is arbitrary, this completes the proof of (II).

References

[1] Itô, K., and McKean, H.P., Diffusion Processes and Their Sample Paths, Springer-Verlag, New York, 1965.

[2] Kawazu, K., and Tanaka,H., On the maximum of a diffusion process in a drifted Brownian environment, Séminaire de Probabilités, LMN 1557, 78–85.

[3] Kawazu, K., and Tanaka,H., A diffusion process in a Brownian environment with drift, to appear in J.Math.Soc.Japan.

[4] Kesten, H., Kozlov, M. V., and Spitzer, F., A limit law for random walk in a random environment, Composito Math. **30** (1975), 145–168.

[5] Prohorov, Yu. V., Convergence of random processes and limit theorems in probability theory, Theor. Probab. Appl. **1** (1956), 157–214.

[6] Solomon, F., Random walks in a random environment, Ann. Probab. **3** (1975),1–31.

The complex story of simple exclusion

S.R.S. Varadhan

Courant Institute of Mathematical Sciences, New York University,
251 Mercer Street, New York, NY 10012, USA

1. Introduction and description of the model

The simple exclusion model serves as perhaps the simplest model of an interacting system of particles that still exhibits some of the complex behavior one expects from an interacting system. The aim of this expository article is to describe, for various simple exclusion models, the different types of results that are known that fall within the framework of hydrodynamic scaling limits.

First we start with a description of the models. The physical space is Z^d, the lattice of integers in d dimensions. We will have particles located at some of the sites i.e. points of Z^d. Any given site can have atmost one particle. The quantity $\eta(x)$ is defined to be equal to 1 if there is a particle at site x and 0 otherwise. The function $\eta : Z^d \to \{0,1\}$ will then completely describe the state of our system and tell us exactly which sites are locations of particles. Of course in this picture the particles are indistinguishable and only the configuration of occupied sites is relevent. The state space then is the set

$$\Omega = \{\eta : Z^d \to \{0,1\}\}$$

of all configurations of particles on Z^d.

The particles move in time by random jumps. Each particle waits for a random exponential time and at the end of this time tries to jump to a new site. The jump size z is randomly chosen from Z^d according to a probability distribution $p(z)$. If the particle is located currently at x, and z is chosen as the jump size, an attempt is made to jump to $x+z$. The attempt is successful and the jump is executed if the site $x+z$ is unoccupied. If the site $x+z$ is occupied the jump is disallowed and the particle starts waiting for a new exponential time. If the jump to $x+z$ is completed the system starts afresh. Of course all the particles are doing this simultaneously and independently of each other. Since we are dealing with continuous time there will be no ties to resolve. However the possibility of having an infinite number of particles requires a careful construction of the process. See the book [5] for details in this regard. One can describe our motion as a Markovian evolution on the state space Ω. We define the operator $\nabla_{x,x'}$ acting on any function f defined on Ω by,

$$(\nabla_{x,x'} f)(\eta) = f(\eta^{x,x'}) - f(\eta) \tag{1.1}$$

The author was supported by NSF grant DMS- 9503419 and ARO grant DAAL03-92-G-0317

where $\eta^{x,x'}$ is the new configuration obtained from η by the rule

$$\eta^{x,x'}(a) = \begin{cases} \eta(a) & \text{if } a \neq x \text{ or } x' \\ \eta(x') & \text{if } a = x \\ \eta(x) & \text{if } a = x'. \end{cases} \tag{1.2}$$

In other words $\nabla_{x,x'}$ interchanges the situation at x and x' leaving things alone at all oter sites. If both the sites x and x' are occupied, or if both are empty, $\nabla_{x,x'}$ does nothing and if only one of the sites is occupied, it moves the particle from the occupied site to the free site. $\nabla_{x,x'}$ is symmetric in x and x' i.e. $\nabla_{x,x'} \equiv \nabla_{x',x}$.

Let $f(\eta)$ be a local function, i.e. a function that depends only on a finite number of coordiantes, that is to say on $\{\eta(x) : x \in F\}$ for some finite subset $F \subset Z^d$. The infinitesimal generator of our Markov Process acts on such local functions and expresses in mathematical terms the verbal description that was provided earlier.

$$(\mathcal{L}f)(\eta) = \sum_{x,y} p(y-x)\eta(x)(1-\eta(y))(\nabla_{x,y}f)(\eta) \tag{1.3}$$

We can assume without loss of generality that $p(0) = 0$.

Since Z^d is infinite we could have some technical problems in proving the existence of the process and in establishing some of the results. To simplify matters let us take a large integer N and consider the finite periodic lattice Z^d_N, of integers modulo N, in every coordinate. We shall make the hypothesis that $p(z)$ is local i.e. $p(z) = 0$ for $z \notin F$ for some fixed finite set F. The definition of $\mathcal{L}$ does not really depend on N in any essential way. Our process then will have a finite, but large, state space of 2^{N^d} points. The problems of hydrodynamic scaling have to do with the behavior of the system over large spatial and temporal scales. Space is scaled by imbedding the lattice Z^d_N in the d-dimensional torus T^d by sending $x \to \frac{x}{N}$ viewed as a point of the unit torus T^d. The lattice spacing then is $\frac{1}{N}$. The time scale is changed by speeding up time by a factor of N^α and this shows up as an extra factor multiplying the infinitesimal generator. The value of α is either 1 or 2 depending on the circumstance. There will be three main classes into which our models will fall depending on the nature of $p(\cdot)$.

Case1. $p(\cdot)$ is symmetric i.e. $p(z) = p(-z)$.

Case2. $p(\cdot)$ is asymmetric, but $\sum zp(z) = 0$.

Case3. $\sum zp(z) = b \neq 0$

No matter what $p(\cdot)$ is, the invariant measures of the process are easy to describe. For any fixed N the total number K of the particles is conserved in our dynamics and for each N and $K \leq N^d$, there will be an inavariant measure $\mu_{N,K}$ for our process. There are $\binom{N^d}{K}$ possible configurations of size

K among N^d sites and $\mu_{N.K}$ is the uniform distribution among them. These are precisely all extremal or ergodic inavariant distributions. As $K, N \to \infty$, with $\frac{K}{N^d} \to \rho$, we get as limits invariant measures μ_ρ for the evolution on Z^d, which are the Bernoulli product measures on Ω with

$$\mu_\rho[\eta(x) = 1] = \rho \quad \text{for each} \quad x \in Z^d \tag{1.4}$$

In the infinite volume case there may be other non translation invariant measures which are also invariant measures for the process. But we do not have to deal with them. The problem of hydrodynamic scaling is to study the behavior of the system in the speeded up time scale especially when it starts from an initial state that is far away from any equilibrium configuration. Our questions will deal with the collective behavior of our system and we will have more or less explicit answers for our questions.

2. The symmetric case: the behavior of the density

For each configuration $\eta(\cdot)$ we can associate the empirical density

$$\nu_N = \frac{1}{N^d} \sum \delta_{\frac{x}{N}} \eta(x) \tag{2.1}$$

and view it as a random measure on T^d. Its total mass is $\frac{K}{N^d}$ and is a constant in time. Suppose that our evolution starts at time 0 from a configuration that may or may not be random and for some nonrandom density $\rho_0(y)$

$$\lim_{N\to\infty} \frac{1}{N^d} \sum \delta_{\frac{x}{N}} \eta(x) = \lim_{N\to\infty} \nu_N(0) = \rho_0(y)dy \tag{2.2}$$

(in probabiliy in the weak topology). Let us also suppose that the time scale has been speeded up by a factor of N^2, which is the appropriate rescaling factor in this case. Then we may ask if for any $t > 0$ (in the new time scale)

$$\lim_{N\to\infty} \frac{1}{N^d} \sum \delta_{\frac{x}{N}} \eta_t(x) = \lim_{N\to\infty} \nu_N(t) = \rho(t,y)dy \tag{2.3}$$

also exists and if so how to calculate $\rho(t,y)$ in terms of $\rho_0(y)$. In particular since $\rho_0(\cdot)$ is deterministic, one expects $\rho(t,\cdot)$ to be deterministic as well. Since weak convergence is tested by integrating against smooth test functions, we need to compute

$$d\left(\frac{1}{N^d} \sum J(\frac{x}{N})\eta_t(x)\right) = d\xi_J(t)$$

By standard Markov Process theory

$$d\xi_J(t) = A_J(t)dt + dM_J(t)$$

where $M_J(t)$ is a martingale. One can make explicit calculations and verify that for each $t > 0$,

$$E\{[M_J(t)]^2\} \to 0 \quad \text{as} \quad N \to \infty$$

The problem then reduces to calculating

$$A_J(t) = G_N(\eta_t(\cdot))$$

where

(2.4)

$$\begin{aligned}
&G_N(\eta(\cdot))\\
&= N^2 \mathcal{L}\left(\frac{1}{N^d}\sum J(\frac{x}{N})\eta(x)\right)\\
&= N^2\frac{1}{N^d}\sum p(x'-x)\eta(x)(1-\eta(x'))[J(\frac{x'}{N}) - J(\frac{x}{N})]\\
&= \frac{N^2}{2N^d}\sum_{x,x'} p(x'-x)[J(\frac{x'}{N}) - J(\frac{x}{N})][\eta(x)(1-\eta(x')) - \eta(x')(1-\eta(x))]\\
&= \frac{N^2}{2N^d}\sum_{x,x'} p(x'-x)[J(\frac{x'}{N}) - J(\frac{x}{N})][\eta(x) - \eta(x')]\\
&\simeq \frac{1}{2N^d}\sum_{x}(\mathcal{A}J)(\frac{x}{N})\eta(x).
\end{aligned}$$

The operator $\mathcal{A}$ is given by

$$(\mathcal{A}J)(y) = \sum C_{i,j}\frac{\partial^2 J}{\partial y_i \partial y_j}(y) = \nabla C \nabla J$$

with

$$C_{i,j} = \sum_z z_i z_j p(z) \quad ; \quad z = (z_1, \cdots, z_d)$$

being the covariance matrix. From this observation it is routine to show that (2.3) is in fact true if (2.2) is valid, and that $\rho(t,y)$ can be obtained by solving the heat equation

$$\frac{\partial \rho}{\partial t} = \frac{1}{2}\nabla C \nabla \rho \tag{2.5}$$

with the initial condition

$$\rho(0,y) = \rho_0(y) \tag{2.7}$$

This calculation was made possible by a lucky break. The nonlinearity in (2.4) canceled out, and that allowed us to sum by parts twice, getting rid of the two powers of N in front of the generator.

3. Weak Asymmetry

Let us change things a little bit. Suppose we destroy the symmetry by adding a term $\frac{1}{N}q(x'-x)$ to $p(x'-x)$. Think of $\frac{1}{N}q$ as a small anti-symmetric modification of p. Then the new $G_N(\eta(\cdot))$ has an additional term

$$(3.1)\qquad \begin{aligned} &\frac{N^2}{N^d}\sum_{x,x'}\eta(x)(1-\eta(x'))\frac{q(x'-x)}{N}[J(\frac{x'}{N})-J(\frac{x}{N})] \\ &\simeq \frac{1}{N^d}\sum_{x,x'}\eta(x)(1-\eta(x'))\langle(\nabla J)(\frac{x}{N})\ ,\ b\ \rangle \end{aligned}$$

where $b=\sum zq(z)$. We are now stuck with the non-linearity. We invoke a general principle that if the density at some point of space and time is $\rho(t,y)$ then the joint distribution of $\{\eta_t(x)\}$ around the macroscopic point y i.e when $\frac{x}{N}\sim y$ will be nearly the Bernoulli measure $\mu_{\rho(t,y)}$. In particular $\eta(x)(1-\eta(x+z))$ can be replaced by $\rho(t,y)(1-\rho(t,y))$, and we end up with the equation

$$(3.2)\qquad \frac{\partial\rho}{\partial t}=\frac{1}{2}\nabla C\nabla\rho-\nabla.(\rho(1-\rho)b)$$

This in fact has been rigorously established in [2].

4. Large Deviations

One of the problems we can raise is the estimation of the probability of large deviations from the hydrodynamical behavior, which is after all a form of law of large numbers. More precisely, if we start from some configuration $\eta(\cdot)$ on Z^d, (which may or may not be random) and follow the empirical measures $\nu_t(\cdot)$ in some time interval, say $[0,T]$, what is the probability that ν_t stays close to an arbitrary given path $\lambda(t)$ in that time interval $[0,T]$? In particular, if it is exponentially small, i.e. $\exp[-N^dI(\lambda(\cdot))+o(N^d)]$, then we should determine the functional $I(\lambda(\cdot))$ explicitly.

First let us suppose that $\lambda(t,dy)=\lambda(t,y)dy$. Otherwise the rate function would be clearly infinite. Typically, there would be an initial rate function $I_0(\lambda)$ that controls the large deviation rate for the initial profile. If the initial conditions were deterministic then $I_0(\lambda)$ would be 0 for the correct macroscopic profile ρ_0 and $+\infty$ for any other profile. We will now concentrate on the large deviation arising out of the dynamical part. One way to generate deviant behavior, is to assume that the model behaves like it had some small asymmetry q that could depend on t and x as well, i.e. $\frac{1}{N}q(t,y,x'-x)$. This is unlikely, but exactly how unlikely, can be measured by relative entropy calculated via a Girsanov type formula. A simple calculation yields the following estimate for the relative entropy H

$$
\begin{aligned}
H &\simeq E\Big\{\frac{1}{2}\int_0^1 \sum_{x,x'} \eta_t(x)(1-\eta_t(x'))\frac{q^2(t,\frac{x}{N},x'-x)}{p(x'-x)}dt\Big\} \\
&\simeq \frac{N^d}{2}\int_0^1\int_{T^d}\rho(t,y)(1-\rho(t,y))e(t,y)dtdy
\end{aligned}
\tag{4.1}
$$

where

$$e(t,y)=\sum_z \frac{q^2(t,y,z)}{p(z)}$$

This will produce a density ρ that will be the solution of

$$\frac{\partial\rho}{\partial t}=\frac{1}{2}\nabla C\nabla\rho-\nabla.\rho(1-\rho)b \tag{4.2}$$

with $b(t,y)=\sum zq(t,y,z)$. It definitely behooves us to minimize $\sum\frac{q^2(z)}{p(z)}$ subject to $\sum zq(z)=b$. An elementary calculation provides the answer as $\langle b\,,C^{-1}\,b\,\rangle$. The best lower bound then is

$$\Psi(b)=\frac{1}{2}\int_0^1\int_{T^d}\rho(t,y)(1-\rho(t,y))\langle b(t,y)\,,C^{-1}\,b(t,y)\rangle dtdy \tag{4.3}$$

Finally the best relative entropy that we can achieve is

$$\inf_{b\in\mathcal{B}_\rho}\Psi(b)$$

where

$$\mathcal{B}_\rho=\{b:\frac{\partial\rho}{\partial t}=\frac{1}{2}\nabla C\nabla\rho-\nabla.\rho(1-\rho)b\} \tag{4.4}$$

One can see that the infimum is attained when b is of the form $b=C\nabla\varphi$ for some φ and the rate function is computed to be

$$I_0(\lambda_0)+\frac{1}{2}\int_0^1\|\frac{\partial\rho}{\partial t}-\frac{1}{2}\nabla C\nabla\rho\|^2_{-1,\rho(t,\cdot)(1-\rho(t,\cdot))C}\;dt \tag{4.5}$$

where

$$\|u\|^2_{-1,A(\cdot)}=\sup_J\Big[2\int u(y)J(y)dy-\int\langle(\nabla J)(y),A(y)(\nabla J)(y)\rangle dy\Big] \tag{4.6}$$

In [2] it was shown that this best possible lower bound is also an upper bound so that (4.5) is the exact rate function. This describes the large deviations as well as the way they arise.

5. Tagged particles in equilibrium and nonequilibrium

In a system with a large number of interacting particles, like in our simple exclusion process, we can tag one particular particle as it interacts and moves amidst the rest of the particles. Let us suppose that we start our system in equlibrium on Z^d, at a density ρ. Let us condition our system to have initially a particle at the site 0, and let us tag that particle. If we watch our tagged particle, we might reasonably expect it to wander diffusively and converge to a Brownian motion after a suitable rescaling of space and time. The limiting Brownian motion will have a covariance matrix $S(\rho)$ depending on the density ρ. That this is infact the case was demonstrated in [3]. The diffusion matrix $S(\rho)$ was shown to be nondegenerate in all cases except the one dimensional nearest neighbor case. In this case, the particles block each other completely, and $S(\rho) \equiv 0$ for all $0 < \rho \leq 1$. In the nonexceptional cases one can see that $S(\rho) \to C$ as $\rho \to 0$ and $S(\rho) \to 0$ as $\rho \to 1$.

We could start far away from equilibrium and still ask how a tagged particle that starts somewhere in this mileu will evolve in time. Let us return to our model on Z_N^d, tag a particle that starts at time 0 from the origin 0. We will assume that the rest of the particles have an initial configuration that has asymptotically an initial density profile $\rho_0(y)$. If we look at the speeded up time scale, as we saw earlier, the density will evolve according to the heat equation (2.5). Let us suppose that, after rescaling time by a factor N^2 and space by N to get the torus T^d, our tagged particle is at a point y on the torus at time t. Then it will find itself surrounded by an environment somewhat similar to the equilibrium environment with density $\rho(t, y)$. The tagged particle might then diffuse like a tagged particle in equilibrium at density $\rho(t, y)$, i.e. with a diffusion coefficient that is equal to $S(\rho(t, y))$. However the description may not be accurate enough and we may have an as yet undetermined first order term or drift that is present. An educated guess for the motion of our tagged particle then is a time dependent diffusion with the backward Kolmogorov generator

$$\mathcal{L}_t = \frac{1}{2}\nabla S(\rho(t,y))\nabla + c(t,y).\nabla \tag{5.1}$$

for some suitable choice of c. The corresponding forward equation is given by

$$\frac{\partial \tilde{\rho}}{\partial t} = \frac{1}{2}\nabla S(\rho(t,y))\nabla\tilde{\rho} - \nabla.c(t,y)\tilde{\rho} \tag{5.2}$$

and the solution at time t will give the marginal distribution at that time of our tagged particle. If we pick at random one of the other particles and tag it, (while untagging our particle) the equation (5.2) does not change. The initial condition for $\tilde{\rho}$ at $t = 0$ bcomes ρ_0 normalized to have total integral 1 on T^d. In fact if the tagged particle is randomly chosen, its density at any time t, is nothing other than $\rho(t,y)$ normalized to have integral 1. In other words, since the equation (5.2) is linear, ρ must satisfy it.

$$\frac{\partial\rho}{\partial t} = \frac{1}{2}\nabla S(\rho(t,y))\nabla\rho - \nabla.c(t,y)\rho \tag{5.3}$$

Now comparing (2.5) and (5.3), we obtain

$$\frac{1}{2}\nabla S(\rho(t,y))\nabla\rho - \nabla.c(t,y)\rho = \frac{1}{2}\nabla C\nabla\rho$$

Continuing on,

$$\nabla.c\rho = \frac{1}{2}(\nabla S(\rho)\nabla\rho - \nabla C\nabla\rho).$$

Perhaps with some luck,

$$c\rho = \frac{1}{2}(S(\rho)\nabla\rho - C\nabla\rho)$$

or

$$c = \frac{1}{2}[S(\rho) - C]\frac{\nabla\rho}{\rho}.$$

One can then reasonably expect the tagged particle in nonequlibrium to behave, after rescaling, like a diffusion with a time dependent backward Kolmogorov generator

$$\mathcal{L}_t = \frac{1}{2}\nabla S(\rho(t,y))\nabla + \frac{1}{2\rho}(C - S(\rho(t,y))\nabla\rho(t,y).\nabla. \tag{5.4}$$

To prove this for every particle in every initial configuration appears to be hard and has not been done yet. If we denote by Q_0 the measure on the path space $C[[0,T];T^d]$ corresponding to our generator (5.4) with initial (sub probability) density $\rho_0(y)$ one may consider the empirical process

$$R_{N,\omega} = \frac{\delta_{y_1(\cdot)} + \delta_{y_2(\cdot)} + \cdots + \delta_{y_K(\cdot)}}{N^d}$$

where $y_i(\cdot)$ are the trajectories of the K particles that are individually tagged. $R_{N,\omega}$ is viewed as a random measure on $D[[0,T];T^d]$ and one can ask if as $N \to \infty$ the measures $R_{N,\omega} \Rightarrow Q_0$. This was answered in the affirmative in [9] and we will return to this later.

6. Several colors and nongradient systems

We may also decide that there are n types of particles distinguished by colors and may want to keep track of the colors while the evolution is colorblind. We define

$$\eta_j : Z_N^d \to \{0,1\}$$

where $\eta_j(x)$ is 1 if there is a particle of color j at site x. Otherwise it is 0. Of course

$$\sum_{j=1}^{n} \eta_j(x) = \eta(x).$$

If we denote the typical element of the state space by

$$\zeta = (\eta_1, \dots, \eta_n)$$

then the n-color generator is given by

$$(6.1) \quad (\mathcal{L}_N^n f)(\zeta) = N^2 \sum_x \sum_z p(z)\eta(x)(1-\eta(x+z)) \left(f(\zeta^{x,x+z}) - f(\zeta)\right)$$

or in a more symmetric form by,
(6.2)

$$(\mathcal{L}_N^n f)(\zeta) = \frac{N^2}{2} \sum_{x,x'} p(x'-x)[\eta(x)(1-\eta(x')) + (1-\eta(x))\eta(x')] \left(f(\zeta^{x,x'}) - f(\zeta)\right)$$

Here $\zeta^{x,x'}$ is the configuration which is defined in a manner similar to the way $\eta^{x,x'}$ was defined in (1.2) by interchanging the situation at x and x' without changing the color of particles, if any, at these sites. The trace of the ζ process on the torus will be a measure valued process with n components

$$\tilde{\nu}_N(t) = \left\{\nu_N^j(t)\right\}$$

where

$$\nu_N^j(t) = \frac{1}{N^d} \sum_x \delta_{\frac{x}{N}} \eta_{j,t}(x)$$

of total masses K_j/N^d, where K_j is the number of particles of color j. The marginal of the empirical process $R_{N,\omega}$ at time t is then $\nu_N(t)$.

Let us define C to be the convex set in R^n consisting of $\tilde{\rho} = (\rho_1, \rho_2, \dots, \rho_n)$ such that $\rho_j \geq 0$ for j=1, 2, ..., n and $\rho = \sum_{j=1}^n \rho_j \leq 1$

We define the following matrices A and χ which are really to be thought of as $n \times n$ matrices corresponing to color types whose entries are $d \times d$ matrices that correspond to coordinates in R^d. If j, j' are two color types then

$$(6.3) \quad \begin{aligned} A_{j,j'} &= A_{j,j'}^{(1)} + A_{j,j'}^{(2)} \\ &= S(\rho)\left(\rho_j \delta_{j,j'} - \frac{\rho_j \rho_{j'}}{\rho}\right) + C\left(\rho_j \rho_{j'} \frac{(1-\rho)}{\rho}\right) \end{aligned}$$

Each entry of $A_{j,j'}$, as a $d \times d$ matrix, is a linear combination of $S(\rho)$ and C with weights that have been given in (6.3). Similarly

$$(6.4) \quad \chi_{j,j'} = I\left(\frac{\delta_{j,j'}}{\rho_j} + \frac{1}{(1-\rho)}\right)$$

where I is the identity matrix of size $d \times d$. Let us note that although χ is undefined when $\rho = 1$, A and $A\chi$ are well defined on C. In [6] it was shown that the empirical measures $\tilde{\nu}_N(\cdot)$ satisfy a law of large numbers. Assuming that the initial configurations, whether they are random are deterministic, satisfy

$$\tilde{\nu}_N(\cdot) \Rightarrow \tilde{\rho}^0(\theta)d\theta$$

in probability, in the weak topology on T^d, it was shown that for all $t > 0$,

$$\tilde{\nu}_N(\cdot) \Rightarrow \tilde{\rho}(t,\theta)d\theta$$

The limit $\tilde{\rho}(t,\theta)$ is characterized as the unique solution of

$$\frac{\partial \tilde{\rho}}{\partial t} = \frac{1}{2}\nabla \cdot A\chi(\tilde{\rho})\nabla\tilde{\rho} \tag{6.5}$$

with initial condition

$$\tilde{\rho}(0,\theta) = \tilde{\rho}^0(\theta) \tag{6.6}$$

The equations (6.5) and (6.6) have the following interpretation. The sum

$$\rho(t,\theta) = \sum_{j=1}^{n} \rho^j(t,\theta)$$

satisfies the heat equation

$$\frac{\partial \rho}{\partial t} = \frac{1}{2}\nabla \cdot C\nabla\rho \tag{6.7}$$

with initial condition

$$\tilde{\rho}(0,\theta) = \rho^0(\theta) = \sum_1^n \rho_j^0(\theta)$$

and, given the solution of (6.7), each component of (6.5) can be recast as

$$\frac{\partial \rho_j}{\partial t} = \mathcal{A}_\rho\, \rho_j$$

with $\rho_j(0,\theta) = \rho_j^0(\theta)$, where

$$\mathcal{A}_\rho f = \frac{1}{2}\nabla \cdot S(\rho(t,\theta))\nabla f + \frac{1}{2}\nabla \cdot [(C - S(\rho(t,\theta))]\,\frac{\nabla\rho(t,\theta)}{\rho(t,\theta)}\, f \tag{6.8}$$

This is compatible with the expectation for the diffusive behavior of the tagged particle keping in mind that coloring is a form of tagging.

The proof proceeds as follows. For the color j, if we compute as we did earlier

$$d\left(\frac{1}{N^d}\sum_x J(\frac{x}{N})\eta_{j,t}(x)\right) = A_J(t)dt + dM_J(t)$$

where $E\{[M_J(t)]^2\} \to 0$ as $N \to \infty$ as before, but now $A_J(t)$ is given by

$$\begin{aligned} A_J(t) &= \frac{N^2}{N^d}\sum_{x,x'} \eta_j(x)(1-\eta(x'))[J(\frac{x'}{N}) - J(\frac{x}{N})]p(x'-x) \\ &\simeq \frac{N}{N^d}\sum_x \eta_j(x)(\nabla J)(\frac{x}{N})\cdot\sum_{x'}(1-\eta(x'))(x'-x)p(x'-x) \end{aligned}$$

Unlike the case of particles with single color we cannot perform summation by parts for a second time and we are saddled with a term of order N. The term does have mean 0. Some methods were developed in [10], for handling such situations referred to as nongradient models. There are several terms, also of order N that are completely negligible because of the rapid time scale. Our term is not one of them. Neither are the density gradients. Let us define the currents that constitute our terms,

$$W_{i,j} = \eta_j(0)\sum_{x'} p(x')(1-\eta(x'))(e_i, x')$$

and the density gradients

$$\xi_{i,j} = (\nabla_{e_i}\eta_j)(0) = \eta_j(e_i) - \eta_j(0)$$

where $\{e_1, \cdots, e_d\}$ is the standard basis for R^d. Relative to any equlibrium $\mu_{\tilde{\rho}}$ for the n color evolution, $W_{i,j}$ can be effectively replaced by a linear combinations of $\{\xi_{i,j}\}$, because the differences are negligible. However the coefficients depend on the density $\tilde{\rho}$. It is proved in [6] that the linear combinations are given by

$$W_{i,j} \sim \sum_{i',j'} [A(\tilde{\rho})\chi(\tilde{\rho})]^{i,j}_{i',j'}\xi_{i',j'}$$

If we do the formal substitution, we can then do another summation by parts and the equation (6.5) emerges. Actual details are rather complex and involve analysis in equlibrium at every $\mu_{\tilde{\rho}}$, and transition to nonequlibrium that involves methods from large deviation theory. This is typical of nongradient systems and was carried out in [6]. The result of Rezakhanlou that we quoted earlier on the behavior of $R_{N.\omega}$ was deduced from the n color behavior. We increase the number of colors indefinitely and color code past history. This way, modulo some difficult compactness estimates proved in [9], one can infer the behavior of $R_{N,\omega}$ from that of $\tilde{\nu}_N(\cdot)$.

7. Large deviations for several colors and empirical processes

We will now describe the large deviation theory for the empirical measure valued process $\tilde{\nu}(\cdot)$. If P_N is the probability distribution of $\tilde{\nu}_N(\cdot)$, it has a standard large deviation property with the following rate function $I_n(\tilde{\rho}(\cdot,\cdot))$ defined for functions

$$\tilde{\rho}(\cdot) : [0,T] \times T^d \to C$$

such that $\int \rho_j(t,\theta)d\theta$ is independent of t for each j. Given a function $q(\theta)$ on T^d taking values in the space of symmetric positive semidefinite $d \times d$ matrices and a function $\tilde{r}(\cdot) : T^d \to R^n$ with mean zero we define

$$\text{(7.1)}\quad \|\tilde{r}\|_{-1,q(\cdot)} = \sup_{\Phi} \left\{ 2\int_{T^d} \langle \tilde{r}(\theta), \Phi(\theta)\rangle \, d\theta - \int_{T^d} \langle \nabla\Phi(\theta), q(\theta)\nabla\Phi(\theta)\rangle \, d\theta \right\}$$

where the supremum is over C^∞ maps from T^d to R^n. We then define

$$I_0^n(\tilde{\rho}(.,.))$$
$$= \sup_{\Phi} \left\{ 2\int_0^T \int_{T^d} \langle (A\chi)(\tilde{\rho})\nabla\tilde{\rho}, \nabla\Phi\rangle dt d\theta - \int_0^T \int_{T^d} \langle \nabla\Phi, (A)(\tilde{\rho})\nabla\Phi\rangle dt d\theta \right\}$$

where the supremum is now taken over C^∞ functions on $[0,T] \times T^d$ with values in R^{nd}. When it makes sense,

$$\text{(7.2)}\qquad \bar{I}^n(\tilde{\rho}) = \int_0^T \int_{T^d} \langle \nabla\tilde{\rho}, (\chi A\chi)(\tilde{\rho})\nabla\tilde{\rho}\rangle \, dt d\theta$$

One can explicitly calculate $\chi A \chi$ and the quantity in (7.2) is finite if and only if

$$\int_0^T \int_{T^d} (1-\rho) \sum_j \frac{|\nabla \rho_j|^2}{\rho_j} d\theta dt \ < \ \infty$$

and

$$\int_0^T \int_{T^d} \frac{|\nabla\rho|^2}{(1-\rho)} d\theta dt \ < \ \infty$$

The dynamical part $I_1^n(\tilde{\rho}(\cdot,\cdot))$ is defined to be $+\infty$ unless $\bar{I}^n(\tilde{\rho}(\cdot,\cdot)) < \infty$ and in such a case

$$I_1^n(\tilde{\rho}(\cdot,\cdot)) = \frac{1}{2}\int_0^T \left\| \frac{\partial\tilde{\rho}}{\partial t} - \frac{1}{2}\nabla\cdot A\chi\nabla\tilde{\rho} \right\|^2_{-1,A(\tilde{\rho}(t,\cdot))} dt$$

If the initial conditions are deterministic and the law of large numbers holds initially with profile $\tilde{\rho}^0(\cdot)$ then we define

$$I_0^n(\tilde{\rho}(0,\cdot)) = \begin{cases} 0 & \text{if } \tilde{\rho}(0,\cdot) = \tilde{\rho}^0(\cdot) \\ +\infty & \text{otherwise.} \end{cases}$$

For other suitable initial conditions we pick the corresponding rate function for the initial profile and this shall be our $I_0^n(\tilde{\rho}(0,\cdot))$. The full rate function is given by

$$I^n(\tilde{\rho}(\cdot,\cdot)) = I_0^n(\tilde{\rho}(0,\cdot)) + I_1^n(\tilde{\rho}\cdot,\cdot)).$$

Large deviation estimates with this rate function have been obtained for dimension ≥ 3 in [7]. The restriction on the dimension is dictated by technical demands, namely Lipschitz continuity of $S(\rho)$ as a function of ρ, a fact that is only availble if $d \geq 3$. The transition from the above result to the large deviation result for the empirical process is in principle straight forward and is a matter of careful calculation and identification. First a superexponential tightness estimate for the empirical process $R_{N,\omega}$ is proved that allows one to interchange limits freely and provides a cover for various technical points along the route. The rate function for the empirical process is then calculated by computing rate functions for finite dimensional projections. The torus T^d is divided into a finite number of cells and the large deviation principle is reduced to a large deviation principle for the proportions of particles that lie in a specified sequence of cells at time points $0 < t_1 < t_2 < \cdots < T$. This is a weak form of tagging and reduces the problem to one with a finite but large number of colors, by assigning different colors to particles with different histories. When the dust settles, it is a matter of identification of the rate function by chasing through the limiting procedure as the number of cells in the partition and the number of inspection times become infinite. This is carried out in [7]. The big surprise is the reemergence of entropy in the final formula and the physical interpretation that the rate function has because of it. This is not evident in the formula for the large deviation rate for n-colors. Finally let us describe the rate functional. Given Q, we can compute the marginal densities $\rho(t,\theta)$ for $0 \leq t \leq T$. There is the initial rate $I_0(\rho(0,\cdot))$ which depends on how the choice of the initial condition is made. The rate function is given by

$$I(Q) = I_0(\rho(0,.)) + \inf_{b \in \mathcal{B}_{\rho(\cdot,\cdot)}} \left\{ \frac{1}{2} \int_0^T \int_{T^d} \rho(1-\rho) \left\langle C^{-1}b, b \right\rangle dt dy + H(Q; P_\rho^b) \right\}$$

Here P_ρ^b is a time dependent diffusion process determined by b and will have one dimensional marginal densities $\rho(t,y)$. The class $\mathcal{B}_{\rho(\cdot,\cdot)}$ is given by (4.4). The diffusion process P_ρ^b is defined for each admissible ρ, and admissible b in $\mathcal{B}_{\rho(\cdot,\cdot)}$ and has for its backward Kolmogorov generator

$$\mathcal{L}_t^{\rho,b} = \frac{1}{2} \nabla S(\rho(t,y)) \nabla + \frac{1}{2\rho} (S(\rho(t,y)) - C) \nabla \rho(t,y).\nabla + (1-\rho) b.\nabla$$

8. The asymmetric mean zero case

The case where $p(\cdot)$ is asymmetric but has the property $\sum_z z p(z) = 0$ is again a nongradient system and is therefore difficult. Even if one is only interested in the simplest of questions, namely the evolution of the particle density in hydrodynamic scales, the analysis is still hard. An added complication is the nonreversible nature of the dynamics due to the asymmetry of $p(.)$. In the application of the large deviation techniques to handle nonequilibrium, as well in establishing some of the properties of the system in equlibrium this causes difficulties. The fact that the mean is zero provides an important control in the form of a bound for the asymmetric part in terms of the symmetric part. At some level it is a bounded perturbation. This was exploited in [12], to show that the analog in this context of the heat equation (2.5) that appears in the symmetric case, is a nonlinear diffusion equation of the form

$$\frac{\partial \rho}{\partial t} = \frac{1}{2}\nabla \bar{C}(\rho)\nabla \rho \tag{8.1}$$

where the diffusion matrix $\bar{C}$ really depends on ρ and can be shown to be strictly larger than the matrix C that would result in the symmetric case provided $0 < \rho < 1$ and $p(.)$ is not symmetric.

9. The case of nonzero mean

The case where $\sum_x zp(z) = b \neq 0$ is qualitatively very different. The motion of the densities is hyperbolic rather than diffusive. The scaling reflects this and time is rescaled by a factor of N instead. A calculation shows

$$d\left(\frac{1}{N^d}\sum_x J(\frac{x}{N})\eta_t(x)\right) = A_J(t)dt + dM_J(t) \tag{9.1}$$

where

$$\begin{aligned} A_J(t) &= \frac{N}{N^d}\sum_{x,x'}\eta(x)(1-\eta(x'))p(x'-x)[J(\frac{x'}{N}) - J(\frac{x}{N})] \\ &\simeq \frac{1}{N^d}\sum_{x,x'}\eta(x)(1-\eta(x'))\left\langle(\nabla J)(\frac{x}{N}), b\right\rangle \\ &\simeq \int_{T^d}\rho(t,y)(1-\rho(t,y))\left\langle(\nabla J)(y), b\right\rangle dy \end{aligned}$$

Here $\rho(t,\cdot)$ is the macroscopic density of the system in the limit at time t. This leads to Burgers' equation for ρ

$$\frac{\partial \rho}{\partial t} + \langle b, \nabla \rho(1-\rho)\rangle = 0 \tag{9.2}$$

This equation in general develops shocks and the physical solution is characterized as the unique solution that satisfies certain entropy conditions. It has been shown that if the initial conditions are suitably chosen, for example, if the initial distribution of particles is a product measure with

$$P[\eta_0(x) = 1] = \rho_0(\frac{x}{N})$$

then for $t > 0$ and as $N \to \infty$ the empirical density converges in probability to the entropy solution of (9.2) with the given initial condition. For the best results in the general case see [8]. More detailed information is available in the 1-dimensional case. In the nearest neighbor totally asymmetric case, i.e.[p(1)=1] one can handle arbitrary initial data as was shown in [11].

Finally one can ask if viscosity corrections to Burgers' equation appear in some manner from the particle model. One can answer in the affirmative in two distinct ways, both for dimensions $d \geq 3$. One can ask for a small correction to Burgers' equation so that the empirical density is approximated by this modified equation to a higher order of accuracy. Another possibility is to start close to equlibrium i.e a small perturbation of constant or uniform initial density. Then this initial perturbation will move with the constant speed of $(1 - 2\rho)b$ in the time scale that was speeded up by N. It is possible that the perturbation dissipates in the time scale N^2 after recentering. And this dissipation should be governed by Burgers' equation with viscosity. Both questions have affirmative answers as was shown respectively in [4] and [1]. The viscosity coefficients are the same in both cases and identification of this coefficient is carried out by techniques developed for nongradient models. The situation is much more complex than in [12], because now the asymmetric part is far more dominant.

References

1. Esposito, R., Marra, R., and Yau, H. T., Diffusive limit of asymmetric simple exclusion, Reviews in Mathematical Physics, Vol 6. No 5A, (1994), 1233-1267.
2. Kipnis, C., Olla, S., and Varadhan, S. R. S., Hydrodynamics and large deviation for simple exclusion process, Comm. Pure Appl. Math. XLII (1989), 115–137.
3. Kipnis, C., and Varadhan, S. R. S., Central limit theorem for additive functionals of reversible Markov processes and application to simple exclusions, Comm. Math. Phys. 104 (1986), 1–19.
4. Landim, Claudio., Olla, S. and Yau, H.T., First order correction for the hydrodynamic limit of asymmetric simple exclusion processes in dimension $d \geq 3$, to appear in Communications in Pure and Applied Mathematics.
5. Ligget, Thomas M. Interacting Particle Systems, Springer-Verlag, New York, (1985).
6. Quastel, J., Diffusion of color in simple exclusion process, Comm. Pure Appl. Math. XLV (1992), 623–680.
7. Quastel, Jeremy.,Rezakhanlou, F., and Varadhan, S. R. S., Large Deviations for the symmetric simple exclusion process.(Preprint) 1995.
8. Rezakhanlou, F., Hydrodynamical limit for attractive particle systems on Z^d, Comm. Math. Phys. 140 (1991), 417–448.
9. Rezakhanlou, F., Propagation of Chaos for Symmetric Simple Exclusions, Communications on Pure and applied mathematics, Vol XLVII, (1994), 943-958.
10. Varadhan, S. R. S., Nonlinear diffusion limit for a system with nearest neighbor interactions, II, Asymptotic Problems in Probability Theory: Stochastic Models and Diffusions

on Fractals, K. D. Elworthy and N. Ikeda (Editors), Pitman Research Notes in Mathematics Series 283, 1991, 75–130.
11. Venkatasubramani, R., Ph.D. Thesis., New York University, 1995.
12. Xu, Lin, Ph.D. Thesis, New York University, 1993.

Lévy's stochastic area formula and Brownian motion on compact Lie groups

Shinzo Watanabe

Department of Mathematics, Graduate School of Science, Kyoto University, Kyoto 606-01, Japan

1. Introduction

Let (W, P) be the d-dimensional Wiener space: W be the space of continuous paths $W = \{w \in C([0,\infty) \to \mathbf{R}^d) | w(0) = 0\}$ and P be the standard d-dimensional Wiener measure on W. Then $w = (w^k(t))_{k=1}^d$ in W is a canonical realization of a d-dimensional Wiener process. Lévy's stochatic area is defined on the two-dimensional Wiener space by Itô's stochastic integral as follows:

$$S(t,w) = \frac{1}{2}\int_0^t w^1(s)dw^2(s) - w^2(s)dw^1(s).$$

Then we have the well-known Lévy's formula:

$$E\left[\exp(\alpha S(1,w))|w(1)=0\right] = \frac{\frac{\alpha}{2}}{\sin\frac{\alpha}{2}} \quad \text{if } |\alpha| < 2\pi. \tag{1.1}$$

In the general d-dimensional case, if $A = (A_{ij}) \in so(d)$, i.e., A is a skew symmetric real $d \times d$-matrix, we have (under a usual convention for the summation)

$$E\left[\exp\left(\frac{1}{2}A_{kl}\int_0^1 w^k(s)dw^l(s)\right)|w(1)=0\right] = \left(\det\left[\frac{1-e^{-A)}}{A}\right]\right)^{-\frac{1}{2}}. \tag{1.2}$$

Note that (1.1) is a particular case of (1.2) when $d = 2$ and $A_{ij} = \alpha\delta_{i1}\delta_{j2} - \alpha\delta_{i2}\delta_{j1}$. On the other hand, (1.2) can be easily deduced from (1.1) if we transform A by $T \in \mathrm{SO}(d)$ so that TA^tT is a block-diagonal form in 2×2 skew symmetric matrices.

The right-hand side (RHS) of (1.2) is just the $\hat{A}$-genus in the Atiyah-Singer index theorem if we substitute for A the curvature form. This connection has been made clear in a probabilistic proof of index theorems by Bismut (cf. [5], [6], [20]). On the other hand, Berline, Getzler and Vergne remarked a striking similarity between the $\hat{A}$-genus and the Jacobian of the exponential map on a compact Lie group. This connection has been studied in Chapter 5 of [4]. These observations clearly suggest that there is some connection between Lévy's formula (1.1) or (1.2) and the formula for Jacobian of exponential map on a compact Lie group. A motivation of this note is to see this connection more directly. In section 2, we study a short-time asymptotic property of the heat kernel on a compact Lie group in two different ways. On one hand, we apply a well-known result in the analysis on Riemannian manifolds that an off-diagonal short-time asymptotic of the heat kernel can be described by the Jacobian of exponential map or Ruse's invariant. If we connect this with the formula for the differential of exponential map in a Lie group, we see that the asymptotic is described by the RHS of (1.2). On the other hand, we study it probabilistically by using a stochastic differential equation (SDE) for Brownian motion on the group. Then we find that the asymptotic is described by the LHS in (1.2). This way of looking at Lévy's formulas (1.1) and (1.2) will perhaps provide us with a new proof or interpretation of these formulas which may be different from the standard proofs by diagonalization (cf. [14]) or by a h-path transform (a change of measure by a Cameron-Martin-Maruyama-Girsanov density, cf. [21], [11]).

In section 3, we give a remark on stochastic moving frames on Riemannian manifolds. As we know, this notion plays an important role in a probabilistic approach to analysis and geometry on vector bundles. A usual stochastic moving frame is realized as a horizontal diffusion on the frame bundle (cf. [13], [16]). Here we modify this by adding a random motion along the fibre so that this modified stochastic moving frame is a Brownian motion on the frame bundle.

Main probabilistic tools in this note are SDEs on manifolds and Brownian motions on Lie groups thereby constructed. We refer to [7] or [13] for details of SDEs on manifolds. We would remark that SDEs on differentiable manifolds have been first introduced by Itô [14] in an intrinsic form although the notion of Stratonovich differentials was not available at that time. We would also remark that Brownian motions on Lie groups have been constructed for the first time by Itô [15] by using the SDE method.

2. Brownian motions and heat kernels on compact Lie groups

Let G be a compact and connected Lie group of dimension d. For simplicity we assume that G is a matrix group, i.e., a multiplication subgroup embedded in the algebra M_N of all $N \times N$ matrices for some N. We use therefore the matrix notation freely; in particular, the right (left) translations and their differentials are denoted simply by the multiplications from the right (left). Sometimes, computations are made in the larger space M_N, being not necessarily confined in the group G; the final conclusion can be stated intrinsically in terms of G, however.

Let $\mathcal{G}$ be the Lie algebra of G and endow on it an Ad(G)-invariant inner product $< *, ** >$. By identifying the tangent space $T_e(G)$ at e ($e =$ the identity) with $\mathcal{G}$ and defining the inner product $(*, **)_{T_g(G)}$ on the tangent space $T_g(G)$ at $g \in G$ by

$$(\xi,\eta)_{T_g(G)} = < g^{-1}\xi, g^{-1}\eta >, \quad \xi,\eta \in T_g(G),$$

G is given a Riemannian structure so that G is a compact Riemannian manifold, indeed, a Riemannian symmetric space. Note that the Riemannian volume is a Haar measure of G. Let $a_1,\ldots,a_d$ be an orthonormal basis (ONB) in $\mathcal{G}$. Then the Laplacian Δ on G is given by

$$\Delta = \sum_{k=1}^{d} L_{a_k}^2 \tag{2.1}$$

where L_a, $a \in \mathcal{G}$, is the left-invariant vector field defined by

$$(L_a f)(g) = \frac{d}{dt} f(g \exp ta)|_{t=0}, \quad f \in C^\infty(G).$$

We note also that, for every $g \in G$ and $a \in \mathcal{G}$, the curve $c(t) = g \exp ta$, $t \in \mathbf{R}$, is a geodesic in G.

The exponential map $a \in \mathcal{G} \to \exp a \in G$ is a diffeomorphism between a neighborhood U in $\mathcal{G}$ of 0 and a neighborhood $\tilde{U}$ in G of the identity e. As usual, $a \in U$ is called the normal coordinate of $g = \exp a \in \tilde{U}$: more generally, for fixed $h \in G$, $a \in U$ is called the normal coordinate around h of $g = \exp a \cdot h \in \tilde{U}h$. The following rather easy fact in Lie group theory plays a key role in this note: The differential of the exponential map at $a \in \mathcal{G}$:

$$d_a \exp : \mathcal{G} \to T_{\exp a}(G)$$

is givev by

$$d_a \exp(b) = \exp a \cdot \frac{1 - e^{-\mathrm{ad}(a)}}{\mathrm{ad}(a)}(b), \quad b \in \mathcal{G} \tag{2.2}$$

where $\mathrm{ad}(a) \in \mathrm{End}(\mathcal{G})$ is defined by $\mathrm{ad}(a)(b) = [a,b] = ab - ba$. Hence Ruse's invariant $\theta(a) = \det(d_a \exp)$ is given by

$$\theta(a) = \det\left(\frac{1 - e^{-\mathrm{ad}(a)}}{\mathrm{ad}(a)}\right) \tag{2.3}$$

and the Riemannian volume dg in the normal coordinate is given by

$$dg = \theta(a)da \tag{2.4}$$

where da is the Lebesgue measure in $\mathcal{G}$.

Let $p(t,g,h), t > 0, g,h \in G$, be the heat kernel on G, i.e. $u(t,g) = \int_G p(t,g,h)f(h)dg$ solves uniquely the intial value problem of heat equation:

$$\frac{\partial u}{\partial t} = \frac{1}{2}\Delta u, \quad u|_{t=0} = f.$$

We see easily that $p(t,g,h) = p(t,e,g^{-1}h) = p(t,e,h^{-1}g)$. Furthermore, $p(t,e,g)$ is invariant under the inner automorphisms: $g \to hgh^{-1}, h \in G$, and under the transformation $g \to g^{-1}$. For any smooth Riemannian manifold M of dimension d, the following aymptotic expansion (the Minakushisudarum-Pleijel expansion) is well-known: if x and y are close enough, then as $t \to 0$,

$$\begin{aligned} p(t,x,y) &= (2\pi t)^{-d/2} \exp\left(\frac{-d(x,y)^2}{2t}\right) \\ &\times \left(c_0(x,y) + tc_1(x,y) + \cdots + t^n c_n(x,y) + O(t^{n+1})\right), \end{aligned}$$

$d(x,y)$ being the Riemannian distance and $c_0(x,y) = \theta_x(y)^{-1/2}$ where

$$\theta_x(y) = |\det(d_y \exp_x)|$$

is Ruse's invariant or the determinant at y of the Jacobi vector field along the geodesic from x to y. (As usual, $\exp_x$ is the exponential map around x and d_y is the differential at y.) Cf. e.g.,[3] or [4] for a standard analytical proof and [8], [12] and [17] for various probabilistic proofs. Combining this with (2.2), we have

Theorem 2.1. *There exists $\kappa > 0$ such that, for any $a \in \mathcal{G}$ with $||a|| \le \kappa$,*

$$p(t,e,\exp a) \sim (2\pi t)^{-d/2} e^{-\frac{||a||^2}{2t}} \left[\det\left(\frac{1-e^{-\mathrm{ad}(a)}}{\mathrm{ad}(a)}\right)\right]^{-1/2} \tag{2.5}$$

as $t \to 0$ uniformly in $a \in \mathcal{G}$ such that $||a|| \le \kappa$.

Although we derived Theorem 2.1 from a general analysis on Riemannian manifolds, we would remark that the asymptotic of the heat kernel can be studied by Lie group theory based on the explicit form of the Laplacian acting on functions invariant under the inner automorphisms.

Example 2.1. Let $G = \mathrm{SO}(3)$ and let a_1, a_2, a_3 be an ONB in $so(3)$ such that

$$[a_1, a_2] = a_3, \quad [a_2, a_3] = a_1, \quad [a_3, a_1] = a_2.$$

For every $g \in G$, there exists unique $\alpha, 0 \le \alpha \le \pi$ such that $hgh^{-1} = \exp \alpha a_1$. We denote $\alpha = \alpha(g)$. For a smooth function $F(\alpha)$ on $[0,\pi]$ such that $f(g) = F(\alpha(g))$ is a smooth function on G, $f(g)$ is invariant under the inner automorphisms of G and, conversely, every such function has this representation. Then setting $j(\alpha) = \sin\alpha/2$, we have (cf.[2])

$$(\Delta f)(g) = \left[\frac{1}{j}\left(\frac{d^2}{d\alpha^2} + \frac{1}{4}\right)(j \cdot F)\right](\alpha(g)).$$

From this, we obtain that

$$p(t,e,g) = (2\pi t)^{-\frac{3}{2}} \exp\left\{\frac{1}{4}t\right\} \sum_{n \in \mathbf{Z}} \exp\left\{-\frac{|\alpha(g) + 2n\pi|^2}{2t}\right\} \left(\frac{\frac{\alpha(g)}{2} + n\pi}{\sin(\frac{\alpha(g)}{2} + n\pi)}\right).$$

Note that Ruse's invariant (2.3) is given by

$$\theta(\alpha a_1) = \left(\frac{2}{\alpha}\sin\frac{\alpha}{2}\right)^2.$$

Now (2.5) is easily concluded from these explicit formulas. This formula for the heat kernel on SO(3) is clearly an analogy of the well-known expression of the heat kernel on SO(2) : by identifying SO(2) with the circle $\mathbf{R}/(2\pi\mathbf{Z})$,

$$p(t,0,\alpha) = (2\pi t)^{-1/2}\sum_{n\in\mathbf{Z}}\exp\left\{-\frac{|\alpha+2n\pi|^2}{2t}\right\}.$$

A generalization of these formulas to the case of semisimple compact Lie groups has been obtained by Eskin (cf. [9], [2], [1]).

Let (W,P) be a d-dimensional Wiener space. Fixing an ONB $a_1,\dots,a_d$ in $\mathcal{G}$ as above, we consider the following SDE on G:

$$dg_t = \sum_{k=1}^{d} L_{a_k}(g_t)\circ dw^k(s), \quad g_0 = e. \tag{2.6}$$

Here $\circ$ denotes the stochastic differential in the Stratonovich sense (cf. [14]). If we define $\bar{w}(t) = \sum_{k=1}^{d} w^k(t)a_k$, then $\bar{w}$ is a $\mathcal{G}$-valued Wiener process and the equation (2.6) can be also written in the matrix notation as follows:

$$dg_t = g_t\circ d\bar{w}(t), \quad g_0 = e. \tag{2.7}$$

The solution g_t is denoted by $g(t,w)$ or $g(t,\bar{w})$. The solution of the SDE with the initial condition $g_0 = g$ is given simply by $g_t = gg(t,\bar{w})$. By (2.1) this is a Brownian motion on G in the usual sense that $u(g_t) - u(g_0) - \int_0^t \frac{1}{2}\Delta u(g_s)ds$ is a martingale for every smooth function u on G.

In the following, we compute the asymptotic of the heat kernel $p(t,e,g)$ directly by a probabilistic method as discussed in [19] or [18] and relate this with a Wiener functional expectation as given by the LHS of (1.2). This, combined with (2.5), provides a (perhaps new) proof of the formula (1.2). For this, we first construct the heat kernel by a probabilistic method of generalized Wiener functional expectations in the Malliavin calculus (cf. [13] or [19] for notions and notations). Let $\delta_g(\cdot)$ be the Dirac delta function on G with pole at g; i.e. the Schwartz distribution on G such that

$$\int_G \delta_g(h)u(h)dh = u(g), \quad u\in C^\infty(G),$$

where dh is the Riemannian volume of G. Then the composite $\delta_g(g(t,\bar{w}))$ can be defined as a generalized Wiener functional and we have

$$p(t,e,g) = E[\delta_g(g(t,\bar{w}))] \tag{2.8}$$

in the sense of generalized expectations in the Malliavin calculus. For $\epsilon > 0$, let $g^\epsilon(t,\bar{w})$ be the solution to the following SDE:

$$dg_t = \epsilon g_t \circ d\bar{w}(t), \quad g_0 = e. \tag{2.9}$$

By the scaling property of Wiener process, (2.8) can be rewritten as

$$p(\epsilon^2, e, g) = E[\delta_g(g^\epsilon(1, \bar{w}))]. \tag{2.10}$$

Next, for $a \in \mathcal{G}$, define $h_a \in C([0,1] \to \mathcal{G})$ by $h_a(t) = at$. We regard h_a as an element in the Cameron-Martin space:

$$H = \{h \in C([0,1] \to \mathcal{G}) \mid h(0) = 0,\ \dot{h} = \frac{dh}{dt} \in L^2([0,1] \to \mathcal{G})\},$$

endowed with the norm $||h|| = ||\dot{h}||_{L^2([0,1]\to\mathcal{G})}$. For each fixed $t \in [0,1]$ and $\epsilon > 0$, $g^\epsilon(t, \bar{w} + h_a/\epsilon)$ is a G-valued Wiener functional and, by embedding G into M_N as above, it is a smooth M_N-valued Wiener functional in the sense of elements in $\mathcal{D}^\infty_{\infty-}(W \to M_N)$. It has an asymptotic expansion in ϵ as $\epsilon \to 0$ in the space $\mathcal{D}^\infty_{\infty-}(W \to M_N)$:

$$g^\epsilon(t, \bar{w} + h_a/\epsilon) = g^{(0)}(t, \bar{w}) + \epsilon g^{(1)}(t, \bar{w}) + \epsilon^2 g^{(2)}(t, \bar{w}) + \cdots.$$

M_N-valued Wiener functionals $g^{(n)}(t, \bar{w}) = g_t^{(n)}(\bar{w})$ are easily computed: $g_t = g^\epsilon(t, \bar{w} + h_a/\epsilon)$ satisfies SDE:

$$dg_t = \epsilon g_t \circ d\bar{w}(t) + g_t a dt, \quad g_0 = e,$$

and hence, we have the following series of equations:

$$dg_t^{(0)} = g_t^{(0)} a dt, \quad g_0^{(0)} = e$$

and

$$dg_t^{(n)} = g_t^{(n-1)} \circ d\bar{w}(t) + g_t^{(n)} a dt, \quad g_0^{(n)} = 0 \quad \text{for } n = 1, 2, \ldots.$$

Hence,

$$g_t^{(0)} = \exp ta \quad \text{and} \quad d[g_t^{(n)} \exp(-ta)] = g_t^{(n-1)} \circ d\bar{w}(t) \exp(-ta).$$

Setting

$$\hat{w}(t) = \int_0^t \mathrm{Ad}(\exp sa) d\bar{w}(s) = \int_0^t \exp sa \cdot d\bar{w}(s) \cdot \exp(-sa),$$

$\hat{w}(t)$ is another $\mathcal{G}$-valued Wiener process and we finally deduce that

$$g_t^{(n)} = \int_0^t \int_0^{t_1} \cdots \int_0^{t_{n-1}} d\hat{w}(t_n) \circ \cdots \circ d\hat{w}(t_1) \cdot \exp ta. \tag{2.11}$$

In particular, we have

(2.12)
$$g^\epsilon(1, \bar{w} + h_a/\epsilon) = \exp a + \epsilon \hat{w}(1) \cdot \exp a + \epsilon^2 \int_0^1 \hat{w}(s) \circ d\hat{w}(s) \cdot \exp a + O(\epsilon^3).$$

Next, we convert the asymptotic expansion (2.12) into the asymptotic expansion in the normal coordinate around $\exp a$. The normal coordinate around $\exp a$ was defined in some neighborhood $\tilde{U} \exp a$ and we extend it to the whole G suitably; the way of this extension is irrelevant to the following computation. In M_N, we have the identity for small $\epsilon > 0$:

$$(e + \epsilon a_1 + \epsilon^2 a_2 + \cdots) \cdot \exp a = \exp\{\epsilon b_1 + \epsilon^2 b_2 + \cdots\} \cdot \exp a$$

under the uniquely determined relation between $a_1, a_2, \ldots$ and $b_1, b_2, \ldots$ given by

$$\begin{aligned} b_1 &= a_1, \quad b_2 = a_2 - a_1^2/2, \\ b_3 &= a_3 - \frac{1}{2}\left\{a_1(a_2 - \frac{a_1^2}{2}) + (a_2 - \frac{a_1^2}{2})a_1\right\} - \frac{a_1^3}{6}, \quad \cdots . \end{aligned}$$

Then (2.12) is written in the normal coordinate around $\exp a$ as

$$\begin{aligned} g^\epsilon(1, \bar{w} + h_a/\epsilon) &= \epsilon \hat{w}(1) + \epsilon^2 (\int_0^1 \hat{w}(s) \circ d\hat{w}(s) - \hat{w}(1)^2/2) + O(\epsilon^3) \\ &= \epsilon \hat{w}(1) + \frac{\epsilon^2}{2} \int_0^1 [\hat{w}(s),\ d\hat{w}(s)] + O(\epsilon^3). \end{aligned}$$

Since θ given by (2.3) satisfies $\theta(0) = 1$, it is easy to see that the Dirac delta function $\delta_{\exp a}(\cdot)$ is given in the normal coordinate around $\exp a$ by $\delta_0(\cdot)$ where δ_0 is the usual Dirac delta function on the Euclidean space $\mathcal{G}$. Then we can evaluate the generalized expectation (2.10) with $g = \exp a$ in the same way as [19] or [18]: Here we give an outline of the calculation only, leaving to [19] or [18] the technical part of justification which consists of the truncation arguments combined with the large deviation method.

By the Cameron-Martin theorem applied to the transformation

$$\bar{w} \to \bar{w} + h_a/\epsilon,$$

$$\begin{aligned} p(\epsilon^2, e, \exp a) &= E[\delta_{\exp a}(g^\epsilon(1, \bar{w}))] \\ &= E\left[\exp\left\{-\frac{\|h_a\|^2}{2\epsilon^2} - \frac{(h_a, \bar{w})}{\epsilon}\right\} \cdot \delta_{\exp a}(g^\epsilon(1, \bar{w} + h_a/\epsilon))\right] \end{aligned}$$

where $(h_a, \bar{w})$ is the generalized Cameron-Martin inner product or Wiener's integral. As is easily seen, $\|h_a\| = \|a\|$ and $(h_a, \bar{w}) = \langle a, \hat{w}(1) \rangle$. Then, writing everything in the normal coordinate around $\exp a$, we see that the above generalized expectation is equal to

$$\exp\left\{-\frac{||a||^2}{2\epsilon^2}\right\} E\left[\exp\left\{-\frac{< a, \hat{w}(1) >}{\epsilon}\right\}\right.$$
$$\left.\times\delta_0\left(\epsilon\{\hat{w}(1)+\frac{\epsilon}{2}\int_0^1[\hat{w}(s), d\hat{w}(s)]+O(\epsilon^2)\}\right)\right]$$
$$= \epsilon^{-d}\exp\left\{-\frac{||a||^2}{2\epsilon^2}\right\}$$
$$\times E\left[\exp\left\{\frac{1}{2} < a, \int_0^1[\hat{w}(s), d\hat{w}(s)] + O(\epsilon) >\right\}\delta_0(\hat{w}(1)+O(\epsilon))\right]$$
$$\sim \epsilon^{-d}\exp\left\{-\frac{||a||^2}{2\epsilon^2}\right\} E\left[\exp\left\{\frac{1}{2} < a, \int_0^1[\hat{w}(s), d\hat{w}(s)] >\right\}\delta_0(\hat{w}(1))\right]$$
$$= \epsilon^{-d}\exp\left\{-\frac{||a||^2}{2\epsilon^2}\right\} E\left[\exp\left\{\frac{1}{2}(\mathrm{ad}(a), \int_0^1 \hat{w}(s)\otimes d\hat{w}(s))_{\mathrm{HS}}\right\}\delta_0(\hat{w}(1))\right]$$
$$= (2\pi\epsilon^2)^{-\frac{d}{2}}\exp\left\{-\frac{||a||^2}{2\epsilon^2}\right\}$$
$$\times E\left[\exp\left\{\frac{1}{2}(\mathrm{ad}(a), \int_0^1 \hat{w}(s)\otimes d\hat{w}(s))_{\mathrm{HS}}\right\}|\hat{w}(1)=0\right].$$

Here we used the following identity which is a consequence of $\mathrm{Ad}(G)$-invariance of the inner product $< *, ** >$ on $\mathcal{G}$: for $a, b, c \in \mathcal{G}$,

$$< a, [b, c] >= - < [b, a], c >=< \mathrm{ad}(a)b, c >= (\mathrm{ad}(a), b\otimes c)_{\mathrm{HS}}$$

where $(*, **)_{\mathrm{HS}}$ is the Hilbert-Schmidt inner product on $\mathcal{G}\otimes\mathcal{G} \cong \mathrm{End}(\mathcal{G})$. Thus, we have obtained

$$p(t, e, \exp a) \sim (2\pi t)^{-d/2} e^{-\frac{||a||^2}{2t}} \times C$$

as $t \to 0$ where the constant C is given by the LHS of (1.2) with $A = \mathrm{ad}(a)$. In particular, if $G = \mathrm{SO}(3)$ and $a = \alpha a_3$ as in Ex.1, then C is given by the LHS of (1.1).

3. Some remarks on stochastic moving frames

Let M be a connected, oriented and compact Riemannian manifold of dimension d. The stochastic moving frame is usually realized as the horizontal diffusion on the bundle $O(M)$ of orthonormal frames (cf. [13], [16]). It may be a hypoelliptic diffusion but is never an elliptic diffusion: Here the term 'elliptic diffusion' means that its generator is an elliptic differential operator. $O(M)$ is a principal fibre bundle over M with the structure group $\mathrm{SO}(d)$ acting from the right and we can define a natural Riemannian structure on $O(M)$. Then we can define another stochastic moving frame as an elliptic diffusion, indeed as a Brownian motion, on $O(M)$. Such a stochastic moving frame has been already introduced, e.g., by Fang [10].

The Riemannian structure is introduced as follows: Let $r = (x, \mathbf{e} = [e_k]) \in O(M)$ where $x \in M$ and $e_k, k = 1, \ldots d$, forms an positively oriented ONB in the tangent space $T_x(M)$. For $\xi = (\xi^\alpha) \in \mathbf{R}^d$ and $\eta = (\eta^\alpha_\beta) \in so(d)$ so that $\eta^\alpha_\beta = -\eta^\beta_\alpha$, a tangent vector $X_{\xi,\eta} \in T_r(O(M))$ is defined by $X_{\xi,\eta} = \dot{c}(0)$ where $c(t)$ is a smooth curve in $O(M)$ determined by $c(t) = \tilde{x}(t) \exp t\eta$. Here $\tilde{x}(t)$ is the horizontal lift starting at r of the geodesic $x(t)$ in M such that $x(0) = x$ and $\dot{x}(0) = \xi^\alpha e_\alpha$. It is clear that

$$X_{\xi,\eta} = X_{\xi,0} + X_{0,\eta} \quad \text{and} \quad T_r(O(M)) = \{X_{\xi,\eta} | \xi \in \mathbf{R}^d, \eta \in so(d)\}$$

so that

$$T_r(O(M)) = H_r(O(M)) \oplus V_r(O(M)) \quad \text{(a direct sum)}$$

where $H_r(O(M)) = \{X_{\xi,0} | \xi \in \mathbf{R}^d\}$ and $V_r(O(M)) = \{X_{0,\eta} | \eta \in so(d)\}$. We introduce a Riemannian norm on $T_r(0(M))$ by

$$||X_{\xi,\eta}||^2 = ||X_{\xi,0}||^2 + ||X_{0,\eta}||^2 = |\xi|^2 + \frac{1}{2}\text{trace } \eta \cdot^t \eta$$

so that $H_r(O(M))$ and $V_r(O(M))$ are orthogonal to each other. Note that $< \eta, \eta' >= \frac{1}{2}\text{trace } \eta \cdot^t \eta'$ is an Ad(SO(d))-invariant inner product on the Lie algebra $so(d)$. It is easy to see that a horizontal lift on $O(M)$ of a geodesic in M and, for $r \in O(M)$ and $\eta \in so(d)$, a vertical curve $r \cdot \exp t\eta$ are all geodesics on the Riemannian manifold $O(M)$.

Let L_k, $k = 1, \ldots d$, be the system of horizontal vector fields on $O(M)$ (cf. [14]). For $a \in so(d)$, let L_a be the vector field on $O(M)$ defined by $L_a(r) = \dot{c}(0)$ where $c(t)$ is a vertical curve defined by $c(t) = r \cdot \exp ta$. Choose an ONB $E_1, \ldots, E_{d(d-1)/2}$ in $so(d)$ with respect to the Ad(SO(d))-invariant inner product defined above. Then

$$\Delta^{(2)}_{O(M)} := \sum_{k=1}^{d} L_k^2 + \sum_{\gamma=1}^{d(d-1)/2} L^2_{E_\gamma} = \Delta^{(1)}_{O(M)} + \sum_{\gamma=1}^{d(d-1)/2} L^2_{E_\gamma}$$

is the Laplacian (i.e. Laplace-Beltrami operator) with respect to the Riemannian structure introduced above while $\Delta^{(1)}_{O(M)} = \sum_{k=1}^d L_k^2$ is the horizontal Laplacian. Let (P, W) be the d-dimensional Wiener space and $(\tilde{W}, \tilde{P})$ be the $d(d-1)/2$-dimensional Wiener space. On the product space $(W \times \tilde{W}, P \times \tilde{P})$, we consider the following three SDEs:

$$dr(t) = \sum_{k=1}^{d} L_k(r(t)) \circ dw^k(t), \quad r(0) = r, \tag{3.1}$$

$$(3.2)\quad dr(t) = \sum_{k=1}^{d} L_k(r(t)) \circ dw^k(t) + \sum_{\gamma=1}^{d(d-1)/2} L_{E_\gamma}(r(t)) \circ d\tilde{w}^\gamma(t), \; r(0) = r,$$

$$(3.3)\qquad dg(t) = \sum_{\gamma=1}^{d(d-1)/2} g(t)E_\gamma \circ d\tilde{w}^\gamma(t), \quad g(0) = e.$$

The unique solutions are denoted by $r^{(1)}(t, r, w)$, $r^{(2)}(t, r, (w, \tilde{w}))$ and $g(t, \tilde{w})$, respectively. Then $r^{(2)}(t)$ is the Brownian motion while $r^{(1)}(t)$ is the horizontal diffusion on $O(M)$. $g(t)$ is the Brownian motion on the structure group $\mathrm{SO}(d)$ starting at e.

If we define another d-dimensional Wiener process $\hat{w}(t) = (\hat{w}^\alpha(t))$ by

$$\hat{w}^\alpha(t) = \sum_{\beta=1}^{\alpha} \int_0^t g_\alpha^\beta(s, \tilde{w}) \circ dw^\beta(s) = \sum_{\beta=1}^{\alpha} \int_0^t g_\alpha^\beta(s, \tilde{w}) \cdot dw^\beta(s),$$

then $\hat{w}$ is also independent of the Wiener process $\tilde{w}$ and it is easy to deduce by the Itô formula that

$$r^{(2)}(t, r, (\hat{w}, \tilde{w})) = r^{(1)}(t, r, w) \cdot g(t, \tilde{w}).$$

Thus we see that the stochastic moving frame in the sense of Brownian motion $O(M)$ can be simply constructed from the stochastic moving frame in the sense of horizontal diffusion on $O(M)$ by the independent action of the Brownian motion on the structure group.

As for the short time asymptotic property of the heat kernel on $O(M)$, we have the following:

Theorem 3.1. *There exists $\kappa > 0$ such that if $r \in O(M)$ and $a \in so(d)$, such that $\|a\| \le \kappa$, then*

$$(3.4)\quad p(t, r, r\exp a)$$

$$\sim \quad (2\pi t)^{-d(d+1)/2} e^{-\frac{\|a\|^2}{2t}} \left[\det\left(\frac{1 - e^{-J_r \cdot a/2}}{J_r \cdot a/2}\right) \det\left(\frac{1 - e^{-\mathrm{ad}(a)}}{\mathrm{ad}(a)}\right)\right]^{-1/2}$$

as $t \to 0$ uniformly in $r \in O(M)$ and $a \in so(d)$ such that $\|a\| \le \kappa$. Here, $J_r \cdot a \in so(d)$ is defined by

$$(J_r \cdot a)_l^k = \sum_{i,j} J_{ijkl}(r) a_j^i$$

where $J_{ijkl}(r) = e_i^s e_j^t e_k^u e_l^v R_{stuv}(x)$ (the scalarization of the Riemann curvature) in the local coordinate $r = (x = (x^i), \mathrm{e} = (e_j^i)) : e_k = e_k^i(\partial/\partial x_i)|_x$.

The proof can be given either by computing geometrically the Jacobi vector field along the vertical geodesic $r\exp ta$, thereby its determinant or Ruse's invariant, as was given by Berline, Getzler and Vergne ([4], Chapter 5, Th.5.4), or by a similar probabilistic proof as above in which the final step is reduced to the formula (1.2).

References

1. M. T. Arede. Géométrie du noyau de la chaleur sur les variétés, Thése de Doctorat de 3ème Cycle, Séptembre 1983, Université d'Aix-Marseille II.
2. A.-I. Benabdallah. Noyau de diffusion sur les espaces homogènes compacts, Bull. Soc. math. France **101**(1973), 265-283.
3. M. Berger, P. Gauduchon and E. Mazet. *Le Spéctre d'une Variété Riemannienne*, **LNM 194**, Springer,1971
4. N. Berline, E. Getzler and M. Vergne. *Heat Kernels and Dirac Operators*, Springer, 1991
5. J.-M. Bismut. The Atiyah-Singer theorems: a probabilistic approach, I. index theorem, II. the Lefschetz fixed point formulas, Jour. Funct. Anal. **57**(1984), 55-99 and 329-348
6. J.-M. Bismut. Formule de localisation et formules de Paul Lévy, Astérisque **157-158**, Société Math. France(1988), 37-58
7. D. K. Elworthy. *Stochastic Differential Equations on Manifolds*, Lecture Note Series **70**, London Math. Soc.,1982
8. D. K. Elworthy and A. Truman. The diffusion equation and classical mechanics: an elementary formula, *Stochastic Processes in Quantum Physics*, **LNP 173**, Springer(1982), 136-146
9. L. D. Eskin. The heat equation and the Weierstrass transform on certain symmetric Riemannian spaces, AMS Translation, **75**(1968), 239-254
10. S. Fang. Rotations et quasi-invariance sur l'espace de chemins, Potential Analysis **4**(1995), 67-77
11. K. Helmes and A. Schwane. Lévy's stochastic area formula in higher dimensions, Jour. Funct. Anal. **54**(1983), 177-192
12. N. Ikeda. Probabilistic methods in the study of asymptotics, *Ecole d'Été de Probabilités de Saint-Flour XVIII-1988*, **LNM 1427**, Springer(1990), 197-325
13. N. Ikeda and S. Watanabe. *Stochastic Diffential Equations and Diffusion Processes*, Second Edition, North-Holland/Kodansha, 1988
14. K. Itô. Stochastic differential equations in a differentiable manifold, Nagoya Math. J. **1**(1950), 35-47
15. K. Itô. Brownian motions in a Lie group, Proc. Japan Acad. **26**(1950), 4-10
16. P. Malliavin. *Géométrie Différentielle Stochastique*, Les Presse de l'Université Montréal, 1978
17. S. A. Molchanov. Diffusion processes and Riemannian geometry, Russian Math. Surveys **30** (1975), 1-63
18. S. Takanobu and S. Watanabe. Asymptotic expansion formulas of the Schilder type for a class of conditional Wiener functional integrations, *Asymptotic problems in probability theory: Wiener functionals and asymptotics, Proc. Taniguchi Symp. 1990*, Pitman Research Notes in Math. Ser. **284**, Longman(1993), 194-241
19. S. Watanabe. Analysis of Wiener functionals (Malliavin calculus) and its applications to heat kernels, Annals of Probab. **15**(1987), 1-39
20. S. Watanabe. Short time asymptotic problems in Wiener functional integration theory. Applications to heat kernels and index theorems, *Stochastic Analysis and Related Topics II, Proc. Silivri 1988*, **LNM 1444**, Springer(1990), 1-62.
21. M. Yor. The law of some Brownian functionals, Proc. ICM, Kyoto 1990, II, Springer(1991), 1105-1112

References

[illegible]

Principal values of Brownian local times and their related topics

Toshio Yamada

Department of Mathematics Ritsumeikan University BKC, Kusatsu, Shiga 525-77, Japan

Introduction

The study of the Cauchy's principal value of Brownian local times has started with Ito,K. and McKean,H.P.Jr. [19]. In this article, we would like to see how their idea combined with a generalization of the Ito formula in stochastic calculus has grown up to be an interesting branch in the study of Brownian motion. After Ito-McKean, Ezawa,H.,Klauder,J.R. and Shepp,A. ([10]) discussed principal values of local times in an investigation of vestigical effects of the singular potential $\lambda V(x) = \lambda|x-c|^{-\alpha}$ where $\lambda \to 0$ in the Feynmann-Kac integral. It seems remarkable that in the early 70's, when few payed attention to these notions, they have been already used for some physical purposes by physicists cooperating with a mathematician. From the end of the 70's to the beginning of the 80's two studies have been published, one by Fukushima,M. [14] and the other by Föllmer,H. [13], and they have played the role of the corner stone in the development of a new branch in stochastic analysis. Föllmer has proposed stochastic analysis of Dirichlet process which is a generalization of semi-martingale. The class of additive functionals of zero energy has been investigated by Fukushima in the scheme of symmetric Hunt processes. These additive functionals can not be treated in the frame of the theory of semi-martingales. The studies have been essentially concerned with generalizations of the Ito-formula. Owing to their propositions, we have rich possibilities to apply the Ito calculus to wider class of processes than that of semi-martingales. Concrete and systematic studies of principal values of Brownian local times have been started with Yor,M. [31] and Yamada,T. ([26] and [27]). In the early of 80's, we showed that principal values of local times can be represented as Hilbert transform or fractional derivatives of local times. Thanks to these analytic tools, we see naturally that some properties such as the Hölder continuity inherit from Brownian local times to their principal values. Fractional derivatives of the local times have arisen naturally in some new limit theorems for occupation times of a linear Brownian motion ([28]). Similar limits theorems in the context of stable processes have been obtained recently by Fitzsimmons, P.J. and Getoor,R.K. ([12]). Related topics have been discussed also by Pitman,J.W. and Yor,M. ([23]). Since principal values of local times do not belong to the class of semi-martingales, stochastic integrals based on them have to be defined in a different way from those based on semi-martingales. In this direction, Nakao,S. [21] and Bertoin,J. [2] contributed new ideas on stochastic integrals and chain rules for functionals of zero energy. These stochastic integrals are closely related with Young-Stieltjes integrals (see Youg, L.C. [33]). In connection with their definition of stochastic integrals, p-variation properties of principal values of the local times have been investigated by Bertoin [6]. He has proposed in [6] also some concrete Ito type formulae. His results on p-variation are generalized by Fitzsimmons, P.J. and Getoor, R.K. ([12]) to the case of local times of a stable process. Nakao,S and Nakane,K are investigating p-variation problems for additive functionals of energy zero of a symmetric Hunt process. They also continue to study Young-Stieltjes integrals based on functionals of zero energy [un published]. Not so much seems to be known about the distributions of concerned processes, but an exception is the paper by Biane,Ph. and Yor,M. [8] in which the joint Fourier-Laplace transform of Cauchy's principal value of Brownian local times has been computed. Their computation is extended by Fitzsimmons and Getoor [11] in the context of a symmetric Lévy process. As a nice application, we cite a paper by Bertoin [3] in which he discussed the spectral theory of strings.

Recently,it has been shown in [29] that any continuous one dimensional Brownian additive functional of locally of zero energy can be represented via convolution type transforms of local times.

In reviewing these results, we are really impressed by a fact that a problem in the excellent book [19] has yielded such a variety of theorems.

1. Hilbert transform of local times

Consider a continuous semi-martingale X_t. The local time L_t^a at a of X_t is defined by the Ito-Tanaka formula

$$|X_t - a| = |X_0 - a| + \int_0^t \operatorname{sgn}(X_s - a)\, dX_s + L_t^a \tag{1-1}$$

There exists a version of L_t^a such that $(a,t) \mapsto L_t^a$ is continuous with respect to t and càdlàg with respect to a. The following occupation time formula holds for any non-negative Borel function $\Phi(x)$ and for all $t \geq 0$:

$$\int_0^t \Phi(X_s)\, d < X >_s = \int_{-\infty}^{\infty} \Phi(a) L_t^a \, da \tag{1-2}$$

In the case where $X_t = B_t$ is a linear Brownian motion, there exists a version of L_t^a such that $a \to L_t^a$ is Hölder continuous of order α as well as uniformly continuous on any compact set, where $0 \leq \alpha < \frac{1}{2}$. Let C_t^a be the continuous additive functional of B_t which corresponds to Cauchy's principal value

$v.p.\frac{1}{x-a}$. More precisely C_t^a is defined by,

(1-3)
$$C_t^a = 2(B_t - a)\log|B_t - a| - 2(B_0 - a)\log|B_0 - a| - 2(B_t - a) + 2(B_0 - a) - 2\int_0^t \log|B_s - a|\, dB_s.$$

Next, we will define the additive functional H_t^a which corresponds to the Hadamard's finite part $p.f.x_+^{-1-\alpha}$, $0 < \alpha < \frac{1}{2}$, by

$$H_t^a(-1-\alpha) = 2\frac{(B_t - a)_+^{1-\alpha}}{(-\alpha)(1-\alpha)} - 2\frac{(B_0 - a)_+^{1-\alpha}}{(-\alpha)(1-\alpha)} - 2\int_0^t \frac{(B_s - a)_+^{-\alpha}}{(-\alpha)}\, dB_s \tag{1-4}$$

Then we have following representations for these functionals (Yamada [26]).

Theorem (1-1). *The following equalities hold in L^2 sense.*

$$C_t = C_t^0 = \lim_{\epsilon \to 0+} \{\int_0^t I_{(-\infty,-\epsilon]}(B_s)\frac{ds}{B_s} + \int_0^t I_{[\epsilon,\infty)}(B_s)\frac{ds}{B_s}\} \tag{1-5}$$

(1-6)
$$H_t(-1-\alpha) = H_t^0(-1-\alpha) = \lim_{\epsilon \to 0+} \{\epsilon^{-\alpha}\frac{L_t^0}{(-\alpha)} + \int_0^t I_{[\epsilon,\infty)}(B_s)B_s^{-1-\alpha}\, ds\}$$

There exists a version of $(t, a, \omega) \mapsto C_t^a(\omega)$ $((t, a, \omega) \mapsto H_t^a(-1-\alpha)(\omega))$ such that it is measurable in three variables (t, a, ω) and continuous in t. Let the function $g(x)$ belong to $L^2(R^1)$. Consider the Hilbert transform of the function g :

$$(\mathcal{H}g)(x) = \frac{1}{\pi}(v.p.\frac{1}{x} * g)(x) \tag{1-7}$$

It is well known that $\mathcal{H}g$ also belongs to $L^2(R^1)$ and the inverse of the operator $\mathcal{H}$ is equal to $-\mathcal{H}$.

We now introduce a certain fractional derivative which corresponds to Hadamard's finite part. Let a function f belong to $L^1(R^1)$ satisfying a global Hölder condition of order β such that $|f(x) - f(y)| \le C(f, \beta)|x - y|^\beta$, $\forall x, y \in R^1$.

Let $0 < \alpha < \beta < 1$. We define

$$D_\pm^\alpha f(x) = \frac{1}{\Gamma(-\alpha)}\int_0^\infty y^{-1-\alpha}[f(x \pm y) - f(y)]\, dy. \tag{1-8}$$

D_-^α and D_+^α are well known one sided fractional derivatives of order α.

The following Theorem is based on Yor [31] and Yamada [27].

Theorem (1-2).

(1) *The functional C_t^a can be represented via Hilbert transform of Brownian local times as follows :*

$$C_t^a = \pi(\mathcal{H}L_t^{\cdot})(a). \tag{1-9}$$

(2) *The fractional derivative of Brownian local times can be represented by the functional H_t^a and by its Hilbert transform as follows*

$$(D_-^{\alpha}L_t^{\cdot})(a) = \frac{1}{\Gamma(-\alpha)}\{\cos(\pi(1+\alpha))H_t^a(-1-\alpha) - \sin(\pi(1+\alpha))\mathcal{H}(H_t^{\cdot}(-1-\alpha))(a)\}. \tag{1-10}$$

2. Limit theorems for occupation times

The process C_t as well as the process H_t arise naturally in certain limit theorems on occupation times. Theorems 2-1 and 2-2 in this section have appeared in Yamada [28]

Theorem (2-1). *Let f be a function which belongs to $L^2(R^1)$. Suppose that the Hilbert transform of the function f, $(\mathcal{H}f)(x) = \frac{1}{\pi}(v.p.\frac{1}{x} * f)(x)$ vanishes out side a compact set in R^1.*

Then the family of stochastic processes

$$t \mapsto \frac{1}{\sqrt{\lambda}}\int_0^{\lambda t} f(B_s)\,ds, \quad \lambda > 0, \tag{2-1}$$

converges in the sense of law on the space of continuous functions to the process

$$t \mapsto \frac{1}{\pi}\Big(\int_{-\infty}^{\infty}(\mathcal{H}^{-1}f)(x)\Big)\cdot C_t^0, \tag{2-2}$$

as $\lambda \to \infty$, where $\mathcal{H}^{-1}$ stands for the inverse transform of $\mathcal{H}$; $\mathcal{H}^{-1}f = -\mathcal{H}f$.

Theorem (2-2). *Suppose that $0 < \alpha < \frac{1}{2}, \alpha < \beta \leq 1$. Let g be a function with compact support satisfying the global Hölder condition of order β, such that, there exists a constant $C > 0$ such that*

$$|g(x) - g(y)| \leq C|x-y|^{\beta}, \quad \forall x, y \in R^1.$$

Put $(-D_-^{\alpha}g)(x) = f(x)$. Then the family of stochastic processes

$$t \mapsto \frac{1}{\lambda^{\frac{1}{2}-\frac{\alpha}{2}}}\int_0^{\lambda t} f(B_s)\,ds, \quad \lambda > 0 \tag{2-3}$$

converges in the sense of law on the space of continuous functions to the process

$$t \mapsto \left(\int_{-\infty}^{\infty} g(x)\,dx\right) \cdot H_t^0(-1-\alpha) \tag{2-4}$$

as $\lambda \to \infty$.

Recently, Fitzsimmons and Getoor [12] have obtained limit theorems for occupation times of one dimensional stable Markov processes which generalize results in the Theorem 2-1 and also in the Theorem 2-2.

3. Stochastic analysis of C_t and H_t

Put $C_t = C_t^0$ and $H_t = H_t^0$. P-variation properties of the process $t \mapsto C_t$ and the process $t \mapsto H_t$, have been studied in depth by Bertoin [6]. The following theorem is due to him.

Theorem (3-1).

(1) *The process* $t \mapsto H_t(-1-\alpha)$ *has bounded p-variation a.s. on* $[0,T]$ *for any* $T > 0$, *if and only if* $p > (1-\alpha)^{-1}$.

(2) *The process* $t \mapsto C_t$ *has bounded p-variation a.s. on* $[0,T]$ *for any* $T > 0$, *if and only if* $p > 1$.

Since a linear Brownian motion $t \mapsto B_t$ has a bounded quadratic variation, it is possible to define a Young-Stieltjes type integral $\int_0^t B_s\,dC_s$ (see Bertoin [2] and also Young [33]). By the definition of the process C_t ,using Ito-type formula for Dirichlet process by Föllmer [13], we see that $\int_0^t B_s\,dC_s = t$. In this direction, Bertoin [6] has proposed the following formula.

Theorem (3-2). *Let f be a function with continuous derivative. Then we have the following formula:*

$$B_t \cdot f(C_t) = \int_0^t f(C_s)\,dB_s + \int_0^t f'(C_s)\,ds. \tag{3-1}$$

Similar formula concerning $H_t(-1-\alpha)$ has been also obtained in [6].

4. Fourier-Laplace transform of the distribution of the Hilbert transform of local times

The paper by Biane and Yor [8] abounds in information on distributions of Brownian functionals . Let C_t be the additive functional of a linear Brownian motion B_t which corresponds to Cauchy's principal value.

Let T be an exponential random variable with expectation q^{-1} and independent of the Brownian motion B_t. We have seen in the Theorem 1-1 that the process C_t can be represented by the Hilbert transform of local times of B_t. The following formula is one of the most striking results in [8].

Theorem (4-1).

$$E[\exp(i\frac{\lambda}{\pi}C_T)] = \mathrm{sech}(\lambda(2q)^{-\frac{1}{2}}),\ \lambda \in R^1. \tag{4-1}$$

In the proof of this formula, the excursion theory combined with some properties of Bessel processes plays an essential role. The formula (4-1) and various related identities hold in the context of a wide class of one dimensional symmetric Lévy processes (see Fitzsimmons and Getoor [11]).

5. Representations of Brownian additive functionals of zero energy

This section is based on Bass [1], Oshima and Yamada [22] and Yamada [29]. The basic references of additive functionals of zero energy is the chapter 5 of the book of Fukushima, Oshima and Takeda [15]

Let $A = A_t = A(t,\omega)$be a continuous additive functional (abbreviated CAF) of d-dimensional brownian motion $B_t = (B_t^{(1)},\ldots,B_t^{(d)})$. The real number $e(A) = \lim_{t\to 0+}\frac{1}{2t}\int_{R^d} E_x[A^2(t,\omega)]$ is called energy of the CAF A_t. If $e(A) = 0$, A is called a CAF of zero energy. We say that a CAF A is locally of zero energy if there exists an increasing sequence of relatively compact finely open sets G_n and a sequence of CAF $A^{(n)}$ of zero energy such that $\lim_{n\to\infty}\tau_{G_n} = \infty$ and $A_t = A_t^{(n)}$, for $0 \le t \le \tau_{G_n} a.s.(P_x)$, for quasi every x, where $\tau_{G_n} = \inf\{t; B_t \notin G_n\}$.

Here, we will introduce a Sobolev space. Let $\mathcal{D}^1_{L^2}(R^1)$ be the Sobolev space of order 1, i.e., $u \in \mathcal{D}^1_{L^2}(R^1)$ if and only if $\frac{du}{dx}$ belongs to $L^2(R^1)$, where derivative is taken in the sense of Schwartz distributions. It's dual space $\mathcal{D}'^{,1}_{L^2}(R^1)$ is characterized as the set of distributions T such that there exist two functions $f, g \in L^2(R^1)$ so that $T = \frac{df}{dx} + g$. Now, we will introduce the CAF associated with a distribution $T \in \mathcal{D}'^{,1}_{L^2}(R^1)$. Let T be written as $T = \frac{df}{dx} + g$, where f and g belong to $L^2(R^1)$. We define the CAF $A_T(t,\omega)$ of a one-dimensional Brownian motion B_t associated with the distribution T as

$$A_T(t,\omega) = 2\{F(B_t) - F(B_0)\} - 2\int_0^t f(B_s)\,dB_s + \int_0^t g(B_s)\,ds, \tag{5-1}$$

where the function F is a primitive function of the function f. In multi-dimensional case, put $\mathcal{D}^1_{L^2}(R^d) = \{u; \frac{\partial u}{\partial x_i} \in L^2(R^d), i = 1,\ldots,d\}$. Let $T = \Delta F + g$ where $F \in \mathcal{D}^1_{L^2}(R^d)$ and $g \in L^2(R^d)$. We will define the CAF A_T of a d-dimensional Brownian motion B_t associated with the distribution T as follows;

$$(5\text{-}2)\quad A_T(t,\omega) = 2\{F(B_t)-F(B_0)\}-2\int_0^t \sum_{i=1}^d \frac{\partial F}{\partial x_i}(B_s)\,dB^{(i)} + \int_0^t g(B_s)\,ds.$$

Fukushima ([14] and [15]) has shown that CAFs defined by (5-1) and also by (5-2) have zero energy. In one dimensional case the class of CAFs locally of zero energy can be characterized as follows (see Oshima and Yamada [22]);

Theorem (5-1). *Let A_T be a CAF locally of zero energy . Then there exists an absolutely continuous function u with derivative $\frac{du}{dx}$ belonging to $L^2_{loc}(R^1)$ such that*

$$(5\text{-}3)\qquad A(t,\omega) = 2\{u(B_t) - u(B_0)\} - 2\int_0^t \frac{du}{dx}(B_s)\,dB_s$$

holds.

Here we will introduce some notation concerning the translation of distributions and CAFs associated to them. Let φ be a function. Put $(\tau_a\varphi)(x) = \varphi(x-a)$. Let T be a distribution. Put $(\tau_a T)(\varphi) = T(\tau_{-a}\varphi)$ for $\varphi \in \mathcal{D}$, where $\mathcal{D}$ is the Schwartz test function space. Put $A^a_T(t,\omega) = A_{\tau_a T}(t,\omega)$. Now we introduce some notation concerning the symmetry. By $\check{f}$ we mean the symmetry with respect to the origin of the function f given by $\check{f}(x) = f(-x)$. Let T be a distribution. By $\check{T}$ we denote the symmetry of T, so that $\check{T}(\varphi) = T(\check{\varphi})$. We now come to state a representation theorem for CAF of zero energy.

Theorem (5-2). *Let T belong to $\mathcal{D}'^1_{L^2}$ such that $T = \frac{df}{dx} + g$, where the functions f and g belong to $L^2(R^1)$. Then for each $t > 0$*

$$(5\text{-}3)\qquad A^a_T(t,\omega) = (\check{T} * L^{\cdot}_t)(a) a.s.$$

holds in the sense of distribution, where $\check{T} * L^{\cdot}_t(a)$ stands for the value at a of the convolution of the distribution $\check{T}$ and the function $\cdot \mapsto L^{\cdot}_t$ in the sense of Schwartz distribution.

As an example, $H^a_t(-1-\alpha)$ is represented by local times as follows (confer Theorem 1-2);

(5-4)

$$\begin{aligned} H^a_t(-1-\alpha) &= (pf.\check{x}_+^{-1-\alpha} * L^{\cdot}_t)(a) \\ &= \int_0^\infty y^{-1-\alpha}(L^{a-y}_t - L^a_t)\,dy = \Gamma(-\alpha) D^\alpha_+(L^{\cdot}_t)(a) \end{aligned}$$

Combining Theorem 5-1 and Theorem 5-2 with a localization method, we can conclude that in the one dimensional case, any CAF locally of zero energy can be represented via convolution type transform of local times (see Yamada [29]).

In the rest of this section, we will discuss a representation problem in multi dimensional case. Let T be a distribution which satisfies the following conditions.

Assumption on T**.** *There exist two functions* F *and* g *such that*

(1) *the partial derivative of* F , $\frac{\partial F}{\partial x_i} = f_i (i = 1, \ldots, d)$ *and the function* g *belong to* L^2 *and moreover they are of compact support.*
(2) $T = \Delta F + g$

We need some notation concerning the Radon transform. The basic reference of the Radon transform is the book of Helgason [17]. See also the book of Gelfand, Graev and Vilenkin [16].

Let $\mathbf{P}^d$ be the space of all hyperplane in R^d. Each $\alpha \in \mathbf{P}^d$ can be written as $\alpha = \{x \in R^d; (x, \xi) = p\}$, where $\xi = (\xi_1, \ldots, \xi_d) \in \mathbf{S}^{d-1}$ and $p \in R^1$. Here,$\mathbf{S}^{d-1}$stands for the unit sphere in R^d i.e., $\sum_{i=1}^{i=d} \xi_i^2 = 1$. Notice that the pair (ξ, p) and $(-\xi, -p)$ give the same hyperplane. Hence $\mathbf{P}^d$ can be identified with $\mathbf{S}^{d-1} \times R^1/ \sim$, where $\sim$ stands for the equivalent relation $(\xi, p) \sim (-\xi, -p)$. The Radon transform of a function $h(x) \in L^1(R^d)$ is defined as a function on $\mathcal{P}^d$ such that $\hat{R}[h](\xi, p) = \int_{(\xi,x)=p} h(x)\, m(dx)$, where $m(dx)$ is the Lebesgue measure on the hyperplane $\{x; (\xi, x) = p\}$. We also define the dual Radon transform of a locally integrable function $\phi(\xi, p)$ on $\mathbf{P}^d$ as a function on R^d such that $\check{R}[\phi](x) = \frac{\Gamma(\frac{d}{2})}{2\sqrt{\pi^d}} \int_{\mathbf{S}^{d-1}} \phi(\xi, (\xi, x))\, \sigma(d\xi)$, where $\sigma(d\xi)$ is the area element of the sphere $\mathbf{S}^{d-1}$.

Now we can state the main theorem in the second half of the section.

Theorem (5-3). *Suppose that the dimension* d *is odd. Let* T *be a distribution satisfying the above Assumption on* T*. Then for* $t \geq 0$,

$$(5\text{-}5) \qquad (-4\pi)^{\frac{d-1}{2}} \frac{\Gamma(\frac{d}{2})}{\Gamma(\frac{1}{2})} A_T^a(t, \omega)$$
$$= \Delta^{\frac{d-1}{2}} \{ -\sum_{k=1}^{k=d} \frac{\partial}{\partial x_k} \check{R}[(\hat{R}[f_k](\xi, \cdot) * L_t^{\cdot}(\xi))(p)](a)$$
$$+ \check{R}[(\hat{R}[g](\xi, \cdot) * L_t^{\cdot}(\xi))(p)](a) \}$$

holds in the sense of distribution.

For the case where the dimension d is even, see Bass [1] and also Yamada [29].

In the case where T is sufficiently smooth function, our representation (5-5) coincides with the result obtained by Bass (see Theorem 4-6 in Bass [1]).

References

1. Bass, R., *Joint continuity and representations of additive functionals of d-dimensional Brownian motion*, Stoch. Proc. Appl. **17** (1984), 211-227.
2. Bertoin, J., *Sur une intégrale pour les processus à α-variation bornée*, Ann.Prob. **17,4** (1989), 1521-1535.

3. Bertoin, J., *Applications de la théorie spectrale des cordes vivrantes aux fonctionelles additives pricipales d'un brownien réfl'echi*, Ann. Inst.Henri Poincaré **25,3** (1989), 307-323.
4. Bertoin, J., *Applications des processus de Dirichlet aux temps locaux et temps locaux d'intersection d'un mouvement brownien*, Prob. Th. Rel. F. **80** (1989), 433-460.
5. Bertoin, J., *Excursion of a $BES_0(d)$ and its drift term* $(0 < d < 1)$, Prob. Th. Rel. F. **84** (1990), 231-250.
6. Bertoin, J., *Complements on Hilbert transform and the fractional derivative of Brownian local times*, J. Math. Kyoto Univ. **30,4** (1990), 651-670.
7. Bertoin, J., *How does a reflected one-dimensional diffusion bounce back ?*, Forum Math. **4** (1992), 549-565.
8. Biane, Ph. and Yor, M., *Valeurs principales associés aux temps locaux brownien*, Bull. Sci. Math. **111** (1987), 23-101.
9. Dellacherie, C. and Meyer, P.A., *Probabilités et potentiel vol.2 , vol.5*, Hermann, Paris, 1980 1992.
10. Ezawa, H.,Klauder, J.R. and Shepp, A., *Vestigial effects of singular potentials in diffusion theory and quantum mechanics*, J. Math. Phys. **16,4** (1975), 783-799.
11. Fitzsimmons, P.J. and Getoor, R.K., *On the distribution of the Hilbert transform of the local time of a symmetric Lévy process*, Ann.Prob. **20,3** (1992), 1484-1497.
12. Fitzsimmons, P.J. and Getoor, R.K., *Limit theorems and variation properties for fractional derivatives of local time of a stable process*, Ann. Inst. Henri Poincaré **28,2** (1992), 311-333.
13. Föllmer, H., *Calcul d'Ito sans probabilités*, S'eminaire de Prob., Lect.Note Math. 850 **15** (1981), 143-150.
14. Fukushima, M., *A decomposition of additive functionals of finite energy*, Nagoya Math. J. **74** (1979).
15. Fukushima, M.,Oshima, Y. and Takeda, M., *Dirichlet forms and symmetric Markov processes*, Walter de Gruyter, Berlin New York, 1994.
16. Gelfand, I.M.,Graev, M.I. and Vilenkin, N.I., *Generalized functions vol. 5*, Academic Press, New York London, 1966.
17. Helgason, S., *The Radon transform*, Birkhäuser, Boston, 1980.
18. Ikeda, N. and Watanabe, S., *Stochastic differential equations and diffusion processes, Second edition*, North Holland / Kodansha, Amsterdam Oxford NewYork Tokyo, 1989.
19. Ito, K. and McKean, H.P.Jr., *Diffusion processes and their sample paths*, Springer Verlag, Berlin Heidelberg New York, 1965.
20. McKean, H.P.Jr. and Tanaka, H., *Additive functionals of Brownian motion path*, Mem. Coll. Sci. Univ. Kyoto **33** (1961), 479-506.
21. Nakao, S., *Stochastic calculus for continuous additive functionals locally of zero energy*, Z. Wahr. **68** (1985), 557-578.
22. Oshima, Y. and Yamada, T., *On some representations of continuous additive functionals locally of zero energy*, J. Math. Soc. Japan **36,2** (1984), 315-339.
23. Pitman, J.W. and Yor, M., *Further asymptotic laws of planer Brownien motion*, Ann. Prob. **17,3** (1989), 965-1011.
24. Revuz, D. and Yor, M., *Continuous martingales and Brownian motion, Second edition*, Springer Verlag, Berlin Heidelberg New York london Tokyo Hong Kong Barcelona Budapest, 1994.
25. Tanaka, H., *Note on continuous additive functionals of 1-dimensional Brownian path*, Z. Wahr. **1** (1963), 251-257.
26. Yamada, T., *On some representations concerning the stochastic integrals*, Prob. Math. Stat. **4,2** (1984), 153-166.
27. Yamada, T., *On the fractional derivative of Brownian local times*, J. Math. Kyoto Univ. **25,1** (1985), 49-58.

28. Yamada, T., *On some limit theorems for occupation times of one dimensional Brownian motion and its continuous additive functionals locally of zero energy*, J. Math. Kyoto Univ. **26,2** (1986), 309-322.
29. Yamada, T., *Representations of continuous additive functionals of zero energy via convolution type transforms of Brownian local times and the Radon transform*, Stochastics and Stoch. Rep. **46,1** (1994), 1-15.
30. Yor, M., *Un exemple de processus qui n'est pas une semi-martingale*, Astérisque **52-53** (1978), 219-221.
31. Yor, M., *Sur la transformée de Hilbert des temps locaux browniens et une extension de la formule d'Ito*, Séminaire de Prob. Lect. Note Math. 920 **16** (1982), 238-247.
32. Yor, M., *The laws of some Brownian functionals*, Proceedings of ICM 1990 Kyoto **2** (1991), 1105-1112.
33. Young, L.C., *An inequality of Hölder type connected with Stieltjes integration*, Acta Math. **67** (1936), 251-282.